ISNM
International Series of
Numerical Mathematics
Vol. 112

Edited by
K.-H. Hoffmann, München
H. D. Mittelmann, Tempe
J. Todd, Pasadena

Numerical Integration IV

Proceedings of the Conference at the Mathematical Researach Institute, Oberwolfach, November 8–14, 1992

Edited by

H. Brass
G. Hämmerlin

Birkhäuser Verlag
Basel · Boston · Berlin

Editors

H. Brass
Institut für Angewandte Mathematik
TU Braunschweig
Pockelsstr. 14
D-3300 Braunschweig

G. Hämmerlin
Mathematisches Institut der
Ludwig-Maximilian-Universität
Theresienstr. 39
D-8000 München

Library of Congress Cataloging-in-Publication Data
Numerical integration IV : proceedings of the conference at the
Mathematical Research Institute at Oberwolfach, November 8–14, 1992
/ edited by H. Brass, G. Hämmerlin
p. cm. – (International series of numerical mathematics :
vol. 112)
ISBN 3-7643-2922-X (acid-free) – ISBN
0-8176-2922-X (acid-free)
1. Numerical integration--Congresses. I. Brass, Helmut.
II. Hämmerlin, G. (Günther), 1928– . III. Mathematisches
Forschungsinstitut Oberwolfach. IV. Title: Numerical integration
4. V. Series: International series of numerical mathematics ; v.
112.
QA299.3.N822 1993
515'.624--dc20

Deutsche Bibliothek Cataloging-in-Publication Data
Numerical integration IV : proceedings of the conference at the
Mathematical Research Institute at Oberwolfach, November 8–
14, 1992 / ed. by H. Brass ; G. Hämmerlin. – Basel ; Boston ;
Berlin : Birkhäuser, 1993
(International series of numerical mathematics ; Vol. 112)
ISBN 3-7643-2922-X (Basel ...)
ISBN 0-8176-2922-X (Boston)
NE: Brass, Helmut [Hrsg.]; Mathematisches Forschungsinstitut
<Oberwolfach>; GT

Camera-ready copy prepared by the authors/editors
Printed on acid-free paper produced from chlorine-free pulp
Cover design: Heinz Hiltbrunner, Basel
Printed in Germany
ISBN 3-7643-2922-X
ISBN 0-8176-2922-X

9 8 7 6 5 4 3 2 1

Contents

Preface vii

List of Participants viii

G. Baszenski and F.-J. Delvos
Multivariate Boolean midpoint rules 1

M. Beckers and R. Cools
A relation between cubature formulae of trigonometric degree and lattice rules 13

T. Bloom, D.S. Lubinsky and H. Stahl
Distribution of points in convergent sequences of interpolatory integration rules: The rates 25

H. Brass
Bounds for Peano kernels 39

R. Cools and H.J. Schmid
A new lower bound for the number of nodes in cubature formulae of degree $4n + 1$ for some circularly symmetric integrals 57

S. Ehrich
On the construction of Gaussian quadrature formulae containing preassigned nodes 67

T.O. Espelid
Integrating singularities using non-uniform subdivision and extrapolation 77

K.-J. Förster
Variance in quadrature – a survey 91

W. Gautschi
Gauss-type quadrature rules for rational functions 111

A. Genz
Subdivision methods for adaptive integration over hyperspheres 131

A. Guessab
Formules de quadrature dans $\mathbb{R}^2$ avec "réseau" minimal de droites 141

S.-Å. Gustafson
Quadrature rules derived from linear convergence acceleration schemes 151

A. Haegemans and P. Verlinden
Construction of fully symmetrical cubature rules of very high degree for the square 167

T. Hasegawa and T. Torii
Numerical integration of nearly singular functions 175

D.B. Hunter and H.V. Smith
Some problems involving orthogonal polynomials 189

P. Köhler
Intermediate error estimates for quadrature formulas 199

F. Locher
Stability tests for linear difference forms 215

J.N. Lyness
The canonical forms of a lattice rule 225

G. Mastroianni and P. Vértesi
Error estimates for product quadrature formulae 241

H. Niederreiter and I.H. Sloan
Quasi-Monte Carlo methods with modified vertex weights 253

G. Nikolov
Gaussian quadrature formulae for splines 267

E. Novak
Quadrature formulas for convex classes of functions 283

F. Peherstorfer
On positive quadrature formulas 297

K. Petras
Quadrature theory of convex functions 315

K. Ritter, G.W. Wasilkowski and H. Woźniakowski
On multivariate integration for stochastic processes 331

C. Schneider
Rational Hermite interpolation and quadrature 349

A. Sidi
A new variable transformation for numerical integration 359

Problems 375

Preface

After the conferences in 1978, 1981 and 1987 (Proceedings in ISNM 45, 57, 85), a fourth symposium on numerical integration was held at Oberwolfach November 8–14, 1992. 46 mathematicians from 16 countries participated. This volume contains 27 refereed papers presented at the conference.

Algorithms for the numerical computation of definite integrals have been proposed for more than 300 years, but practical considerations have lead to problems of ever increasing complexity, so that, even with current computing speeds, numerical integration may be a difficult task. High dimension and complicated structure of the region of integration and singularities of the integrand are the main sources of difficulties.

The construction of new rules was therefore a central theme of the conference. Another line of research attempts to study the set of all rules of a class with regard to the error or another characteristic. The theory of information-based complexity introduced new concepts and results into the discussion. The widening of the notion of optimality to the average case setting seems to be of great interest. As it has been the case in the former conferences, the Gaussian rule and its relatives have continued to be important objects of study.

Many open problems were discussed in Oberwolfach, showing liveliness and actuality of the theory, nine of these problems could be given precise formulations; they are collected at the end of this volume.

The editors' thanks go to all contributors and participants who made the conference a success; to the management of the Institute with its unique atmosphere; to the Birkhäuser Verlag for the possibility of publishing the volume in the well-known ISNM series and to the co-workers of the editors for assistance in editorial work.

H. Brass
Braunschweig

G. Hämmerlin
München

List of participants of the conference

Prof. Dr. Julius Albrecht
Institut für Mathematik
TU Clausthal
Erzstr. 1
W-3392 Clausthal-Zellerfeld 1
Germany

Prof. Dr. Borislav Bojanov
Dept. of Mathematics
University of Sofia
Boul. James Boucher 5
1126 Sofia
Bulgaria

Prof. Dr. Helmut Braß
Institut für Angewandte Mathematik
TU Braunschweig
Pockelsstr. 14
W-3300 Braunschweig
Germany

Dr. Ronald Cools
Department of Computer Science
Katholieke Universiteit Leuven
Celestijnenlaan 200 A
B-3001 Heverlee-Leuven
Belgium

Prof. Dr. Franz-Jürgen Delvos
Fachbereich 6 Mathematik
Universität Gesamthochschule Siegen
Hölderlinstr. 3
W-5900 Siegen
Germany

Prof. Dr. Elise de Doncker
Computer Science Department
Western Michigan University
Kalamazoo MI 49008-5021
USA

Sven Ehrich
Institut für Mathematik
Universität Hildesheim
Marienburger Platz 22
W-3200 Hildesheim
Germany

Prof. Dr. Terje O. Espelid
Department of Informatics
University of Bergen
Hoyteknologisenteret
N-5020 Bergen
Norway

Prof. Dr. Heinz Fiedler
Abteilungen für Mathematik
Universität Ulm
Helmholtzstr. 18
W-7900 Ulm
Germany

Prof. Dr. Klaus-Jürgen Förster
Institut für Mathematik
Universität Hildesheim
Marienburger Platz 22
W-3200 Hildesheim
Germany

Prof. Dr. Luigi Gatteschi
Dipartimento di Matematica
Universita di Torino
Via Carlo Alberto, 10
I-10123 Torino
Italy

Prof. Dr. Walter Gautschi
Department of Computer Sciences
Computer Science Building 164C
Purdue University
West Lafayette, IN 47907
USA

Prof. Dr. Alan Genz
School of EE and Computer science
Washington State University
Pullman WA 99164-3113
USA

Prof. Dr. David B. Hunter
Department of Mathematics
University of Bradford
GB-Bradford, Yorkshire BD7 1DP
U.K.

Dr. Allal Guessab
Département de Mathématiques
Université de Pau, Appl. URA 1204
B.P. 290
Avenue de l'Université
F-64000 Pau
France

Prof. Dr. Kurt Jetter
Fachbereich Mathematik
Universität-GH Duisburg
Postfach 10 16 29
Lotharstr. 65
W-4100 Duisburg 1
Germany

Prof. Dr. Sven-Åke Gustafson
HSR Box 2557
Ullandhaug
N-4004 Stavanger
Norway

Dr. Peter Köhler
Institut für Angewandte Mathematik
TU Braunschweig
Pockelsstr. 14
W-3300 Braunschweig
Germany

Prof. Dr. Ann Haegemans
Department of Computer Science
Katholieke Universiteit Leuven
Celestijnenlaan 200 A
B-3001 Heverlee-Leuven
Belgium

Prof. Dr. Franz Locher
Fachbereich Mathematik
Fernuniversität Gesamthochschule
Postfach 940
W-5800 Hagen 1
Germany

Prof. Dr. Günther Hämmerlin
Mathematisches Institut
Universität München
Theresienstr. 39
W-8000 München 2
Germany

Prof. Dr. Doron S. Lubinsky
Department of Mathematics
University of Witwatersrand
P. O. Box
Wits 2050
South Africa

Prof. Dr. Takemitsu Hasegawa
Department of Information Science
Faculty of Engineering
Fukui University
Fukui 910
Japan

Dr. James N. Lyness
Mathematics and Computer Science
Division
Argonne National Laboratory
9700 South Cass Avenue
Argonne , IL 60439-4844
USA

Prof. Dr. Claus Schneider
Fachbereich Mathematik
Universität Mainz
Saarstr. 21
Postfach 3980
W-6500 Mainz 1
Germany

Prof. Dr. Frank Stenger
680 Terrace Hills Drive
Salt Lake City UT 84103
USA

Prof. Dr. Avram Sidi
Computer Science Department
TECHNION
Israel Institute of Technology
Haifa 32000
Israel

Prof. Dr. Hans Strauß
Institut für Angewandte Mathematik
Universität Erlangen
Martensstr. 3
W-8520 Erlangen
Germany

Prof. Dr. Ian H. Sloan
Dept. of Mathematics
The University of New South Wales
P. O. Box 1
Kensington N. S. W. 2033
Australia

Prof. Dr. Grzegorz W. Wasilkowski
Comp. Science Department
University of Kentucky
915 Paterson
Lexington , KY 40506-0027
USA

Prof. Dr. Harry V. Smith
22 Hodgson Avenue
GB- Leeds LS 17 8PQ
U.K.

Prof. Dr. Henryk Woźniakowski
Institute of Applied Mathematics
University of Warsaw
ul. Banacha 2
02-097 Warszawa
Poland

Prof. Dr. Tor Sørevik
Department of Informatics
University of Bergen
Hoyteknologisenteret
N-5020 Bergen
Norway

Prof. Dr. Karl Zeller
Mathematisches Institut
Universität Tübingen
Auf der Morgenstelle 10
W-7400 Tübingen 1
Germany

Prof. Dr. Giuseppe Mastroianni
Universitá degli Studi della
Basilicata
Dipartimento di Matematica
Via Nazario Sauro 85
I-85100 Potenza
Italy

Prof. Dr. Harald Niederreiter
Inst. für Informationsverarbeitung
Österreichische Akademie
der Wissenschaften
Sonnenfelsgasse 19
A-1010 Wien
Austria

Athanassios Nikolis
Mathematisches Institut
Universität München
Theresienstr. 39
W-8000 München 2
Germany

Dr. Geno Nikolov
Dept. of Mathematics
University of Sofia
Boul. James Boucher 5
1126 Sofia
Bulgaria

Dr. Sotirios E. Notaris
1 Xenokratous Street
10675 Athen
Greece

Dr. Erich Novak
Mathematisches Institut
Universität Erlangen
Bismarckstr. 1 1/2
W-8520 Erlangen
Germany

Prof. Dr. Franz Peherstorfer
Institut für Mathematik
Johannes Kepler Universität
A-4040 Linz
Austria

Dr. Knut Petras
Institut für Angewandte Mathematik
TU Braunschweig
Pockelsstr. 14
W-3300 Braunschweig
Germany

Wolfgang Prock
Mathematisches Institut
Universität München
Theresienstr. 39
W-8000 München 2
Germany

Prof. Dr. Gazi Ibadur Rahman
Dept. of Mathematics and Statistics
University of Montréal
C. P. 6128, Succ. A
Montréal, P. Q. H3C 3J7
Canada

Prof. Dr. Gerhard Schmeißer
Mathematisches Institut
Universität Erlangen
Bismarckstr. 1 1/2
W-8520 Erlangen
Germany

Prof. Dr. Hans Joachim Schmid
Mathematisches Institut
Universität Erlangen
Bismarckstr. 1 1/2
W-8520 Erlangen
Germany

Multivariate Boolean Midpoint Rules

Günter Baszenski and Franz-Jürgen Delvos

Abstract Boolean methods of interpolation have been applied to the construction of bivariate integration formulas (Delvos 1990). These formulas are comparable with lattice rules of multivariate numerical integration (Sloan 1985, 1987). In this paper we will construct Boolean midpoint rules for multivariate numerical integration of arbitrary dimensions which are based on the ideas of multivariate Boolean interpolation and which extend results of Delvos (1990).

1. Multivariate product midpoint rules

Let d be a positive integer. Then

$$U^d = \{(x_1, \ldots, x_d) \in \mathbf{R}^d : 0 \le x_j \le 1 \quad (j = 1, \ldots, d)\}$$

denotes the d-variate unit cube and $\mathcal{C}(U^d)$ is the algebra of continuous complex-valued functions $f : U^d \to \mathbf{C}$. Next we introduce

$$n_k = 2^{k-1} \quad (k = 1, 2, \ldots).$$

Given $f \in \mathcal{C}(U^d)$ its d-variate *product midpoint sum* is defined by

$$S(k_1, \ldots, k_d)[f] = \sum_{\mu_1=1}^{n_{k_1}} \cdots \sum_{\mu_d=1}^{n_{k_d}} f\left(\frac{2\mu_1 - 1}{2n_{k_1}}, \ldots, \frac{2\mu_d - 1}{2n_{k_d}}\right).$$

The number of function evaluations for computing $S(k_1,\dots,k_d)[f]$ is denoted by $|S(k_1,\dots,k_d)|$. It is obvious that $|S(k_1,\dots,k_d)| = n_{k_1}\cdots n_{k_d} = 2^{k_1+\dots+k_d-d}$.

The d-variate *product midpoint rule* is then given by

$$M(k_1,\dots,k_d)[f] = \frac{1}{n_{k_1}\cdots n_{k_d}} \cdot S(k_1,\dots,k_d)[f].$$

It is used to approximate the d-variate integral

$$I_d[f] = \int_0^1 \dots \int_0^1 f(x_1,\dots,x_d)\,dx_1\dots dx_d.$$

The error $M(k_1,\dots,k_d)[f] - I_d[f]$ is discussed for functions $f \in \mathcal{C}(U^d)$ which are also elements of the Korobov space $\mathcal{E}^\alpha(U^d)$ with $\alpha > 1$. The Fourier coefficients of f are given by

$$a(m_1,\dots,m_d)[f] =$$
$$\int_0^1 \dots \int_0^1 f(x_1,\dots,x_d)\exp\bigl(-2\pi i(m_1x_1+\dots+m_dx_d)\bigr)\,dx_1\dots dx_d$$

with $(m_1,\dots,m_d) \in \mathbf{Z}^d$. If the Fourier coefficients satisfy

$$a(m_1,\dots,m_d)[f] = \mathcal{O}(|m_1\cdots m_d|^{-\alpha}) \quad (|m_1|,\dots,|m_d| \to \infty)$$

then f ist an element of the Korobov space $\mathcal{E}^\alpha(U^d)$. Since $\alpha > 1$, f is represented by its absolutely convergent Fourier series:

$$f(x_1,\dots,x_d) =$$
$$\sum_{m_1=-\infty}^{\infty} \dots \sum_{m_d=-\infty}^{\infty} a(m_1,\dots,m_d)[f]\exp\bigl(2\pi i(m_1x_1+\dots+m_dx_d)\bigr).$$

The discrete orthogonality of the exponentials yields the relation

$$M(k_1,\dots,k_d)[f] = \sum_{r_1=-\infty}^{\infty} \dots \sum_{r_d=-\infty}^{\infty} a(r_1n_{k_1},\dots,r_dn_{k_d})[f]\cdot(-1)^{r_1+\dots+r_d}. \quad (1.1)$$

Recall that

$$a(0,\dots,0)[f] = I_d[f] = \int_0^1 \dots \int_0^1 f(x_1,\dots,x_d)\,dx_1\dots dx_d.$$

We introduce different types of multiple series to obtain an alternative representation of the remainder $M(k_1,\ldots,k_d)[f]-I_d[f]$.

Let $w=(w_1,\ldots,w_d)\in\{0,1\}^d$. We define

$$R^{(w_1,\ldots,w_d)}_{(k_1,\ldots,k_d)}[f]= \sum_{w_1r_1\neq 0}\cdots\sum_{w_dr_d\neq 0} a(w_1r_1n_{k_1},\ldots,w_dr_dn_{k_d})[f]\cdot(-1)^{w_1r_1+\cdots+w_dr_d} \tag{1.2}$$

For every $w\in\{0,1\}^d$ there are unique positive integers u_j and s such that $\{j:1\leq j\leq d, w_j=1\}=\{u_1,\ldots,u_s\}$ and $1\leq u_1<\ldots<u_s\leq d$. We therefore also use the notation

$$R^{(w_1,\ldots,w_d)}_{(k_1,\ldots,k_d)}[f]=R(k_{u_1},\ldots,k_{u_s})[f].$$

Proposition 1.1 The remainder of the multivariate midpoint rule possesses the series representation

$$M(k_1,\ldots,k_d)[f]-I_d[f]=\sum_{(w_1,\ldots,w_d)\neq(0,\ldots,0)} R^{(w_1,\ldots,w_d)}_{(k_1,\ldots,k_d)}[f]. \tag{1.3}$$

Proof Relation (1.3) is an immediate consequence of (1.1) and (1.2). ■

The series representation is used to establish the asymptotic behavior of the error. Let

$$|w|_1=|w_1|+\cdots+|w_d|$$

be the ℓ_1-norm of w. Note that $|w|_1$ is the number of ones in w. Taking into account the definition of $n_{k_1},\ldots,n_{k_d}$ and $R^{(w_1,\ldots,w_d)}_{(k_1,\ldots,k_d)}[f]$ we obtain

$$R^{(w_1,\ldots,w_d)}_{(k_1,\ldots,k_d)}[f]=\mathcal{O}(2^{-(w_1k_1+\cdots+w_dk_d)\alpha})\quad(k_1,\ldots,k_d\to\infty). \tag{1.4}$$

With $R^{(w_1,\ldots,w_d)}_{(k_1,\ldots,k_d)}[f]=R(k_{u_1},\ldots,k_{u_s})[f]$ this relation obtains the form

$$R(k_{u_1},\ldots,k_{u_s})[f]=\mathcal{O}(2^{-(k_{u_1}+\cdots+k_{u_s})\alpha})\quad(k_{u_1},\ldots,k_{u_s}\to\infty). \tag{1.5}$$

We consider the special case $k_j = q \quad (j = 1, \ldots, d)$. In this case (1.4) becomes

$$R^{(w_1,\ldots,w_d)}_{(q,\ldots,q)}[f] = \mathcal{O}(2^{-|w|_1 q\alpha}) \quad (q \to \infty). \tag{1.6}$$

Proposition 1.2 If f is an element of the Korobov space $\mathcal{E}^\alpha(U^d)$ then the remainder in the multivariate product midpoint rule satisfies the asymptotic relation

$$M(q, \ldots, q)[f] - I_d[f] = \mathcal{O}(2^{-q\alpha}) \quad (q \to \infty). \tag{1.7}$$

Proof Relation (1.7) follows immediately from (1.6) and (1.3). ■

2. Multivariate Boolean midpoint rules

To obtain an asymptotic error estimate of the form $\mathcal{O}(2^{-q\alpha})$ via the multivariate product midpoint rule $M(q, \ldots, q)[f]$ it is necessary to evaluate f at $2^{dq-d} = \mathcal{O}(2^{dq})$ points. It is well known that this number can be reduced drastically by using efficient lattice rules for multivariate numerical integration, see Sloan (1985), Sloan (1987). We will construct multivariate Boolean midpoint rules which possess similar properties and which are easy to generate.

The basic tools for constructing multivariate Boolean midpoint rules are multivariate Boolean midpoint sums which are defined by

$$S^{d;q}[f] = \sum_{k_1+\cdots+k_d=q} S(k_1, \ldots, k_d)[f]$$

where

$$S(k_1, \ldots, k_d)[f] = \sum_{\mu_1=1}^{n_{k_1}} \cdots \sum_{\mu_d=1}^{n_{k_d}} f\left(\frac{2\mu_1 - 1}{2n_{k_1}}, \ldots, \frac{2\mu_d - 1}{2n_{k_d}}\right).$$

The number of function evaluations for $S(k_1, \ldots, k_d)[f]$ is

$$|S(k_1, \ldots, k_d)| = 2^{q-d}$$

since $k_1 + \cdots + k_d = q$.

Recall the combinatorial result concerning partitions

$$\left|\{(k_1,\ldots,k_d) \in \mathbf{N}^d : k_1 + \cdots + k_d = q\}\right| = \binom{q-1}{d-1}. \tag{2.1}$$

Thus the number of function evaluations for $S^{d;q}[f]$ is

$$|S^{d;q}| = \binom{q-1}{d-1} 2^{q-d} = \mathcal{O}(q^{d-1}2^q). \tag{2.2}$$

Next we define the semi-normalized Boolean midpoint sums

$$M^{d;q}[f] = 2^{-q+d} S^{d;q}[f] = \sum_{k_1+\cdots+k_d=q} M(k_1,\ldots,k_d)[f].$$

Motivated by the formula of multivariate Boolean interpolation (see Delvos 1982) we introduce the *multivariate Boolean midpoint rule*:

$$H^{d;q}[f] = \sum_{j=0}^{d-1} (-1)^j \binom{d-1}{j} M^{d;q-j}[f].$$

It follows from (2.1) that the number of multivariate product midpoint rules generating $M^{d;q-j}[f]$ is $\binom{q-j-1}{d-1}$. From (2.2) we get that the number of function evaluations for $M^{d;q-j}[f]$ satisfies the asymptotic relation $\mathcal{O}(q^{d-1}2^q)$. Thus the number of function evaluations for $H^{d;q}[f]$ is also $\mathcal{O}(q^{d-1}2^q)$:

$$|H^{d;q}| = \mathcal{O}(q^{d-1}2^q) \quad (q \to \infty). \tag{2.3}$$

If $M(k_1,\ldots,k_d)[f] = I_d[f]$ (which is true for $f = 1$) it follows that

$$H^{d;q}[f] = \sum_{j=0}^{d-1} (-1)^j \binom{d-1}{j} \binom{q-j-1}{d-1} I_d[f].$$

It was shown in Baszenski-Delvos (1991) using a result of Netto (1901) (see page 252) that

$$\sum_{j=0}^{d-1}(-1)^j \binom{d-1}{j}\binom{q-j-1}{d-1} = 1$$

which implies that $H^{d;q}[f]$ is a reasonable approximation of the d-variate integral

$$I_d[f] = \int_0^1 \cdots \int_0^1 f(x_1, \ldots, x_d)\, dx_1 \ldots dx_d.$$

Next we will investigate the remainder $H^{d;q}[f] - I_d[f]$ of the multivariate Boolean midpoint rule in greater detail. We have

$$\begin{aligned} H^{d;q}[f] - I_d[f] &= \sum_{j=0}^{d-1}(-1)^j \binom{d-1}{j} M^{d;q-j}[f] - I_d[f] \\ &= \sum_{j=0}^{d-1}(-1)^j \binom{d-1}{j} \sum_{k_1+\cdots+k_d=q-j} (M(k_1, \ldots, k_d)[f] - I_d[f]). \end{aligned}$$

Now the remainder representation (1.3) of multivariate product midpoint rules is applicable:

$$H^{d;q}[f] - I_d[f] = \\ \sum_{j=0}^{d-1}(-1)^j \binom{d-1}{j} \sum_{k_1+\cdots+k_d=q-j} \quad \sum_{(w_1,\ldots,w_d)\neq(0,\ldots,0)} R^{(w_1,\ldots,w_d)}_{(k_1,\ldots,k_d)}[f]. \tag{2.4}$$

Let us introduce the sums

$$R^s_{(k_1,\ldots,k_d)}[f] = \sum_{|w|_1=s} R^{(w_1,\ldots,w_d)}_{(k_1,\ldots,k_d)}[f] \quad (1 \le s \le d). \tag{2.5}$$

Inserting (2.5) into (2.4) we get

$$H^{d;q}[f] - I_d[f] = \sum_{s=1}^{d}\sum_{j=0}^{d-1}(-1)^j \binom{d-1}{j} \sum_{k_1+\cdots+k_d=q-j} R^s_{(k_1,\ldots,k_d)}[f]. \tag{2.6}$$

Next we investigate the sums

$$\sum_{j=0}^{d-1}(-1)^j\binom{d-1}{j}\sum_{k_1+\cdots+k_d=q-j} R^s_{(k_1,\ldots,k_d)}[f]. \tag{2.7}$$

Typical for such a sum is the expression

$$T(1,\ldots,s)[f]=\sum_{j=0}^{d-1}(-1)^j\binom{d-1}{j}\sum_{k_1+\cdots+k_d=q-j} R^{(1,\ldots,1,0,\ldots,0)}_{(k_1,\ldots,k_s,k_{s+1},\ldots,k_d)}[f].$$

In view of

$$R^{(w_1,\ldots,w_d)}_{(k_1,\ldots,k_d)}[f]=R(k_{u_1},\ldots,k_{u_s})[f]$$

we have

$$T(1,\ldots,s)[f]=\sum_{j=0}^{d-1}(-1)^j\binom{d-1}{j}\sum_{k_1+\cdots+k_d=q-j} R(k_1,\ldots,k_s)[f].$$

The other terms in (2.7) are

$$T(u_1,\ldots,u_s)[f]=\sum_{j=0}^{d-1}(-1)^j\binom{d-1}{j}\sum_{k_1+\cdots+k_d=q-j} R(k_{u_1},\ldots,k_{u_s})[f] \quad (1\le u_1<\ldots<u_s\le d). \tag{2.8}$$

Thus we have

$$\sum_{j=0}^{d-1}(-1)^j\binom{d-1}{j}\sum_{k_1+\cdots+k_d=q-j} R^s_{(k_1,\ldots,k_d)}[f]=\sum_{1\le u_1<\ldots<u_s\le d} T(u_1,\ldots,u_s)[f].$$

The remainder formula (2.6) obtains the following form

$$H^{d;q}[f]-I_d[f]=\sum_{s=1}^{d}\sum_{1\le u_1<\ldots<u_s\le d} T(u_1,\ldots,u_s)[f]. \tag{2.9}$$

The sums $T(u_1,\ldots,u_s)[f]$ have been studied in Baszenski-Delvos (1991), Prop. 1. It was shown there (see relation (9)) that

$$T(u_1,\ldots,u_s)[f]=\sum_{t=1}^{s}(-1)^{s-t}\binom{s-1}{t-1}\sum_{k_{u_1}+\cdots+k_{u_s}=q-d+t} R(k_{u_1},\ldots,k_{u_s})[f] \quad (1\le u_1<\ldots<u_s\le d). \tag{2.10}$$

Proposition 2.1 If f is an element of the Korobov space $\mathcal{E}^\alpha(U^d)$ then the remainder of the multivariate Boolean midpoint rule satisfies the asymptotic relation

$$H^{d;q}[f] - I_d[f] = \mathcal{O}(q^{d-1}2^{-q\alpha}) \quad (q \to \infty). \tag{2.11}$$

Proof Recall from (1.5) that

$$R(k_{u_1}, \ldots, k_{u_s})[f] = \mathcal{O}(2^{-(k_{u_1}+\cdots+k_{u_s})\alpha}) \quad (k_{u_1}, \ldots, k_{u_s} \to \infty).$$

Since $k_{u_1} + \cdots + k_{u_s} = q - d + t \quad (1 \leq t \leq s \leq d)$ we get

$$R(k_{u_1}, \ldots, k_{u_s})[f] = \mathcal{O}(2^{-q\alpha}) \quad (q \to \infty). \tag{2.12}$$

It follows from

$$\left|\{(k_{u_1}, \ldots, k_{u_s}) \in \mathbf{N}^s : k_{u_1} + \cdots + k_{u_s} = q - d + t\}\right| = \binom{q-d+t-1}{s-1} = \mathcal{O}(q^{d-1})$$

and (2.10) that

$$T(u_1, \ldots, u_s)[f] = \mathcal{O}(q^{d-1}2^{-q\alpha}) \quad (q \to \infty). \tag{2.13}$$

Now the asymptotic error relation (2.11) is a consequence of (2.9). ∎

It follows from (2.3) and (2.11) that the number of point evaluations for the multivariate Boolean midpoint rule satisfies the asymptotic relation

$$\left|H^{d;q}\right| = \mathcal{O}(q^{d-1}2^q) \quad (q \to \infty)$$

while the asymptotic error behaviour is given by

$$H^{d;q}[f] - I_d[f] = \mathcal{O}(q^{d-1}2^{-q\alpha}) \quad (q \to \infty).$$

This is in great contrast to the multivariate product midpoint rule where we have the relations

$$M(q, \ldots, q)[f] - I_d[f] = \mathcal{O}(2^{-q\alpha}) \quad (q \to \infty),$$

$$\left|M(q, \ldots, q)\right| = \mathcal{O}(2^{dq}) \quad (q \to \infty).$$

3. A numerical example

An example for the bivariate case has been presented in Delvos (1990). Here we will consider the case of $d = 3$ to 7 variables. The d-variate Boolean midpoint rules are in these cases given by

$$
\begin{aligned}
H^{3;q}[f] &= M^{3;q}[f] - 2M^{3;q-1}[f] + M^{3;q-2}[f] \\
H^{4;q}[f] &= M^{4;q}[f] - 3M^{4;q-1}[f] + 3M^{4;q-2}[f] - M^{4;q-3}[f] \\
H^{5;q}[f] &= M^{5;q}[f] - 4M^{5;q-1}[f] + 6M^{5;q-2}[f] - 4M^{5;q-3}[f] + M^{5;q-4}[f] \\
H^{6;q}[f] &= M^{6;q}[f] - 5M^{6;q-1}[f] + 10M^{6;q-2}[f] - 10M^{6;q-3}[f] + 5M^{6;q-4}[f] \\
&\quad - M^{6;q-5}[f] \\
H^{7;q}[f] &= M^{7;q}[f] - 6M^{7;q-1}[f] + 15M^{7;q-2}[f] - 20M^{7;q-3}[f] + 15M^{7;q-4}[f] \\
&\quad - 6M^{7;q-5}[f] + M^{7;q-6}[f]
\end{aligned}
$$

The function $f(x_1, \ldots, x_d)$ to be integrated is taken as

$$f(x_1, \ldots, x_d) = \frac{1}{1 + \left((2x_1 - 1)\cdots(2x_d - 1)\right)^2}.$$

Its integral over $U^d = [0,1]^d$ can be computed explicitly by expanding $f(x_1, \ldots, x_d)$ into a geometric series. The values are:

$$
\begin{aligned}
&I_3[f] = 0.9689461, \quad I_4[f] = 0.9889446, \quad I_5[f] = 0.9961578 \\
&I_6[f] = 0.9986852, \quad I_7[f] = 0.9995545.
\end{aligned}
$$

The periodic extension of $f(x_1, \ldots, x_d)$ is continuous. This implies that $f(x_1, \ldots, x_d)$ is an element of the Korobov space $\mathcal{E}^2(U^d)$. The following table illustrates the validity of the asymptotic error estimate.

q	$H^{3;q}[f]$	$H^{3;q}[f] - I_3[f]$
5	1.000000000	-3.10539e-02
6	0.984615385	-1.56692e-02
7	0.973989259	-5.04311e-03
8	0.970007508	-1.06136e-03
9	0.969119608	-1.73462e-04
10	0.968970289	-2.41430e-05
11	0.968947284	-1.13750e-06
12	0.968945027	1.11974e-06
13	0.968945473	6.73133e-07
14	0.968945870	2.76650e-07
15	0.968946048	9.87463e-08
16	0.968946114	3.27037e-08
17	0.968946136	1.03370e-08
18	0.968946143	3.16186e-09
19	0.968946145	9.45846e-10
20	0.968946146	2.76443e-10

q	$H^{4;q}[f]$	$H^{4;q}[f] - I_4[f]$
7	1.000000000	-1.10554e-02
8	0.996108949	-7.16440e-03
9	0.992296956	-3.35240e-03
10	0.990099616	-1.15506e-03
11	0.989223001	-2.78449e-04
12	0.988983998	-3.94464e-05
13	0.988942672	1.87987e-06
14	0.988940662	3.88935e-06
15	0.988942592	1.95928e-06
16	0.988943772	7.79614e-07
17	0.988944273	2.78703e-07
18	0.988944458	9.34522e-08
19	0.988944522	2.99851e-08
20	0.988944542	9.34296e-09

q	$H^{5;q}[f]$	$H^{5;q}[f] - I_5[f]$
9	1.000000000	-3.84217e-03
10	0.999024390	-2.86656e-03
11	0.997811116	-1.65329e-03
12	0.996920909	-7.63081e-04
13	0.996436734	-2.78905e-04
14	0.996233293	-7.54651e-05
15	0.996169391	-1.15630e-05
16	0.996156229	1.59918e-06
17	0.996155751	2.07707e-06
18	0.996156792	1.03581e-06
19	0.996157419	4.08939e-07
20	0.996157683	1.45126e-07

q	$H^{6;q}[f]$	$H^{6;q}[f] - I_6[f]$
11	1.000000000	-1.31478e-03
12	0.999755919	-1.07070e-03
13	0.999390266	-7.05044e-04
14	0.999072179	-3.86956e-04
15	0.998864432	-1.79210e-04
16	0.998754125	-6.89024e-05
17	0.998705884	-2.06617e-05
18	0.998689056	-3.83404e-06
19	0.998684869	3.52974e-07
20	0.998684499	7.23264e-07

q	$H^{7;q}[f]$	$H^{7;q}[f] - I_7[f]$
13	1.000000000	-4.45492e-04
14	0.999938969	-3.84461e-04
15	0.999832198	-2.77690e-04
16	0.999725589	-1.71082e-04
17	0.999646096	-9.15881e-05
18	0.999597300	-4.27920e-05
19	0.999571683	-1.71746e-05
20	0.999560132	-5.62389e-06

References

Baszenski, G., and Delvos, F.-J. (1991), Remainders for Boolean interpolation. In: Progress in Approximation Theory, Eds. P. Nevai, A. Pinkus, pp. 31–41. Academic Press.

Delvos, F.-J. (1982), *d*-variate Boolean interpolation. J. Approx. Theory **34**, pp. 99–114.

Delvos, F.-J. (1990), Boolean methods for double integration. Math. Comp. **55**, pp. 683–692.

Netto, E. (1901), Lehrbuch der Kombinatorik. Chelsey, New York.

Sloan, I. H. (1985), Lattice methods for multiple integration. J. Comput. Appl. Math. **12–13**, pp. 131–143.

Sloan, I. H. (1987), Lattice methods for multiple integration: Theory, Error analysis and examples. SIAM J. Numer. Anal. **24**, pp. 116–128.

Authors' addresses:

Prof. Dr. Günter Baszenki
Fachhochschule Dortmund
Fachbereich Nachrichtentechnik
Sonnenstraße 96–100
4600 Dortmund
Germany

Prof. Dr. Franz-Jürgen Delvos
Fachbereich Mathematik I
Universität GH Siegen
Hölderlinstraße 3
5900 Siegen
Germany

International Series of Numerical Mathematics, Vol. 112, © 1993 Birkhäuser Verlag Basel

A relation between cubature formulae of trigonometric degree and lattice rules

Marc Beckers and Ronald Cools

Abstract. Part of this paper contains a summary of results on cubature formulae of trigonometric degree that appeared in the Russian literature. By proving a condition for the weights, we show that such formulae can be approached with the tools used to construct cubature formulae of algebraic degree. We also approach them from the field of lattice rules. A new family of cubature formulae of trigonometric degree with the lowest possible number of points is presented.

1 Trigonometric polynomials

A trigonometric monomial in the variables $x_1, x_2, \ldots, x_n$ is a function of the form

$$f : \mathbb{C}^n \to \mathbb{C} : (x_1, x_2, \ldots, x_n) \mapsto e^{2\pi i \alpha_1 x_1} . e^{2\pi i \alpha_2 x_2} \ldots e^{2\pi i \alpha_n x_n} \tag{1}$$

where $\alpha_1, \alpha_2, \ldots, \alpha_n \in \mathbb{Z}$ and $i^2 = -1$.
The degree of a monomial (1) is $\sum_{j=1}^{n} |\alpha_j|$. The space of all trigonometric polynomials in n variables is denoted by $\mathbb{T}^n$. The subspace of $\mathbb{T}^n$ with only the polynomials of degree at most d is denoted by $\mathbb{T}^n_d$.
Let $\tau(n, d)$ denote the number of monomials (1) in n variables of degree d. Using combinatorics one can derive (see Mysovskikh 1987) the identity

$$\begin{aligned} \tau(n,0) &= 1 \\ \tau(n,d) &= \sum_{j=1}^{n} \binom{n}{j} \binom{d-1}{j-1} 2^j \text{ , for } d \geq 1, \end{aligned}$$

where $\binom{a}{b} = 0$ if $b > a$.
Let $t(n, d)$ denote the number of monomials (1) in n variables of degree at most d. Then

$$t(n,d) = \dim \mathbb{T}^n_d = \sum_{j=0}^{d} \tau(n,j) = \sum_{j=0}^{n} \binom{n}{j} \binom{d}{j} 2^j$$

(see Mysovskikh 1987). The following equalities are interesting and not too difficult to derive.

$$t(n,d) = t(d,n) \tag{2}$$

$$\tau(n,d) + \tau(n-1,d) = 2t(n-1,d) \tag{3}$$

$$n\tau(d,n) = d\tau(n,d)$$

$$2t(n-1,d-1) = \tau(n,d) - \tau(n-1,d)$$

Lemma 1.1 *The number of monomials (1) of degree at most d, with the same parity as d is* $\frac{\tau(d+1,n)}{2}$.

Proof.

a) $d=0$: There is only 1 monomial of degree 0 and $\frac{\tau(1,n)}{2} = 1$.

b) $d=1$: There are 2n monomials of degree 1 and

$$\frac{\tau(2,n)}{2} = \frac{1}{2}\left(\binom{2}{1}\binom{n-1}{0}2 + \binom{2}{2}\binom{n-1}{1}4\right) = 2n$$

c) Assume the lemma is true for $d-1$.

The number of monomials of degree $\leq d$ with the same parity as d

= the number of monomials of degree $\leq d$

minus

the number of monomials of degree $\leq d-1$ with the same parity as $d-1$

= $t(n,d) - \frac{\tau(d,n)}{2}$

= $t(d,n) - \frac{\tau(d,n)}{2}$ (because of (2))

= $\left(\frac{\tau(d+1,n)}{2} + \frac{\tau(d,n)}{2}\right) - \frac{\tau(d,n)}{2}$ (because of (3))

= $\frac{\tau(d+1,n)}{2}$.

□

2 Cubature formulae of trigonometric degree

2.1 What is it?

We are concerned with integrals

$$I[f] = \int_0^1 \cdots \int_0^1 f(x_1,\ldots,x_n)dx_1\ldots dx_n = \int_{[0,1]^n} f(\mathbf{x})d\mathbf{x}. \tag{4}$$

(In the Russian literature, the region $[0,2\pi]^n$ is used.) It is often desirable to approximate $I[f]$ by a weighted sum of function values

$$I[f] \simeq Q[f] := \sum_{j=1}^{N} w_j f(\mathbf{x}^{(j)}) \tag{5}$$

with $w_j \in \mathbb{R}$ and $\mathbf{x}^{(j)} \in [0,1]^n$. If $n=1$ then Q is called a *quadrature formula*. If $n \geq 2$ then Q is called a *cubature formula*.

Definition 2.1 *A cubature formula Q (5) for an integral I (4) has trigonometric degree d if $Q[f] = I[f], \forall f \in \mathbb{T}_d^n$ and if there exists a $g \in \mathbb{T}_{d+1}^n$ such that $Q[g] \neq I[g]$.*

Consider the following group of linear transformations:

$$\begin{aligned} \mathcal{G}_c \ := \ & \{(x_1, x_2, \ldots, x_n) \mapsto (x_1, x_2, \ldots, x_n), \\ & (x_1, x_2, \ldots, x_n) \mapsto (x_1 + \tfrac{1}{2}, x_2 + \tfrac{1}{2}, \ldots, x_n + \tfrac{1}{2})\}. \end{aligned}$$

An integral I with an integrand with period 1 with respect to each variable separately and a cubature formula Q are called *centrally symmetric* if they are invariant with respect to $\mathcal{G}_c$. At first sight, this is a strange definition. One should realise that this symmetry for the integrals and cubature formulae we are currently interested in, has similar consequences as the more familiar central symmetry for cubature formulae of algebraic degree. Indeed, for trigonometric monomials f we know that if $\deg(f)$ is odd then $f(\mathbf{x}) = -f(\mathbf{x} + (\frac{1}{2}, \ldots, \frac{1}{2}))$ and if $\deg(f)$ is even then $f(\mathbf{x}) = f(\mathbf{x} + (\frac{1}{2}, \ldots, \frac{1}{2}))$. Furthermore, if f is an odd degree monomial , then $I[f] = 0$ and a centrally symmetric cubature formula is exact for such monomials.

2.2 How many knots do they have?

The analogous of Tchakalov's theorem (1957) for trigonometric polynomials is presented by Mysovskikh (1987) .

Theorem 2.1 *For the integral I (4) there exists a cubature formula Q (5) which is exact for all trigonometric polynomials of degree at most d and whose $N = \dim \mathbb{T}_d^n = t(n,d)$ knots belong to $[0,1]^n$.*

The analogous of the traditional lower bound has been proved by Mysovskikh (1987) .

Theorem 2.2 *The number of knots of a cubature formula (5) which is exact for all trigonometric polynomials of degree at most d satisfies $N \geq \dim \mathbb{T}_{\lfloor d/2 \rfloor}^n = t(n, \lfloor d/2 \rfloor)$.*

Let G_k be the span of trigonometric monomials of degree at most k with the same parity as k.

Theorem 2.3 *The number of knots N of a cubature formula (5) for the integral (4) which is exact for all trigonometric polynomials of degree at most $d = 2k + 1$ satisfies $N \geq 2 \dim G_k = \tau(k+1, n)$.*

Proofs of this theorem are given by Noskov (1985) and Mysovskikh (1987) .

2.3 What about the weights?

In this section we will illustrate that cubature formulae of trigonometric degree can be approached with the same tools as cubature formulae of algebraic degree.
An integral I defines an inner product

$$(\phi, \psi) = I[\overline{\phi} \cdot \psi]. \tag{6}$$

Let $\mathbb{F}$ be a subspace of $\mathbb{T}^n$. Choose the polynomials $\phi_1(\mathbf{x}), \phi_2(\mathbf{x}), \ldots \in \mathbb{F}$ so that $\phi_i(\mathbf{x})$ is I-orthogonal to $\phi_j(\mathbf{x})$, $\forall j < i$, and $(\phi_i(\mathbf{x}), \phi_i(\mathbf{x})) = 1$. Note that the trigonometric monomials form an orthonormal sequence $\{\phi_j\}_{j=1}^{\infty}$. We order the sequence such that $\deg(\phi_j) \leq \deg(\phi_{j+1})$.
For a given $k \in \mathbb{N}$ we set $s := \dim(\mathbb{F} \cap \mathbb{T}_k^n)$ and

$$K(\mathbf{x}, \mathbf{y}) := \sum_{j=1}^{s} \overline{\phi_j(\mathbf{x})} \cdot \phi_j(\mathbf{y}). \tag{7}$$

$K(\mathbf{x}, \mathbf{y})$ is a polynomial in $2n$ variables of degree $\leq 2k$. K is a *reproducing kernel* in the space $\mathbb{F} \cap \mathbb{T}_k^n$ with the inner product (6) :

$$\text{if } f \in \mathbb{F} \cap \mathbb{T}_k^n \text{ then } f(\mathbf{a}) = (f(\mathbf{x}), K(\mathbf{x}, \mathbf{a})) = \sum_{j=1}^{s} \phi_j(\mathbf{a}) \cdot I[f(\mathbf{x})\overline{\phi_j}(\mathbf{x})]. \tag{8}$$

Using the inner product (6) and the reproducing kernel (7), we can prove the equivalent of Theorem 2 of Cools and Haegemans (1988) for cubature formulae of trigonometric degree.

Theorem 2.4 *Let $\mathcal{G}$ be a transformation group. Let $\mathbb{F}$ be a subspace of $\mathbb{T}^n$ such that if ψ_1 and $\psi_2 \in \mathbb{F} \Rightarrow \psi_1.\psi_2$ is invariant with respect to $\mathcal{G}$. Let V_k be the set of common zeros of all elements of $\mathbb{F} \cap \mathbb{T}_k^n$ and $Q_{\mathcal{G}}(\mathbf{y})[f]$ the average of the function values over all points of the $\mathcal{G}$-orbit of $\mathbf{y}$. If*

$$Q[f] := \sum_{j=1}^{\alpha} w_j Q_{\mathcal{G}}(\mathbf{y}^{(j)})[f] + \sum_{j=1}^{\beta} v_j Q_{\mathcal{G}}(\mathbf{z}^{(j)})[f] \tag{9}$$

is an invariant cubature formula of degree $\geq 2k$ with respect to $\mathcal{G}$ for the integral I with $\mathbf{y}^{(j)} \notin V_k$, $\mathbf{z}^{(j)} \in V_k$ and if $\alpha \leq s := \dim(\mathbb{F} \cap \mathbb{T}_k^n)$ then $w_j > 0$, $\forall j \in \{1, 2, \ldots, \alpha\}$.

Proof. Let $K(\mathbf{x}, \mathbf{y})$ be the reproducing kernel (7) in the space $\mathbb{F} \cap \mathbb{T}_k^n$,
$F_j(\mathbf{x}) := K(\mathbf{x}, \mathbf{y}^{(j)}) \in \mathbb{T}_k^n, j = 1, 2, \ldots, \alpha$, and A the symmetric matrix

$$A = \begin{pmatrix} F_1(\mathbf{y}^{(1)}) & \cdots & F_1(\mathbf{y}^{(\alpha)}) \\ \vdots & & \vdots \\ F_\alpha(\mathbf{y}^{(1)}) & \cdots & F_\alpha(\mathbf{y}^{(\alpha)}) \end{pmatrix}. \tag{10}$$

A^{-1} exists, otherwise there is a polynomial of degree $\leq k$ that becomes zero at all knots of the cubature formula and that is not possible for a cubature formula of degree $\geq 2k$. From (8) follows that $F_l(\mathbf{y}^{(j)}) = I[F_l(\mathbf{x})F_j(\mathbf{x})]$ with $F_l(\mathbf{x})F_j(\mathbf{x})$ a polynomial of degree $\leq 2k$, so the cubature formula (9) can be used:

$$F_l(\mathbf{y}^{(j)}) = \sum_{m=1}^{\alpha} w_m Q(\mathbf{y}^{(m)})[F_l F_j] + \sum_{m=1}^{\beta} v_m Q(\mathbf{z}^{(m)})[F_l F_j].$$

Because the polynomials $F_l F_j$ are invariant with respect to $\mathcal{G}$ and $\mathbf{z}^{(j)} \in V_k$

$$F_l(\mathbf{y}^{(j)}) = \sum_{m=1}^{\alpha} w_m F_l(\mathbf{y}^{(m)}) F_j(\mathbf{y}^{(m)}).$$

Thus

$$A = A \begin{pmatrix} w_1 & & 0 \\ & \ddots & \\ 0 & & w_\alpha \end{pmatrix} A^T \Longleftrightarrow A^{-1} = \begin{pmatrix} w_1 & & 0 \\ & \ddots & \\ 0 & & w_\alpha \end{pmatrix}$$

$$\Longleftrightarrow A = \begin{pmatrix} 1/w_1 & & 0 \\ & \ddots & \\ 0 & & 1/w_\alpha \end{pmatrix}. \tag{11}$$

Comparing (10) and (11) gives

$$\begin{aligned} F_l(\mathbf{y}^{(j)}) = K(\mathbf{y}^{(j)}, \mathbf{y}^{(l)}) &= 0 \quad if \ l \neq j \\ &= \tfrac{1}{w_j} \quad if \ l = j. \end{aligned}$$

This gives $w_j = (K(\mathbf{y}^{(j)}, \mathbf{y}^{(j)}))^{-1}$ and because $\mathbf{y} \notin V_k$, $w_j > 0$ for $j = 1, 2, \ldots, \alpha$. □

Corollary 2.1 *If $\mathcal{G}$ is such that the orthonormal basis of $\mathbb{F}$ contains only trigonometric monomials, then the weights w_j, $j \in \{1, 2, \ldots, \alpha\}$ in (9) are all equal to $1/s$.*

Proof. From the proof of Theorem 2.4 we know that

$$w_l^{-1} = K(\mathbf{y}^{(l)}, \mathbf{y}^{(l)}) = \sum_{j=1}^{s} \overline{\phi_j(\mathbf{y}^{(l)})} \cdot \phi_j(\mathbf{y}^{(l)}) = \sum_{j=1}^{s} |\phi_j(\mathbf{y}^{(l)})|^2.$$

If ϕ_j is a trigonometric monomial then $|\phi_j| = 1$. Consequently

$$w_l^{-1} = \sum_{j=1}^{s} 1 = s.$$

□

Corollary 2.2 *If a cubature formula of degree $2k$ has $t(n,k)$ knots, then all weights are equal.*

Proof. In this case $\mathcal{G}$ is the group with one element. Let $\mathbb{F} := \mathbb{T}^n$, then $V_k = \emptyset$ and $s = t(n,k)$. Theorem 2.4 now gives the result. □

Corollary 2.3 *If a central symmetric cubature formula of degree $2k+1$ has $\tau(k+1,n)$ knots, then all weights are equal.*

Proof. In this case $\mathcal{G} := \mathcal{G}_c$. Let $\mathbb{F}$ be the space of all trigonometric monomials of the same parity as k, then $V_k = \emptyset$ and $s = \frac{\tau(k+1,n)}{2}$.
The minimal number of knots in a cubature formula of degree $2k+1$ for a central symmetric integral is, according to Theorem 2.3, $\tau(k+1,n)$. This means $\frac{\tau(k+1,n)}{2}$ generators of 2 points. Theorem 2.4 gives the result. □

2.4 What is known?

In this section we give an overview of the cubature formulae of trigonometric degree that we found in the literature.

By $\{\mathbf{a}\}$ we denote the vector each of whose components is the fractional part of the corresponding component of $\mathbf{a}$, lying in $[0,1)$.

Known minimal cubature formulae that illustrate Theorem 2.2 and Corollary 2.2 are:

- $n = 1$, $d = 2k$, $N = 2k+1$, $k \in \mathbb{N}$: the shifted rectangular rule (Mysovskikh 1985)
- $n = 3$, $d = 2$, $N = 7$: (Noskov 1988)

$$Q[f] = \frac{1}{7}\sum_{j=1}^{7} f\left(\left\{\frac{j}{7}, \frac{3j}{7}, \frac{9j}{7}\right\}\right)$$

Known minimal cubature formulae that illustrate Theorem 2.3 and Corollary 2.3 are:

- $n \in \mathbb{N}$, $d = 1$, $N = 2$: (Mysovskikh 1990)

$$Q[f] = \frac{1}{2}(f(x_1,\ldots,x_n) + f(x_1+\tfrac{1}{2},\ldots,x_n+\tfrac{1}{2})) \text{ with } x_j \in [0,\tfrac{1}{2})$$

- $n \in \mathbb{N}$, $d = 3$, $N = 4n$, (Noskov 1988)

$$Q[f] = \frac{1}{4n}\sum_{j=1}^{4n} f\left(\left\{\frac{j}{4n}, \frac{3j}{4n}, \ldots, \frac{(2n-1)j}{4n}\right\}\right)$$

- $n = 1$, $d = 2k+1$, $N = 2k+2$, $k \in \mathbb{N}$: the shifted rectangular rule (Mysovskikh 1985)
- $n = 2$, $d = 2k+1$, $N = 2(k+1)^2$, $k \in \mathbb{N}$: the body-centred cubic lattice (Reztsov 1990)

$$Q[f] = \frac{1}{2(k+1)^2}\sum_{m=0}^{2k+1}\sum_{j=0}^{k} f\left(\frac{2j+m}{2(k+1)}, \frac{m}{2(k+1)}\right)$$

- $n = 2$, $d = 3$, $N = 8$: (Noskov 1988)

$$Q[f] = \frac{1}{8}\sum_{m,j=0}^{1}\left[f\left(\frac{m}{2}, \frac{j}{2}\right) + f\left(\frac{m}{2}+\frac{1}{4}, \frac{j}{2}+\frac{1}{4}\right)\right]$$

- $n = 2$, $d = 3$, $N = 8$: (Noskov 1988)

$$Q[f] = \frac{1}{8}\sum_{j=1}^{8} f\left(\left\{\frac{j}{8}, \frac{3j}{8}\right\}\right)$$

- $n = 3$, $d = 5$, $N = 38$: (Frolov 1977 and Noskov 1988)

$$Q[f] = \frac{1}{38}\sum_{j=1}^{38} f\left(\left\{\frac{j}{38}, \frac{7j}{38}, \frac{27j}{38}\right\}\right)$$

  The formula presented by Frolov (1977) is, apart from some typographical errors, geometrically equivalent with the one given by Noskov (1988).

3 A short course on lattice rules

In the previous section we proved that minimal formulae of trigonometric degree are equal weight formulae and this property also holds for lattice rules. Observe that all published formulae of trigonometric degree with a minimal number of knots, are lattice rules.

Definition 3.1 *A multiple integration lattice L in $\mathbb{R}^n$ is a subset of $\mathbb{R}^n$ which is discrete and closed under addition and subtraction and which contains $\mathbb{Z}^n$ as a subset.*

A *lattice rule* for approximating $I[f]$ is a cubature formula where the N knots are the points of a multiple integration lattice L that lie in $[0,1]^n$ and the weights are all equal to $1/N$.

Definition 3.2 *The dual of the lattice L is $L^\perp := \{\mathbf{r} \in \mathbb{R}^n : \mathbf{r}\cdot\mathbf{x} \in \mathbb{Z}, \forall \mathbf{x} \in L\}$.*

The dual of a multiple integration lattice is a subset of $\mathbb{Z}^n$. It plays an important role in the error representation. Assume that f can be expanded into an absolutely convergent multiple Fourier series:

$$f(\mathbf{x}) = \sum_{\mathbf{r}\in\mathbb{Z}^n} a(\mathbf{r}) e^{2\pi i \mathbf{r}\cdot\mathbf{x}}, \qquad \mathbf{x} \in I_n$$

with

$$a(\mathbf{r}) = \int_{I_n} e^{-2\pi i \mathbf{r}\cdot\mathbf{x}} f(\mathbf{x}) d\mathbf{x} .$$

Theorem 3.1 *Let L be a multiple integration lattice. Then the corresponding lattice rule Q has an error*

$$Q[f] - I[f] = \sum_{\mathbf{r}\in L^\perp}{}' a(\mathbf{r}),$$

where the prime indicates that the $\mathbf{r} = \mathbf{0}$ term is to be omitted from the sum.

Proof. See Sloan and Kachoyan (1987) . □

A popular construction criterion for lattice rules is the *Zaremba index* (1972) . The Zaremba index is defined as

$$\rho(Q) := \min_{\substack{\mathbf{r} \neq \mathbf{0} \\ \mathbf{r} \in L^\perp}} (\bar{r}_1 \bar{r}_2 \ldots \bar{r}_n) \text{ with } \bar{r}_j := \begin{cases} 1 & \text{if } r_j = 0 \\ |r_j| & \text{if } r_j \neq 0. \end{cases}$$

The trigonometric degree of a lattice rule can also be seen from the dual lattice. The *trigonometric degree* is

$$\delta(Q) := \min_{\substack{\mathbf{r} \neq \mathbf{0} \\ \mathbf{r} \in L^\perp}} \left(\sum_{j=1}^{n} |r_j| \right) - 1 .$$

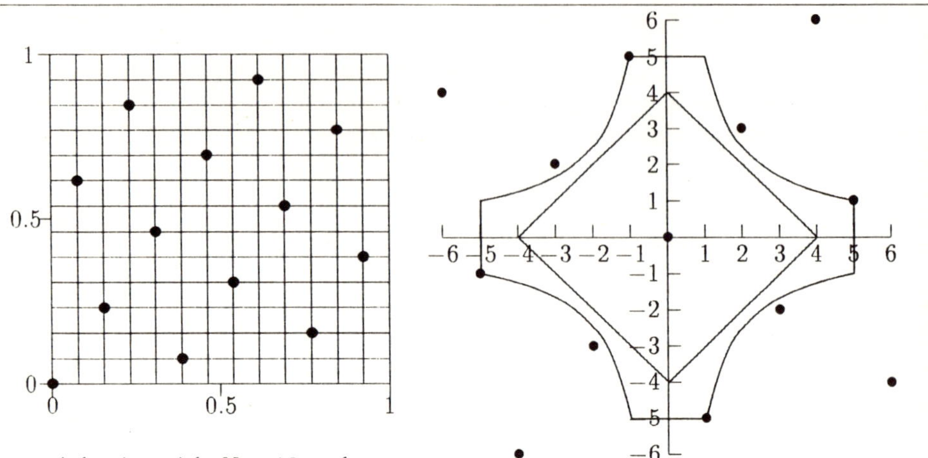

A lattice with $N = 13$ and generator $(\frac{1}{13}, \frac{8}{13})$

The dual lattice

In the left part of the figure, we present the Fibonacci lattice rule with 13 points (see section 4.2). The right part is a picture of the dual lattice. The trigonometric degree depends on the biggest octahedron that does not contain a point of the dual lattice. It is 4 in this example. The Zaremba index depends on the biggest Zaremba-cross with a point of the dual lattice on its boundary. In this example it is 5.

4 Illustrations

4.1 Minimal cubature formulae for 2-dimensions

First, we will study the following lattice rules:

$$Q[f] = \frac{1}{N} \sum_{j=0}^{N-1} f\left(\frac{j}{N}, \frac{j(2k+1)}{N}\right) \ , \ N = 2(k+1)^2 \tag{12}$$

and

$$Q[f] = \frac{1}{N} \sum_{j=0}^{N-1} f\left(\frac{j}{N}, \frac{j(2k+3)}{N}\right) \ , \ N = 2(k+1)^2 .$$

These rules are geometrically equivalent (Beckers 1992), so we only have to study one of them.

Theorem 4.1 *Let $k \in \mathbb{N}$. The cubature formula (12) has trigonometric degree $2k+1$. The Zaremba index of this cubature formula is also $2k+1$.*

Proof. The dual lattice that corresponds to the lattice rule (12) is

$$L^{\perp} = \left\{ \mathbf{r} \in \mathbb{Z}^2 : \mathbf{r} \cdot \mathbf{p} = 0 \bmod N \ \text{ with } \ \mathbf{p} = (1, 2k+1) \right\} . \tag{13}$$

The formula has at least degree $2k+1$ if the points of the dual lattice except the origin, satisfy $|r_1| + |r_2| \geq 2k + 2$. So, we have to show that

$$r_1 + (2k+1)r_2 \neq pN \text{ for } |r_1| + |r_2| \leq 2k+1 \text{ and } p \in \mathbb{Z}$$

or

$$|r_1 + (2k+1)r_2| \neq pN \text{ for } |r_1| + |r_2| \leq 2k+1 \text{ and } p \in \mathbb{N}.$$

Suppose there exists a (r_1, r_2) such that

$$|r_1 + (2k+1)r_2| = pN \text{ for } |r_1| + |r_2| \leq 2k+1 \text{ and } p \in \mathbb{N}.$$

Because $|r_1| + |r_2| \leq 2k + 1$

$$|r_1 + (2k+1)r_2| \leq (2k+1)^2 < 2N.$$

So $p = 0$ or $p = 1$.

1. $p = 0$:
   This means that $r_1 + (2k+1)r_2 = 0$ thus $r_1 = -r_2(2k+1)$. This is impossible because $0 \leq |r_2| \leq 2k+1$ and $0 \leq |r_1| + |r_2| \leq 2k+1$.

2. $p = 1$:

   (a) $r_1 + (2k+1)r_2 = N = 2k^2 + 4k + 2$:
   In this case $r_2 = k+1$ because $(k+1)(2k+1) < N < (k+2)(2k+1)$. Consequently $r_1 = k+1$. This is not allowed because in that case $|r_1|+|r_2| = 2k+2 > 2k+1$

   (b) $r_1 + (2k+1)r_2 = -N$: Analogous to the previous item, one can show that this is also impossible.

It is already mentioned that $(k+1, k+1)$ belongs to the dual lattice. Hence the degree is lower than $2k+2$. This concludes the first part of the proof.
The Zaremba index of the formula is $2k+1$ if the points of the dual lattice except the origin, satisfy $\bar{r}_1\bar{r}_2 \geq 2k+1$. So, we have to show that

$$r_1 + (2k+1)r_2 \neq pN \text{ for } \bar{r}_1\bar{r}_2 \leq 2k \text{ and } p \in \mathbb{Z}$$

or

$$|r_1 + (2k+1)r_2| \neq pN \text{ for } \bar{r}_1\bar{r}_2 \leq 2k \text{ and } p \in \mathbb{N}.$$

Suppose there exists a (r_1, r_2) such that

$$|r_1 + (2k+1)r_2| = pN \text{ for } \bar{r}_1\bar{r}_2 \leq 2k \text{ and } p \in \mathbb{N}.$$

Because $\bar{r}_1\bar{r}_2 \leq 2k$

$$|r_1 + (2k+1)r_2| \leq 2k(2k+1) < 2N.$$

So $p = 0$ or $p = 1$.

1. $p = 0$:
   This means that $r_1 + (2k+1)r_2 = 0$ thus $r_1 = -r_2(2k+1)$. This is impossible because $0 \leq |r_2| \leq 2k$ and $0 \leq \bar{r}_1\bar{r}_2 \leq 2k$.

2. $p = 1$:

   (a) $r_1 + (2k+1)r_2 = N = 2k^2 + 4k + 2$:
   In this case $r_2 = k+1$ because $(k+1)(2k+1) < N < (k+2)(2k+1)$. Consequently $r_1 = k+1$. This is not allowed because in that case $\bar{r}_1\bar{r}_2 = (k+1)^2 > 2k$

   (b) $r_1 + (2k+1)r_2 = -N$:Analogous to the previous item, one can show that this is also impossible.

A point of the dual lattice that is on the boundary of the Zaremba-cross, is $(2k+1, -1)$. □

4.2 Fibonacci lattice rules

Let F_k be the k-th Fibonacci number, defined by $F_1 = 1$, $F_2 = 1$ and $F_k = F_{k-1} + F_{k-2}$ for $2 < k \in \mathbb{N}$. Consider the following lattice rules:

$$Q[f] = \frac{1}{F_k} \sum_{j=0}^{F_k - 1} f\left(\frac{j}{F_k}, \frac{jF_{k-1}}{F_k}\right) \tag{14}$$

and

$$Q[f] = \frac{1}{F_k} \sum_{j=0}^{F_k - 1} f\left(\frac{j}{F_k}, \frac{jF_{k-2}}{F_k}\right).$$

These rules are geometrically equivalent, so we only have to study one of them.

Theorem 4.2 *Let $m \geq 2$. If $k = 2m+1$ then the cubature formula (14) has trigonometric degree $F_{m+2} - 1$.*

Proof. The cubature formula (14) with $k = 2m+1$ is a lattice rule with generator matrix of the corresponding lattice:

$$M = \begin{pmatrix} \frac{1}{F_{2m+1}} & \frac{F_{2m}}{F_{2m+1}} \\ 0 & 1 \end{pmatrix}.$$

The rows of M generate the lattice. A generator matrix of the corresponding dual lattice is (see Sloan and Kachoyan 1987)

$$B = (M^{-1})^T = \begin{pmatrix} F_{2m+1} & 0 \\ -F_{2m} & 1 \end{pmatrix}.$$

The matrix

$$C = \begin{pmatrix} F_m & F_{m+1} \\ -F_{m-1} & -F_m \end{pmatrix}$$

is an integer unimodular matrix:

$$\det(C) = -F_m^2 + F_{m+1}F_{m-1} = (-1)^m F_1 = (-1)^m.$$

Let

$$A = \begin{pmatrix} F_m & (-1)^{m+1}F_{m+1} \\ F_{m+1} & (-1)^m F_m \end{pmatrix}$$

then

$$CA = \begin{pmatrix} F_m & F_{m+1} \\ -F_{m-1} & -F_m \end{pmatrix} \begin{pmatrix} F_m & (-1)^{m+1}F_{m+1} \\ F_{m+1} & (-1)^m F_m \end{pmatrix}$$

$$= \begin{pmatrix} F_m^2 + F_{m+1}^2 & (-1)^{m+1}F_{m+1}F_m + (-1)^m F_m F_{m+1} \\ -F_{m-1}F_m - F_m F_{m+1} & (-1)^m F_{m+1}F_{m-1} + (-1)^{m+1}F_m^2 \end{pmatrix} = B$$

because

$$\begin{aligned} F_m^2 + F_{m+1}^2 &= F_{2m+1} \\ F_m(F_{m-1} + F_{m+1}) &= F_{2m} \\ (-1)^{m-1}(F_m^2 - F_{m+1}F_{m-1}) &= F_1 = 1. \end{aligned}$$

So, A is another generator matrix for the dual lattice. Thus the dual lattice contains the points

$$((j_1F_m + j_2F_{m+1}), (-1)^m(-j_1F_{m+1} + j_2F_m)) \text{ with } j_1, j_2 \in \mathbb{Z}.$$

The cubature formula has trigonometric degree $F_{m+2} - 1$ if

$$|j_1F_m + j_2F_{m+1}| + |-j_1F_{m+1} + j_2F_m| \geq F_{m+2} \quad \forall j_1, j_2 \in \mathbb{Z} \text{ with } j_1 \text{ or } j_2 \neq 0. \tag{15}$$

If $j_1 = 0$ then $|j_1F_m + j_2F_{m+1}| + |-j_1F_{m+1} + j_2F_m| = |j_2|(F_{m+1} + F_m) \geq F_{m+2}$.

If $j_2 = 0$ then $|j_1F_m + j_2F_{m+1}| + |-j_1F_{m+1} + j_2F_m| = |j_1|(F_m + F_{m+1}) \geq F_{m+2}$.
If $j_1j_2 > 0$ then $|j_1F_m + j_2F_{m+1}| + |-j_1F_{m+1} + j_2F_m| \geq |j_1F_m + j_2F_{m+1}| \geq F_{m+2}$.
If $j_1j_2 < 0$ then $|j_1F_m + j_2F_{m+1}| + |-j_1F_{m+1} + j_2F_m| \geq |-j_1F_{m+1} + j_2F_m| \geq F_{m+2}$.
If ($j_1 = 1$ and $j_2 = 0$) or ($j_1 = 0$ and $j_2 = 1$) then (15) becomes an equality. □

Theorem 4.3 *If $k = 2m$, the cubature formula (14) has trigonometric degree $2F_m - 1$.*

Proof. The proof of this theorem is analogous to the proof of Theorem 4.2. This time, one can use

$$A = \begin{pmatrix} F_m & (-1)^m F_m \\ F_{m-1} & (-1)^{m+1}F_{m+1} \end{pmatrix}$$

as an alternative generator matrix of the dual lattice. □

References

M. Beckers (1992), *Numerical integration in high dimensions.* PhD thesis, Katholieke Universiteit Leuven.

R. Cools and A. Haegemans (1988), Why do so many cubature formulae have so many positive weights ? *BIT*, 28:792–802.

K.K. Frolov (1977), On the connection between quadrature formulas and sublattices of the lattice of integral vectors. *Dokl. Akad. Nauk SSSR*, 232:40–43. (Russian) Soviet Math. Dokl. 18: 37–41 (English).

I.P. Mysovskikh (1985) Quadrature formulae of the highest trigonometric degree of accuracy. *Zh. vychisl. Mat. mat. Fiz.*, 25:1246–1252. (Russian) *U.S.S.R. Comput. Maths. Math. Phys.* 25: 180–184 (English).

I.P. Mysovskikh (1987), On cubature formulas that are exact for trigonometric polynomials. *Dokl. Akad. Nauk SSSR*, 296:28–31. (Russian) *Soviet Math. Dokl.* 36: 229–232, 1988 (English).

I.P. Mysovskikh (1990), On the construction of cubature formulas that are exact for trigonometric polynomials. In A. Wakulicz, editor, *Numerical Analysis and Mathematical Modelling*, volume 24 of *Banach Center Publications*, pages 29–38. PWN - Polish Scientific Publishers, Warsaw. (Russian).

M.V. Noskov (1985), Cubature formulae for the approximate integration of periodic functions. *Metody Vychisl. Vyp.*, 14:15–23. (Russian).

M.V. Noskov (1988), Cubature formulae for the approximate integration of functions of three variables. *Zh. vychisl. Mat. mat. Fiz.*, 28:1583–1586. (Russian) *U.S.S.R. Comput. Maths. Math. Phys.* 28: 200-202 (English).

A. Reztsov (1990), Gauss-type cubature formulas with an asymptotically minimal number of nodes. *Mat. Zametki*, 48:151–152. (Russian).

I.H. Sloan and P.J. Kachoyan (1987) Lattice mathods for multiple integration: theory, error analysis and examples. *SIAM J. Numer. Anal.*, 24:116–128.

M.V. Tchakaloff (1957), Formules de cubatures mécaniques à coefficients non négatifs. *Bull. des Sciences math.*, 2e *série*, 81:123–134.

S.K. Zaremba (1972), La méthode des "bons treillis" pour le calcul des intégrales multiples. In S.K. Zaremba, editor, *Applications of Number Theory to Numerical Analysis*, pages 39–116. Academic Press.

Authors' addresses:

Marc Beckers
Alcatel Bell Telephone Research Centre
Access Networks Department
Francis Wellesplein 1
B-2018 Antwerpen, Belgium
mbec@ra.alcatel.be

Ronald Cools
Department of Computer Science
Katholieke Universiteit Leuven
Celestijnenlaan 200 A
B-3001 Leuven, Belgium
Ronald.Cools@cs.kuleuven.ac.be

Distribution of Points in Convergent Sequences of Interpolatory Integration Rules: The Rates

Tom Bloom, Doron S. Lubinsky and Herbert B. Stahl

Abstract. Suppose that we are given a convergent sequence of interpolatory integration rules $\{I_n\}_{n=1}^{\infty}$. What can we say about the *distribution* of the points in the rules? We review previous results, which show that *half* the points in the rules behave like zeros of appropriate orthogonal polynomials, and half may be arbitrarily distributed. In the case of the interval $(-1,1)$, this usually means that half the points have *arcsin distribution*. We also present and prove a new result relating the *rate* at which half the points converge to the arcsin distribution, and the *size of the weights* in the rule I_n.

1. Introduction and Results

Let $w : [-1,1] \to [0,\infty)$ be a non–negative integrable function, and with

$$\int_{-1}^{1} w(x)dx > 0.$$

We call w a *weight*. Associated with this weight we may consider integration rules

$$I_n[f] := \sum_{j=1}^{n} w_{jn} f(x_{jn}), \tag{1.1}$$

with *points* x_{jn} in $[-1,1]$. The rules are called *interpolatory* if they integrate exactly polynomials of degree $\leq n-1$, that is

$$I_n[P] = \int_{-1}^{1} Pw, \quad P \in \mathcal{P}_{n-1}, \tag{1.2}$$

where $\mathcal{P}_m$ denotes the set of polynomials of degree $\leq m$.

Suppose the rules converge for all continuous $f : [-1,1] \to \mathbb{R}$, so

$$\lim_{n\to\infty} I_n[f] = \int_{-1}^{1} fw. \tag{1.3}$$

What can then be said about the *points* x_{jn} and *weights* w_{jn}? A classical result of Polya (see for example, Davis and Rabinowitz 1984) asserts that (1.3) is true for every continuous f iff the *norms*

$$\|I_n\| := \sum_{j=1}^{n} |w_{jn}| \tag{1.4}$$

satisfy

$$\sup_n \|I_n\| < \infty. \tag{1.5}$$

Of course, this is a trivial application of the uniform boundedness principle of functional analysis, but in quadrature theory is still referred to as the Polya–Steklov theorem.

This is a result about the weights. What can we say about the points? Surprisingly this had not been answered in general, and recent papers of the authors attended to it (see Bloom et al. 1992a,b,c). For quadrature rules I_n with all non-negative weights, that is

$$w_{jn} \geq 0, \quad 1 \leq j \leq n, \tag{1.6}$$

Bernstein had answered the question in the 1930's, see Bernstein (1936). We thank K.J. Förster for bringing his work to our attention; A more convenient reference for Bernstein's work on positive quadratures is the book of H. Brass (1977), see also Förster (1982).

But before discussing Bernstein's work, let us try and develop some intuition about the problem. The Newton–Cotes rules, which place their points equally spaced in $[-1,1]$, do not even converge for all functions analytic on $[-1,1]$. Since the proportion of such points in any subinterval $[a,b]$ of $[-1,1]$ is $\frac{1}{2}(b-a)$ as the order of the Newton–Cotes rules $\to \infty$, these points are called *uniformly distributed* in $[-1,1]$. So uniform distribution is not a good idea, at least for interpolatory rules.

We know the Gauss rules I_n are good rules: They are interpolatory, they have positive weights, their norms satisfy (1.5), and so they converge on all continuous functions. What distribution do their points have? For weights w positive almost everywhere on $(-1,1)$, they have *arcsin distribution*. Roughly speaking, this means that they behave like the zeros $\cos\left(j-\frac{1}{2}\right)\frac{\pi}{n}$, $1 \leq j \leq n$, of the nth Chebyshev polynomial $T_n(x)$: More or less spaced $\frac{1}{n}$ apart well inside $[-1,1]$, and spaced $\frac{1}{n^2}$ apart near ± 1.

More formally, if $-1 \leq a < b \leq 1$ and $N_n[a,b]$ denotes the number of x_{jn}, $1 \leq j \leq n$, lying in $[a,b]$, then arcsin distribution asserts that

$$\begin{aligned}\lim_{n\to\infty} N_n[a,b]/n &= \int_a^b \frac{dx}{\pi\sqrt{1-x^2}} \\ &= \frac{1}{\pi}\left[\arcsin(b) - \arcsin(a)\right],\end{aligned} \tag{1.7}$$

for every such a, b. So the *proportion* of integration points lying in $[a, b]$, is given by $\frac{1}{\pi}(\arcsin(b) - \arcsin(a))$, as $n \to \infty$.

In particular, the Gauss rules for Jacobi weights have arcsin distribution, and so do the Clenshaw–Curtis rules. So an initial guess was that convergent interpolatory sequences of integration rules always have arcsin distribution. Some further evidence for this guess comes from the theory of Lagrange interpolation.

Recall that if $L_n[f]$ denotes the Lagrange interpolation polynomial to f at x_{jn}, $1 \le j \le n$, then the rule I_n is called interpolatory because

$$I_n[f] = \int_{-1}^{1} L_n[f]w,$$

for every function $f\colon [-1, 1] \to \mathbb{R}$. So the quadrature convergence (1.3) is equivalent to

$$\lim_{n\to\infty} \int_{-1}^{1} \{L_n[f] - f\}w = 0. \tag{1.8}$$

If we insert mod signs in (1.8) and demand that

$$\lim_{n\to\infty} \int_{-1}^{1} |L_n[f] - f|w = 0, \tag{1.9}$$

for every continuous $f\colon [-1, 1] \to \mathbb{R}$, then classical results of Walsh, Fekete, Erdös and Turan (see Erdös and Turan, 1940b) indeed tell us that the points x_{jn} have arcsin distribution. But, as we shall see, (1.8) does *not* imply arcsin distribution. So what a difference a mod sign makes!

The correct intuition is provided by Gauss quadrature, suitably interpreted. Suppose that n is even, and let us consider the Gauss quadrature rule $I_{n/2}$ involving $n/2$ points and weights:

$$I_{n/2}[P] = \sum_{j=1}^{n/2} w_{jn} P(x_{jn}) = \int_{-1}^{1} Pw,$$

degree $(P) \le 2\left(\frac{n}{2}\right) - 1 = n - 1$. Let us choose any $n/2$ additional points x_{jn}, $n/2 + 1 \le j \le n$, and set the corresponding weights $w_{jn} := 0$. Then we obtain a rule I_n with n points and weights and that is interpolatory. *Half* the points in the rule (the $\frac{n}{2}$ Gauss points) are zeros of an orthogonal polynomial of degree $\frac{n}{2}$ for w, and half (the $\frac{n}{2}$ dummy points) are *arbitrary*. We shall see that in a sense, this is the general case.

The results of Bernstein for positive quadratures fit into this pattern: If I_n is an interpolatory quadrature rule with non–negative weights, that is, satisfies (1.1), (1.2) and (1.6), then half the points in I_n interlace the zeros of the orthogonal polynomial of degree $\langle n/2 \rangle$ for w. (Here, and in the sequel, $\langle x \rangle$ denotes the greatest integer $\le x$). More precisely, if s and t

are adjacent zeros of this orthogonal polynomial, then the interval $[s, t]$ contains at least one of the integration points. So for positive quadratures, at least half the points have the same distribution as zeros of orthogonal polynomials.

Bernstein's methods involved properties of positive linear functionals. These do not extend to general non-positive quadratures. We shall see that for general non-positive quadratures the answer is, in a distributional sense, the same, but the methods of proof necessary involve potential theory.

We have already seen that one way to describe distribution involves $N_n[a,b]/n$, the proportion of integration points of I_n in $[a,b]$. An equivalent way involves *counting measures.* Let δ_x denote a Dirac delta or unit mass at x. The *zero counting measure* corresponding to I_n is

$$\mu_n := \frac{1}{n}\sum_{j=1}^{n} \delta_{x_{jn}}, \tag{1.10}$$

that is a probability measure assigning mass $\frac{1}{n}$ to each integration point x_{jn}. (By a probability measure, we mean a non-negative Borel measure with total mass 1.) For $S \subset [-1,1]$,

$$\begin{aligned}\mu_n(S) &= \frac{1}{n} \times \text{ Number of } x_{jn} \text{ in } S\\ &= \text{proportion of points from } I_n \text{ in } S.\end{aligned}$$

In particular

$$\mu_n[a,b] = \int_a^b d\mu_n = N_n[a,b]/n.$$

Thus (1.7) can be recast as

$$\lim_{n\to\infty} \int_a^b d\mu_n(x) = \int_a^b \frac{dx}{\pi\sqrt{1-x^2}},$$

$\forall -1 \le a < b \le 1$, which easily leads to

$$\lim_{n\to\infty} \frac{1}{n}\sum_{j=1}^{n} f(x_{jn}) = \lim_{n\to\infty} \int_a^b f(x)d\mu_n(x) = \int_a^b f(x)\frac{dx}{\pi\sqrt{1-x^2}}$$

$\forall$ continuous $f: [-1,1] \to \mathbb{R}$. We write

$$d\mu_n(x) \xrightarrow{\star} \frac{dx}{\pi\sqrt{1-x^2}},$$

and say that $d\mu_n$ *converges weakly* to $\frac{dx}{\pi\sqrt{1-x^2}}$ as $n \to \infty$.

In the general case, the full sequence of zero counting measures $\{d\mu_n\}_{n=0}^{\infty}$ need not converge weakly. So we have to describe the distribution of the integration points in terms of weak limits of ***subsequences***. We shall say that a probability measure ν is a *weak limit of a subsequence* of $\{d\mu_n\}_{n=1}^{\infty}$ if there exists an infinite sequence $\mathcal{N}$ of positive integers such that

$$\lim_{\substack{n\to\infty \\ n\in\mathcal{N}}} \int_{-1}^{1} f\,d\mu_n = \int_{-1}^{1} f\,d\nu \tag{1.11}$$

for every continuous $f\colon [-1,1] \to \mathbb{R}$, and we write

$$d\mu_n \xrightarrow{\star} d\nu,\ n\to\infty,\ n\in\mathcal{N}.$$

A theorem of Helly (really a special case of the Banach–Alaoglu theorem/weak–$\star$–compactness of functional analysis) asserts that every sequence of probability measures on $[-1,1]$ contains a weakly convergent subsequence. So we can describe the distribution of points of $\{I_n\}_{n=1}^{\infty}$ in terms of the set of all weak limits of subsequences of $\{d\mu_n\}_{n=1}^{\infty}$.

We are now ready to state one of the main results of Bloom et al (1992a). We use the notation (1.1), (1.2), (1.4) and (1.10).

Theorem 1.1. Let w be a weight positive a.e. on $[-1,1]$.

(a) Let $\{I_n\}_{n=1}^{\infty}$ be a sequence of interpolatory integration rules associated with w satisfying

$$\lim_{n\to\infty} \|I_n\|^{1/n} = \lim_{n\to\infty} \left(\sum_{j=1}^{n} |w_{jn}|\right)^{1/n} = 1. \tag{1.12}$$

Then every weak limit of a subsequence of the corresponding counting measures $\{d\mu_n\}_{n=1}^{\infty}$ has the form

$$\frac{1}{2}\,\frac{dx}{\pi\sqrt{1-x^2}} + \frac{1}{2}d\nu(x), \tag{1.13}$$

where ν is a probability measure on $[-1,1]$.

(b) Conversely, given any probability measure ν on $[-1,1]$ there exist interpolatory rules $\{I_n\}_{n=1}^{\infty}$ associated with w with all weights positive, that is,

$$w_{jn} > 0,\ 1 \le j \le n, \tag{1.14}$$

satisfying (1.12), and moreover, exhibiting the convergence (1.3) for every continuous $f\colon [-1,1] \to \mathbb{R}$, and for which,

$$d\mu_n(x) \xrightarrow{\star} \frac{1}{2}\,\frac{dx}{\pi\sqrt{1-x^2}} + \frac{1}{2}d\nu(x),\ n\to\infty. \tag{1.15}$$

Of course if a sequence of interpolatory rules converges on every continuous function, then Polya's result (1.5) trivially implies (1.12). More generally, sequences of rules whose norms

$||I_n||$ grow no faster than a power of n (say rules converging on all functions with a fixed number of derivatives), will satisfy (1.12), and so half their points will have arcsin distribution.

An alternative way to formulate (a) is to say that

$$\liminf_{n\to\infty} N_n[a,b]/n \geq \frac{1}{2}\int_a^b \frac{dx}{\pi\sqrt{1-x^2}},$$

$\forall -1 \leq a < b \leq 1$; The weak convergence in (b) may be reformulated as

$$\lim_{n\to\infty} N_n[a,b]/n = \frac{1}{2}\int_a^b \frac{dx}{\pi\sqrt{1-x^2}} + \frac{1}{2}\int_a^b d\nu(x),$$

$\forall -1 \leq a < b \leq 1$.

It is possible to generalize Theorem 1.1 to rules that integrate polynomials of degree $m_n (\neq n-1)$ exactly: The following result is the main result, Theorem 1.3, of Bloom et al (1992a):

Theorem 1.2. Let w be a weight positive a.e. on $[-1,1]$.

(a) Let $\{I_n\}_{n=1}^{\infty}$ be a sequence of integration rules of the form (1.1), with

$$I_n[P] = \int_{-1}^{1} Pw, \quad \text{degree } (P) \leq m_n, \tag{1.16}$$

some non–negative integers m_n, and

$$\alpha :=\frac{1}{2}\lim_{n\to\infty} m_n/n > 0, \tag{1.17}$$

and let the norms $\{||I_n||\}_{n=1}^{\infty}$ satisfy (1.12). Then every weak limit of a subsequence of the corresponding counting measures $\{d\mu_n\}_{n=1}^{\infty}$ has the form

$$\alpha\frac{dx}{\pi\sqrt{1-x^2}} + (1-\alpha)d\nu(x), \tag{1.18}$$

where ν is a probability measure on $[-1,1]$.

(b) Conversely, given any probability measure ν on $[-1,1]$, there exist rules $\{I_n\}_{n=1}^{\infty}$ of the form (1.1), with the precision (1.16), with all weights positive as in (1.14), with norms satisfying (1.12), and with the convergence (1.3) holding for every continuous function, and for which

$$d\mu_n(x) \xrightarrow{\star} \alpha\frac{dx}{\pi\sqrt{1-x^2}} + (1-\alpha)d\nu(x),\ n\to\infty. \tag{1.19}$$

In the interpolatory case $m_n = n-1$ and so $\alpha = \frac{1}{2}$. For Gauss rules associated with w, $m_n = 2n-1$ and so $\alpha = 1$; Then the "arbitrary" piece $d\nu$ drops out of (1.18), and we recover the classical result of Erdös and Turan (1940b) that the Gauss points, that is the zeros of the orthogonal polynomials for w, have arcsin distribution.

There are analogous results for interpolatory integration rules associated with weights on $\mathbb{R}$. Again, convergent interpolatory integration rules have half their points distributed like zeros of appropriate orthogonal polynomials, and half arbitrarily distributed. Here the arcsin distribution is replaced by a Nevai–Ullman distribution, and the integration points are first scaled to $[-1,1]$ before forming a (now contracted) zero counting measure. See Bloom et al. (1992b).

Theorems 1.1(a) and 1.2(a) assert that a suitable proportion of the integration points converge to the arcsin distribution. However, it is possible to say somewhat more: One can give a *rate* at which the points converge. To do this, we need a suitable notion of *distance* or *discrepancy* between distributions. Given a probability measure ν on $[-1,1]$, we define its *discrepancy from the arcsin distribution* to be

$$\text{Discr}\,(\nu, \text{arcsin}) := \sup_{-1\leq a<b\leq 1}\left|\int_a^b d\nu(x) - \int_a^b \frac{dx}{\pi\sqrt{1-x^2}}\right|. \tag{1.20}$$

We shall estimate discrepancy using a recent breakthrough due to H.P. Blatt (1992), as improved by V. Totik (1992). Classical results of Erdös and Turan showed how to do this for general polynomials; Blatt's breakthrough applies to polynomials with all simple zeros, giving substantially better results.

Our rate will involve the norms of the integration rules, but also involves a Nikolskii–type quantity

$$\tau_n := \sup_{P\in\mathcal{P}_n\setminus\{0\}}\left(\max_{x\in[-1,1]}|P(x)|\Big/\left\{\int_{-1}^{1} P^2(x)w(x)dx\right\}^{1/2}\right). \tag{1.21}$$

It is known (see Stahl and Totik 1992, p.91) that for w positive a.e. in $[-1,1]$,

$$\lim_{n\to\infty} \tau_n^{1/n} = 1. \tag{1.22}$$

However, much more can be said for special weights: For example, for the Jacobi weight $w(x) = (1-x)^\alpha(1+x)^\beta$,

$$C_1 \leq \tau_n/n^{\max\left\{\frac{1}{2},\alpha+1,\beta+1\right\}} \leq C_2,\ n\geq 1, \tag{1.23}$$

where C_1, C_2 are positive constants independent of n (see, for example Nevai 1979, 1986).

Recall that in Theorem 1.2, asymptotically αn of the points have arcsin distribution. Theorem 1.3 shows that $m(n)/2 = \alpha n(1+o(1))$ points converge at a certain rate:

Theorem 1.3. Assume the hypotheses of Theorem 1.2(a), and let

$$\rho_n : = n\tau_n \|I_n\|^{1/2},\ n \geq 1. \tag{1.24}$$

Then for $n \geq 1$, we can choose

$$\sigma(n) : = \langle m_n/2 \rangle \tag{1.25}$$

integration points from I_n, say x_{jn}, $1 \leq j \leq \sigma(n)$, such that if

$$\nu_n : = \frac{1}{\sigma(n)} \sum_{j=1}^{\sigma(n)} \delta_{x_{jn}}, \tag{1.26}$$

then for $n \geq 1$,

$$\text{Discr}\,(\nu_n, \text{arcsin}) \leq C \left(\frac{\log C\rho_n}{n} \right) \log \left(\frac{n}{\log C\rho_n} \right), \tag{1.27}$$

where C is independent of n.

Note that our hypothesis (1.12), and (1.22), ensure that

$$\lim_{n \to \infty} \frac{\log C\rho_n}{n} = 0.$$

Hence the right–hand side of (1.27) converges to 0 as $n \to \infty$, and Theorem 1.3 implies Theorem 1.2(a), but in a more quantitative form.

If τ_n and $\|I_n\|$ grow no faster than a power of n (which is true if w is a Jacobi weight, and the rules converge on all continuous functions), then (1.27) becomes

$$\text{Discr}\,(\nu_n, \text{arcsin}) \leq C_1 \frac{(\log(n+1))^2}{n},\ n \geq 1. \tag{1.28}$$

When $m_n = 2n - 1$, we are dealing with zeros of orthogonal polynomials, and a result of Blatt (1992, p.254) gives (1.28) for quite general w.

We prove the theorem in section 2. We close this section with a note of caution: Distribution of points and *spacing of points* are related, but only loosely. Thus our results neither contain, nor imply, a result of Brass (1977, p.95) which asserts that for a convergent sequence of interpolatory rules associated with the Legendre weight $w \equiv 1$, necessarily for large enough n,

$$\max_{1 \leq j \leq n-1} \left| \arccos x_{j+1,n} - \arccos x_{jn} \right| < \frac{4 \log n}{n}.$$

Here the points have been arranged in increasing/decreasing order (and there was earlier related material of Locher and Zeller).

2. Proofs

We shall use the methods of potential theory, but in a fairly simple form: The difficult side of the potential theory is hidden in the proof of the lemmas. Given a compact set $E \subset \mathbb{C}$, and $\sigma \geq 1$, we say $z_1 \ldots z_\sigma \in E$ are *σth Fekete points* for E if

$$\prod_{1 \leq j < k \leq \sigma} |z_k - z_j| = \max_{y_1 \ldots y_\sigma \in E} \prod_{1 \leq j < k \leq \sigma} |y_k - y_j|.$$

See Chapter 16 of Hille (1962) for an accessible treatment of Fekete points. One of their notable features is that they are good points for Lagrange interpolation:

Lemma 2.1. Let $z_1 \ldots z_\sigma$ be σth Fekete points for a compact set E containing at least σ points, and let

$$q(z) := \prod_{j=1}^{\sigma} (z - z_j),$$

and

$$\ell_j(z) := \frac{q(z)}{q'(z_j)(z - z_j)}, \; 1 \leq j \leq \sigma.$$

Then

(a) $|\ell_j(z)| \leq 1, \; z \in E, \; 1 \leq j \leq \sigma.$

(b) $\max_{z \in E} |q(z)| \leq \operatorname{diam}(E) \min_{1 \leq j \leq \sigma} |q'(z_j)|$, where diam (E) denotes the diameter of E.

(c) If $E \subset [-1, 1]$, then

$$\max_{z \in E} |q(z)| \leq (\sigma + 1) 2^{-\sigma + 1}.$$

Proof. (a) See Theorem 16.2.4, page 273 of Hille (1962).

(b) From (a), we obtain

$$|q(z)| \leq |z - z_j| \, |q'(z_j)|, \; z \in E, \; 1 \leq j \leq \sigma,$$

and then the result follows.

(c) Let

$$T_\sigma(x) = \cos(\sigma \arccos x) = 2^{\sigma - 1} x^\sigma + \cdots$$

denote the Chebyshev polynomial of degree σ. We apply the Lagrange interpolation formula to the polynomial $T_\sigma(x)/2^{\sigma-1} - q(x)$, which has degree at most $\sigma - 1$:

$$T_\sigma(x)/2^{\sigma-1} - q(x) = \sum_{j=1}^{\sigma} (T_\sigma(z_j)/2^{\sigma-1}) \ell_j(x).$$

Applying our estimate of (a), yields, for $x \in E$,

$$\begin{aligned}|q(x)| &\le |T_\sigma(x)|/2^{\sigma-1} + \sum_{j=1}^{\sigma} |T_\sigma(z_j)|/2^{\sigma-1} \\ &\le 2^{-\sigma+1} + \sigma 2^{-\sigma+1}. \ \blacksquare\end{aligned}$$

Lemma 2.2. Let

$$q(x) := \prod_{j=1}^{\sigma} (x - z_j)$$

be a polynomial with all zeros simple, and in $[-1,1]$. Assume that

$$\max_{x\in[-1,1]} |q(x)| \le A/2^\sigma$$

and

$$|q'(z_j)| \ge 1/(B2^\sigma),\ 1 \le j \le \sigma.$$

Let

$$\chi := \max\{A, B, \sigma\} \le e^{\sigma/2}$$

and

$$\nu_\sigma := \frac{1}{\sigma}\sum_{j=1}^{\sigma} \delta_{z_j}.$$

Then

$$\text{Discr}\,(\nu_\sigma, \arcsin) \le C\frac{\log\chi}{\sigma}\log\left(\frac{\sigma}{\log\chi}\right).$$

Here C is an absolute constant.

Proof. This is one of the main results of Totik (1992). ■

We turn to the

Proof of Theorem 1.3. Let $E_n := \{x_{1n}, x_{2n}, \ldots, x_{nn}\}$ be the points in I_n. Choose a $\sigma(n)$-th Fekete set of points for E_n (note that necessarily $m_n \le 2n-1$, so $\sigma(n) \le n-1$) and let us denote them by x_{jn}, $1 \le j \le \sigma(n)$. Let

$$q_n(x) := \prod_{j=1}^{\sigma(n)} (x - x_{jn}).$$

Since q_n^2 has degree $2\sigma(n) \le m_n$, the precision (1.16) and the definition (1.21) of τ_n give

$$\begin{aligned}\max_{x\in[-1,1]}|q_n(x)| &\le \tau_{\sigma(n)}\left\{\int_{-1}^{1} q_n^2 w\right\}^{1/2}\\ &\le \tau_{\sigma(n)}\left\{I_n(q_n^2)\right\}^{1/2}\\ &\le \tau_{\sigma(n)}\|I_n\|^{1/2}\max_{x\in E_n}|q_n(x)|\\ &\le \tau_{\sigma(n)}\|I_n\|^{1/2} 2\min_{1\le j\le\sigma(n)}|q'(x_{jn})|,\end{aligned} \tag{2.1}$$

by Lemma 2.1(b). Since q_n is monic of degree $\sigma(n)$,

$$\max_{x\in E}|q_n(x)| \ge 2^{-\sigma(n)+1},$$

so

$$\min_{1\le j\le n}|q_n'(x_{jn})| \ge 1/\left\{2^{\sigma(n)}\tau_n\|I_n\|^{1/2}\right\} =: 1/\{B_n 2^{\sigma(n)}\}.$$

Here we have also used $\tau_{\sigma(n)} \le \tau_n$. Next, Lemma 2.1(c) and (2.1) give

$$\begin{aligned}\max_{x\in[-1,1]}|q_n(x)| &\le \tau_n\|I_n\|^{1/2}(\sigma(n)+1)2^{-\sigma(n)+1}\\ &\le \tau_n\|I_n\|^{1/2}\sigma(n)2^{-\sigma(n)+2} =: A_n/2^{\sigma(n)},\end{aligned}$$

for n large enough. Now

$$\tau_n \ge 1\Big/\left\{\int_{-1}^{1} w\right\}^{1/2},$$

and

$$\|I_n\| = \sum_{j=1}^{n}|\lambda_{jn}| \ge \sum_{j=1}^{n}\lambda_{jn} = \int_{-1}^{1} w,$$

so

$$\tau_n\|I_n\|^{1/2} \ge 1.$$

Then for n large enough, our condition (1.17) ensures that

$$\chi_n : = \max\{A_n, B_n, n\}$$

satisfies

$$8 \ge \chi_n/\{n\tau_n\|I_n\|^{1/2}\} = \chi_n/\rho_n \ge 2\alpha.$$

Our hypothesis (1.12) and (1.22) show then that

$$\lim_{n\to\infty}\chi_n^{1/\sigma(n)} = 1.$$

Hence for large enough n, Totik's Lemma 2.2, ensures that ν_n, defined by (1.26), satisfies

$$\text{Discr }(\nu_n, \arcsin) \le C\frac{\log\chi_n}{\sigma(n)}\log\left(\frac{\sigma(n)}{\log\chi_n}\right).$$

Since the function $t \to t \log \frac{1}{t}$ is increasing in $(0, e^{-1})$ we obtain (1.27) for n large enough. Since

$$\text{Discr}\,(\nu_n, \arcsin) \leq 2,$$

we obtain (1.27) for all n by making C large enough. ■

We remark finally that for Fekete points associated with more general sets, Blatt and Mhaskar (1992) estimated the discrepancy, and generalized Totik's result to compact sets with smooth boundaries.

References

Bernstein S. (1936) Sur la formule de quadrature approchée de Tchebycheff. Comptes Rendus Acad. Sci. Paris 203, 1305–1306.

Blatt H.P. (1992) On the Distribution of Simple Zeros of Polynomials. J. Approx. Theory 69, 250–268.

Blatt H.P. and Mhaskar H.N. (1992) A General Discrepancy Theorem, manuscript.

Bloom T., Lubinsky D.S., and Stahl H.B. (1992a) What Distributions of Points are Possible for Convergent Sequences of Interpolatory Integration Rules? Constructive Approxn., 9, 41–58.

Bloom, T., Lubinsky D.S., and Stahl H.B. (1992b) Distribution of Points for Convergent Interpolatory Integration Rules on $(-\infty, \infty)$, Constructive Approxn., 9, 59–82.

Bloom, T., Lubinsky D.S., and Stahl H.B. (1992c) Interpolatory Integration Rules and Orthogonal Polynomials with Varying Weights, to appear.

Brass H. (1977) Quadraturverfahren. Vandenhoeck & Ruprecht, Göttingen.

Davis P.J. and Rabinowitz P. (1984) Methods of Numerical Integration, 2nd edn. Academic Press, New York.

Erdös P. and Turan P. (1940a) On The Uniformly–Dense Distribution of Certain Sequences of Points. Annals of Mathematics. 41, 162–173.

Erdös P. and Turan P. (1940b) On Interpolation III. Interpolatory Theory of Polynomials. Annals of Mathematics. 41, 510–553.

Förster K.-J. (1982) A Comparison Theorem for Linear Functionals and its Application in Quadrature, in Numerical Integration, ed. by G. Hämmerlin, ISNM 57, Birkhäuser-Verlag, 66–76.

Hille E. (1962) Analytic Function Theory, Volume II, 2nd edn. Chelsea, New York.

Nevai P. (1979) Orthogonal Polynomials. Memoirs of the American Math. Soc. 18, No. 213.

Nevai P. (1986) Geza Freud, Orthogonal Polynomials and Christoffel Functions. A Case Study. J. Approx. Theory 48, 3–167.

Stahl H.B. and Totik V. (1992) General Orthogonal Polynomials. Encyclopaedia of Mathematics and its Applications, Vol. 43. Cambridge University Press, Cambridge.

Totik V. (1992) Distribution of Simple Zeros of Polynomials, to appear in Acta Mathematica (Sweden).

Authors' addresses:

T. Bloom,
Dept. of Mathematics,
University of Toronto,
Toronto, Ontario,
Canada M5S 1A1.

D.S. Lubinsky,
Dept. of Mathematics,
Witwatersrand University,
P.O. Wits 2050,
South Africa.

H. Stahl,
Technische Fachhochschule,
Berlin/FB2,
Luxemburgerstrasse 20,
D-1000 Berlin 65,
Germany.

Bounds for Peano kernels

Helmut Brass

1. Introduction

Let I denote a fixed given functional on $C[-1,1]$ of the type

$$I[f] := \int_{-1}^{1} f(x)w(x)\,dx \quad \text{with} \quad w \in L^1,\ w \geq 0,\ \int_{-1}^{1} w(x)\,dx > 0.$$

The numerical computation of $I[f]$ is often done with help of "quadrature rules" Q, these are functionals of the form

$$Q[f] := \sum_{\nu=1}^{n} a_\nu f(x_\nu), \qquad -1 \leq x_1 < x_2 < \ldots < x_n \leq 1.$$

The error is the functional $R := I - Q$. The degree of Q is the number $\deg Q := \sup\{m : R[\mathcal{P}_m] = 0\}$, where $\mathcal{P}_m$ denotes the space of polynomials of degree $\leq m$. The most interesting quadrature rules are the Gaussian rules Q_n^{G}, which are characterized as rules with n evaluation points $x_1^{\mathrm{G}}, \ldots, x_n^{\mathrm{G}}$ and degree $2n-1$.

The "Peano kernel" of a quadrature rule Q is defined by

$$K_s(\xi) := \frac{1}{(s-1)!} R\left[(\cdot - \xi)_+^{s-1}\right], \quad s = 1, 2, \ldots, \deg Q + 1,$$

where we have used the function $(\cdot)_+^s$, which is given by

$$(u)_+^s = \left(\frac{u+|u|}{2}\right)^s \quad \text{if} \quad s > 0, \quad (u)_+^0 = \begin{cases} 1 & u > 0 \\ 1/2 & u = 0 \\ 0 & u < 0 \end{cases}.$$

An important tool for the discussion of R is the Peano representation

(1.1) $$R[f] = \int_{-1}^{1} f^{(s)}(x) K_s(x)\,dx, \quad s = 1, 2, \ldots, \deg Q + 1,$$

holding for all f such that $f^{(s-1)}$ is absolutely continuous (e.g. Brass (1977)). An immediate consequence of (1.1) is

$$|R[f]| \leq \|K_s\| \int_{-1}^{1} |f^{(s)}(x)|\, dx$$

($\|\cdot\|$ means the sup norm throughout this paper) or, equivalently

$$|R[f]| \leq \|K_s\| \operatorname{Var} f^{(s-1)}, \tag{1.2}$$

where $\operatorname{Var} g$ denotes the total variation of g. Actually (1.2) holds for any f with $\operatorname{Var} f^{(s-1)} < \infty$. This generality does not seem to be quite trivial, we shall give a proof below.

Error bounds of the type (1.2) are not as well known as bounds of the type

$$|R[f]| \leq \|f^{(s)}\| \int_{-1}^{1} |K_s(x)|\, dx \tag{1.3}$$

(likewise an immediate consequence of (1.1)), but there are good reasons to prefer (1.2) to (1.3). (1.2) can be applied to a much larger class of functions than (1.3) (e.g. $f(x) = \sqrt{1-x^4}$ and $s = 1$) and if both can be applied, then the resulting bounds are of the same order for many important quadrature rules. To be more precise in this last point: For the Gaussian rules the order of $\|K_s\|$ and $\int_{-1}^{1} |K_s(x)|\, dx$ as function of the number n of evaluation points is n^{-s} in both cases, at least if we make some assumptions on w; and if we replace the Gaussian rules by the respective optimal rules, we have the same situation.

There is no special difficulty in the computation of $\|K_s\|$ for any given rule Q, but we are interested in the asymptotics of $\|K_s\|$ for sequences $Q_1, Q_2, \ldots$ of quadrature rules with increasing number of evaluation points ("quadrature methods"). Our main result is an explicitly given upper bound for $\|K_s\|$, holding for a large class of quadrature methods, which is asymptotically best possible for the Gaussian method. Our main tool is a result of Ganelius (1956) concerning one-sided approximation of the Bernoulli functions.

2. The results

Theorem 1 *If $f^{(s)}$ is of bounded variation, then we have*

$$|R[f]| \leq \|K_{s+1}\| \operatorname{Var} f^{(s)}, \quad (s = 0, 1, \ldots, \deg Q).$$

It is an immediate consequence of the definition of K_{s+1}, that this bound is best possible.

If $s = 0$, then f is not necessarily continuous, but I and Q are defined for all functions of bounded variation. It seems rather natural to have a theorem for all these functions, but the inclusion of noncontinuous functions makes the proof more difficult.

A quadrature rule Q is called positive if all coefficients a_ν are positive. We are interested in the class $\mathbf{Q}_m$ of quadrature rules

$$\mathbf{Q}_m := \{Q \,:\, Q \text{ is positive and } \deg Q \geq m\}.$$

There is no assumption on the number of evaluation points of the elements of $\mathbf{Q}_m$.

The Bernoulli function B_s is defined by

$$B_s(x) := -2 \sum_{\mu=1}^{\infty} \frac{\cos(2\pi\mu x - \frac{1}{2}\pi s)}{(2\pi\mu)^s}. \tag{2.1}$$

The restriction of B_s to $]0,1[$ is identical with the Bernoulli polynomial (e.g. Hämmerlin/Hoffmann (1989)).

A further particular function which is used in the formulation of the theorems is h:

$$h(x) := \sqrt{1-x^2}.$$

Theorem 2 *If $\|wh\| < \infty$, then we have for any Peano kernel K_s of a quadrature rule $Q \in \mathbf{Q}_m$*

$$\|K_s\| \leq \|wh\|(2\pi)^s \|B_s\| \frac{(m-s+1)!}{(m+1)!}.$$

The special case $s = 1$ reads

$$\|K_1\| \leq \|wh\| \frac{\pi}{m+1}. \tag{2.2}$$

Here is no improvement possible, for we have equality if $w(x) = (1-x^2)^{-1/2}$ for the Gaussian rule (if m is odd) or the Radau rule (if m is even).

For the important weight function $w \equiv 1$ Förster/Petras (1991) have given a result very similar to (2.2) and by an explicit determination of $\|K_1\|$ for the Gaussian rule they could show that an improvement of (2.2) would only be possible by a factor

$1+O(m^{-1})$. Our next theorem shows that we have the same situation (but replace $1+O(m^{-1})$ by $1+o(1)$) for any continuous weight function with $w>0$ a.e.

Theorem 3 *Let* wh *be continuous and*
if s *is odd* : *let* $w>0$ a.e.,
if s *is even* : *let there exist a polynomial* $q \not\equiv 0$ *such that* $w(x) \geq q^2(x)$.

Then we have

$$\lim_{m\to\infty} m^s \sup\left\{\|K_s\| : Q\in \mathbf{Q}_m\right\} = \lim_{m\to\infty} m^s \left\| K_s\left(Q^{\mathrm{G}}_{\lceil\frac{m+1}{2}\rceil}\right)\right\|$$
$$= (2\pi)^s \|B_s\| \|wh^s\|.$$

Petras (1992 b) has proved this theorem for the special weight functions $w_\lambda(x) := (1-x^2)^\lambda$, $\lambda > -1/2$, but our method is not a generalization of his method. Using these weight functions in Theorem 2 we see that this theorem can not be sharpened in the asymptotic sense for any s.

Our next theorem gives localized upper and lower bounds for K_s:

Theorem 4 *Let* wh *be continuous. Then*

$$\overline{\lim_{m\to\infty}}\, m^s \sup\left\{K_s(\xi) : Q\in \mathbf{Q}_m\right\} \leq (2\pi)^s h^s(\xi) w(\xi) \sup_x B_s(x)$$

and

$$\underline{\lim_{m\to\infty}}\, m^s \inf\left\{K_s(\xi) : Q\in \mathbf{Q}_m\right\} \geq (2\pi)^s h^s(\xi) w(\xi) \inf_x B_s(x).$$

Both relations hold uniformly in ξ.

This theorem allows to replace $\|wh\|$ in theorem 2 by $\|wh^s\|$ (and so to be more in conformity with Theorem 3), if we are willing to weaken the assertion from an inequality to an asymptotic inequality.

Petras (1992 b) proved Theorem 4 for a narrower class of weight functions with the sign of equality instead of our signs $\geq$ and $\leq$, but without the assertion of uniformity.

Our last result holds for a more specialized situation, but gives a rather good bound.

Theorem 5 *Let* $w\equiv 1$, *then*

$$|K_1(\xi)| \leq \frac{\pi}{m}\sqrt{1-\xi^2} + \frac{2\ln m + 3}{m^2}$$

holds for all $Q \in \mathbf{Q}_{m-1}$.

The first theorem of this type was given by DeVore/Scott (1984) and improved by Petras (1992 a), our theorem is asymptotically best possible if $\xi \in]-1,1[$.

3. Proof of Theorem 1

If f is of bounded variation on $[-1,1]$, then there are two increasing functions f_1 and f_2 such that

$$f = f_1 - f_2, \quad \operatorname{Var} f = \operatorname{Var} f_1 + \operatorname{Var} f_2.$$

The second equality does not seem to be known generally, but the proof can be given without difficulty using

$$f_1 = \frac{1}{2}(v+f), \quad f_2 = \frac{1}{2}(v-f); \quad v(x) := \operatorname{Var}^{[-1,x]} f.$$

So we can suppose that f is increasing.

(i) Let $s = 1$.
We have

$$R[f] = \int_{-1}^{1} f(u)w(u)\,du - \sum_{\nu=1}^{n} a_\nu f(x_\nu)$$

$$= \sum_{\nu=0}^{n} f(x_{\nu+1})\Big(\int_{x_\nu}^{x_{\nu+1}} w(u)\,du - a_{\nu+1}\Big) + \sum_{\nu=0}^{n} \int_{x_\nu}^{x_{\nu+1}} (f(u) - f(x_{\nu+1}))w(u)\,du,$$

where $a_{n+1} := 0$, $x_{n+1} := 1$, $x_0 := -1$. A partial summation leads to

$$R[f] = \sum_{\nu=0}^{n} (f(x_{\nu+1}) - f(x_\nu))\Big(\int_{x_\nu}^{1} w(u)\,du - a_{\nu+1} - \ldots - a_n\Big)$$

$$+ \sum_{\nu=0}^{n} \int_{x_\nu}^{x_{\nu+1}} (f(u) - f(x_{\nu+1}))w(u)\,du,$$

and by an application of the second mean value theorem we obtain

$$\int_{x_\nu}^{x_{\nu+1}} (f(u) - f(x_{\nu+1}))w(u)\,du = (f(x_\nu) - f(x_{\nu+1})) \int_{x_\nu}^{\eta_\nu} w(u)\,du,$$

where $\eta_\nu \in [x_\nu, x_{\nu+1}]$. So we have

$$R[f] = \sum_{\nu=0}^{n} (f(x_{\nu+1}) - f(x_\nu))\Big(\int_{\eta_\nu}^{1} w(u)\,du - a_{\nu+1} - \ldots - a_n\Big)$$

and this means

$$|R[f]| \leq (f(1) - f(-1))\|K_1\|.$$

(ii) Let $s > 1$ (but $s \leq \deg Q$)
Starting with the Peano representation we obtain by an application of the second mean value theorem

$$R[f] = \int_{-1}^{1} f^{(s)}(u)K_s(u)\,du \;=\; f^{(s)}(-1)\int_{-1}^{\eta} K_s(u)du + f^{(s)}(1)\int_{\eta}^{1} K_s(u)\,du.$$

We use now $\int_{-1}^{1} K_s(u)\,du = 0$ and $K'_{s+1} = -K_s$ and get

$$\begin{aligned} R[f] &= f^{(s)}(-1)\int_{-1}^{\eta} K_s(u)\,du - f^{(s)}(1)\int_{-1}^{\eta} K_s(u)\,du \\ &= (f^{(s)}(-1) - f^{(s)}(1))(-K_{s+1}(\eta)), \end{aligned}$$

and the assertion follows.

4. On one-sided approximation

Denote by $\mathcal{T}_m$ the space of trigonometric polynomials of order m. The main tools of our proofs are the following quantities:

$$\underset{\sim}{E}_m[q;f] := \inf\left\{\int_{-\pi}^{\pi} (f(y) - t(y))q(y)\,dy \;:\; t \in \mathcal{T}_m,\, t \leq f\right\},$$

$$\hat{E}_m[q;f] := \inf\left\{\int_{-\pi}^{\pi} (t(y) - f(y))q(y)\,dy \;:\; t \in \mathcal{T}_m,\, t \geq f\right\}.$$

The following relations are more or less obvious: If f and g are 2π periodic and $\alpha \in \mathbb{R}$, then

(4.1) $$\underset{\sim}{E}_m[q;f+g] \leq \underset{\sim}{E}_m[q;f] + \underset{\sim}{E}_m[q;g],$$

(4.2) $$\underset{\sim}{E}_m[q;\alpha f] = \alpha \underset{\sim}{E}_m[q,f] \qquad \text{if} \quad \alpha \geq 0,$$

(4.3) $$\underset{\sim}{E}_m[q;-f] = \hat{E}_m[q;f],$$

(4.4) $$\underset{\sim}{E}_m[q;f] \leq 4\pi\|q\|\|f\|,$$

(4.5) $$\underset{\sim}{E}_m(q;f] \leq \|q\|\underset{\sim}{E}_m[1;f],$$

(4.6) $$\underset{\sim}{E}_m[1;f(\alpha \pm \cdot)] = \underset{\sim}{E}_m[1;f]$$

and we may permute $\underset{\sim}{E}_m$ and $\hat{E}_m$.

By no means obvious is the following theorem of Ganelius (1956)

$$E_{m-1}\left[1; B_s\left(\frac{\cdot}{2\pi}\right)\right] = -\frac{2\pi}{m^s}\inf_x B_s(x), \tag{4.7}$$

$$\hat{E}_{m-1}\left[1; B_s\left(\frac{\cdot}{2\pi}\right)\right] = \frac{2\pi}{m^s}\sup_x B_s(x). \tag{4.8}$$

If $s = 1$, then the trigonometric polynomial of best one-sided L^1-approximation is given by

$$G_1(x) = \frac{\alpha_o}{2} + \sum_{\nu=1}^{m-1} \alpha_\nu \cos \nu x + \beta_\nu \sin \nu x,$$

$$\alpha_\nu = -m^{-2}(m-\nu), \quad \beta_\nu = -(\pi m)^{-1} - m^{-2}(m-\nu)\operatorname{ctg} \pi\nu m^{-1}$$

for (4.7) and by $\hat{G}_1(x) = -G_1(-x)$ for (4.8).

If $s > 1$, then the extremal polynomials are given by

$$G_s(x) = \sum_{\nu=0}^{m-1}\left(B_s\left(\frac{\theta_\nu}{2\pi}\right)F(x-\theta_\nu) + (2\pi)^{-1}B_{s-1}\left(\frac{\theta_\nu}{2\pi}\right)D(x-\theta_\nu)\right),$$

$$F(x) := \left(\frac{\sin m\frac{x}{2}}{m \sin\frac{x}{2}}\right)^2, \quad D(x) := 2m^{-2}\left(\sin m\frac{x}{2}\right)^2 \operatorname{ctg}\frac{x}{2}, \quad \theta_\nu = \frac{2\pi\nu + A_s}{m},$$

where A_s is determined by $B_s(\frac{A_s}{2\pi}) = \inf_x B_s(x)$ for (4.7) and by $B_s(\frac{A_s}{2\pi}) = \sup_x B_s(x)$ for (4.8).

The connection between one-sided approximation and the Peano kernel is described in the following lemma:

Lemma 1 *Let* $d_s(y) := (\cos y - \cos\delta)_+^s$ *and* $\tilde{w}(y) := w(\cos y)|\sin y|$. *Then we have for all* $Q \in \mathbf{Q}_m$,

$$-\frac{1}{2}\hat{E}_m[\tilde{w}; d_s] \le s!\,K_{s+1}(\cos\delta) \le \frac{1}{2}E_m[\tilde{w}; d_s], \quad s = 0, 1, \ldots, m;\ \delta \in [0, \pi]. \tag{4.9}$$

Proof: If $p \in \mathcal{P}_m$ and $p \le f$ holds (pointwise), then we have

$$R[f] = R[f-p] = I[f-p] - Q[f-p] \le I[f-p]$$

because of the positivity of Q. This leads to

$$R[f] \le \inf\left\{\int_{-1}^{1}(f(x) - p(x))w(x)\,dx\ :\ p \in \mathcal{P}_m, p \le f\right\},$$

and if we apply this inequality to $f(x) := (x-\xi)_+^s$, we get

$$s!\,K_{s+1}(\xi) \le \inf\left\{\int_{-1}^{1} ((x-\xi)_+^s - p(x))w(x)\,dx \,:\, p\in\mathcal{P}_m, p \le (\cdot-\xi)_+^s\right\}.$$

If we now substitute $x = \cos y$ in the integral and put $\xi = \cos\delta$, we obtain

$$\begin{aligned} s!\,K_{s+1}(\cos\delta) &\le \inf\left\{\int_{0}^{\pi} (d_s(y) - p(\cos y))\tilde{w}(y)\,dy \,:\, p\in\mathcal{P}_m,\ p(\cos\cdot)\le d_s\right\} \\ &= \frac{1}{2}E_m[\tilde{w}; d_s]. \end{aligned}$$

In the last step we used that the infimum is the same if taken over all of $\mathcal{T}_m$ or only over the even polynomials of $\mathcal{T}_m$.

We have proven the upper bound in (4.9) and evidently we can apply the same argument for the proof of the lower bound.

5. Two lemmata

Lemma 2 *If* $g_s(y) := (\cos y - \cos\delta)^s$, $(s = 1, 2, \ldots)$ *then*

(5.1) $$g_s^{(\nu)}(\delta) = 0, \quad \nu = 0, 1, \ldots, s-1,$$

(5.2) $$g_s^{(s)}(\delta) = (-1)^s s!(\sin\delta)^s,$$

(5.3) $$|g_s^{(\nu)}(\delta)| \le S_\nu^{(s)} := (-1)^s \sum_{\kappa=0}^{s} (-1)^\kappa \binom{s}{\kappa}\kappa^\nu, \quad \nu = 0, 1, \ldots$$

(5.4) $$|g_s^{(\nu)}(\delta)| \le \frac{4}{7}S_\nu^{(s)}, \quad s = 2, 4, \ldots, \quad \nu = s+1, s+2, \ldots$$

Proof: The numbers $S_\nu^{(s)}$ are the s-th differences of the function $f_\nu(x) := x^\nu$ at the point $x = 0$, that is

$$S_\nu^{(s)} = \Delta^s f_\nu(0), \qquad (\Delta f := f(\cdot+1) - f).$$

Therefore we have

(5.5) $$S_\nu^{(s)} = 0, \qquad \nu = 0, 1, \ldots, s-1.$$

By an application of Leibniz's rule we obtain

(5.6) $$g_{s+1}^{(\lambda)} = (g_s g_1)^{(\lambda)} = \sum_{\sigma=0}^{\lambda}\binom{\lambda}{\sigma} g_s^{(\sigma)} g_1^{(\lambda-\sigma)}.$$

From this we derive

$$(5.7) \qquad |g_{s+1}^{(\lambda)}(\delta)| \le \sum_{\sigma=0}^{\lambda-1} \binom{\lambda}{\sigma} |g_s^{(\sigma)}(\delta)|.$$

Using (5.7) we can easily prove (5.3) by induction beginning with $s = 1$. (5.1) is a special case of (5.3), see (5.5).

Now (5.6) leads to

$$g_{s+1}^{(s+1)}(\delta) = (s+1) g_s^{(s)}(\delta)(-\sin\delta),$$

this relation gives (5.2).

For the proof of (5.4) we use induction too, but this time we apply Leibniz's rule in the form

$$g_{s+2}^{(\lambda)} = (g_s g_2)^{(\lambda)} = \sum_{\sigma=0}^{\lambda} \binom{\lambda}{\sigma} g_s^{(\sigma)} g_2^{(\lambda-\sigma)},$$

hence

$$(5.8) \qquad |g_{s+2}^{(\lambda)}(\delta)| \le \sum_{\sigma=0}^{\lambda-2} \binom{\lambda}{\sigma} |g_s^{(\sigma)}(\delta)|\, |g_2^{(\lambda-\sigma)}(\delta)|.$$

To begin the induction, we compute

$$\begin{aligned} g_2^{(4\nu+1)}(\delta) &= (-2^{4\nu}+1)\sin 2\delta \\ g_2^{(4\nu+2)}(\delta) &= (-2^{4\nu+2}+2)\cos^2\delta + 2^{4\nu+1} \\ g_2^{(4\nu+3)}(\delta) &= (2^{4\nu+2}-1)\sin 2\delta \\ g_2^{(4\nu+4)}(\delta) &= (2^{4\nu+4}-2)\cos^2\delta - 2^{4\nu+3} \end{aligned}$$

that is

$$|g_2^{(\nu)}(\delta)| \le 2^{\nu-1} - \frac{1-(-1)^\nu}{2}$$

and we conclude

$$(5.9) \qquad |g_2^{(\nu)}(\delta)| \le \frac{4}{7}(2^\nu - 2) = \frac{4}{7} S_\nu^{(2)}, \qquad \nu = 3, 4, \ldots$$

To do the induction step, we use in the right hand side of (5.8) the induction hypothesis for the terms $g_s^{(\sigma)}(\delta)$ with $\sigma > s$, (5.9) for the term $g_2^{(\lambda-s)}$ with $\sigma = s$ and (5.3) for all other terms. By virtue of (5.5) we get

$$|g_{s+2}^{(\lambda)}(\delta)| \le \frac{4}{7} \sum_{\sigma=0}^{\lambda-2} \binom{\lambda}{\sigma} S_\sigma^{(s)} S_{\lambda-\sigma}^{(2)}.$$

Finally we have to verify the identity

$$\sum_{\sigma=0}^{\lambda-2} \binom{\lambda}{\sigma} S_\sigma^{(s)} S_{\lambda-\sigma}^{(2)} = S_\lambda^{(s+2)}$$

and the proof is done.

Lemma 3 *Let* $d_s(y) := (\cos y - \cos\delta)_+^s$ *and* $g_s(y) = (\cos y - \cos\delta)^s$. *Then for any* $r \in \{1, 2, \ldots\}$ *there exists a* $t_r \in \mathcal{T}_s$ *such that*

$$\left| d_s(y) - \sum_{\lambda=s}^{s+r-1} (-1)^{\lambda+1}(2\pi)^\lambda \left(B_{\lambda+1}\left(\frac{\delta - y}{2\pi}\right) + B_{\lambda+1}\left(\frac{\delta + y}{2\pi}\right)\right) g_s^{(\lambda)}(\delta) - t_r(y) \right|$$
$$\leq 4(s+1)(\frac{s}{s+1})^{s+r} 2^s$$

holds.

Proof: The result is quite simple if $s = 0$, so we may suppose $s \geq 1$. If f has period 2π, then

$$f(y) = \frac{1}{2\pi}\int_{-\pi}^{\pi} f(u)du - (2\pi)^{s-1}\int_{-\pi}^{\pi} B_s\left(\frac{y-u}{2\pi}\right) f^{(s)}(u)\, du$$

holds, this is a well-known special case of the Euler-Maclaurin sum formula. We apply this formula to $f = d_s$ and get

$$d_s(y) = \frac{1}{2\pi}\int_{-\delta}^{\delta} g_s(u)\, du - (2\pi)^{s-1}\int_{-\delta}^{\delta} B_s\left(\frac{y-u}{2\pi}\right) g_s^{(s)}(u)\, du.$$

An r-fold partial integration leads to

$$\begin{aligned} d_s(y) &= \frac{1}{2\pi}\int_{-\delta}^{\delta} g_s(u)\, du \\ &+ \sum_{\lambda=s}^{s+r-1} (-1)^{\lambda+1}(2\pi)^\lambda \left(B_{\lambda+1}\left(\frac{\delta-y}{2\pi}\right) + B_{\lambda+1}\left(\frac{\delta+y}{2\pi}\right)\right) g_s^{(\lambda)}(\delta) \\ &- (2\pi)^{s+r-1}\int_{-\delta}^{\delta} B_{s+r}\left(\frac{y-u}{2\pi}\right) g_s^{(s+r)}(u)\, du. \end{aligned} \tag{5.10}$$

We observe now

$$\left| B_{s+r}\left(\frac{y-u}{2\pi}\right) + 2\sum_{\nu=1}^{s} \frac{\cos\left(\nu(y-u) - \frac{s+r}{2}\pi\right)}{(2\nu\pi)^{s+r}} \right| \leq \frac{2}{(2\pi)^{s+r}} \sum_{\nu=s+1}^{\infty} \frac{1}{\nu^{s+r}}$$
$$\leq \frac{4}{(2\pi)^{s+r}} \frac{1}{(s+1)^{s+r-1}}.$$

By Bernstein's inequality $\|g_s^{(s+r)}\| \leq s^{s+r}\|g_s\|$ holds, hence we may write

$$(2\pi)^{s+r-1}\int_{-\delta}^{\delta} B_{s+r}\left(\frac{y-u}{2\pi}\right) g_s^{(s+r)}(u)\,du = \tilde{t}_r + \sigma_r,$$

where $\tilde{t}_r \in \mathcal{T}_s$ and

$$|\sigma_r| \leq (2\pi)^{s+r-1}2\delta\frac{4}{(2\pi)^{s+r}}\frac{1}{(s+1)^{s+r-1}}s^{s+r}\|g_s\| \leq 4(s+1)\left(\frac{s}{s+1}\right)^{s+r}2^s.$$

This relation inserted in (5.10) yields the lemma.

6. Proof of Theorem 2

We shall prove an upper bound for $K_s(\xi)$, the same method can be used to construct a lower bound, combination of both bounds leads to Theorem 2.

We start with lemma 1 and substitute d_s by its approximation from lemma 3. An application of (4.1) and (4.4) leads to

$$\begin{aligned} &s!K_{s+1}(\cos\delta)\\ &\leq \frac{1}{2}\sum_{\lambda=s}^{s+r-1} E_m\left[\tilde{w};(-1)^{\lambda+1}(2\pi)^{\lambda}\left(B_{\lambda+1}\left(\frac{\delta-\cdot}{2\pi}\right)+B_{\lambda+1}\left(\frac{\delta+\cdot}{2\pi}\right)\right)\right]g_s^{(\lambda)}(\delta)\\ &\quad + 2\pi\|\tilde{w}\|4(s+1)\left(\frac{s}{s+1}\right)^{s+r}2^s. \end{aligned}$$

In this inequality we let $r \to \infty$. Then we apply (5.2), (4.2) and (4.3) and use the symbol E_m for $\max\{\underset{\sim}{E}_m, \hat{E}_m\}$:

$$\begin{aligned} s!K_{s+1}(\cos\delta) \leq &\frac{1}{2}(2\pi)^s s!(\sin\delta)^s\hat{E}_m\left[\tilde{w};B_{s+1}\left(\frac{\delta-\cdot}{2\pi}\right)+B_{s+1}\left(\frac{\delta+\cdot}{2\pi}\right)\right]\\ &+\frac{1}{2}\sum_{\lambda=s+1}^{\infty}(2\pi)^{\lambda}|g_s^{(\lambda)}(\delta)|E_m\left[\tilde{w};B_{\lambda+1}\left(\frac{\delta-\cdot}{2\pi}\right)+B_{\lambda+1}\left(\frac{\delta+\cdot}{2\pi}\right)\right]. \end{aligned}$$

Combining (4.1), (4.5), (4.6) and the theorem of Ganelius, we obtain

$$E_m\left[\tilde{w};B_{\lambda+1}\left(\frac{\delta-\cdot}{2\pi}\right)+B_{\lambda+1}\left(\frac{\delta+\cdot}{2\pi}\right)\right] \leq 2\|\tilde{w}\|\frac{2\pi}{(m+1)^{\lambda+1}}\|B_{\lambda+1}\|. \tag{6.1}$$

Thus we arrive at

$$\begin{aligned} s!K_{s+1}(\cos\delta) \leq &\frac{1}{2}(2\pi)^s s!(\sin\delta)^s\hat{E}_m\left[\tilde{w};B_{s+1}\left(\frac{\delta-\cdot}{2\pi}\right)+B_{s+1}\left(\frac{\delta+\cdot}{2\pi}\right)\right]\\ &+\|\tilde{w}\|\sum_{\lambda=s+1}^{\infty}\left(\frac{2\pi}{m+1}\right)^{\lambda+1}\|B_{\lambda+1}\||g_s^{(\lambda)}(\delta)|. \end{aligned} \tag{6.2}$$

Using (5.3) and (5.4) we obtain
(6.3)

$$\sum_{\lambda=s+1}^{\infty}\left(\frac{2\pi}{m+1}\right)^{\lambda+1}\|B_{\lambda+1}\||g_s^{(\lambda)}(\delta)| \le c_s \sum_{\lambda=s+1}^{\infty}\left(\frac{2\pi}{m+1}\right)^{\lambda+1}\|B_{\lambda+1}\|S_\lambda^{(s)},$$

where we have put

$$c_s = \begin{cases} 1 & \text{if } s \text{ is odd,} \\ \frac{4}{7} & \text{if } s \text{ is even.} \end{cases}$$

Now let s be an odd number, then we have for $\lambda \ge s$

$$(2\pi)^{s+1}\|B_{s+1}\| = 2\sum_{\nu=1}^{\infty}\frac{1}{\nu^{s+1}} \ge (2\pi)^{\lambda+1}\|B_{\lambda+1}\|.$$

If s is even, then we have for $\lambda \ge s+1$

$$(2\pi)^{s+1}\|B_{s+1}\| \ge (2\pi)^{s+1}\left|B_{s+1}\left(\frac{1}{4}\right)\right| \ge 2\left(1-\frac{1}{3^{s+1}}\right) \ge \frac{52}{27}$$

$$\ge \frac{4}{7}\cdot\frac{\pi^4}{45} = \frac{4}{7}(2\pi)^4\|B_4\| \ge \frac{4}{7}(2\pi)^{\lambda+1}\|B_{\lambda+1}\|.$$

Inserting these inequalities in (6.3) gives us

$$(6.4)\quad \sum_{\lambda=s+1}^{\infty}\left(\frac{2\pi}{m+1}\right)^{\lambda+1}\|B_{\lambda+1}\||g_s^{(\lambda)}(\delta)| \le (2\pi)^{s+1}\|B_{s+1}\|\sum_{\lambda=s+1}^{\infty}\frac{S_\lambda^{(s)}}{(m+1)^{\lambda+1}}.$$

We return now to (6.2). For the first summand of the right hand side we use (6.1) and $|\sin\delta| \le 1$, for the sum we use (6.4). We obtain

$$s!K_{s+1}(\cos\delta) \le \|\tilde{w}\|(2\pi)^{s+1}\|B_{s+1}\|\sum_{\lambda=s}^{\infty}\frac{S_\lambda^{(s)}}{(m+1)^{\lambda+1}}$$

$$= \|\tilde{w}\|(2\pi)^{s+1}\|B_{s+1}\|\sum_{\lambda=0}^{\infty}\frac{S_\lambda^{(s)}}{(m+1)^{\lambda+1}}$$

$$= \|\tilde{w}\|(2\pi)^{s+1}\|B_{s+1}\|\sum_{\kappa=0}^{s}(-1)^{\kappa+s}\binom{s}{\kappa}\sum_{\lambda=0}^{\infty}\frac{\kappa^\lambda}{(m+1)^{\lambda+1}}$$

$$= \|\tilde{w}\|(2\pi)^{s+1}\|B_{s+1}\|\sum_{\kappa=0}^{s}(-1)^{\kappa+s}\binom{s}{\kappa}\frac{1}{m+1-\kappa}$$

$$= \|\tilde{w}\|(2\pi)^{s+1}\|B_{s+1}\|\frac{s!}{(m+1)m(m-1)\dots(m+1-s)}.$$

The theorem is proved.

7. Proof of Theorem 4

We begin with (6.2). Using the bounds for the infinite sum in (6.2) which we have obtained in the proof of Theorem 2, we can write

$$K_{s+1}(\cos\delta) \tag{7.1}$$
$$\leq \frac{1}{2}(2\pi)^s(\sin\delta)^s \hat{E}_m\left[\tilde{w}\,;\,B_{s+1}\left(\frac{\delta-\cdot}{2\pi}\right)+B_{s+1}\left(\frac{\delta+\cdot}{2\pi}\right)\right]+O(m^{-s-2}).$$

We use now the best one-sided approximations of $B_{s+1}(\frac{\delta\pm\cdot}{2\pi})$ for the weight function $w\equiv 1$, which are determined by Ganelius as approximating functions for the general weight function. So we have

$$\hat{E}_m\left[\tilde{w}\,;\,B_{s+1}\left(\frac{\delta-\cdot}{2\pi}\right)+B_{s+1}\left(\frac{\delta+\cdot}{2\pi}\right)\right]$$
$$\leq\int_{-\pi}^{\pi}\left(\hat{G}_{s+1}(\delta-u)-B_{s+1}\left(\frac{\delta-u}{2\pi}\right)+\hat{G}_{s+1}(\delta+u)-B_{s+1}\left(\frac{\delta+u}{2\pi}\right)\right)\tilde{w}(u)\,du$$
$$=2\int_{-\pi}^{\pi}\left(\hat{G}_{s+1}(v)-B_{s+1}\left(\frac{v}{2\pi}\right)\right)\tilde{w}(\delta-v)\,dv.$$

If we could prove that for any 2π periodic and continuous $\tilde{w}$

$$\int_{-\pi}^{\pi}\left(\hat{G}_{s+1}(v)-B_{s+1}\left(\frac{v}{2\pi}\right)\right)\tilde{w}(\delta-v)\,dv \tag{7.2}$$
$$=\tilde{w}(\delta)\frac{2\pi}{(m+1)^{s+1}}\sup_x B_{s+1}(x)+o(m^{-s-1})$$

holds uniformly in δ, then (half of) our theorem would follow immediately.

We introduce an operator H_m, defined on the space of 2π periodic and continuous functions by

$$H_m[f]:=\left(2\pi\sup_x B_{s+1}(x)\right)^{-1}(m+1)^{s+1}\int_{-\pi}^{\pi}\left(\hat{G}_{s+1}(v)-B_{s+1}\left(\frac{v}{2\pi}\right)\right)f(\cdot-v)\,dv$$

and (7.2) would follow from

$$\lim_{m\to\infty}H_m[f]=f. \tag{7.3}$$

For the proof of (7.3) we use the theorem of Korovkin (see e.g. DeVore (1972)) which says that a relation like (7.3) holds for all f, if the operators H_m are positive (which is the case) and (7.3) holds for $f_0\equiv 1$ (this is the result of Ganelius), for $f_1(x)=\sin x$ and for $f_2(x)=\cos x$. For the last two conditions it is sufficient to verify

$$\lim_{m\to\infty}(m+1)^{s+1}\int_{-\pi}^{\pi}\left(\hat{G}_{s+1}(v)-B_{s+1}\left(\frac{v}{2\pi}\right)\right)\cos v\,dv=2\pi\sup_x B_{s+1}(x)$$

and

$$\lim_{m\to\infty}(m+1)^{s+1}\int_{-\pi}^{\pi}\left(\hat{G}_{s+1}(v)-B_{s+1}\left(\frac{v}{2\pi}\right)\right)\sin v\,dv=0.$$

This can be done with a somewhat lengthy but straightforward computation.

8. Proof of Theorem 5

We use the same method as used for the proof of Theorem 4, but things are simpler, for we have the identity

$$d_0(y)=-\left(B_1\left(\frac{\delta-y}{2\pi}\right)+B_1\left(\frac{\delta+y}{2\pi}\right)\right)+const.$$

We can replace (7.1) by

$$K_1(\cos\delta)\le\frac{1}{2}\hat{E}_{m-1}\left[|\sin\cdot|\,;\,B_1\left(\frac{\delta-\cdot}{2\pi}\right)+B_1\left(\frac{\delta+\cdot}{2\pi}\right)\right]$$

and using again Ganelius' extremal functions, we obtain

$$\begin{aligned}K_1(\cos\delta)&\le\int_{-\pi}^{\pi}\left(\hat{G}_1(v)-B_1\left(\frac{v}{2\pi}\right)\right)|\sin(\delta-v)|\,dv\\&\le\sin\delta\int_{-\pi}^{\pi}\left(\hat{G}_1(v)-B_1\left(\frac{v}{2\pi}\right)\right)dv+\int_{-\pi}^{\pi}\left(\hat{G}_1(v)-B_1\left(\frac{v}{2\pi}\right)\right)|\sin v|\,dv\\&=\frac{\pi}{m}\sin\delta+\frac{4}{m^2}\begin{cases}(1+\frac{1}{3}+\frac{1}{5}+\ldots+\frac{1}{m-2}+\frac{1}{2}\frac{1}{m}) & m\text{ odd}\\(1+\frac{1}{3}+\frac{1}{5}+\ldots+\frac{1}{m-1}) & m\text{ even}.\end{cases}\end{aligned}$$

The theorem follows.

9. Proof of Theorem 3

Theorem 2 shows that it is sufficient to prove

$$\underline{\lim}_{m\to\infty}\, m^s\left\|K_s\left(Q^{\mathrm{G}}_{\lceil\frac{m+1}{2}\rceil}\right)\right\|\ge(2\pi)^s\|B_s\|\,\|wh^s\|.$$

We make the hypothesis that there is a sequence $m_1, m_2, \ldots$ such that we have

$$\lim_{\nu\to\infty} m_\nu^s\left\|K_s\left(Q^{\mathrm{G}}_{\lceil\frac{m_\nu+1}{2}\rceil}\right)\right\|<(2\pi)^s\|B_s\|\,\|wh^s\| \tag{9.1}$$

and this will lead us to a contradiction.

As an abbreviation for $\lceil \frac{m_\nu+1}{2} \rceil$ we shall write n and for $K_s(Q_n^{\mathrm{G}})$ we shall write K_s. We give the proof only for $s \geq 2$; if $s = 1$, then there are some minor changes necessary.

K_s has (under our assumptions on w) $2n - s + 2$ zeros $-1 = \xi_0^{(s)} < \xi_1^{(s)} < \ldots < \xi_{2n-s+1}^{(s)} = 1$. K_s changes its sign at all interior zeros, if not, we could construct a $p \in \mathcal{P}_{2n-1}$ with $R[p] > 0$. From the theorem of Rolle and $K_l' = -K_{l-1} (l = 3, 4, \ldots 2n-1)$ we conclude

$$\xi_\kappa^{(l-1)} < \xi_\kappa^{(l)} < \xi_{\kappa+1}^{(l-1)}, \qquad \kappa = 1, 2, \ldots, 2n - l$$

hence

$$\xi_\kappa^{(2)} < \xi_\kappa^{(l)} < \xi_{\kappa+l-2}^{(2)}, \qquad l = 3, 4, \ldots, 2n-1. \tag{9.2}$$

K_2 is strictly convex on $[x_\kappa^{\mathrm{G}}, x_{\kappa+1}^{\mathrm{G}}]$, therefore K_2 has not more than two zeros on this interval, this means

$$x_\kappa^{\mathrm{G}} < \xi_{2\kappa-1}^{(2)} < \xi_{2\kappa}^{(2)} < x_{\kappa+1}^{\mathrm{G}}. \tag{9.3}$$

It follows from the relations (9.2) and (9.3) that the systems

$$(\xi_1^{(l)}, \xi_3^{(l)}, \ldots, \xi_{2\lceil \frac{2n-l}{2} \rceil - 1}^{(l)}) \quad \text{and} \quad (\xi_2^{(l)}, \xi_4^{(l)}, \ldots, \xi_{2\lceil \frac{2n-l}{2} \rceil}^{(l)})$$

have for $n \to \infty$ the same asymptotic distribution as the system $(x_1^{\mathrm{G}}, \ldots, x_n^{\mathrm{G}})$.

We shall write now ξ_ν instead of $\xi_\nu^{(s)}$. We have

$$\begin{aligned}
\int_{-1}^{1} |K_s(x)|\, dx &= \left| \sum_{\nu=0}^{2n-s} (-1)^\nu \int_{\xi_\nu}^{\xi_{\nu+1}} K_s(x)\, dx \right| \\
&= \left| \sum_{\nu=0}^{2n-s} (-1)^\nu (K_{s+1}(\xi_\nu) - K_{s+1}(\xi_{\nu+1})) \right| \\
&= 2 \left| \sum_{\mu=1}^{\lfloor \frac{2n-s}{2} \rfloor} K_{s+1}(\xi_{2\mu}) - \sum_{\mu=1}^{\lceil \frac{2n-s}{2} \rceil} K_{s+1}(\xi_{2\mu-1}) \right|.
\end{aligned}$$

Denote by c^l and c^u continuous functions such that we have

$$c^l(x) \leq m_\nu^{s+1} K_{s+1}(x) \leq c^u(x)$$

for all $\nu > \nu_0$. Suppose now that K_{s+1} has maxima at the points $\xi_2, \xi_4, \ldots$. Then we obtain

$$m_\nu^{s+1} \int_{-1}^{1} |K_s(x)|\, dx \le 2\Big(\sum_{\mu=1}^{\lfloor \frac{2n-s}{2} \rfloor} c^u(\xi_{2\mu}) - \sum_{\mu=1}^{\lceil \frac{2n-s}{2} \rceil} c^l(\xi_{2\mu-1}) \Big).$$

Our assumption on w implies that the x_μ^{G} have (asymptotically) a "trigonometric equidistribution", that means

$$\lim_{n\to\infty} \frac{1}{n} \sum_{\nu=1}^{n} f(x_\nu^{\mathrm{G}}) = \frac{1}{\pi} \int_{-1}^{1} \frac{f(x)}{\sqrt{1-x^2}}\, dx$$

(e.g. Freud (1969) p. 134) holds for $f \in C[-1,1]$, and since the systems $\xi_{2\mu}$ and $\xi_{2\mu-1}$ have the same distribution, we conclude

$$\overline{\lim_{\nu\to\infty}}\, m_\nu^s \int_{-1}^{1} |K_s(x)|\, dx \le \frac{1}{\pi} \int_{-1}^{1} \frac{c^u(x) - c^l(x)}{\sqrt{1-x^2}}\, dx.$$

But the limit on the left hand side is explicitly known (Petras (1988), Brass (1993)), so we obtain

$$(2\pi)^s I[h^s] \int_0^1 |B_s(x)|\, dx \le \frac{1}{\pi} \int_{-1}^{1} \frac{c^u(x) - c^l(x)}{\sqrt{1-x^2}}\, dx.$$

We use now Theorem 4 to construct a bound for the right hand side, but we observe that because of (9.1) one of the functions c^u and $-c^l$ may be lowered on a set of positive measure (continuity!). We get

$$(2\pi)^s I[h^s] \int_0^1 |B_s(x)|\, dx < \frac{1}{\pi} (2\pi)^{s+1} I[h^s] \big(\sup_x B_{s+1}(x) - \inf_x B_{s+1}(x) \big)$$

or

$$\int_0^1 |B_s(x)|\, dx < 2\big(\sup_x B_{s+1}(x) - \inf_x B_{s+1}(x) \big).$$

But this is false, we have equality here. Hence we have our contradiction.

References

Brass, H. (1977), *Quadraturverfahren*, Vandenhoeck und Ruprecht, Göttingen.

Brass, H. (1993), Asymptotically optimal error bounds for quadrature rules of given degree. *(To appear in Math. of Computation).*

DeVore, R.A. (1972), *The approximation of continuous functions by positive linear operators.* Springer Verlag, Berlin, Heidelberg, New York.

DeVore, R.A. / Scott, L.R. (1984), Error bounds for Gaussian quadrature and weighted-L^1 polynomial approximation. *SIAM J. Numer. Anal. 21*, 400–412.

Förster, K.-J. / Petras, K. (1991), Error estimates in Gaussian quadrature for functions of bounded variation. *SIAM J. Numer. Anal. 28*, 880–889.

Freud, G. (1969), *Orthogonale Polynome*, Birkhäuser Verlag, Basel, Stuttgart.

Ganelius, T. (1956), On one-sided approximation by trigonometric polynomials. *Math. Scand. 4*, 247–258.

Hämmerlin, G. / Hoffmann, K.-H. (1989), *Numerische Mathematik*, Springer Verlag, Berlin, Heidelberg, New York.

Petras, K. (1988), Asymptotic behaviour of Peano kernels of fixed order. In: H. Brass/ G. Hämmerlin (eds.): *Numerical integration III*. Birkhäuser Verlag, Basel, Boston, Berlin, 186–198.

Petras, K. (1992a), Error bounds for Gaussian and related quadrature and applications to r-convex functions. *SIAM J. Numer. Anal. 29*, 578–585.

Petras, K. (1992b), One sided L1-Approximation and bounds for Peano kernels. In: Espelid, T.O., Genz, A. (eds.), *Numerical Integration.* Kluwer Academic Publishers, Dordrecht/Boston/London, 165–174.

Helmut Brass
Institut für Angewandte Mathematik
Technische Universität Braunschweig
Pockelsstr. 14
W-3300 Braunschweig, Germany

A new lower bound for the number of nodes in cubature formulae of degree 4 n + 1 for some circularly symmetric integrals

Ronald Cools and Hans Joachim Schmid

Abstract. For two classes of integrals it will be shown that the number of nodes of cubature formulae of degree $4n+1, n > 1$, will not attain Möller's lower bound. Thus in these cases that bound has to be increased by 1.

1. Introduction

Consider an integral I such that

$$I[f] = \int\int_R w(\sqrt{x^2+y^2})f(x,y)dx\,dy = \int_a^b w(\tau)\tau(\int_0^{2\pi} f(\tau\cos\theta, \tau\sin\theta)d\theta)d\tau, \quad (1)$$

where τ, θ are polar coordinates and $w(\tau)$ is positive for $a < \tau < b$. The integral I has circular symmetry.

A cubature formula Q is an approximation of I of the form

$$Q[f] = \sum_{j=1}^{N} C_j f(x_j, y_j) \simeq I[f].$$

The cubature formula Q is of degree d for I if the identity $Q[f] = I[f]$ holds whenever f is a polynomial of degree $\leq d$, but does not hold for at least one polynomial of degree $d+1$. The number of knots N in such an approximation for (1) of odd degree $2k-1$ satisfies the lower bound by Möller (1976), i. e.,

$$N \geq \frac{k(k+1)}{2} + \lfloor\frac{k}{2}\rfloor. \quad (2)$$

For circularly symmetric integrals Verlinden and Cools (1992) have derived necessary and sufficient conditions such that cubature formulae of degree $4n+1$ attain Möller's lower bound. In their paper they showed that these conditions do not hold for some integrals of circular symmetry and some $n \in \mathbb{N}$. This result will be extended to all $1 < n \in \mathbb{N}$ for the following integrals

$$\int_{\mathbb{R}^2} f(x,y)(x^2+y^2)^{\alpha-1}e^{-x^2-y^2}dx\,dy, \alpha > 0, \tag{3}$$

and

$$\int_{\Omega} f(x,y)(x^2+y^2)^{\alpha}(1-x^2-y^2)^{\beta}dx\,dy,\ \Omega = \{(x,y) \in \mathbb{R}^2 : x^2+y^2 \leq 1\}, \alpha,\beta > -1. \tag{4}$$

Thus cubature formulae of degree $4n+1$, $n \geq 2$, for both integrals need at least $2(n+1)^2$ nodes.

2. Necessary and sufficient conditions

In this section we describe the main result of Verlinden and Cools (1992). A one-dimensional integral J is associated with a two-dimensional circular symmetric integral I , cf. (1), in the following way. Let $\rho = x^2 + y^2$ and let J be the one-dimensional integral

$$J[f(\rho)] \quad = \quad \int_{a^2}^{b^2} \pi w(\sqrt{\rho})\rho f(\rho)d\rho. \tag{5}$$

So, with the integral (3) corresponds

$$J_\alpha[f] = \int_0^\infty f(\rho)\rho^\alpha e^{-\rho}d\rho,\ \alpha > 0$$

and with the integral (4) corresponds

$$J_{\alpha,\beta}[f] = \int_0^1 f(\rho)\rho^{\alpha+1}(1-\rho)^\beta d\rho,\quad \alpha,\beta > -1.$$

Let $p_d(\rho)$ denote the orthogonal polynomial of degree d w.r.t. the integral J and let $\rho_1, \ldots, \rho_k$ be the zeros of $p_k(\rho)$.

Theorem 1 *The following propositions are equivalent:*

1. *There exists a cubature formula of degree* $4n+1$, *attaining Möller's lower bound, for the integral (1).*
2. $(-1)^1\frac{p_{n-1}(\rho_1)}{\rho_1^{n+1}} = (-1)^2\frac{p_{n-1}(\rho_2)}{\rho_2^{n+1}} = \cdots = (-1)^n\frac{p_{n-1}(\rho_n)}{\rho_n^{n+1}}$,
   where $\rho_1 < \rho_2 < \cdots < \rho_n$ *are the successive zeros of* $p_n(\rho)$.
3. *The polynomials* ρ^{2n+2} *mod* $p_n(\rho)$ *and* $p_{n-1}^2(\rho)$ *mod* $p_n(\rho)$ *are equal up to a constant factor.*

For a proof see Verlinden and Cools (1992).
The third proposition provides a practical test for the existence of a cubature formula attaining Möller's lower bound. The test does not require the numerical, approximate computation of the zeros of the polynomial $p_n(\rho)$, but involves only algebraic manipulations. However, the test is restricted to a given integral and fixed n. In this way it was proved by Verlinden and Cools (1992) that Möller's lower bound cannot be attained for some integrals and some $n \in \mathbb{N}$. It is the aim of this paper to present more general applications of Theorem 1.

3. A necessary condition

Let

$$I_\alpha[f] = \int_\Omega f(\rho)\rho^\alpha \omega(\rho)d\rho, \ \Omega \subseteq [0,\infty), \ \alpha \geq 0,$$

be a strictly positive linear functional. The orthogonal polynomials with respect to I_α will be denoted by

$$p_0^\alpha, p_1^\alpha, p_2^\alpha, \ldots.$$

They are normalized such that

$$p_i^\alpha(\rho) = \rho^i + \text{lower order terms.}$$

For a fixed n we will consider the following condition

$$\sigma(-1)^i \rho_i^{n+1} = p_{n-1}^\alpha(\rho_i), \ i = 1, 2, \ldots, n, \tag{6}$$

where σ is a real number and $\rho_1 < \rho_2 < \ldots < \rho_n$ are the roots of p_n^α. This is condition 2 of Theorem 1 in a slightly modified form. We will show that under the assumption of equation (6) the orthogonal polynomial p_n^α can be factored into two quasi-orthogonal polynomials which belong to the functional $I_{n+\alpha+1}$.
The Gaussian quadrature formula of degree $2n-1$ for I_α generated by the roots of p_n^α is of the form

$$Q_n[f] = \sum_{i=1}^{n} C_i f(\rho_i).$$

In particular we obtain

$$\sum_{i=1}^{n} C_i \rho_i^{n+j+1} = I_\alpha[\rho^{n+j+1}], \ j = 0, 1, \ldots, n-2, \tag{7}$$

and

$$\sum_{i=1}^{n} C_i \rho_i^j p_{n-1}^\alpha(\rho_i) = 0, \ j = 0, 1, \ldots, n-2.$$

Inserting (6) we get

$$\sum_{i=1}^{n} C_i (-1)^i \rho_i^{n+j+1} = 0, \quad \text{hence} \ \sum_{i \ even} C_i \rho_i^{n+j+1} = \sum_{i \ odd} C_i \rho_i^{n+j+1}.$$

Thus (7) can be written as

$$2\sum_{i\ \ even} C_i\rho_i^{n+1}\rho_i^j = I_\alpha[\rho^{n+j+1}] = I_{n+\alpha+1}[\rho^j],\ j=0,1,\ldots,n-2,$$

or as

$$2\sum_{i\ \ odd} C_i\rho_i^{n+1}\rho_i^j = I_\alpha[\rho^{n+j+1}] = I_{n+\alpha+1}[\rho^j],\ j=0,1,\ldots,n-2.$$

For $n \geq 2$ this implies that the ρ_i, i even, are the nodes of a positive quadrature formula of degree $n-2$ for $I_{n+\alpha+1}$. Similarly the ρ_i, i odd, generate such a formula. Hence condition (6) implies that p_n^α can be factored in the following way.

$$p_n^\alpha = \begin{cases} (p_r^{n+\alpha+1} + \gamma_1 p_{r-1}^{n+\alpha+1})(p_r^{n+\alpha+1} + \gamma_2 p_{r-1}^{n+\alpha+1}), \text{ if } n = 2r, \\ p_r^{\alpha+n+1}(p_{r+1}^{n+\alpha+1} + \gamma_1 p_r^{n+\alpha+1} + \gamma_2 p_{r-1}^{n+\alpha+1}), \text{ if } n = 2r+1, \end{cases} \tag{8}$$

where γ_1, γ_2 are real parameters. Let

$$p_n^\alpha(\rho) = \sum_{\nu=0}^{n} a_\nu^{n,\alpha}\rho^\nu$$

and $I_\alpha[p_n^\alpha p_n^\alpha]$ be given. A first condition on γ_1, $\gamma_1\gamma_2$, respectively, will be obtained by applying $I_{n+\alpha+1}$ to both sides of equation (8).
Since $I_{n+\alpha+1}[p_n^\alpha] = I_\alpha[\rho^{n+1}p_n^\alpha] = -a_n^{n+1,\alpha}I_\alpha[p_n^\alpha p_n^\alpha]$ we find

$$-a_n^{n+1,\alpha}I_\alpha[(p_n^\alpha)^2] = \begin{cases} \gamma_1 I_{n+\alpha+1}[(p_r^{n+\alpha+1})^2], \text{ if } n = 2r+1, \\ I_{n+\alpha+1}[(p_r^{n+\alpha+1})^2] + \gamma_1\gamma_2 I_{n+\alpha+1}[(p_{r-1}^{n+\alpha+1})^2], \text{ if } n = 2r \end{cases} \tag{9}$$

3.1 n is odd. If $n = 1$, then condition (6) holds, i.e. the lower bound by Möller will be attained. For $n = 2r+1 > 1$ we study the product

$$p_{2r+1}^\alpha = p_r^{2r+\alpha+2}(p_{r+1}^{2r+\alpha+2} + \gamma_1 p_r^{2r+\alpha+2} + \gamma_2 p_{r-1}^{2r+\alpha+2}).$$

Comparing the highest non-trivial coefficients on both sides we find

$$\gamma_1 = a_{2r}^{2r+1,\alpha} - a_r^{r+1,n+\alpha+1} - a_{r-1}^{r,n+\alpha+1}. \tag{10}$$

3.2 n is even. If $n = 2$ the necessary condition will be satisfied. In this case, however, condition 2 of Theorem 1 is equivalent to

$$a_0^{2,\alpha}(a_1^{2,\alpha} - 2a_0^{2,\alpha}) = a_0^{1,\alpha}a_1^{2,\alpha}[(a_1^{2,\alpha})^2 - 3a_0^{2,\alpha}], \tag{11}$$

which will be studied separately. We consider for $n = 2r \geq 4$ the polynomial

$$p_{2r}^\alpha - (p_r^{2r+\alpha+1} + \gamma_1 p_{r-1}^{2r+\alpha+1})(p_r^{2r+\alpha+1} + \gamma_2 p_{r-1}^{2r+\alpha+1}) \tag{12}$$

which must be the zero-polynomial in order to satisfy the necessary condition (8). The coefficients of this polynomial can be computed using the following Maple-script.

```
# file ecompare --- procedure ecompare: extract the n first non-
# trivial highest degree coefficients of P(2r,alpha)-P(r,2r+alpha+1)^2
#           -s*P(r,2r+alpha+1)*P(r-1,2r+alpha+1)-t*P(r-1,2r+alpha+1)^2
ecompare:= proc(n)
local f, p, q, i, z; f:= 1; p:= 1; q:= x; v:= array(1..n);
for i from 1 to n do
     f:= f + a(2*r,alpha,2*r-i) * x^i;
     p:= p + a(r,2*r+alpha+1,r-i) * x^i;
     q:= q + a(r-1,2*r+alpha+1,r-1-i) * x^(i+1); od;
z:= taylor(p^2 + s * p * q + t * q^2 -f,x,n+1);
for i from 1 to n do
     v[i]:= collect(expand(coeff(z,x,i)),[s,t]); # coefficient 2r - i
od; procname; end;
```

Setting the highest coefficients to zero we find

$$\gamma_1 + \gamma_2 = a_{2r-1}^{2r,\alpha} - 2a_{r-1}^{r,\alpha+n+1} \tag{13}$$

and

$$\gamma_1\gamma_2 = a_{2r-2}^{2r,\alpha} - 2a_{r-2}^{r,n+\alpha+1} - (a_{r-1}^{r,n+\alpha+1} + a_{r-2}^{r-1,n+\alpha 1})(\gamma_1 + \gamma_2) - a_{r-1}^{r,n+\alpha+1} a_{r-1}^{r,n+\alpha+1}. \tag{14}$$

Finally, in addition,

$$\begin{aligned} 0 = \; & (a_{r-2}^{r,2r+\alpha+1} + a_{r-1}^{r,2r+\alpha+1} a_{r-2}^{r-1,2r+\alpha+1} + a_{r-3}^{r-1,2r+\alpha+1})(\gamma_1 + \gamma_2) \\ & +2a_{r-3}^{r,2r+\alpha+1} + 2a_{r-1}^{r,2r+\alpha+1} a_{r-2}^{r,2r+\alpha+1} + 2a_{r-2}^{r-1,2r+\alpha+1} \gamma_1\gamma_2 - a_{2r-3}^{2r,\alpha} \end{aligned} \tag{15}$$

These results will be applied in the remaining part of the paper to prove

Theorem 2 *Let N be the number of nodes of a cubature formula of degree $4n+1, n > 1$, for the integrals (3) and (4). Then*

$$N \geq 2(n+1)^2.$$

Cubature formulae that attain this bound are known for degree 9 ($n = 2$). For the integral (3) with $\alpha = 1$, a formula is constructed by Haegemans and Piessens (1977). For the integral (4) with $\alpha = \beta = 0$, a formula is constructed by Piessens and Haegemans (1975).

4. Generalized Laguerre polynomials

For $\alpha > 0$ we consider

$$I_\alpha[f] = \int_0^\infty f(\rho)\rho^\alpha e^{-\rho} d\rho \tag{16}$$

and the associated orthogonal polynomials

$$L_n^\alpha(\rho) = \sum_{\nu=0}^{n} a_\nu^{n,\alpha} \rho^\nu, \text{ where } a_\nu^{n,\alpha} = (-1)^{n+\nu} \frac{n!}{\nu!} \binom{n+\alpha}{n-\nu}.$$

The coefficients can be computed via

```
# file laguerre
# compute the m-th coefficient of Laguerre(n,mu)
# note that a(n,mu,m) contains the factor 0, if m is negative.
a:= proc(n,mu,m) local i, j, sum; sum:= 1;
for j from 1 to n - m do i:= n-j; sum:= sum*(-(i+1)*(mu+i+1)/j); od;
sum; end;
```

In particular we obtain

$$\begin{aligned} a_n^{n,\alpha} &= 1, & a_{n-2}^{n,\alpha} &= n(n-1)(n+\alpha)(n+\alpha-1)/2, \\ a_{n-1}^{n,\alpha} &= -n(n+\alpha), & a_{n-3}^{n,\alpha} &= n(n-1)(n-2)(n+\alpha)(n+\alpha-1)(n+\alpha-2)/6. \end{aligned}$$

The corresponding moments are of the form

$$I_\alpha[L_n^\alpha L_n^\alpha] = n!(n+\alpha)!.$$

4.1 n is odd. In the odd case, i. e. $n = 2r+1 > 1$, the necessary condition (10) is of the form

$$\gamma_1 = a_{2r}^{2r+1,\alpha} - a_r^{r+1,2r+\alpha+2} - a_{r-1}^{r,2r+\alpha+2} = 2(r+1)^2. \tag{17}$$

From (9) we get

$$\gamma_1 = \frac{(2r+2)!(2r+\alpha+2)!}{r!(3r+\alpha+2)!}.$$

Hence the identity $(2r+1)!(2r+\alpha+2)! = (3r+\alpha+2)!(r+1)!$ is necessary for the factorization. For $\alpha > 0$, this holds for $r = 0$ alone.

4.2 n is even. If $n = 2$ then equation (11) does not hold, which can be easily checked. For $n = 2r \geq 4$, we can compute $\gamma_1 + \gamma_2$ and $\gamma_1\gamma_2$ from (13) and (14). In addition, we get one further equation from (15). Furthermore, the discriminant of γ_1 and γ_2 must be positive. The computation can be checked by the following Maple-script which calls the procedures `ecompare` and `laguerre`.

```
lprint(' Laguerre even, (alpha)');
read(ecompare): ecompare(3): read(laguerre):
lprint(' s = gamma_1+gamma_2 from (13)'); s:= factor(solve(v[1],s));
lprint(' t = gamma_1*gamma_2 from (14)'); t:= factor(solve(v[2],t));
lprint(' discriminant for s and t'); discr:= factor(s^2 - 4*t);
lprint(' From (15) we get'); u:= factor(v[3]):
lprint('0 = ',u); lprint(' Hence'); alpha:= solve(u,alpha);
lprint(' Inserting this into the discriminant we obtain'); discr;
```

Thus $\gamma_1 + \gamma_2 = 2r(r+1)$ and $\gamma_1\gamma_2 = r^2(2r^2 - \alpha)$. Furthermore, from (15),

$$0 = r^2(r^2-1)(2r^2-6r-3\alpha-2).$$

Hence $\alpha := 2(r^2-3r-1)/3$. Now the discriminant is of the form

$$-4r^2(r^2-2r-1-\alpha) = -4r^2(r^2-1)/3$$

This is negative for $r > 1$. Consequently the necessary condition does not hold.

5. Shifted Jacobi Polynomials

We consider

$$I_{\alpha,\beta}[f] = \int_0^1 f(\rho)\rho^{\alpha+1}(1-\rho)^{\beta}d\rho, \quad \alpha, \beta > -1.$$

The associated orthogonal polynomials are known as shifted Jacobi polynomials,

$$G_n^{\alpha+\beta+2,\alpha+2}(\rho) = \sum_{\nu=0}^{n} a_\nu^{n,\alpha,\beta}\rho^\nu,$$

where

$$a_\nu^{n,\alpha,\beta} = (-1)^{n-\nu}\binom{n}{\nu}\frac{\Gamma(n+\alpha+2)\Gamma(n+\nu+\alpha+\beta+2)}{\Gamma(\nu+\alpha+2)\Gamma(2n+\alpha+\beta+2)}.$$

The coefficients can be computed via

```
# file jacobi
# procedure to compute the m-th coefficient of Jacobi(n,mu,beta),
# note that a(n,mu,m) contains the factor 0, if m is negative.
a:= proc(n,mu,m) local i, j, sum; sum:= 1;
for j from 1 to n - m do
    i:= n-j; sum:= sum*(-(i+1)*(mu+i+2)/(j*(mu+beta+n+i+2))); od;
sum; end;
```

In particular we obtain

$$a_n^{n,\alpha,\beta} = 1,\ a_{n-1}^{n,\alpha,\beta} = -n\frac{n+\alpha+1}{2n+\alpha+\beta+1},\ a_{n-2}^{n,\alpha,\beta} = \frac{n(n-1)(n+\alpha+1)(n+\alpha)}{2(2n+\alpha+\beta+1)(2n+\alpha+\beta)},$$
$$a_{n-3}^{n,\alpha,\beta} = -\frac{n(n-1)(n-2)(n+\alpha+1)(n+\alpha)(n+\alpha-1)}{6(2n+\alpha+\beta+1)(2n+\alpha+\beta)(2n+\alpha+\beta-1)},$$

and

$$I_{\alpha,\beta}[G_n^{\alpha+\beta+2,\alpha+2}G_n^{\alpha+\beta+2,\alpha+2}] = \frac{n!\,\Gamma(n+\alpha+2)\Gamma(n+\alpha+\beta+2)\Gamma(n+\beta+1)}{(2n+\alpha+\beta+2)\Gamma^2(2n+\alpha+\beta+2)}.$$

This can be computed via

```
# file ll
# procedure to evaluate Int(Jacobi(n,mu,beta)^2)
ll:= proc(n,mu)
GAMMA(n+1)*GAMMA(n+mu+2)*GAMMA(n+mu+beta+2)*GAMMA(beta+n+1)
   / ((2*n+mu+beta+2)*GAMMA(2*n+mu+beta+2)*GAMMA(2*n+mu+beta+2));
end;
```

5.1 n is odd. Let us start by considering the odd case, $n = 2r + 1 > 1$. From equation (10) we obtain

$$\gamma_1 = \frac{2(r+1)(4r^2 + \beta r + 6r + r\alpha + \beta + 1)}{(4r + \alpha + \beta + 3)(4r + \alpha + \beta + 5)}$$

and from equation (9) we get

$$\gamma_1 = \frac{(2r+2)!(\alpha + 2r + 3)!(2r + \alpha + \beta + 2)!(\beta + 2r + 1)!}{(4r + \alpha + \beta + 5)r!(3r + \alpha + 3)!(3r + \alpha + \beta + 3)!(r + \beta)!}.$$

Thus the identity

$$\begin{aligned} & \frac{(2r+1)!(2r+\alpha+3)!(2r+\alpha+\beta+2)!(2r+\beta+1)!(4r+\alpha+\beta+3)}{r!(3r+\alpha+3)!(3r+\alpha+\beta+3)!(r+\beta)!} \\ = \quad & 4r^2 + \beta r + 6r + \alpha r + \beta + 1 \end{aligned}$$

is necessary for the factorization. Rewriting it as

$$\begin{aligned} & \frac{(r+1)\dots(2r+1)(r+\beta+1)\dots(2r+\beta+1)(4r+\alpha+\beta+3)}{(2r+\alpha+4)\dots(3r+\alpha+3)(2r+\alpha+\beta+3)\dots(3r+\alpha+\beta+3)} \\ = \quad & 4r^2 + (\alpha+\beta+6)r + \beta + 1 \end{aligned}$$

we obtain

$$(3r + \alpha + \beta + 3)(4r^2 + \beta r + 6r + r\alpha + \beta + 1) \leq (r+1)(r+\beta+1)(4r+\alpha+\beta+3).$$

Hence $r(8r^2 + (6\alpha + 2\beta + 19)r + (\alpha+3)(\alpha+\beta+4) - 1) \leq 0$ is necessary for the factorization. It holds for $r = 0$ alone.

5.2 n is even. If $n = 2$ then equation (11) does not hold. For $n = 2r \geq 4$, we can compute $\gamma_1 + \gamma_2$ and $\gamma_1\gamma_2$ from (13) and (14). In addition, we get two further equations from (15) and (9). The computations can be checked by the following Maple-script which calls the procedures `ecompare`, `jacobi` and `ll`.

```
lprint(` Jacobi even, (alpha,beta), r > 1`);
read(ecompare); ecompare(3): read(jacobi); read(ll);
lprint(` s = gamma_1 + gamma_2 from (13)`); s:= factor(solve(v[1],s));
lprint(` t = gamma_1 * gamma_2 from (14)`); t:= factor(solve(v[2],t));
lprint(` from equation (15): f1 = 0, where`);
f1:= factor(v[3]); f1:= numer(f1)/(2*r^2*(r-1)):
lprint(` we obtain S =  0, where`); S:= taylor(f1,r,10);
lprint(` condition (9) (multiplied by the factor hh = hh1 = hh2)`);
lprint(` can be transformed to`);
h:= 4 * r + alpha + beta + 1:
hh:= (h-1)*(h+1)*(h+2)*GAMMA(h+1)^2 / (GAMMA(r+1)*
    GAMMA(3*r+alpha+2)*GAMMA(3*r+alpha+beta+2)*GAMMA(r+beta)):
```

```
hh1:= (h-1)*(h+2)*GAMMA(h+2)^2 / ((h+1)*GAMMA(r+1)*
    GAMMA(3*r+alpha+2)*GAMMA(3*r+alpha+beta+2)*GAMMA(r+beta)):
hh2:= (h-1)*h^2*(h+1)*(h+2)*GAMMA(h)^2 / (GAMMA(r+1)*
    GAMMA(3*r+alpha+2)*GAMMA(3*r+alpha+beta+2)*GAMMA(r+beta)):
z:= simplify(- ll(2*r,alpha)*a(2*r+1,alpha,2*r)* hh,GAMMA):
z1:= simplify(ll(r,2*r+alpha+1) * hh1, GAMMA):
z2:= simplify(t*ll(r-1,2*r+alpha+1)*hh2,GAMMA): zz:= factor(z1+z2):
lprint(' Thus (9 bis)'); z; T:= taylor(zz,r,8): lprint('  = '); T;
lprint(' Hence we obtain the inequality: 0 > u, where');
u:= T + S - (r+1)*(r+2)*(r+beta)*(4*r+alpha+beta);
lprint(' i.e.: 0 > v, where'); v:= taylor(expand(u),r,6);
```

From (13) and (14) we get

$$\begin{aligned}\gamma_1+\gamma_2 &= \frac{2r(4r^2+(\alpha+\beta+3)r+\beta)}{(4r+\alpha+\beta+1)(4r+\alpha+\beta+2)},\\ \gamma_1\gamma_2 &= \frac{r^2\,W(r;\alpha,\beta)}{(4r+\alpha+\beta)(4r+\alpha+\beta+1)(4r+\alpha+\beta+2)^2},\end{aligned}$$

where

$$\begin{aligned}W(r;\alpha,\beta) \quad &= 32r^4+4(4\alpha+4\beta+5)r^3+(2(\alpha+\beta)^2-3\alpha+5\beta-28)r^2\\ &-2(\alpha^2+3\alpha\beta+8\alpha+8\beta+12)r\\ &-(\alpha^2\beta+\alpha\beta^2+\alpha^2+6\alpha\beta+\beta^2+4\alpha+8\beta+4).\end{aligned}$$

From (15) we obtain

$$0=\frac{2r^2(r-1)S(r;\alpha,\beta)}{3(4r+\alpha+\beta-1)(4r+\alpha+\beta)^2(4r+\alpha+\beta+1)^2},$$

where

$$\begin{aligned}S(r;\alpha,\beta) \;=\; &32r^5+(16(\alpha+\beta)-20)r^4+(2(\alpha+\beta)^2-21(\alpha+\beta)-45)r^3\\ &-(4(\beta^2+\alpha^2)+20\alpha\beta+11\alpha+44\beta-11)r^2\\ &-(3\alpha\beta(\alpha+\beta)+13\alpha\beta-\alpha^2+11\beta^2-12\alpha+9\beta-19)r\\ &-(3\alpha\beta^2-\alpha^2-2\alpha\beta+5\beta^2-4\beta-4\alpha-3).\end{aligned}$$

Thus (15) can be simplified to

$$S(r;\alpha,\beta)=0. \tag{18}$$

From condition (9) we obtain

$$\frac{(2r+\alpha+\beta+1)!(2r+\beta)!(4r+\alpha+\beta)(2r+1)!(2r+\alpha+2)!}{r!(3r+\alpha+1)!(3r+\alpha+\beta+1)!(r+\beta-1)!}=T(r;\alpha,\beta),$$

where

$$\begin{aligned} T(r;\alpha,\beta) &= 32r^5 + (16(\alpha+\beta)+28)r^4 + (2(\alpha+\beta)^2 - \alpha + 7\beta - 18)r^3 \\ &\quad -(2\alpha^2 + 6\alpha\beta + 12\alpha + 4\beta + 24)r^2 \\ &\quad -(\alpha\beta(\alpha+\beta) - 2\beta^2 + 3\alpha + \beta + 6)r \\ &\quad +\alpha\beta(\alpha+\beta) + 2\beta(\alpha+\beta). \end{aligned}$$

This is equivalent to

$$\frac{(4r+\alpha+\beta)(r+1)\ldots(2r+1)(r+\beta)\ldots(2r+\beta)}{(2r+\alpha+\beta+2)\ldots(3r+\alpha+\beta+1)(2r+\alpha+3)\ldots(3r+\alpha+1)} = T(r;\alpha,\beta).$$

Thus we obtain the inequality

$$(r+1)(r+2)(r+\beta)(4r+\alpha+\beta) > T(r;\alpha,\beta).$$

Subtracting equation (18) we obtain the following necessary condition for the factorization

$$\begin{aligned} 0 &> T(r;\alpha,\beta) - (r+1)(r+2)(r+\beta)(4r+\alpha+\beta) - S(r;\alpha,\beta) \\ &= 44r^4 + (19\alpha + 23\beta + 15)r^3 + (2\alpha^2 + 3\beta^2 + 13\alpha\beta - 4\alpha + 25\beta - 43)r^2 \\ &\quad +(2\alpha\beta(\alpha+\beta) - \alpha^2 + 10\beta^2 + 10\alpha\beta - 17\alpha - 2\beta - 25)r \\ &\quad +\alpha^2\beta + 4\alpha\beta^2 - \alpha^2 - 2\alpha\beta + 5\beta^2 - 4\alpha - 4\beta - 3. \end{aligned}$$

For $r \geq 2$, however, the right hand side is positive for all α, β under consideration, so the factorization does not hold.

References

Haegemans, A. and R. Piessens (1977), Construction of cubature formulas of degree seven and nine for symmetric planar regions. *SIAM J. Numer. Anal.*, 14:492–508.

Möller, H. M. (1976), Kubaturformeln mit minimaler Knotenzahl. *Numer. Math.*, 25:185–200.

Piessens, R. and A. Haegemans (1975), Cubature formulas of degree nine for symmetric planar regions. *Math. Comp.*, 29:810-815.

Verlinden, P. and R. Cools (1992), On cubature formulae of degree $4k+1$ attaining Möller's lower bound for integrals with circular symmetry. *Numer. Math.*, 61:395–407.

Authors' addresses:

Ronald Cools
Dept. of Computer Science
Katholieke Universiteit Leuven
Celestijnenlaan 200 A
B-3001 Leuven
Ronald.Cools@cs.kuleuven.ac.be

Hans Joachim Schmid
Mathematisches Institut
Universität Erlangen-Nürnberg
Bismarckstraße 1 1/2
DW-8520 Erlangen
schmid@mi.uni-erlangen.de

International Series of Numerical Mathematics, Vol. 112, © 1993 Birkhäuser Verlag Basel

On the Construction of Gaussian Quadrature Formulae containing Preassigned Nodes

Sven Ehrich

Abstract An iterative method is proposed for the construction of Gaussian quadrature formulae that contain preassigned nodes. The method is locally convergent of order two. It can be implemented in a way that every iteration step involves a number of arithmetical operations that depends quadratically on the number of nodes, and it provides a posteriori error estimates. A comparison is made with known methods. Numerical examples show the suitability of the method for some of the possible applications.

1. Construction by Fixed Point Iteration

We will consider the approximation of the weighted integral $I_\rho[f] := \int_a^b \rho(x)f(x)\,dx$ of f over a finite or infinite interval (a,b), where ρ is assumed continuous and non-negative in (a,b), and all moments $M_k = I_\rho[p_k]$, $p_k(x) = x^k$, $k = 0,1,2,\ldots$ are assumed to exist. For given real numbers $\eta_1 < \ldots < \eta_m$, $m \geq 0$, let there exist real distinct numbers $\xi_1,\ldots,\xi_n$, pairwise different from $\eta_1,\ldots,\eta_m$, such that the interpolatory quadrature formula

$$Q^G_{n,m}[f] := \sum_{\nu=1}^{m} \alpha_\nu f(\eta_\nu) + \sum_{\mu=1}^{n} \beta_\mu f(\xi_\mu) \tag{1}$$

has polynomial degree of exactness

$$\deg(Q^G_{n,m}) \geq 2n+m-1, \tag{2}$$

i.e., integrates exactly all polynomials of degree $\leq 2n+m-1$. Then (1) will be called a *Gaussian quadrature formula for the preassigned nodes* $\eta_1,\ldots,\eta_m$. It is easy to see that a Gaussian quadrature formula for given $\eta_1,\ldots,\eta_m$ is unique. Examples of nonexistence have been found for the special case of Kronrod quadrature (cf., among others, Gautschi and Notaris 1988a). The aim of this paper is to propose an iterative method to compute the nodes $\xi_1\ldots,\xi_n$ numerically. We place special emphasis on the a posteriori verification of the numerical results, which is strongly motivated by an example in Gautschi (1983), where two approximations for $Q^G_{15,0}$ with respect to $\rho(x) = e^{-x^3/3}$, $x \in (0,\infty)$, are discussed which agree in only about two decimal digits but both satisfy the seemingly reasonable test of reproducing the moments

$$\sum_{\nu=1}^{m} \alpha_\nu\, p_k(\eta_\nu) + \sum_{\mu=1}^{n} \beta_\mu\, p_k(\xi_\mu) - M_k = 0, \qquad k = 0,1,2,\ldots,2n+m-1. \tag{3}$$

This test was therefore regarded as unreliable, and alternative heuristic tests were proposed on the basis of three-term recurrence relations. It is well known that these exist for Gaussian formulas without preassigned nodes, but not necessarily for Kronrod quadrature (cf. Monegato 1979). On the other hand, numerical results have shown that for hitherto used methods the computation of Gaussian quadrature formulas containing preassigned nodes may even be numerically more unstable than without preassigned nodes (cf. Gautschi and Notaris 1988b).

In the following, assume that the Gaussian formula (1) exists for given $\eta_1, \dots, \eta_m$. Let $\Omega(x) := \prod_{\nu=1}^{m}(x-\eta_\nu)$, $\omega(x) = \prod_{\mu=1}^{n}(x-\xi_\mu)$, $\omega^{[i]}(x) := \prod_{\mu=1,\mu\neq i}^{n}(x-\xi_\mu)$, $\omega^{[i,j]}(x) := \prod_{\mu=1,\mu\neq i,\mu\neq j}^{n}(x-\xi_\mu)$, $\omega^{[j,i]} := \omega^{[i,j]}$, $1 \le i < j \le n$. It is well known (cf. Krylov 1962, section 9.1), that for (2) it is necessary and sufficient that $Q_{n,m}^G$ be interpolatory and that ω be orthogonal with respect to $\rho \cdot \Omega$ to all polynomials of degree $\le n-1$, i.e., the values $\xi_1, \dots, \xi_n$ are a solution of the nonlinear system

$$\Phi_o(X) := \begin{pmatrix} \int_a^b \rho(x)\Omega(x)\omega(x)\omega^{[1]}(x)\,dx \\ \vdots \\ \int_a^b \rho(x)\Omega(x)\omega(x)\omega^{[n]}(x)\,dx \end{pmatrix} = 0. \tag{4}$$

The weights $\alpha_1, \dots, \alpha_m, \beta_1, \dots, \beta_n$ in (1) are then given by

$$\alpha_\nu = (\Omega^{[\nu]}(\eta_\nu)\omega(\eta_\nu))^{-1} \int_a^b \rho(x)\Omega^{[\nu]}(x)\omega(x)\,dx, \tag{5}$$

$$\beta_\mu = (\Omega(\xi_\mu))^{-1} \left(\omega^{[\mu]}(\xi_\mu)\right)^{-2} \int_a^b \rho(x)\Omega(x)\left(\omega^{[\mu]}(x)\right)^2 dx. \tag{6}$$

Without loss of generality we can assume that

$$\beta_\mu \neq 0 \qquad \text{for all} \quad \mu = 1, \dots, n, \tag{7}$$

since otherwise n could be replaced by a smaller number. The nonlinear system (4) may be solved by Newton's rootfinding method, or alternatively by the fixed point method defined by

$$X_0 \in I\!R^n, \qquad X_k = \Phi_E(X_{k-1}) := \begin{pmatrix} \phi_1(X_{k-1}) \\ \vdots \\ \phi_n(X_{k-1}) \end{pmatrix}, \tag{8}$$

where

$$\phi_i(X) = \frac{\int_a^b \rho(x)\Omega(x)\left(\zeta^{[i]}(x)\right)^2 x\,dx}{\int_a^b \rho(x)\Omega(x)(\zeta^{[i]}(x))^2\,dx}, \tag{9}$$

and for $X = (x_1, \dots, x_n) \in I\!R^n$ we define $\zeta(x) = \prod_{\nu=1}^{n}(x-x_\nu)$. Note that in view of (7) the denominator does not vanish near the fixed point. In the following section we will prove that this method is locally and quadratically convergent to the solution $X^* := (\xi_1, \dots, \xi_n)$. The computation of the respective weighted integrals of polynomials can be done exactly if the necessary moments are known, or approximately, in which case however the iteration function and its fixed point may change. A discussion of implementation details will be given in section 4.

For heuristic error bounds, let $\|\cdot\|_V$ be a norm on $\mathbb{R}^n$ with compatible matrix norm $\|\cdot\|_M$. Let $X \in \mathbb{R}^n$ be such that

$$\lambda := \|\Phi_E'(X)\|_M < 1. \tag{10}$$

It is well known that under the assumption $\|\Phi_E(X) - X^*\|_V \le \lambda\|X - X^*\|_V$ there holds the classical a posteriori error bound

$$\|\Phi_E(X) - X^*\|_V \le \frac{\lambda}{1-\lambda}\|\Phi_E(X) - X\|_V. \tag{11}$$

2. Convergence of the fixed point method Φ_E

An application of Newton's method to the system (4) may be expected to have local and quadratic convergence. The following theorem states that (8) possesses similar convergence properties.

Theorem 1. *The iterative procedure*

$$X_0 \in \mathbb{R}^n; \qquad X_k = \Phi_E(X_{k-1}), \quad k \in \mathbb{N},$$

defined in (8) and (9) is locally convergent of order two.

Note that by (9) an equality $x_i = x_j$ implies $\phi_i(X) = \phi_j(X)$, so that global convergence is not possible.

Proof of Theorem 1. The ϕ_i defined by (9) are rational functions in $x_1, \ldots, x_n$. Therefore we can form continuous partial derivatives of ϕ_i with respect to $x_1, \ldots, x_n$ of arbitrary order at the fixed point. Since ϕ_i in (9) is independent of x_i, we have

$$\frac{\partial \phi_i}{\partial x_i}(X) = 0 \tag{12}$$

for all $X \in \mathbb{R}^n$. For $i \neq j$ we have by elementary transformations

$$\frac{\partial \phi_i}{\partial x_j}(X) = -2 + 2\,\frac{\int_a^b \rho(x)\Omega(x)\left(\zeta^{[i]}(x)\right)^2 (x - x_j)\,dx \,\int_a^b \rho(x)\Omega(x)\zeta^{[i]}(x)\zeta^{[i,j]}(x)\,dx}{\left(\int_a^b \rho(x)\Omega(x)\left(\zeta^{[i]}(x)\right)^2\,dx\right)^2}. \tag{13}$$

Evaluating the integrals with the quadrature formula (1) at $X = X^*$, we find

$$\begin{aligned}\frac{\partial \phi_i}{\partial x_j}(X^*) &= -2 + 2\,\frac{\beta_i\Omega(\xi_i)\left(\omega^{[i]}(\xi_i)\right)^2(\xi_i - \xi_j)\,\beta_i\Omega(\xi_i)\omega^{[i]}(\xi_i)\omega^{[i,j]}(\xi_i)}{\left(\beta_i\,\Omega(\xi_i)\left(\omega^{[i]}(\xi_i)\right)^2\right)^2} &(14)\\ &= 0. &(15)\end{aligned}$$

In view of (12) and (15), any norm of the Jacobian matrix vanishes at the fixed point, and the assertion follows from a well-known theorem on fixed point methods. q.e.d.

3. Comparison with known Methods

Several methods exist for the case $m = 0$, $\Omega \equiv 1$ without preassigned nodes (for a survey, cf. Gautschi 1979). An often-used method for the important special case of Kronrod and Patterson extensions (Patterson 1968) is to expand ω into a given polynomial basis and to compute the coefficients with the help of the orthogonality property of ω. Then, $\xi_1, \dots, \xi_n$ are obtained by a suitable rootfinding method. An alternative method takes advantage of a characterization of $\xi_1, \dots, \xi_n$ as eigenvalues of a Hessenberg matrix, which is tridiagonal in the special case $m = 0$, $\Omega \equiv 1$ (cf. Kautsky and Elhay 1984). A third method proposed by Caliò et al. (1986) is to apply Newton's method to the nonlinear system

$$\Phi_{CGM}(X) := \begin{pmatrix} \sum_{\nu=1}^{m} \alpha_\nu \, p_0^*(\eta_\nu) & + \sum_{\mu=1}^{n} \beta_\mu \, p_0^*(\xi_\mu) & - M_0^* \\ \vdots & \vdots & \vdots \\ \sum_{\nu=1}^{m} \alpha_\nu p_{2n+m-1}^*(\eta_\nu) & + \sum_{\mu=1}^{n} \beta_\mu \, p_{2n+m-1}^*(\xi_\mu) & - M_{2n+m-1}^* \end{pmatrix} = 0, \tag{16}$$

in the unknowns $X = (\xi_1, \dots, \xi_n, \alpha_1, \dots, \alpha_m, \beta_1, \dots, \beta_n) \in \mathbb{R}^{2n+m}$, where $\{p_0^*, p_1^*, \dots, p_{2n+m-1}^*\}$ is a basis of $\mathcal{P}_{2n+m-1}$, and $M_k^* = I_\rho[p_k^*]$. For $p_k^* = p_k$ and $M_k^* = M_k$, this method uses the moment reproducing test (3), since Newton's procedure halts if $\Phi_{CGM}(X_k) = 0$ within the range of numerical precision. For $\rho \equiv 1$ and choosing p_k^* as the k-th Legendre polynomial, it has been observed in Gautschi and Notaris (1988b) that Newton's method for (16) deteriorates for subsequent Patterson extensions and in a special example diverges even if the starting values are taken very close to the results published by Patterson. Motivated by a convergence theorem of Ostrowski (1966, p. 187), Gautschi and Notaris further analyzed this case by examining a norm of the inverse of the Jacobian matrix of the system (16). Denoting the iteration function of Newton's method, applied to $\Phi(X) = 0$, with $N[\Phi]$, we obtain from Taylor's theorem with center X^* (cf. Ortega and Rheinboldt 1970, NR 3.3-3 and E 10.2-6)

$$\begin{aligned} \|N[\Phi_{CGM}](X) - X^*\|_2 \;\le\; & \frac{1}{2} \|(\Phi'_{CGM}(X^*))^{-1}\|_F \|\Phi''_{CGM}(X^*)\|_F \, \|X - X^*\|_2^2 \\ & + O(\|X - X^*\|_2^3), \end{aligned} \tag{17}$$

$$\begin{aligned} \|N[\Phi_o](X) - X^*\|_2 \;\le\; & \frac{1}{2} \|(\Phi'_o(X^*))^{-1}\|_F \|\Phi''_o(X^*)\|_F \, \|X - X^*\|_2^2 \\ & + O(\|X - X^*\|_2^3) \end{aligned} \tag{18}$$

and

$$\|\Phi_E(X) - X^*\|_2 \le \frac{1}{2} \|\Phi''_E(X^*)\|_F \, \|X - X^*\|_2^2 + O(\|X - X^*\|_2^3), \tag{19}$$

where $\|\Phi''_{CGM}(X^*)\|_F$ resp. $\|\Phi''_o(X^*)\|_F$ is the square root of the sum of the squares of all second derivatives.

Gautschi and Notaris (1988b) showed

Lemma 1 (Gautschi and Notaris). *Let $\rho \equiv 1$ and let p_k^* in (16) be the normalized Legendre polynomial of degree k; let $m = N$, $n = N+1$ and let $p_m^*(\eta_\nu) = 0$, $\nu = 1, \dots, m$. Then there holds*

$$\|(\Phi'_{CGM}(X^*))^{-1}\|_F = \left(\int_a^b \phi_N(t)\,dt\right)^{\frac{1}{2}}, \tag{20}$$

where the polynomial ϕ_N is defined by

$$\phi_N = \sum_{\nu=1}^{N} A_\nu^2 + \sum_{\mu=1}^{N+1}\left(B_\mu^2 + \frac{1}{\beta_\mu^2}C_\mu^2\right), \tag{21}$$

and A_ν, B_μ, C_μ are elementary Hermite interpolation polynomials of degree $3N+1$ satisfying

$$\begin{array}{lll} A_\nu(\eta_\lambda) = \delta_{\nu\lambda}; & A_\nu(\xi_\mu) = 0; & A'_\nu(\xi_\mu) = 0; \\ B_\mu(\eta_\lambda) = 0; & B_\mu(\xi_\kappa) = \delta_{\mu\kappa}; & B'_\mu(\xi_\kappa) = 0; \\ C_\mu(\eta_\lambda) = 0; & C_\mu(\xi_\kappa) = 0; & C'_\mu(\xi_\kappa) = \delta_{\mu\kappa}. \end{array} \tag{22}$$

Using Lemma 1, the norm of the inverse of the Jacobian matrix of (16) can be computed if the nodes $\xi_1, \dots, \xi_n$ and the weights $\beta_1, \dots, \beta_n$ are known. In the following two Lemmas, we will give similar representations for Newton's method applied to (4) and for the fixed point method Φ_E, respectively.

Lemma 2. *For $\Phi_o : I\!R^n \to I\!R^n$ defined in (4), there hold*

$$\|(\Phi'_o(X^*))^{-1}\|_F = \left(\sum_{i=1}^{n}\left(\beta_i\Omega(\xi_i)(\omega'(\xi_i))^2\right)^{-2}\right)^{\frac{1}{2}} \tag{23}$$

and

$$\|\Phi''_o(X^*)\|_F = 2\left(\sum_{i=1}^{n}\sum_{j=1,\ j\neq i}^{n} 3\left(\beta_i\Omega(\xi_i)\frac{(\omega'(\xi_i))^2}{\xi_i - \xi_j}\right)^2\right)^{\frac{1}{2}}. \tag{24}$$

Lemma 3. *For $\Phi_E : I\!R^n \to I\!R^n$ defined in (8) and (9), there holds*

$$\|\Phi''_E(X^*)\|_F = 2\left(\sum_{i=1}^{n}\sum_{j=1,j\neq i}^{n}\left(\frac{\beta_j\,\Omega(\xi_j)\,(\omega'(\xi_j))^2}{\beta_i\,\Omega(\xi_i)\,(\omega'(\xi_i))^2}\right)^2\right)^{\frac{1}{2}}. \tag{25}$$

In view of (17), (18) and (19), the preceding Lemmas can be used for a heuristic analysis of the convergence of the methods that are proposed in section 1; see Example 2 in the following section.

Proof of Lemma 2. For $X \in I\!R^n$, let $\zeta(x) = \prod_{\nu=1}^n (x - x_\nu)$ and

$$\phi_o^{(i)}(X) := \int_a^b \rho(x)\Omega(x)\zeta(x)\zeta^{[i]}(x)\,dx. \tag{26}$$

Then

$$\frac{\partial \phi_o^{(i)}}{\partial x_i}(X) = -\int_a^b \rho(x)\Omega(x)\left(\zeta^{[i]}(x)\right)^2 dx, \tag{27}$$

and for $i \neq j$ the orthogonality property of ω implies

$$\frac{\partial \phi_o^{(i)}}{\partial x_j}(X^*) = -2\int_a^b \rho(x)\Omega(x)\omega(x)\omega^{[i,j]}(x)\,dx = 0. \tag{28}$$

Evaluating (27) at the fixed point X^* by the quadrature formula (1), we find

$$\frac{\partial \phi_o^{(i)}}{\partial x_i}(X^*) = -\beta_i\,\Omega(\xi_i)\left(\omega'(\xi_i)\right)^2, \tag{29}$$

leading to (23). For $k \neq i$, $k \neq j$ we have

$$\frac{\partial^2 \phi_o^{(i)}}{\partial x_k \partial x_j}(X^*) = \frac{\partial^2 \phi_o^{(i)}}{\partial x_i \partial x_i}(X^*) = 0 \tag{30}$$

because of the orthogonality property of ω and since the right hand side of (27) is independent of x_i. For $j \neq i$ we have

$$\frac{\partial^2 \phi_o^{(i)}}{\partial x_i \partial x_j}(X^*) = \frac{\partial^2 \phi_o^{(i)}}{\partial x_j \partial x_i}(X^*) = \frac{\partial^2 \phi_o^{(j)}}{\partial x_i \partial x_i}(X^*) = 2\,\beta_i\,\Omega(\xi_i)\,\frac{\left(\omega'(\xi_i)\right)^2}{\xi_i - \xi_j}. \tag{31}$$

Combining the equal terms, one obtains

$$\sum_{i=1}^n \sum_{j=1}^n \sum_{k=1}^n \left(\frac{\partial^2 \phi_o^{(i)}}{\partial x_k \partial x_j}(X^*)\right)^2 = \sum_{i=1}^n \sum_{j=1, j\neq i}^n 3\left(2\,\beta_i \Omega(\xi_i)\frac{(\omega'(\xi_i))^2}{\xi_i - \xi_j}\right)^2. \quad \text{q.e.d.} \tag{32}$$

Proof of Lemma 3. In view of (12) and (13), an elementary calculation yields

$$\frac{\partial^2 \phi_i}{\partial x_i \partial x_i}(X^*) = \frac{\partial^2 \phi_i}{\partial x_i \partial x_j}(X^*) = \frac{\partial^2 \phi_i}{\partial x_j \partial x_i}(X^*) = \frac{\partial^2 \phi_i}{\partial x_k \partial x_j}(X^*) = 0, \tag{33}$$

where $i \neq j$, $k \neq i$ and $k \neq j$, while for $i \neq j$ we have

$$\begin{aligned}
\frac{\partial^2 \phi_i}{\partial x_j \partial x_j}(X) \;=\; & 2\left(\int_a^b \rho(x)\Omega(x)\left(\zeta^{[i]}(x)\right)^2 dx\right)^{-4} \\
& \cdot\Bigg[\left(\int_a^b \rho(x)\Omega(x)\left(\zeta^{[i]}(x)\right)^2 dx\right)^2 \\
& \quad \cdot\Bigg(-\int_a^b \rho(x)\Omega(x)\left(\zeta^{[i,j]}(x)\right)^2 dx \int_a^b \rho(x)\Omega(x)\left(\zeta^{[i]}(x)\right)^2 (x-\xi_j)\,dx \\
& \qquad -3\int_a^b \rho(x)\Omega(x)\zeta^{[i]}(x)\zeta^{[i,j]}(x)\,dx \int_a^b \rho(x)\Omega(x)\left(\zeta^{[i]}(x)\right)^2 dx\Bigg) \\
& +4\int_a^b \rho(x)\Omega(x)\zeta^{[i]}(x)\zeta^{[i,j]}(x)\,dx \int_a^b \rho(x)\Omega(x)\left(\zeta^{[i]}(x)\right)^2 (x-\xi_j)\,dx \\
& \quad \int_a^b \rho(x)\Omega(x)\zeta^{[i]}(x)\zeta^{[i,j]}(x)\,dx \int_a^b \rho(x)\Omega(x)\left(\zeta^{[i]}(x)\right)^2 dx\Bigg].
\end{aligned} \tag{34}$$

Evaluating (34) at the fixed point X^* by the quadrature formula (1) yields, after some simplifications,

$$\frac{\partial^2 \phi_i}{\partial x_j \partial x_j}(X^*) = 2\,\frac{\beta_j\,\Omega(\xi_j)(\omega^{[j]}(\xi_j))^2}{\beta_i\,\Omega(\xi_i)(\omega^{[i]}(\xi_i))^2}. \qquad \text{q.e.d.} \tag{35}$$

4. Implementation and Numerical Results

In the following we will suggest that the total expense of the method need not be much more than for the existing methods for Gaussian formulas without preassigned nodes. In particular, the number of arithmetical operations for every iteration step can asymptotically be bounded by $O((2n+m)^2)$.

If the moments M_k are given, $k = 0, 1, 2, \ldots, 2n+m-1$, the integrals in (9) may be evaluated exactly with the help of an arbitrary interpolatory formula

$$Q_{2n+m}[f] = \sum_{\nu=1}^{2n+m} a_\nu\, f(z_\nu) \tag{36}$$

or a Gaussian quadrature formula without preassigned nodes and using a total number of $\lceil (2n+m)/2 \rceil$ nodes. The latter may be constructed by the methods in Gautschi (1979). An exact computation of (8) leads to $O(n\,(2n+m)^2)$ arithmetical operations. Then, however, many values are computed repeatedly, and a slight modification of the method leads to the improved expense of $O((2n+m)^2)$. More precisely, this is the number of arithmetical operations needed for the computation of $\Omega(z_\nu)(\zeta(z_\nu))^2$ if $\zeta(z_\nu) \neq 0$ and $\Omega(z_\nu)(\zeta^{[\mu]}(z_\nu))^2$ if $z_\nu = x_\mu$, $\zeta(x_\mu) = 0$. If these intermediate values are saved, then $O(n\,(2n+m))$ additional arithmetical operations are needed for the iteration step. If additionally the intermediate values $\Omega(z_\nu)\zeta^{[\mu]}(z_\nu)$ for $\Omega(z_\nu) \neq 0$ and $\zeta(z_\nu) \neq 0$, respectively $\Omega^{[\mu]}(z_\nu)\zeta(z_\nu)$ if $z_\nu = \xi_\mu$ and $\Omega(\xi_\mu) = 0$ are saved, only $O(n(2n+m))$ further arithmetical operations are needed for the calculation of the weights using (5) and (6). The Jacobi matrix (10) for the a posteriori bounds (11) can be computed by using the values that are available from the previous calculations and $O(n^2(2n+m))$ further arithmetical operations.

The computations for the following examples have been carried out using the *Mathematica* (Wolfram 1991) software package.

Example 1. We again refer to the example in Gautschi (1983) that was already mentioned as motivation for this paper. The problem is to compute $Q^G_{15,0}$ for $\rho(x) = e^{-x^3/3}$, $x \in (0, \infty)$. In Gautschi (1983), the two different approximations X_{Lee} of Lee (1980) and X_{Gautschi} of Gautschi are compared. Our recomputed values X_{recomp} are contained in Table 1. The error control using (11) confirmed that the results are correct in all decimal places. Rounded to 16 decimal places, the values agree with X_{Gautschi}, except for two digits (underlined). The values of (10) at the different approximations then are $\|\Phi'_E(X_{\text{Lee}})\|_\infty = 1.9$, $\|\Phi'_E(X_{\text{Gautschi}})\|_\infty = 2.8\,(-3)$ and $\|\Phi'_E(X_{\text{recomp}})\|_\infty = 4.6\,(-13)$.

i	ξ_i^{recomp}
1	1. 92976 53896 38693 21070 66546 21897 71158 03766 21336 (−2)
2	1. 00659 91422 26748 54082 77208 83572 61299 65139 54832 (−1)
3	2. 42846 83666 94404 03159 74741 13090 23168 61245 32418 (−1)
4	4. 38764 29468 78456 35042 26034 29255 66966 33598 92655 (−1)
5	6. 78796 50369 04373 33344 01869 00057 92898 02181 84697 (−1)
6	9. 52262 04715 09191 48473 28909 20380 41616 39199 94131 (−1)
7	1. 24916 53111 41012 30353 35075 75847 83679 59309 18828
8	1. 56152 63581 96974 94506 68720 68690 41539 61318 46754
9	1. 88394 66912 23344 33305 46481 31869 63579 95536 74390
10	2. 21359 55701 64660 98177 76516 30842 02292 91207 81393
11	2. 55002 33783 08306 89289 15829 70122 27553 97269 81412
12	2. 89520 86150 30500 29126 13914 15609 43096 41944 99800
13	3. 25436 82224 16161 68420 33398 67973 79248 17760 64867
14	3. 63904 56911 97642 63322 00833 64467 32129 62868 26487
15	4. 08080 54150 15806 76533 09911 58715 16341 60776 68216

Table 1. *Nodes of* $Q_{15,0}^G$ *for* $\rho(x) = e^{-x^3/3}$, $x \in (0, \infty)$

Example 2. For the second example the first four Kronrod resp. Patterson extensions of $Q_{3,0}^G$ are computed. This example was considered before in Gautschi and Notaris (1988b). Newton's method for (16) was observed to deteriorate for subsequent extensions and to diverge for the fourth. In Table 2, values from Gautschi and Notaris (1988b, Table II) are compared with the respective values for Newton's method applied to $\Phi_o(X) = 0$ from Lemma 2. A further column contains the values from Lemma 3 for the method Φ_E.

QF	m	A_1	A_2	T_{CGM}	B_1	B_2	T_o	C
Q_7^{GK}	3	1.6 (0)	2.8 (2)	4.5 (2)	1.3 (2)	7.4 (−1)	9.4 (1)	4.2 (0)
Q_{15}^{PAT2}	7	1.5 (0)	2.3 (3)	3.5 (3)	1.1 (7)	1.4 (−3)	1.7 (4)	1.1 (1)
Q_{31}^{PAT3}	15	7.9 (0)	1.7 (4)	1.3 (5)	7.4 (19)	1.7 (−9)	1.2 (11)	8.3 (1)
Q_{63}^{PAT4}	31	9.6 (5)	1.3 (5)	1.2 (11)	2.0 (49)	1.3 (−21)	2.6 (28)	2.2 (1)

$A_1 = \|(\Phi'_{CGM}(X^*))^{-1}\|_F$, $A_2 = \|\Phi''_{CGM}(X^*)\|_F$,
$B_1 = \|(\Phi'_o(X^*))^{-1}\|_F$, $B_2 = \|\Phi''_o(X^*)\|_F$,
$T_{CGM} = \|(\Phi'_{CGM}(X^*))^{-1}\|_F \cdot \|\Phi''_{CGM}(X^*)\|_F$, $T_o = \|(\Phi'_o(X^*))^{-1}\|_F \cdot \|\Phi''_o(X^*)\|_F$,
$C = \|\Phi''_E(X^*)\|_F$

Table 2. *Convergence analysis of different methods for an example of repeated Patterson extensions*

The difference in magnitude of the values for the third and fourth extension in column A_1 was considered as a possible explanation for the divergence of Newton's method applied to (16). Unlike for the Newton methods, the numerical values for the method Φ_E in column C do not increase monotonically for subsequent extensions. The numerical examples suggest to try the method Φ_E for this problem, especially since the existence of this extension has not yet been proved theoretically and may be questionable in view of the numerical results in Gautschi and Notaris (1988b). Working with 50 digits of precision and taking the values published by Patterson (1968) as starting values, the method converged in five iterations. Rounded to the precision used there, the values agree with those of Patterson (1968).

Example 3. We compute $Q^G_{20,0} = \sum_{\mu=1}^{20} \beta_\mu f(\xi_\mu)$ with respect to the weight function

$$\rho(x) = E_1(x) = \int_1^\infty e^{-xt}\, dt/t, \qquad x \in (0,\infty). \tag{37}$$

Approximations have been given before in Gautschi (1968) and in Danloy (1973). As Danloy noted, these approximations differ slightly. The recomputed values X_{recomp} and the respective weights in Table 3 are checked by the error control (11). They agree with the values of Danloy within the range of numerical precision used there, except for the last digit of β_6^{Danloy} and $\beta_{19}^{\text{Danloy}}$, respectively.

i	ξ_i^{recomp}	β_i^{recomp}
1	4. 15731 01868 41544 93369 19 (−2)	3. 30068 38813 62959 23499 17 (−1)
2	2. 74239 64018 07681 51133 63 (−1)	3. 35018 80062 05731 99532 78 (−1)
3	7. 35213 02463 28344 52316 89 (−1)	2. 02727 08984 17723 33575 29 (−1)
4	1. 43646 48205 72385 35269 50	9. 06794 13781 93196 44895 56 (−2)
5	2. 38684 23639 04846 41304 99	3. 11926 47504 37371 16255 84 (−2)
6	3. 59494 93861 67574 33530 40	8. 30681 68246 06230 48913 05 (−3)
7	5. 07042 04547 99353 42659 69	1. 70519 55214 34001 20466 10 (−3)
8	6. 82474 52200 81413 82277 68	2. 67192 38528 63930 48163 41 (−4)
9	8. 87199 45661 22991 99799 07	3. 15225 25750 86263 81518 02 (−5)
10	1. 12296 31310 20926 75369 91 (1)	2. 75116 43641 99170 39294 13 (−6)
11	1. 39195 56095 01768 92598 00 (1)	1. 73736 44964 63384 74662 42 (−7)
12	1. 69695 73318 81456 88445 62 (1)	7. 71970 10710 39650 85130 15 (−9)
13	2. 04155 65090 75899 98121 80 (1)	2. 32856 51935 81019 50246 01 (−10)
14	2. 43048 83607 90409 02555 08 (1)	4. 54955 05953 07018 17824 40 (−12)
15	2. 87019 53556 27872 06624 27 (1)	5. 40355 19330 41503 95633 65 (−14)
16	3. 36981 99818 80280 48644 97 (1)	3. 56730 02730 34955 66228 29 (−16)
17	3. 94313 66594 44715 45028 82 (1)	1. 14479 49856 57473 46316 40 (−18)
18	4. 61284 47518 32138 71206 05 (1)	1. 43415 82363 95024 31292 65 (−21)
19	5. 42229 68044 88160 23373 87 (1)	4. 63374 05629 14596 02235 03 (−25)
20	6. 48259 44247 29580 88297 67 (1)	1. 36239 85719 61005 19207 55 (−29)

Table 3. *Nodes and Weights of* $Q^G_{20,0}$ *for* $\rho(x) = E_1(x) = \int_1^\infty e^{-xt}\, dt/t,\ x \in (0,\infty)$

The values of the norms of the Jacobian matrices for the different approximations are $\|\Phi'_E(X_{\text{Gautschi}})\|_\infty = 3.0\ (-6)$ and $\|\Phi'_E(X_{\text{Danloy}})\|_\infty = 1.0\ (-12)$.

5. Acknowledgement

The author would like to thank the referee for the helpful comments, which led to a considerable improvement in the readability of this paper.

References

Brass, H. (1977), *Quadraturverfahren*, Vandenhoeck and Ruprecht, Göttingen.

Caliò, F., Gautschi, W. and Marchetti, E. (1986), *On Computing Gauss-Kronrod Formulae*, Math. Comp. 47, pp. 639-650.

Danloy, B. (1973), *Numerical Construction of Gaussian Quadrature Formulas for* $\int_0^b(-Log\, x)\, x^\alpha\, f(x)\, dx$ *and* $\int_0^\infty E_m(x)\, f(x)\, dx$, Math. Comp. 27, pp. 861-869.

Gautschi, W. (1968), *Algorithm 331 – Gaussian Quadrature formulas*, Comm. ACM 11, pp. 432-436.

Gautschi, W. (1979), *On generating Gaussian quadrature rules*, in *Numerische Integration*, G. Hämmerlin, ed., Intern. Ser. Num. Math. 45, Birkhäuser, Basel, pp. 147-154.

Gautschi, W. (1983), *How to check and how not to check Gaussian quadrature formulae*, BIT 23, pp. 209-216.

Gautschi, W. and Notaris, S. E. (1988a) *An Algebraic Study of Gauss-Kronrod Quadrature Formulae for Jacobi Weight Functions*, Math. Comp. 51, pp. 231-248.

Gautschi, W. and Notaris, S. E. (1988b), *Newton's Method and Gauss-Kronrod Quadrature*, in *Numerical Integration III*, H. Brass and G. Hämmerlin, eds., Intern. Ser. Num. Math. 85, Birkhäuser, Basel, pp. 60-71.

Kautsky, J. and Elhay, S. (1984), *Gauss Quadratures and Jacobi Matrices for Weight Functions Not of One Sign*, Math. Comp. 43, pp. 543-550.

Krylov, V. I. (1962), *Approximate Calculation of Integrals*, translated by A. H. Stroud, MacMillan, New York.

Lee, S.-Y. (1980), *The Inhomogeneous Airy functions* $Gi(z)$ *and* $Hi(z)$, J. Chem. Phys. 72, pp. 332-336.

Monegato, G. (1979), *An Overview of Results and Questions Related to Kronrod Schemes*, in *Numerische Integration*, G. Hämmerlin, ed., Intern. Ser. Num. Math. 45, Birkhäuser, Basel, pp. 231-240.

Ortega, J. M. and Rheinboldt, W. C. (1970), *Iterative Solution of Nonlinear Equations in Several Variables*, Academic Press, Orlando.

Ostrowski, A. M. (1966), *Solutions of Equations and Systems of Equations*, Second Edition, Academic Press, New York.

Patterson, T. N. L. (1968), *The Optimum Addition of Points to Quadrature Formulae*, Math. Comp. 22, pp. 847-856.

Wolfram, S. (1991), *Mathematica*, Second Edition, Addison Wesley Publishing Company.

Sven Ehrich, Institut für Mathematik
Universität Hildesheim, Marienburger Platz 22, D–31141 Hildesheim
Federal Republic of Germany
e-mail: ehrich@informatik.uni-hildesheim.de

Integrating Singularities using Non-uniform Subdivision and Extrapolation

Terje O. Espelid

Abstract A new approach to the computation of approximations to multidimensional integrals over an n-dimensional hyper-rectangular region, when the integrand is singular, is described. This approach is based on a non-uniform subdivision of the region of integration and the technique fits well to the subdivision strategy used in many adaptive algorithms. The strategy can be applied to vertex singularities, line singularities and more general subregion singularities. The technique turns out to have good numerical stability properties.

1. Introduction

In 1976 Lyness [9] published the first paper which addresses the problem of finding the error functional expansion in multidimensional quadrature with a singular integrand function. Knowing at least the existence of such expansions is essential in order to compute such integrals effectively. Since 1976 several related papers have appeared, some giving expansions for different regions and others giving expansions for different types (or combinations) of singular behavior [3, 4, 5, 8, 10, 11, 12, 13, 14]. A common feature for these expansions is that they are based on a *uniform* subdivision of the region, applying the same rule on each subregion.

In a recent paper Espelid (1992) [7] describes a new idea applying extrapolation on a sequence of estimates produced through a *non-uniform* subdivision of the initial region. This approach can be used on problems having vertex singularities associated with homogeneous functions. The idea presented in [7] is generally applicable to any dimension and any region which can be subdivided into subregions of the same form, e. g. hypercubes, simplices, etc.. Furthermore, the technique can be applied to internal point singularities as well. In order to apply this technique one has to know at least the position of the singular point.

In this paper we will extend this new approach to the integration of singular integrands to include problems with line singularities in 2 dimensions, line or face singularities in 3 dimensions and subregion (line, face , etc.) singularities in n-dimensions. In order to achieve this generalization we have to restrict the region of integration to an n-dimensional hyperrectangle.

This paper is organized as follows: in the next section we present the problem and develop the basic error expansion. Then we describe the new subdivision approach and finally we give some examples and concluding remarks.

2. Homogeneous functions: basic error expansion

A function $f(\mathbf{x})$, where $\mathbf{x} \in R^n$, is said to be homogeneous of degree α (about the origin) if

$$f(\lambda \mathbf{x}) = \lambda^\alpha f(\mathbf{x}) \text{ for } \forall\, \lambda > 0.$$

We will use the notation introduced by Lyness (1976), [9], and denote such a function $f_\alpha(\mathbf{x})$. Note the following simple rules: $f_\alpha f_\beta$ is of homogeneous of degree $\alpha + \beta$ and $(f_\alpha)^\beta$ is of homogeneous of degree $\alpha\beta$.

We will discuss n-dimensional integration, however we will assume that the singularity is caused by a homogeneous function about the origin which involves exactly s of the n variables, with $1 \leq s \leq n$. To simplify the presentation assume that the variables are numbered such that these s variables are $x_1, x_2, \ldots, x_s$.

Observe that we may use an affine transformation to transform the given rectangular region on to the unit n-cube. Any homogeneous function will still be homogeneous after such a transformation and its degree will be invariant. We know furthermore that any given cubature rule can be transformed as well and its polynomial degree will be invariant too. Thus, choosing the n-dimensional unit hypercube, C_n, as the region of integration represents no restriction

$$C_n : 0 \leq x_i \leq 1 \text{ for } i = 1, 2, \ldots, n.$$

Define

$$I(f) = \int_{C_n} f(\mathbf{x}) d\mathbf{x} = \int_0^1 \int_0^1 \cdots \int_0^1 f(x_1, x_2, \ldots, x_n)\, dx_1\, dx_2 \ldots dx_n. \tag{1}$$

We will discuss how to compute numerical estimates to integration problems of type (1), where the function involved, f, is a product of a homogeneous function $f_\alpha(x_1, x_2, \ldots, x_s)$ and a function $g(\mathbf{x})$ which is regular in C_n. The function $f_\alpha(x_1, x_2, \ldots, x_s)$ is assumed to have a point singularity at the origin of C_s due to the homogeneous property. Define $\mathbf{x}_0 = (0, 0, \ldots, 0, x_{s+1}, x_{s+2}, \ldots, x_n)$ then

$$\begin{cases} f(\mathbf{x}) = f_\alpha(x_1, x_2, \ldots, x_s)\, g(\mathbf{x}), \text{with } \alpha > -s. \\ g(\mathbf{x}_0) \not\equiv 0, \ \mathbf{x}_0 \in C_n. \end{cases}$$

Furthermore we assume that the origin is the only point in C_s where f_α is non-analytic. We give a few examples to illustrate the type of singular problems we will deal with.

1) A line singularity in C_2 due to a point singularity in C_1 ($s = 1$),

$$\int_0^1 \int_0^1 x^\alpha g(x,y)\, dxdy \text{ with } \alpha > -1.$$

2) A line singularity in C_3 due to a point singularity in C_2 ($s = 2$),

$$\int_0^1 \int_0^1 \int_0^1 (x+y)^\alpha g(x,y,z)\, dxdydz \text{ with } \alpha > -2.$$

3) A face singularity in C_3 due to a point singularity in C_1 ($s = 1$),

$$\int_0^1 \int_0^1 \int_0^1 x^\alpha g(x,y,z)\, dxdydz \text{ with } \alpha > -1.$$

4) A point singularity in C_3 ($s = 3$), with $r = \sqrt{x^2 + y^2 + z^2}$,

$$\int_0^1 \int_0^1 \int_0^1 r^\alpha\, g(x,y,z)\, dxdydz \text{ with } \alpha > -3.$$

Example 4) is the only one that can be handled by the technique described in [7] directly. However, all of these problems can be dealt with by that technique after an appropriate transformation. E. g., in example 1) we may apply an inverse Duffy (1982) transformation, [6]; replacing y with t using $y = t/x$ gives

$$\int_0^1 \Big(\int_0^x x^{\alpha-1} g(x, t/x)\, dt \Big)\, dx \text{ with } \alpha > -1.$$

Thus integrating a function over a triangle having a vertex singularity at the origin of degree $\alpha - 1$. An additional difficulty has been introduced due to the transformation: t/x is a homogeneous function of degree 0 around the origin, however this will be handled by the described technique quite easily. It will, however, have a slightly negative effect on the performance of the method. We will give an example of such an approach in Section 5.

In order to deal with example 1) directly we may choose a product of two classical rules: a Gauss-Legendre rule in the y direction where the function is smooth and a Gauss-Jacobi rule in the other direction.

Alternatively we may subdivide the original square in m^2 equal squares and then use the same rule Q over all these squares. Denoting the compound rule $Q^{(m)}$ we get, Sidi (1983) [13] and Lyness and de Doncker-Kapenga (1987) [11],

$$Q^{(m)}(f) \sim I(f) + \sum_{\ell \geq 0} \frac{A_\ell}{m^{\alpha+2+\ell}} + \sum_{\ell \geq 0} \frac{C_\ell \log m}{m^{\alpha+2+\ell}} + \sum_{\ell \geq 1} \frac{B_\ell}{m^\ell}. \tag{2}$$

For many integrals, some of the coefficients in (2) vanish e.g., the C-coefficients vanish if α is non-integer or if a special rule Q is used; some of the B-coefficients may vanish depending on the degree of precision and symmetry of Q. Based on (2) we may now compute $Q^{(m)}$ for different values of m and then use extrapolation to improve these approximations. This last approach does not need information about where the line singularity is, while the Gauss-Legendre/Gauss-Jacobi approach is based on that information.

We will, in what follows, present an alternative approach. Just as in [7] we will base this on a *non-uniform* subdivision of the region of integration combined with extrapolation. The non-uniform aspect resembles the basic idea with the product rules: to treat the problem differently in the different directions of integration. Assume that we know (a) which variables are involved in the singularity and (b) the position of at least one vertex where the function is singular (the origin in this presentation). Cut this region into $s+1$ new subregions by dividing in two halves the region containing the singularity orthogonal to one of the s directions. Continuing this until all s directions have been divided exactly once we have: one region containing the singularity and s regions where the function is supposed to be well behaved. Considering these s subdivisions as one step we may do j such steps. The hyper-rectangle containing the singularity is now $H_n^{(s)}(h) = [0,h]^s \times [0,1]^{n-s}$, with $h = 1/2^j$. Define

$$I_{H_n^{(s)}(h)}(f) = \int_0^1 \cdots \int_0^1 (\int_0^h \cdots \int_0^h f(x_1, x_2, \ldots, x_n)\, dx_1 \ldots dx_s) dx_{s+1} \ldots dx_n. \quad (3)$$

Suppose that both $\mathbf{x}_0$ and $\mathbf{x}$ are points in C_n. Expand $g(\mathbf{x})$ in a Taylor series around $\mathbf{x}_0$ with p basic terms and a remainder term r

$$\begin{aligned} g(\mathbf{x}) &= g(\mathbf{x}_0) + \textstyle\sum_{i=1}^s x_i \frac{\partial g}{\partial x_i}(\mathbf{x}_0) + \frac{1}{2}\sum_{i,j=1}^s x_i x_j \frac{\partial^2 g}{\partial x_i \partial x_j}(\mathbf{x}_0) + \ldots \\ &+ \textstyle\frac{1}{(p-1)!}\sum_{i_1,i_2,\ldots,i_{p-1}=1}^s x_{i_1} x_{i_2} \cdots x_{i_{p-1}} \frac{\partial^{p-1} g}{\partial x_{i_1}\partial x_{i_2}\ldots\partial x_{i_{p-1}}}(\mathbf{x}_0) + r. \end{aligned}$$

Multiply this expression by $f_\alpha(x_1, x_2, \ldots, x_s)$ and integrate over $H_n^{(s)}(h)$. Changing integration variables $x_i = h\ v_i$, $i = 1, 2, \ldots, s$, in each integration term moves all integrals to C_n and gives

$$I_{H_n^{(s)}(h)}(f_\alpha g) = c_0\ h^{\alpha+s} + \sum_{\ell=1}^{p-1} c_\ell\ h^{\alpha+s+\ell} + O(h^{\alpha+s+p}), \quad (4)$$

with

$$\begin{cases} c_0 = \int_{C_n} f_\alpha(x_1, x_2, \ldots, x_s) g(\mathbf{x}_0) d\mathbf{x}, \\ c_\ell = \frac{1}{\ell!}\int_{C_n} f_\alpha(x_1, x_2, \ldots, x_s) \sum_{i_1,\ldots,i_\ell=1}^s x_{i_1} \cdots x_{i_\ell} \frac{\partial^\ell g}{\partial x_{i_1}\ldots\partial x_{i_\ell}}(\mathbf{x}_0)\, d\mathbf{x}, \\ \ell = 1, 2, \ldots p-1. \end{cases}$$

Here we have used the fact that the term ℓ involves functions which are of homogeneous degree $\alpha + \ell$ in C_s about the origin.

Suppose furthermore that we use a fixed cubature rule Q on a hyper-rectangle, H, based on L evaluation points

$$Q_H(f) = \sum_{i=1}^L w_i\ f(\mathbf{x}_i).$$

Assume that $\sum_{i=1}^L w_i$ equals the volume of hyperrectangle H (Q has degree of precision at least 0) and that the evaluation points $\mathbf{x}_i$ have been translated to H. Applying this rule on f over $H_n^{(s)}(h)$ then gives

$$Q_{H_n^{(s)}(h)}(f_\alpha g) = b_0\ h^{\alpha+s} + \sum_{\ell=1}^{p-1} b_\ell\ h^{\alpha+s+\ell} + O(h^{\alpha+s+p}), \quad (5)$$

where

$$\begin{cases} b_0 = Q_{C_n}(f_\alpha(x_1, x_2, \ldots, x_s)g(\mathbf{x}_0)), \\ b_\ell = \frac{1}{\ell!}Q_{C_n}(f_\alpha(x_1, x_2, \ldots, x_s)\sum_{i_1,\ldots,i_\ell=1}^{s} x_{i_1}\cdots x_{i_\ell}\frac{\partial^\ell g}{\partial x_{i_1}\ldots\partial x_{i_\ell}}(\mathbf{x}_0)), \\ \ell = 1, 2, \ldots p-1. \end{cases}$$

Here we have used the fact that the term ℓ involves functions which are of homogeneous degree $\alpha + \ell$ in C_s and that applying Q over $H_n(h)$ is a simple h-scaling of C_n in the first s variables. Thus the volume and the weights w_ℓ must be scaled by a h^s factor. By subtracting (4) from (5) we get the error expansion

$$E_n^{(s)}(h) = Q_{H_n^{(s)}(h)}(f_\alpha g) - I_{H_n^{(s)}(h)}(f_\alpha g) = \sum_{\ell=0}^{p-1} a_\ell\, h^{\alpha+s+\ell} + O(h^{\alpha+s+p}), \tag{6}$$

with

$$a_\ell = b_\ell - c_\ell,\ \ell = 0, 1, \ldots, p-1.$$

b_ℓ is the numerical estimate produced by the rule Q of the integral c_ℓ and the functions involved will become smoother with increasing values of ℓ. Therefore it is reasonable to expect $|a_\ell| >> |a_{\ell+1}|$, at least as long as $\alpha + s + \ell \leq \text{degree}(Q)$.

The extrapolation scheme will be based the error expansion (6). In fact, we assume that the error given in (6) will be the major error source in the estimate of $I(f)$. The error contribution from the rest of the original region C_n has to be kept under control and we will show how it will influence the extrapolation process and the final global estimate of $I(f)$.

3. The Series with Tail-correction Approach

We will now use the error expansion developed in the previous section combined with the series with tail-correction approach presented in [7]. Suppose that we use a strategy which repeatedly subdivides the subregion which is considered to represent the most difficult part of the integration problem (1). Each time the subregion containing the singularity needs subdivision we divide this in $s+1$ new subregions following the procedure described in the previous section. Thus, at any time, the collection of subregions will contain only one hyperrectangle containing the singularity, namely $H_n^{(s)}(h)$. At a given time we may have replaced this hyperrectangle in the collection k times. Define (with $h_i = 1/2^i$)

$$\begin{cases} I_i = I_{H_n^{(s)}(h_i)}(f_\alpha g),\ i = 0, 1, 2, \ldots, k, \\ U_i = I_{i-1} - I_i,\ i = 1, 2, \ldots, k, \\ S_k = \sum_{i=1}^{k} U_i,\ \hat{S}_k = \sum_{i=1}^{k} \hat{U}_i, \end{cases} \tag{7}$$

where $\hat{U}_i$ is an approximation to U_i. The elements in the sequence $U_1, U_2, \ldots$ are by definition (7) integrals over a sequence of subregions, $L_n^{(s)}(h) = [h, 2h]^s \times [0, 1]^{n-s}$ for a given value of h. $L_n^{(s)}(h)$ is similar to the L-shaped regions described by Lyness (1976) [9] when $s = n$ (that is: a corner singularity in the n-cube).

We illustrate, in Figure 1, two subregion sequences in 2 dimensions:

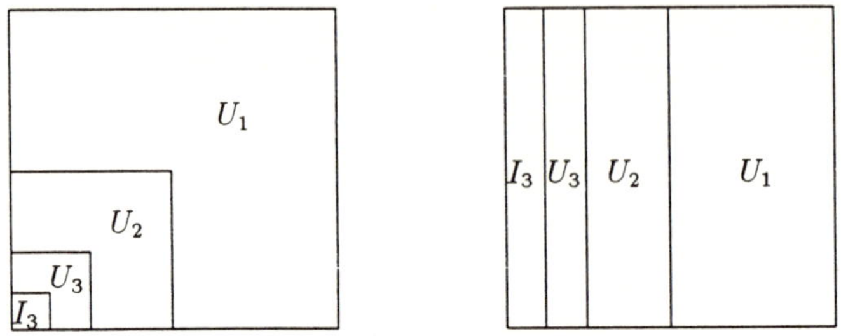

Figure 1. 2-D (a) corner and (b) line singularities: U and I-sequences.

Using the definition of U_i we get $S_k = I_0 - I_k$. Thus S_k is an approximation to I_0 with error $-I_k$. We can use the expansion for $I_{H_n^{(s)}(h)}$ with $h = h_k = 1/2^k$ given in (4)

$$S_k - I_0 = -c_0 h_k^{\alpha+s} - c_1 h_k^{\alpha+s+1} - \cdots - c_{p-1} h_k^{\alpha+s+p-1} - O(h_k^{\alpha+s+p}).$$

In practice we have to approximate U_i, e. g.: we have suggested to divide this region into s rectangles in Section 2, however we may need to subdivide further to achieve the necessary precision. $\hat{S}_k$ is an approximation to I_0 giving

$$\hat{S}_k - I_0 = \sum_{i=1}^{k}(\hat{U}_i - U_i) - c_0 h_k^{\alpha+s} - c_1 h_k^{\alpha+s+1} - \cdots - c_{p-1} h_k^{\alpha+s+p-1} - O(h_k^{\alpha+s+p}). \quad (8)$$

In [7] we discuss to use (8) as a basis for extrapolation. The conclusion is that correcting the tail of the series gives an improved approximation to a low cost and that this actually is to be preferred. Based on this we present the tail correction strategy here: using the rule Q to estimate $I_i = I_{H_n^{(s)}(h_i)}$, then $Q_i = Q_{H_n^{(s)}(h_i)}(f_\alpha g)$ is an estimate of the tail $I_i = \sum_{\ell > i} U_\ell$. Define

$$T_{i0} = Q_i + \sum_{\ell=1}^{i} \hat{U}_\ell,\ i = 0, 1, 2, \ldots, k.$$

T_{i0} is an approximation to I_0 and we have according to (6)

$$T_{i0} - I_0 = \sum_{\ell=1}^{i} E_{U_\ell} + \sum_{\ell=0}^{p-1} a_\ell h_i^{\alpha+s+\ell} + O(h_i^{\alpha+s+p}), \quad (9)$$

where $E_{U_\ell} = \hat{U}_\ell - U_\ell$. By standard linear extrapolation we may use (9) to eliminate the k first terms in the sum of h-powers. Define

$$n_1 = 2^{\alpha+s} - 1,\ n_{j+1} = 2n_j + 1,\ j = 1, 2, 3, \ldots,$$

and then

$$T_{ij} = T_{i,j-1} + (T_{i,j-1} - T_{i-1,j-1})/n_j, i = 1, 2, \ldots, k,\ j = 1, 2, 3, \ldots, i. \quad (10)$$

We may put the T_{ij} in a standard extrapolation tableau as follows

$$\begin{array}{llll} T_{00} & & & \\ T_{10} & T_{11} & & \\ \vdots & \vdots & \ddots & \\ T_{k0} & T_{k1} & \cdots & T_{kk} \end{array}$$

This gives after j extrapolation steps in row k

$$T_{kj} = I_0 + \sum_{\ell=j}^{p-1} a_\ell^{(j)} h_k^{\alpha+s+\ell} + O(h_k^{\alpha+s+p}) + \sum_{\ell=1}^{k}(1 - \beta_{k+1-\ell,j})E_{U_\ell}. \qquad (11)$$

We refer the reader to [7] for more details about these β−coefficients. The important point to make is that these coefficients may be computed and make it possible to keep track of the effect of not knowing each U_ℓ exactly. We may avoid that these errors ruin the extrapolation process by simply deciding that if the effect becomes disturbing, then compute a better approximation to at least one these U_ℓs.

We face two different alternatives if we are not satisfied with the global approximation: (A) Compute a better value for $\hat{U}_\ell$ and update the T-tableau. This can easily be done directly in the last row of this tableau, estimate the new errors and finally decide what to do next. (B) If we are not satisfied with the extrapolation error then increase k: put $h_{k+1} = h_k/2$, compute $\hat{U}_{k+1}$ and Q_{k+1} and create a new row in the T-tableau.

In [7] we observed that T_{kk} can be written

$$T_{kk} = \sum_{i=1}^{k} \gamma_i^{(k)} \hat{U}_i + \sum_{i=0}^{k} \delta_i^{(k)} Q_i,$$

and defined the condition number for T_{kk} as

$$\tau_s^{(k)} = (1 - \frac{1}{2^s}) \sum_{i=1}^{k} |\gamma_i^{(k)}| (\frac{1}{2^s})^{i-1} + \sum_{i=0}^{k} |\delta_i^{(k)}| (\frac{1}{2^s})^i.$$

Observe that the volumes associated with the estimates $\hat{U}_i$ and Q_i enter this expression. We can compute $\tau_s^{(k)}$ once the exponent sequence is known, ([7]):

s / k	1	2	3	4	7	10
1	5.83	6.92	6.30	4.49	1.55	1.07
2	5.83	4.28	3.13	1.80	1.02	1.00
3	5.83	2.96	2.15	1.24	1.00	1.00
4	5.83	2.30	1.81	1.09	1.00	1.00
5	5.83	1.97	1.67	1.04	1.00	1.00

Table 1 τ-table for $s + \alpha = 1/2$.

The extrapolation procedure is remarkably stable and increasing the number of extrapolation steps has a positive effect on the methods stability. $H_n^{(s)}(h)$ has volume h^s

and this volume will decrease in each step with a factor $\frac{1}{2^s}$. This decrease in volume counter-effects the influence by the coefficients $\gamma_i^{(k)}$ and $\delta_i^{(k)}$ on the condition number.

4. Integrating logarithmic singularities

Now suppose that the function f is a product of a homogeneous function f_α, a logarithmic function $\ln g_\beta$ and a function $g(\mathbf{x})$ which is regular in C_n. The functions $f_\alpha(x_1, x_2, \ldots, x_s)$ and $g_\beta(x_1, x_2, \ldots, x_s)$ are both assumed to have a point singularity at the origin of C_s due to the homogeneous property.

$$f(\mathbf{x}) = f_\alpha(x_1, x_2, \ldots, x_s)\,(\ln g_\beta(x_1, x_2, \ldots, x_s))\,g(\mathbf{x}), \text{with } \alpha > -s.$$

Furthermore we assume that origin is the only point in C_s where f_α and g_β are non-analytic. Repeating the line of arguments that led to (4), (5) and (6) we get the error expansion

$$E_n^{(s)}(h) = Q_{H_n^{(s)}(h)}(f) - I_{H_n^{(s)}(h)}(f) = \sum_{\ell=0}^{p-1}(\tilde{a}_\ell + \hat{a}_\ell \ln h)\; h^{\alpha+s+\ell} + O(h^{\alpha+s+p}\ln h), \quad (12)$$

where we have

$$\begin{cases} \tilde{a}_\ell = \tilde{b}_\ell - \tilde{c}_\ell, \; \ell = 0, 1, 2, \ldots, p-1. \\ \tilde{b}_0 = Q_{C_n}(f_\alpha(\ln g_\beta) g(\mathbf{x}_0)), \\ \tilde{b}_\ell = \frac{1}{\ell!}\, Q_{C_n}(f_\alpha(\ln g_\beta) \sum_{i_1,\ldots,i_\ell=1}^{s} x_{i_1}\cdots x_{i_\ell} \frac{\partial^\ell g}{\partial x_{i_1}\ldots\partial x_{i_\ell}}(\mathbf{x}_0)), \; \ell = 1, 2, \ldots p-1. \\ \tilde{c}_0 = \int_{C_n} f_\alpha(\ln g_\beta) g(\mathbf{x}_0) d\mathbf{x}, \\ \tilde{c}_\ell = \frac{1}{\ell!}\int_{C_n} f_\alpha(\ln g_\beta) \sum_{i_1,\ldots,i_\ell=1}^{s} x_{i_1}\cdots x_{i_\ell} \frac{\partial^\ell g}{\partial x_{i_1}\ldots\partial x_{i_\ell}}(\mathbf{x}_0)\, d\mathbf{x}, \; \ell = 1, 2, \ldots p-1, \end{cases}$$

and

$$\begin{cases} \hat{a}_\ell = \hat{b}_\ell - \hat{c}_\ell, \; \ell = 0, 1, 2, \ldots, p-1. \\ \hat{b}_0 = \beta\, Q_{C_n}(f_\alpha g(\mathbf{x}_0)), \\ \hat{b}_\ell = \frac{\beta}{\ell!}\, Q_{C_n}(f_\alpha \sum_{i_1,\ldots,i_\ell=1}^{s} x_{i_1}\cdots x_{i_\ell} \frac{\partial^\ell g}{\partial x_{i_1}\ldots\partial x_{i_\ell}}(\mathbf{x}_0)), \; \ell = 1, 2, \ldots p-1. \\ \hat{c}_0 = \beta \int_{C_n} f_\alpha g(\mathbf{x}_0) d\mathbf{x}, \\ \hat{c}_\ell = \frac{\beta}{\ell!}\int_{C_n} f_\alpha \sum_{i_1,\ldots,i_\ell=1}^{s} x_{i_1}\cdots x_{i_\ell} \frac{\partial^\ell g}{\partial x_{i_1}\ldots\partial x_{i_\ell}}(\mathbf{x}_0)\, d\mathbf{x}, \; \ell = 1, 2, \ldots p-1. \end{cases}$$

The extrapolation scheme will be based the error expansion (12). Using linear extrapolation just as before: it is well known that considering each h-exponent to appear twice will remove both the constant and the $\ln h$ term in front of the $h^{\alpha+s+\ell}$. We will give one example where we apply this idea in the next section.

5. Numerical Examples

In this section we will demonstrate, on a few examples in 2 and 3 dimensions, how efficient this approach may be. We will for this illustrative purpose apply the highest degree basic rule Q implemented in DCUHRE [2] (Berntsen et al. 1991) and furthermore not subdivide any of the rectangles which does not contain the singularity any further.

1) In the first example we look at a problem with a line singularity in C_2 combined with a nice function $g(x,y)$

$$\int_{C_2} x^{-1/2} e^{2x+y}\, dxdy \approx 8.1255\ 96316\ 47 \tag{13}$$

We give the Tail tableau in Table 2. The number of function evaluations (nfe) is given in the left column.

nfe				
65	8.13780021352			
195	8.13431250231	8.12589242260		
325	8.13179057440	8.12570210185	8.12559801197	
455	8.12998733056	8.12563391482	8.12559662210	8.12559632365
585	8.12870513608	8.12560964477	8.12559637104	8.12559631712
	8.12559631649			

Table 2. The Tail extrapolation tableau T_{ij} on (13).

DCUHRE uses in this example a degree 13 rule Q based on 65 evaluation points. Thus, dividing in two pieces and computing $\hat{U}_k$ and Q_k in each step cost 130 function evaluations per step.

As mentioned in Section 2 we can deal with this problem with the technique given in [7]. Using the transformation $y = t/x$ in (13) we get

$$\int_0^1 \left(\int_0^x x^{-3/2} e^{2x+t/x}\, dt \right) dx \approx 8.1255\ 96316\ 47 \tag{14}$$

and applying the tail approach described in [7] gives

nfe				
37	7.65148426982			
185	7.79175999005	8.13041553629		
333	7.89003397527	8.12728836320	8.12557805545	
481	7.95920320233	8.12619248842	8.12559313459	8.12559637265
629	8.00800024677	8.12580673327	8.12559575677	8.12559631985
	8.12559631473			
777	8.04246510858	8.12567064538	8.12559621644	8.12559631515
	8.12559631469	8.12559631469		

Table 3. The Tail extrapolation tableau T_{ij} on (14).

In this example a 37 point rule of degree 13 is used on each triangle (DCUTRI, Berntsen and Espelid 1992 [1]) and we divide each triangle into 4 subtriangles in each step by connection the midpoints of the 3 sides. Thus the cost per step is $4 \times 37 = 148$ evaluation points.

We get an impression of the efficiency of these two approaches by plotting, in Figure 2, the true error for the diagonals in these two tableaus versus the cost of computing these diagonals elements.

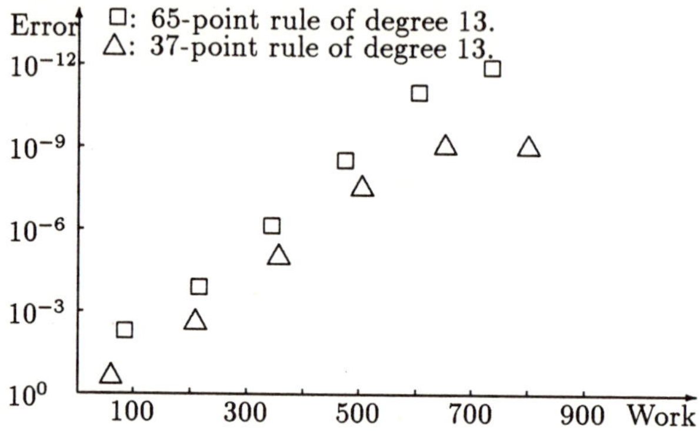

Figure 2. The true error in the diagonal elements in Table 2 (□) and Table 3 (△) versus the number of function values used to produce these elements.

Both techniques demonstrate fast convergence, however they will both converge to a "wrong" answer. Figure 2 indicates that the error in $\hat{U}_\ell$ is $O(10^{-8})$ in the triangle version and much smaller in the rectangle approach. This difference in behavior probably stems from the Hølder discontinuity introduced by the transformation. Thus it looks like the direct treatment of the integration problem (13) is to be preferred. In order to achieve a similar accuracy using adaptive codes, e. g. DCUTRI on (14) or DCUHRE on (13), we need in both cases at least $\approx$ 20,000 function evaluations.

2) In the next example we integrate a face singularity in C_3

$$\int_{C_3} x^{-1/2} e^{x+xy+z/3}\, dxdydz \approx 4.4191\ 59656\ 80 \qquad (15)$$

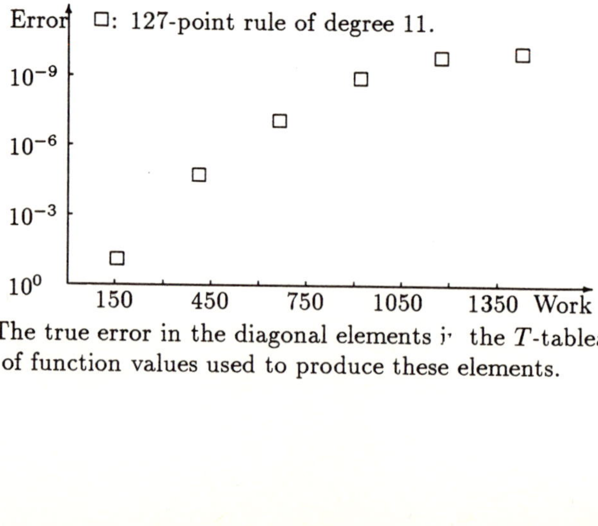

Figure 3. The true error in the diagonal elements in the T-tableau versus the number of function values used to produce these elements.

In this example the error in $\hat{U}_1$ is $O(10^{-10})$ and we see that after 4-5 extrapolation steps we reach this limit.

3) In this example we integrate a face singularity in C_3 created by a logarithmic singularity and a homogeneous function

$$-\int_{C_3} x^{-1/2} \ln x \; e^{x+xy+z/3} \, dxdydz \approx 5.8401\,12318\,46 \qquad (16)$$

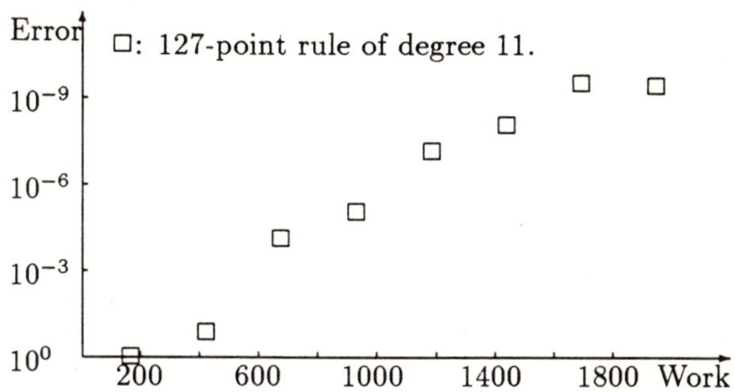

Figure 4. The true error in the diagonal elements in the T-tableau versus the number of function values used to produce these elements.

In this example the error in $\hat{U}_1$ is $O(10^{-9})$ and we see that after 6 extrapolation steps we reach this limit.

4) In the last example we integrate a line singularity in 3 dimensions

$$\int_{C_3} (x+y)^{-1/2} \, e^{x+xy+z/3} \, dxdydz \approx 2.7878925361\ldots \qquad (17)$$

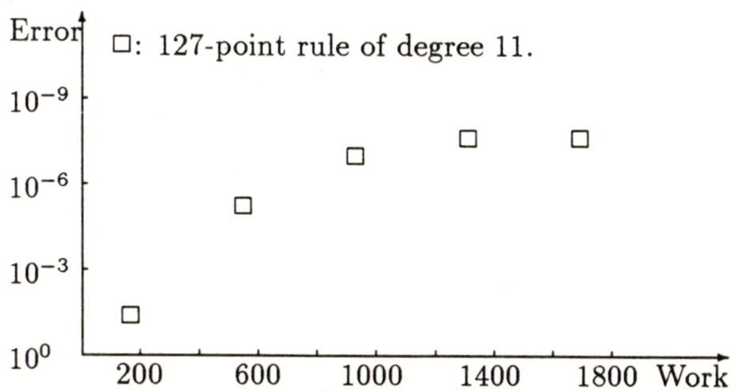

Figure 5. The true error in the diagonal elements in the T-tableau versus the number of function values used to produce these elements.

In this example the error in $\hat{U}_1$ is $O(10^{-8})$ and we see that after 3 extrapolation steps we reach this limit.

6. Conclusions

The power of this technique is clearly demonstrated through these examples. Using a high degree rule gives good accuracy in the computation of $\hat{U}_\ell$ and will in addition give small constants a_i in the error expansion which in turn will reduce the number of extrapolation steps needed. In practice it may be difficult to choose the best rule Q to use and it seems natural to combine an adaptive strategy and extrapolation.

It is essential though, in order to apply the technique, that we have all information about the singularity in C_n. We have used that knowledge in our examples creating two sequences $\hat{U}_1, \hat{U}_2, \hat{U}_3 \ldots$ and $Q_1, Q_2, Q_3, \ldots$. The $\hat{U}_\ell$'s are estimates to integrals over a sequence of non-overlapping regions *all of the same form, but non of them containing the singularity.* Similarly, the Q_ℓ's are estimates to integrals, *over overlapping regions, all of the same form and containing the singularity.*

Having more than one singularity in a region complicates the integration problem considerably. If we can separate these singularities, that is: subdivide the region such that each subregion has at most one singularity the problem can be handled. If this is not the case then further research is needed in order to handle such problems by this non-uniform subdivision technique.

Extrapolation is, as we have seen, the key to the success of both the uniform and non-uniform approaches. We have, in our examples, used the value of α combined with linear extrapolation. If we do not know the value of α, only the position of the singular point, then a pre-computation of function values in selected points along a straight line ending in a singular point, combined with extrapolation, will provide accurate estimates of α. This seems to be a good way to compute α and then use this value in the described linear extrapolation. One would expect such an approach to be at least as efficient as many of the non-linear extrapolation techniques available which do not need a value of α and therefore can be applied directly on the Tail-sequence created by the Q- and U-sequences.

Finally the non-uniform approach fits well with the general strategy one finds in adaptive codes and therefore it is reasonable to expect that these codes can be modified to handle these types of singular integrands quite well.

References

[1] J. Berntsen and T.O. Espelid. DCUTRI: An Algorithm for Adaptive Cubature over a Collection of Triangles. Accepted for publication in ACM Trans. Math. Softw., 1992.

[2] J. Berntsen, T.O. Espelid, and A. Genz. Algorithm 698: DCUHRE: An Adaptive Multidimensional Integration Routine for a Vector of Integrals. *ACM Trans. Math. Softw.*, 17(4):452– 456, 1991.

[3] E. de Doncker. New Euler-Maclaurin expansions and their application to quadrature over the s-dimensional Simplex. *Math. Comp.*, 33:1003–1018, 1979.

[4] E. de Doncker. Numerical integration and asymptotic expansions., 1980. Ph.D. thesis, Katholieke Universiteit, Leuven, Belgium.

[5] E.de Doncker-Kapenga. Asymptotic expansions and their applications in numerical quadrature. In P. Keast and G. Fairweather, editors, *Proceedings of the NATO Advanced Research Workshop on Numerical Integration*, pages 207–218. D. Reidel Publishing Company, 1987.

[6] M. G. Duffy. Quadrature over a pyramid or cube of integrands with a singularity at a vertex. *SIAM J. Numer. Anal.*, 19:1260–1262, 1982.

[7] T. O. Espelid. On integrating vertex singularities using extrapolation. Submitted for publication, 1992.

[8] J.N. Lyness. Applications of extrapolation techniques to multidimensional quadrature of some integrand functions with a singularity. *J. Comp. Phys.*, 20(3):346–364, 1976.

[9] J.N. Lyness. An error functional expansion for n-dimensional quadrature with an integrand function singular at a point. *Math. Comp.*, 30(133):1–23, 1976.

[10] J.N. Lyness. On handling singularities in finite elements. In T. O. Espelid and A. Genz, editors, *Numerical Integration, Recent Developments, Software and Applications*, NATO ASI Series C: Math. and Phys. Sciences Vol. 357, pages 219–233, Dordrecht, The Netherlands, 1992. Kluwer Academic Publishers.

[11] J.N. Lyness and E. de Doncker-Kapenga. On quadrature error expansions - part 1. *J. Comput. and Appl. Math.*, 17:131–149, 1987.

[12] J.N. Lyness and G Monegato. Quadrature error functional expansions for the simplex when the integrand function has singularities at vertices. *Math. Comp.*, 34(149):213–225, 1980.

[13] A. Sidi. Euler-maclaurin expansions for integrals over triangles of functions having algebraic/logaritmic singularities along an edge. *Journal of Approximation Theory*, 39(1):39–53, 1983.

[14] P. Verlinden and A. Haegemans. An error expansion for cubature with an integrand with homogeneous boundary singularities. Report TW 172, Dept. of Comp. Sci., Katholieke Universiteit Leuven, 1992.

Author's address

Dr. Terje O. Espelid
Department of Informatics
University of Bergen
5020 Bergen, Norway.
terje@ii.uib.no

Variance in Quadrature – a Survey

Klaus-Jürgen Förster

Abstract. If the function values have random error, then it is of practical interest that the quadrature formula Q_n has a small variance $Var(Q_n)$. The question of small variance and high algebraic degree was first considered by Chebyshev in 1874. Since that time, several investigations on this subject can be found in literature. The purpose of this paper is to give a survey of the results and open problems in this field.

1. Introduction

Let w be a fixed nonnegative weight function on $[-1, 1]$. We consider the linear functional I,

$$(1.1) \qquad I[f] = \int_{-1}^{1} f(x)\, w(x)\, dx , \qquad I[p_0] > 0 ,$$

where here and in the following p_ν denotes the monomial given by $p_\nu(x) = x^\nu$. For the numerical approximation of I mostly *quadrature formulas* Q_n are used. These are linear functionals of type

$$(1.2) \qquad Q_n[f] = \sum_{\nu=1}^{n} a_{\nu,n}\, f(x_{\nu,n}) , \qquad a_{\nu,n} \in \mathbb{R} , \qquad x_{1,n} \le x_{2,n} \le \ldots \le x_{n,n} .$$

A quadrature formula Q_n is called *positive*, if all weights $a_{\nu,n}$ are nonnegative. We say that a sequence $(Q_{n_i})_{i=1}^{\infty}$ of quadrature formulas Q_{n_i} is a *quadrature method*, if $n_i < n_{i+1}$ for all i. By R_n we denote the *remainder term* of the quadrature formula Q_n, $R_n = I - Q_n$. The associated *algebraic degree* $deg(R_n)$ is defined by

$$(1.3) \qquad deg(R_n) = \sup \{ r \mid R_n[p_\nu] = 0 \quad \text{for every} \quad \nu = 0, 1, \ldots, r \} .$$

The investigation of the "quality" of a quadrature formula Q_n means, in particular, the investigation of the behaviour of the remainder term R_n in many different classes of functions. But if the function values $f(x_{\nu,n})$ have random error, then, under certain statistical assumptions, it is also of interest (cf., e.g., Krylov 1962, p. 71, Allasia 1976/77 and Katz 1984) that the quadrature formula Q_n has a small variance $Var(Q_n)$,

$$(1.4) \qquad Var(Q_n) = \sum_{\nu=1}^{n} (a_{\nu,n})^2 .$$

Requiring convergence at least for constant functions, by the Cauchy-Schwarz inequality we have $\lim_{i\to\infty} \inf\; n_i\; Var(Q_{n_i}) \ge (I[p_0])^2$. In particular, if $deg(R_n) \ge 0$, then

(1.5) $$n\,Var(Q_n) \quad \geq \quad (I[p_0])^2 \,,$$

where the equality sign in (1.5) is valid if and only if Q_n is a *Chebyshev-type formula*, i.e., $a_{1,n} = a_{2,n} = \ldots = a_{n,n} = I[p_0]/n$.

The study of quadrature formulas Q_n with small variance $Var(Q_n)$ and high algebraic degree $deg(R_n)$ began with a classical paper of Chebyshev (1874). He has proved, for the so-called *Chebyshev weight function* w_0 given by $w_0(x) = (1-x^2)^{-1/2}$, that all Gaussian formulas are Chebyshev-type formulas, and he has remarked explicitely that these formulas are minimizing the variance $Var(Q_n)$ among all Q_n with $deg(R_n) \geq 0$. Since Chebyshev's paper, many results on Chebyshev-type formulas have been obtained, see Section 5 and, in particular, for a detailed discussion the survey paper of Gautschi (1976). For most of the investigated weight functions w, existence results for Chebyshev-type methods (Q_{n_i}) having "high" algebraic degree in the sense of

(1.6) $$\lim_{i\to\infty}\ \inf\ deg(R_{n_i})/n_i \quad > \quad 0$$

are negative. Therefore, requiring small variance and high algebraic degree, for such weight functions the use of Chebyshev-type methods seems to be not of great interest. On the other hand, starting with papers of Sard (1949) and Ostrowski (1959), several investigations have shown that (1.6) may be possible, if we require "small" variance in the sense of

(1.7) $$\lim_{i\to\infty}\ \sup\ n_i\,Var(Q_{n_i}) \quad < \quad \infty$$

instead of the minimal variance of Chebyshev-type formulas. In the following, we attempt to review advances in the described field, where for the sake of brevity we restrict the statements to functionals I of type (1.1).

2. Various examples for the variance of quadrature formulas

At first, we consider the variance of three well-known quadrature methods of interpolatory type, where Q_n is called an *interpolatory quadrature formula*, if $deg(R_n) \geq n-1$ holds.

2.1. Gauss-type formulas. Of all quadrature methods that have aquired attention in literature, Gaussian quadrature seems to be the most investigated (for a survey, see Gautschi 1981). A quadrature formula Q_n is called a *Gaussian formula* Q_n^G, if for its remainder term R_n^G the inequality $deg(R_n^G) \geq 2n-1$ holds. The existence and uniqueness of the Gaussian formulas Q_n^G are well-known and their nodes $x_{\nu,n}^G$ are the zeros of the corresponding orthonormal polynomial P_n (cf., e.g, Szegö 1939, p. 47 ff). Using the *Christoffel function* $\lambda_n(w,\cdot)$,

(2.1) $$\lambda_n(w,\cdot) \quad = \quad \Big(\sum_{\nu=0}^{n-1} P_\nu^2\Big)^{-1} \,,$$

then, for every quadrature formula Q_n with $deg(R_n) \geq 2n-2$, it follows from $a_{\nu,n} = \lambda_n(w, x_{\nu,n})$ (cf. Freud 1969, p. 21 ff) that

(2.2) $$Var(Q_n) \quad = \quad Q_n[\lambda_n(w,\cdot)] \,.$$

Equality (2.2) now can be used for further estimates. For example, let w be continuous and positive in a neighbourhood of $[a,b] \subset (-1,1)$. Then, we have

$$(2.3) \qquad \lim_{n\to\infty} n\,\lambda_n(w,x) = \pi\, w(x)\,\sqrt{1-x^2}$$

uniformly for $x \in [a,b]$ (see Freud 1969, p. 248 ff and, for more general results, Nevai 1979, 1986). By (2.2) and (2.3), therefore it is an obvious conjecture that the equality

$$(2.4) \qquad \lim_{n\to\infty} n\, Var(Q_n^G) = \pi \int_{-1}^{1} w^2(x)\,\sqrt{1-x^2}\,dx$$

holds for many weight functions, if the integral on the right hand side exists. For weight functions w, which are continuous in $(-1,1)$ and which satisfy

$$(2.5) \qquad 0 < w(x) < c\,(1-x^2)^{-3/4+\delta}\,, \qquad c>0,\quad \delta>0,\quad x\in(-1,1)\,,$$

equality (2.4) has been proven in [55]. For more general weight functions the question is still open.

2.1.1. Gaussian formulas for ultraspherical weight functions. An elementary representation for $Var(Q_n^G)$ for all Gaussian formulas Q_n^G seems to be known explicitely only for special cases of the ultraspherical weight functions w_λ,

$$(2.6) \qquad w_\lambda(x) = (1-x^2)^{\lambda-1/2}\,, \qquad \lambda > -1/2\,.$$

From the representation of the associated Gaussian weights it directly follows that

$$(2.7) \qquad Var(Q_n^G) = \tfrac{\pi^2}{n} \text{ for } \lambda=0 \text{ and } n\geq 1\,, \qquad Var(Q_n^G) = \tfrac{3\pi^2}{8(n+1)} \text{ for } \lambda=1 \text{ and } n\geq 2\,,$$

$$(2.8) \qquad Var(Q_n^G) = \tfrac{35\pi^2}{128(n+2)}\left\{1+\tfrac{8}{7(n+1)\,(n+3)}\right\} \text{ for } \lambda=2 \text{ and } n\geq 3\,.$$

For $\lambda > -1/4$ and each $n\in\mathbb{N}$, using results stated in [54], Förster & Petras (1992) have proved upper and lower bounds for $Var(Q_n^G)$ which are asymptotically sharp in the sense of (2.4). For example, for the Legendre case $\lambda = \frac{1}{2}$ the obtained bounds are the following:

$$(2.9) \qquad \tfrac{\pi^2}{2n+1}\left\{1-\tfrac{1}{(2n+1)^2}\right\} \leq Var(Q_n^G) \leq \tfrac{\pi^2}{2n+1} \qquad \text{for } n\geq 3\,.$$

If $\lambda\in(-\frac{1}{2},-\frac{1}{4}]$, then the variance of the Gaussian rule (Q_n^G) is not small in the sense of (1.7),

$$(2.10) \qquad Var(Q_n^G) = 2\pi\,\{\ln n + \mathcal{O}(1)\}/n \qquad \text{for } \lambda=-1/4\,,$$

$$(2.11) \qquad 0 < \lim_{n\to\infty}\inf n^{4\lambda+2}\, Var(Q_n^G) \leq \lim_{n\to\infty}\sup n^{4\lambda+2}\, Var(Q_n^G) < \infty \quad \text{for } \lambda < -1/4\,.$$

2.2. Polya formulas. The Polya formula Q_n^{Pol} is the interpolatory quadrature formula which uses the Chebyshev nodes $x_{\nu,n} = -\cos((2\nu-1)\pi/2n)$. From a more general result of Chawla & Ramakrishnan (1976, see also Section 3.2.1) it follows that

$$(2.12) \qquad Var(Q_n^{Pol}) \leq \frac{\pi}{n}\int_{-1}^{1} w^2(x)\,\sqrt{1-x^2}\,dx \qquad \text{for every } n\in\mathbb{N}\,,$$

$$(2.13) \qquad \lim_{n\to\infty} n\, Var(Q_n^{Pol}) = \pi\int_{-1}^{1} w^2(x)\,\sqrt{1-x^2}\,dx$$

with the only additional assumption on w that the integral on the right hand side exists. Comparing (2.4), (2.5) and (2.13), for a great class of weight functions we have a similar asymptotical behaviour of the variance of the Gaussian and the Polya method. For the quadrature formulas of Filippi and Clenshaw-Curtis (for definitions see Brass 1977, p. 116 f) similar results can be obtained.

2.3. Newton-Cotes formulas. The Newton-Cotes formula Q_n^{NC} is the well-known interpolatory quadrature formula which uses the equispaced nodes $x_{\nu,n}^{NC} = -1 + 2\frac{\nu-1}{n-1}$. For the standard weight function $w \equiv 1$, using methods given in section IV.2 of the monograph of Brass (1977), a rough estimation proves the lower bound $Var(Q_n^{NC}) > 2^{2n}/(n^6\sqrt{n}\,\ln^4 n)$ for every $n > 1$. This inequality shows that $Var(Q_n^{NC}) > 10^{45}$ for $n \geq 100$ and, in particular, that $Var(Q_n^{NC})$ increases exponentially as n tends to infinity.

3. Variance of quadrature formulas with fixed nodes

In this section, we consider fixed nodes $x_{1,n} < x_{2,n} < \ldots < x_{n,n}$. For given algebraic degree s we ask for the variance of quadrature formulas Q_n using these nodes $x_{\nu,n}$.

3.1. Upper bounds for the variance. At first, we require divided differences, where the *divided difference* $dvd(x_{1,n},\ldots,x_{n,n})$ is a functional of type (1.2) satisfying $dvd(x_{1,n},\ldots,x_{n,n})[p_\nu] = 0$ for all $\nu = 0,1,\ldots,n-2$ and $dvd(x_{1,n},\ldots,x_{n,n})[p_{n-1}] = 1$. It is well-known that this divided difference exists and is uniquely determined by the nodes $x_{\nu,n}$, if the latter are pairwise distinct. If $deg(R_n) \geq s$ and $n \geq s+2$, then the quadrature formula Q_n^c,

$$(3.1)\quad Q_n^c[f] = Q_n[f] + c\, dvd(x_{1,n},\ldots,x_{s+2,n}) =: Q_n[f] + c\sum_{\nu=1}^{s+2} b_{\nu,s+2}\, f(x_{\nu,n}),\quad c \in \mathbb{R},$$

also has $deg(R_n^c) \geq s$. Therefore, from the only assumption that a quadrature formula Q_n with known nodes $x_{1,n} < x_{2,n} < \ldots < x_{n,n}$ has a given algebraic degree $s < n-1$, no upper bound for $Var(Q_n)$ can be deduced.

3.1.1. Positive quadrature formulas. If we restrict consideration to positive quadrature formulas Q_n, then from the boundedness of the norm of Q_n trivially the boundedness of $Var(Q_n)$ follows. Investigating upper bounds of $Var(Q_n)$ for all positive quadrature formulas Q_n with algebraic degree $deg(R_n) \geq s$, we even can restrict consideration to all positive interpolatory quadrature formulas Q_{s+1} which use as nodes a subset of $\{x_{1,n},\ldots,x_{n,n}\}$. The proof can be sketched in a few lines. We assume that all weights $a_{1,n},\ldots,a_{s+2,n}$ of a given positive quadrature formula Q_n with $deg(R_n) \geq s$ and $n \geq s+2$ are positive. Using (3.1) we have

$$(3.2)\qquad h(c) := Var(Q_n^c) = \sum_{\nu=1}^{s+2}(a_{\nu,n} + c\, b_{\nu,s+2})^2 + \sum_{\nu=s+3}^{n}(a_{\nu,n})^2,\quad h''(c) = 2\sum_{\nu=1}^{s+2}(b_{\nu,s+2})^2.$$

It is well-known that the sign of the weights $b_{\nu,s+2}$ of a divided difference alternates. First, let $h'(0) \geq 0$. Then, there exists a unique $\tilde{c} > 0$ such that $Q_n^{\tilde{c}}$ is positive but one of the associated weights $a_{\nu,n}^{\tilde{c}}$ ($\nu = 1,\ldots,s+2$) is zero. From the strict convexity of h it follows that $Var(Q_n^{\tilde{c}}) = h(\tilde{c}) > h(0) = Var(Q_n)$. For $h'(0) < 0$ an analogous argumentation holds. Repeated application of this method shows that there exists a positive interpolatory quadrature formula $\bar{Q}_{s+1}$ using as nodes a subset of $\{x_{1,n},\ldots,x_{n,n}\}$ with $Var(\bar{Q}_{s+1}) > Var(Q_n)$.

3.2. Minimum variance formulas. For given algebraic degree $s \in \mathbb{N}_0$, $s < n$, we say that a quadrature formula Q_n^* is a (s,n) *minimum variance formula for the nodes* $\{x_{1,n}, x_{2,n},\ldots,x_{n,n}\}$, if Q_n^* has the smallest variance under all quadrature formulas Q_n using the same nodes $x_{1,n}, x_{2,n},\ldots,x_{n,n}$ and having algebraic degree $deg(R_n) \geq s$. First examples of (s,n) minimum variance formulas for equispaced nodes have been computed by Sard (1949). By the papers of Wilson (1970a) and Chawla & Ramakrishnan (1976) we have the following results: (s,n) minimum variance formulas Q_n^* for fixed nodes $\{x_{1,n}, x_{2,n}, \ldots, x_{n,n}\}$ always exist and are uniquely determined. Consider the inner product

$$(3.3) \qquad (f,g) \;=\; \sum_{\nu=1}^{n} f(x_{\nu,n})\, g(x_{\nu,n})$$

with the associated norm $||f|| = (f,f)^{1/2}$. Let $P_i^* \in \Pi_i$ denote the polynomials being orthonormal with respect to the inner product (3.3). Then the weights $a_{\nu,n}^*$ and the variance $Var(Q_n^*)$ of the (s,n) minimum variance formula Q_n^* for the nodes $\{x_{1,n}, x_{2,n}, \ldots, x_{n,n}\}$ are given by

$$(3.4)\ a_{\nu,n}^* \;=\; \sum_{i=0}^{s} I[P_i^*]\, P_i^*(x_{\nu,n}) \quad \text{for} \quad \nu = 1,2,\ldots,n\,, \qquad Var(Q_n^*) \;=\; \sum_{i=0}^{s} (I[P_i^*])^2\,.$$

The construction of (s,n) minimum variance formulas using generalized inverses and inversion of numerical differentation formulas was investigated by Duris (1971).

3.2.1. Chebyshev nodes. For the Chebyshev nodes $x_{\nu,n} = -\cos((2\nu-1)\pi/(2n))$ the corresponding orthonormal polynomials P_i^* are given by

$$(3.5) \qquad P_0^* \;=\; n^{-1/2}\, T_0\,, \qquad P_i^* \;=\; (2/n)^{1/2}\, T_i \qquad \text{for} \quad i = 1,2,\ldots,n-1\,,$$

where T_ν denotes the Chebyshev polynomial of the first kind of degree ν. Therefore, we have for these (s,n) minimum variance formulas Q_n^*,

$$(3.6)\ a_{\nu,n}^* \;=\; \tfrac{2}{n} \sum_{i=0}^{s}{}' \, I[T_i]\, T_i(x_{\nu,n}) \quad \text{for} \quad \nu = 1,2,\ldots,n\,; \qquad Var(Q_n^*) \;=\; \tfrac{2}{n} \sum_{i=0}^{s}{}' \, (I[T_i])^2\,,$$

see (Chawla & Ramakrishnan 1976). For $s = n-1$ we obtain the variance of the Polya formula Q_n^{Pol}. Using Bessel's inequality, we directly obtain the results stated in Section 2.2. Results on the norm and the weights of these Clenshaw-Curtis-type quadrature methods have been proved by Sloan & Smith (1978).

3.2.2. Equispaced nodes. A detailed investigation of (s,n) minimum variance formulas with equispaced nodes has been given by Chawla & Ramakrishnan (1976). Using results of Wilson (1970b), numerical investigations of Mencke (1990) for $w \equiv 1$ show the somewhat surprising result that for some special choices of s and n, for which positive quadrature formulas Q_n with equispaced nodes and $deg(R_n) \geq s$ exist, the associated (s,n) minimum variance formulas are nonpositive.

3.3. Computation of the variance of interpolatory quadrature formulas. Let Q_n be an interpolatory quadrature formula with pairwise distinct nodes $x_{\nu,n}$. Then

$$(3.7)\quad a := \begin{pmatrix} a_{1,n} \\ a_{2,n} \\ \vdots \\ a_{n,n} \end{pmatrix} = X^{-1}\, m, \quad m := \begin{pmatrix} I[p_0] \\ I[p_1] \\ \vdots \\ I[p_{n-1}] \end{pmatrix}, \quad X := \begin{pmatrix} 1 & 1 & \cdots & 1 \\ x_{1,n} & x_{2,n} & \cdots & x_{n,n} \\ \vdots & \vdots & \ddots & \vdots \\ x_{1,n}^{n-1} & x_{2,n}^{n-1} & \cdots & x_{n,n}^{n-1} \end{pmatrix}.$$

Therefore, we obtain

$$(3.8)\ Var(Q_n) \;=\; a^T a \;=\; m^T\, S^{-1}\, m\,, \quad S := \begin{pmatrix} s_0 & s_1 & \cdots & s_{n-1} \\ s_1 & s_2 & \cdots & s_n \\ \vdots & \vdots & \ddots & \vdots \\ s_{n-1} & s_n & \cdots & s_{2n-2} \end{pmatrix}, \quad s_\mu := \sum_{\nu=1}^{n} x_{\nu,n}^{\mu}\,.$$

Now, using Newton's identities (see Section 5.1.1), the variance $Var(Q_n)$ of an interpolatory quadrature formula Q_n can be computed directly from the coefficients $b_{\nu,n}$ of its node polynomial $h(Q_n, \cdot)$, $h(Q_n, x) = \Pi_{\nu=1}^{n}(x - x_{\nu,n}) =: \sum_{\nu=0}^{n} b_{\nu,n} x^{n-\nu}$, and from the moments $I[p_\nu]$. If all these coefficients $b_{\nu,n}$ and moments $I[p_\nu]$ are rational numbers, then the computation remains rational and therefore also $Var(Q_n)$ is a rational number. For the weight function $w \equiv 1$, we obtain as a somewhat surprising consequence that the variance $Var(Q_n^G)$ of every Gaussian formula Q_n^G is a rational number, although most of the associated Gaussian nodes and weights are irrational.

4. Variance of quadrature formulas with free nodes

4.1. Estimates for the variance of convergent quadrature rules. Let $(Q_{n_i})_{i=1}^{\infty}$ be a quadrature method which is convergent for every continuous function and let $y \in \mathbb{R}$ be fixed, where y is not a node of any of the quadrature formulas Q_{n_i}. Then, the quadrature method $(\tilde{Q}_{m_i})$ with $\tilde{Q}_{m_i}[f] = Q_{n_i}[f] + c(f(y + \frac{1}{n}) - f(y))$ and fixed $c \in \mathbb{R}$ is also convergent for every continuous function and we obtain $Var(\tilde{Q}_{m_i}) \geq c^2$. Therefore, from the only assumption that (Q_{n_i}) converges for every continuous function, we can only deduce the boundedness of $(Var(Q_{n_i}))_{i=1}^{\infty}$.

A much sharper result is valid, if we additionally assume that the quadrature method $(Q_{n_i})_{i=1}^{\infty}$ is positive: Using an idea of Fejér (1933), we have $\lim_{i\to\infty} \max\{a_{\nu,n_i} \mid \nu = 1, 2, \ldots, n_i\} = 0$. An application of the rough estimate $Var(Q_n) \leq \max\{a_{\nu,n} \mid \nu = 1, 2, \ldots, n\}\, Q_n[p_0]$ shows for every positive convergent method $(Q_{n_i})_{i=1}^{\infty}$ that

$$\lim_{i\to\infty} Var(Q_{n_i}) = 0 . \tag{4.1}$$

4.2. Estimates for the variance of nonpositive quadrature formulas. From the only assumption that a quadrature formula Q_n has a given algebraic degree $deg(R_n) = s < n - 1$, no upper bound for $Var(Q_n)$ can be deduced, see the method given in Section 3.1. Furthermore, assuming for Q_n a high algebraic degree $deg(R_n) = s \geq n - 1$, results on upper bounds for the variance may be not satisfactory as well: Let $n > 1$, let $y \in [-1, 1]$ be not a node of the Gaussian formula Q_{n-1}^G and let b be an arbitrary negative number. Then, by a result in [50], there exists a (unique) quadrature formula Q_n with $deg(R_n) \geq 2n - 3$ and all nodes contained in $[-1, 1]$, which has a node in y and the associated weight b. It follows that, for every fixed $n > 1$,

$$\sup\{Var(Q_n) \mid deg(R_n) \geq 2n - 3, \quad -1 \leq x_{1,n} \leq \ldots \leq x_{n,n} \leq 1\} = \infty . \tag{4.2}$$

Note that $deg(R_n) > 2n - 3$ implies the positivity of Q_n. Therefore, in the following we restrict consideration to positive quadrature formulas, more precisely, we consider estimates for

$$V_{\sup}(s, n) := \sup\{Var(Q_n) \mid deg(R_n) \geq s, \quad Q_n \text{ is positive}\} , \tag{4.3}$$

$$V_{\inf}(s, n) := \inf\{Var(Q_n) \mid deg(R_n) \geq s, \quad Q_n \text{ is positive}\} . \tag{4.4}$$

4.3. Upper estimates for the variance of positive quadrature formulas. Using the Gaussian formula Q_m^G, it is shown in [52] that, for the variance $Var(Q_n)$ of every positive quadrature formula Q_n with $deg(R_n) \geq 2m - 1$,

$$Var(Q_n) \leq 2\, Var(Q_m^G) . \tag{4.5}$$

For functionals I resp. $\tilde{I}$ with weight functions w resp. $\tilde{w}$, this basic result easily can be used for comparison results as follows. Let Q_m^G be the Gaussian formula for I and let $\tilde{Q}_n$ be a positive quadrature formula for $\tilde{I}$ with $deg(\tilde{R}_n) \geq 2m - 1$. If $\tilde{w} \leq w$ on $(-1, 1)$, then

$$(4.6) \qquad Var(\tilde{Q}_n) \quad \leq \quad 2\,Var(Q_m^G) \;-\; (I[p_0] - \tilde{I}[p_0])^2/n \;.$$

Applying results on the variance of Gaussian formulas for ultraspherical weight functions (see Section 2.1.1), by (4.6) we obtain, for weight functions w satisfying

$$(4.7) \qquad w(x) \quad \leq \quad c(1-x^2)^{-3/4+\delta}\;, \qquad c > 0,\; \delta > 0,\; x \in (-1,1)\;,$$

that all positive quadrature methods $(Q_{n_i})_{i=1}^{\infty}$, which have high algebraic degree in the sense of (1.6), have small variance in the sense of (1.7). On the other hand, if

$$(4.8) \qquad w(x) \quad \geq \quad c(1-x^2)^{-3/4}\;, \qquad c > 0,\; x \in (-1,1)\;,$$

then at least the Gaussian method does not have a small variance in the sense of (1.7). In the latter case, note that other positive quadrature methods with high algebraic degree may have small variance. E.g., inequality (4.8) holds for the weight function w given in (5.19) below, but this weight function w admits a Chebyshev-type formula Q_n with $deg(R_n) \geq n$ for every $n \in \mathbb{N}$.

For weight functions w which are continuous in $(-1,1)$ and which satisfy (4.7), inequality (4.5) can be improved asymptotically: Since for every positive quadrature formula Q_n with $deg(R_n) \geq 2m - 2 > 0$ the inequality $a_{\nu,n} \leq \lambda_m(w, x_{\nu,n})$ holds (cf., e.g., Karlin & Studden 1966), we can apply the method demonstrated in Section 2.1. Using this argument it is proved in [55] that, for every positive quadrature method $(Q_{n_i})_{i=1}^{\infty}$ with $\lim_{i\to\infty} 1/deg(R_{n_i}) = 0$,

$$(4.9) \qquad \lim_{i\to\infty}\;\sup\; deg(R_{n_i})\; Var(Q_{n_i}) \quad \leq \quad 2\,\pi \int_{-1}^{1} w^2(x)\,\sqrt{1-x^2}\;dx\;.$$

4.3.1. Ultraspherical weight functions. By comparison of inequality (4.9) with equality (2.4) we see that, for the weight functions w considered there, the variance of the Gaussian formula Q_m^G is asymptotically worst under all positive quadrature formulas Q_n with $deg(R_n) \geq 2m-1$, if m tends to infinity. For ultraspherical weight funcions w_λ, this result can be improved for certain parameters λ under the assumption that all nodes of Q_n are contained in the interval $[-1,1]$. Then, for every positive quadrature formula Q_n with $deg(R_n) \geq 2m - 1$,

$$(4.10) \qquad Var(Q_n) \quad \leq \quad Var(Q_m^G) \qquad \text{for } \;\lambda \in (-1/2, 0],\; m \in \mathbb{N}\;,$$

$$(4.11) \qquad Var(Q_n) \quad < \quad Var(Q_m^G) \;+\; \tfrac{1}{2}\left(\tfrac{\pi}{m}\right)^2 \qquad \text{for } \;\lambda \in (0, 1/2],\; m \in \mathbb{N}\;,$$

see [45], where also extensions of the type (4.6) are given. The equality sign in (4.10) holds if and only if $Q_n = Q_m^G$. For the Legendre case $\lambda = 1/2$, by a comparison with the Lobatto formula Q_{m+1}^{Lo} it can be shown that a result of type (4.10) is not valid (cf. [55]).

4.3.2. Structure of positive quadrature formulas of given algebraic degree having high variance. As stated in Section 3.1, for every (strict) positive quadrature formula Q_n with $deg(R_n) = s < n-1$, there exists a positive interpolatory quadrature formula $\bar{Q}_{s+1}$ with $\{\bar{x}_{1,s+1}, \ldots, \bar{x}_{s+1,s+1}\} \subset \{x_{1,n}, \ldots, x_{n,n}\}$ and $Var(\bar{Q}_{s+1}) > Var(Q_n)$. Therefore, if for a positive quadrature formula Q_m^+ with $deg(R_m^+) \geq s$ the equality

$$(4.12) \qquad Var(Q_m^+) \quad = \quad \sup_{n\in\mathbb{N}} V_{\sup}(s,n)$$

holds, then it follows that $m \le s+1$. Up to special cases (see (4.10)), general existence results for such positive quadrature formulas Q_m^+ being extremal in the sense of (4.12) are unknown. If such an extremal formula Q_m^+ exists, there seems to be computional evidence that then even $m \le [(s+3)/2]$.

4.4. Lower estimates for the variance of positive quadrature formulas. In this section, we ask for estimates for $V_{\inf}(s,n)$ defined in (4.4). First, we have (cf. [46])

$$(4.13) \quad V_{\inf}(s,n) \quad \le \quad (I[p_0])^2/(n-[s/2]) \qquad \text{for each } s \in \mathbb{N} \text{ and each } n > [s/2] \; .$$

Substituting $s = n$ we obtain for each $n \in \mathbb{N}$ that there exists a positive quadrature formula Q_n with $deg(R_n) \ge n$ which has at most twice of the minimal possible variance of a Chebyshev-type formula with n nodes. Furthermore, from (4.13) it follows that

$$(4.14) \qquad \lim_{n\to\infty} n\, V_{\inf}(s,n) \quad = \quad (I[p_0])^2 \; , \qquad \text{if} \quad \lim_{n\to\infty} s/n \quad = \quad 0 \; ,$$

showing in this case of "small" algebraic degree the existence of a positive quadrature method which has asymptotically the smallest possible variance.

For hight algebraic degree in the sense of $\lim_{n\to\infty} \inf\, s/n \ge d > 0$ and $w > 0$ a.e. in $(-1,1)$, a nontrivial lower asymptotical estimate for $V_{\inf}(s,n)$ can be obtained by comparison of w with the Chebyshev weight function $w_0(x) = (1-x^2)^{-1/2}$, if there exists an interval $(a,b) \subseteq [-1,1]$ such that

$$(4.15) \qquad w(x) \quad < \quad d\, I[p_0]\, w_0(x)/(2\pi) \qquad \text{a.e. in} \quad (a,b) \; .$$

By a generalization of a method introduced by Kahaner (1969) we have the estimate (cf. [46])

$$(4.16) \qquad \lim_{n\to\infty} \inf\, n\, V_{\inf}(s,n) \quad \ge \quad \left\{1 + \frac{(B-A)^2}{B(1-B)}\right\} (I[p_0])^2 \quad > \quad (I[p_0])^2 \; ,$$

$$\text{where} \quad A := \frac{1}{I[p_0]} \int_a^b w(x)\, dx \; , \qquad B := \frac{d}{2\pi} \int_a^b w_0(x)\, dx \; , \qquad \lim_{n\to\infty} \inf\, s/n \ge d > 0 \; .$$

Under the conditions given above, a comparison with (4.13) shows that (4.16) is unimprovable for every $d \in (0,2)$, since $\sup\left\{1 + \frac{(B-A)^2}{B(1-B)} \mid w > 0 \text{ a.e. in } [-1,1]\right\} = 1/(1-d/2)$.

4.4.1. Results for the weight function $w \equiv 1$. For the standard weight function $w \equiv 1$ sharper estimates have been proven in [51]. E.g., for interpolatory quadrature formulas, the upper bound $n\, V_{\inf}(n-1,n) < 1.0625\, (I[p_0])^2$ is valid for every $n \in \mathbb{N}$, while the lower bound $n\, V_{\inf}(n-1,n) > 1.0325\, (I[p_0])^2$ holds for every $n \ge 100$.

4.5. Existence and structure of minimum variance formulas with free nodes. We say that a quadrature formula Q_n^- is a (s,n) *minimum variance formula with free nodes*, if $deg(R_n^-) \ge s$ and

$$(4.18) \qquad Var(Q_n^-) \quad = \quad \inf\, \{\, Var(Q_n) \mid deg(R_n) \ge s \,\} \; .$$

If Q_n^- is not a Chebyshev-type formula, then results on the existence, uniqueness or structure of such formulas Q_n^- seem to be rare. For $s = n-1$, Kahaner (1968, 1969) has proved the existence of Q_n^- under the additional assumption that all nodes are contained in $[-1,1]$. He also has pointed out that several methods for the numerical computation of such formulas are instable, since some of the nodes seem to coalesce. The case $s = n$ has been considered by Salkauskas (1971) under the following additional restrictions on the quadrature formulas: Q_n is positive, the nodes lie in $(-1, 1)$ and are pairwise distinct. He has shown that, if a (n, n) minimum variance formula exists, then it is of Chebyshev-type. For nonnegative symmetric weight functions w, Engels & Merschen (1984, 1987) have proved a necessary condition for $(2n - 2, n)$ minimum variance formulas with free nodes, which is fullfilled for the Gaussian formula Q_n^G.

5. Chebyshev-type formulas

We say that a quadrature formula Q_n is a (s, n) *Chebyshev-type formula*, if Q_n is of type $Q_n[f] = (I[p_0]/n)\sum_{\nu=1}^{n} f(x_{\nu,n})$ and $deg(R_n) \geq s$. By $C(s, n)$ we denote the set of all (s, n) Chebyshev-type quadrature formulas for w. If $Q_n \in C(n, n)$ and if all nodes are contained in $[-1, 1]$, then Q_n is called a *Chebyshev formula in the strict sense*. In the following, let

$$d_n \quad := \quad \sup \{ \, s : \; C(s, n) \neq \emptyset \, \} \, . \tag{5.1}$$

Here, we are interested in Chebyshev-type formulas because of their property that they minimize $Var(Q_n)$ under all quadrature formulas Q_n with $deg(R_n) \geq 0$. With regard to the open problems stated in Section 4.5, in the following we consider mostly results on the existence, structure and construction of these special "minimum variance formulas with free nodes" for high algebraic degree in the sense of (1.6). For the progress which has been made in Chebyshev-type quadrature in many other fields of mathematical interest, we refer to Section 6 and, in particular, to the survey article of Gautschi (1976).

5.1. Construction of Chebyshev-type formulas.

5.1.1. Newton's identities. In the following we denote by $h(Q_n, \cdot)$ the so-called node polynomial of the quadrature formula Q_n,

$$h(Q_n, x) \quad = \quad x^n + \sum_{\nu=1}^{n} b_{\nu,n} x^{n-\nu} \quad := \quad \prod_{\nu=1}^{n} (x - x_{\nu,n}) \, . \tag{5.2}$$

Consider Newton's identities which, for n arbitrary complex number $z_1, z_2, \ldots, z_n$, yield

$$\left\{ \begin{array}{lll} s_1 + c_1 & = & 0 \\ s_2 + c_1 s_1 + 2c_2 & = & 0 \\ \quad \vdots & & \\ s_{n-1} + c_1 s_{n-2} + c_2 s_{n-3} + \ldots + c_{n-2} s_1 + (n-1) c_{n-1} & = & 0 \, , \\ s_{\gamma+n} + c_1 s_{\gamma+n-1} + c_2 s_{\gamma+n-2} + \ldots + c_{n-1} s_{\gamma+1} + c_n s_\gamma & = & 0 \qquad (\gamma = 0, 1, 2, \ldots) \end{array} \right. \tag{5.3}$$

$$\Leftrightarrow \quad s_\mu \;=\; \sum_{\nu=1}^{n} (z_\nu)^\mu \quad (\mu = 1, 2, \ldots) \, , \quad g(z) \;=\; z^n + \sum_{\nu=1}^{n} c_\nu z^{n-\nu} \;:=\; \prod_{\nu=1}^{n} (z - z_\nu) \tag{5.4}$$

(cf., e.g., Perron (1932), p. 150 ff). Substituting $s_\mu := n \, I[p_\mu]/I[p_0]$, from (5.4) it follows that the coefficients $b_{1,n}, b_{2,n}, \ldots, b_{s,n}$ of all node polynomials $h(Q_n, \cdot)$ with $Q_n \in C(s, n)$ are identical. Conversely, every $h(Q_n, \cdot)$ of type (5.2) with these coefficients $b_{1,n}, b_{2,n} \ldots, b_{s,n}$ and only real roots represents a quadrature formula $Q_n \in C(s, n)$. In particular, if $Q_n \in C(n, n)$, the uniqueness of Q_n follows immediately. Starting with Radau (1880) several authors have used Newton's identities to prove or to disprove the existence of (s, n) Chebyshev-type formulas.

5.1.2. Expansion method. Consider the polynomial F_n defined by

$$(5.5)\quad F_n(z) = \exp(H_n(z)) + \mathcal{O}(z^{-1}), \quad H_n(z) := \frac{n}{I[p_0]} \int_{-1}^{1} \ln(z-x)w(x)dx, \; z \in \mathbb{C}, \; |z| \to \infty.$$

Let Q_n be a Chebyshev-type formula, let $h(Q_n, \cdot)$ be the associated node polynomial and let $m \in \mathbb{N}$. Then $Q_n \in C(n-1+m, n)$ holds if and only if $F_n(z) = \exp(H_n(z)) + \mathcal{O}(z^{-m})$ and $F_n = h(Q_n, \cdot)$. This follows by expansion of $R_n[f_z]$ in powers of $1/z$ for $|z|$ large, where $f_z(x) = 1/(z-x)$ (cf., e.g. Krylow 1962, p. 181 ff, Peherstorfer 1988b). Representation (5.5) has been used first by Chebyshev (1874) and has been proved to be a powerful tool in several fields of Chebyshev quadrature. Using Faber polynomials, for the polynomials F_n Peherstorfer (1988b) has obtained the helpful recursion relation

$$(5.6)\quad \begin{aligned} F_{n+1}(x) &= x\,F_n(x) - \sum_{k=0}^{n-1} c_k\,F_{n-k}(x) - (n+1)\,c_n, \qquad F_0(x) = 1, \\ (n+1)\,c_n &= \int_{-1}^{1} x\,F_n(x)\,w(x)\,dx \qquad \text{for every } n \in \mathbb{N}_0. \end{aligned}$$

5.2. Structure of Chebyshev-type formulas. If $s \geq n$, then the set $C(s,n)$ contains at most one element, see Section 5.1.1. If $s < n$, then it also may be possible that $C(s,n)$ contains only one element. E.g., for the ultraspherical weight function w_1, $C(9,12)$ contains only one element and this also holds for $C(8,12)$, see [53]. Förster & Ostermeyer (1986) have proved that if $C(s,n)$ with $s < n$ contains only one element, then this Chebyshev-type formula has at most $s-1$ pairwise distinct nodes. On the other hand, for most of the investigated examples, a nonempty set $C(s,n)$ with $s < n$ has more than one element (cf., e.g., Barnhill et al. 1969, Gautschi & Yanagiwara 1974, Anderson & Gautschi 1975). If $C(s,n)$ contains more than one element, then there exists an infinite number of Chebyshev-type formulas in $C(s,n)$ which have n pairwise distinct nodes (see [53]). Therefore, if there exists a Chebyshev formula in $C(s,n)$ with at least s pairwise distinct nodes, then there exist infinitely many Chebyshev-type formulas in $C(s,n)$ having n pairwise distinct nodes. Using a different approach, the latter result has also been proved by Kuijlaars (1992a). If $s \leq d_n < n$, then the set $C(s,n)$ always contains a formula having at most s pairwise distinct nodes. This has been proved first by Anderson & Gautschi (1975) for $s = d_n$ and later in [53] for $s \leq d_n$. Finally, let us remark that even (n,n) Chebyshev-type formulas may have multiple nodes or nodes outside the interval $[-1,1]$, see the various examples in the appendix of [43].

5.2.1. E-extremal Chebyshev-type formulas. Let $1 < s \leq d_n$. A more detailed description of the structure of $C(s,n)$ can be given by use of *E-extremal* Chebyshev-type formulas being introduced in [53]. We call a formula $Q_n \in C(s,n)$ *E-minimal (E-maximal)* in $C(s,n)$ and denote it by $Q_n^{\min s}$ ($Q_n^{\max s}$) if

$$(5.7)\quad R_n^{\min s}[p_{s+1}] = \min\{R_n[p_{s+1}] \mid Q_n \in C(s,n)\},$$

$$(5.8)\quad R_n^{\max s}[p_{s+1}] = \max\{R_n[p_{s+1}] \mid Q_n \in C(s,n)\}.$$

In $C(s,n)$ there exists one and only one E-minimal (E-maximal) formula and this formula has at most s distinct nodes. These two E-extremal formulas can be characterised by a special arrangement of their multiple nodes, see [53]. Moreover, for every $r \in (R_n^{\min s}[p_{s+1}], R_n^{\max s}[p_{s+1}])$ there exist formulas $\tilde{Q}_n$ and $\bar{Q}_n$ in $C(s,n)$ with $\tilde{R}_n[p_{s+1}] = \bar{R}_n[p_{s+1}] = r$ such that (i) $\tilde{Q}_n$ has n pairwise distinct nodes and (ii) $\bar{Q}_n$ has at most $s+1$ distinct nodes. In particular, if $Q_n^{\min s} = Q_n^{\max s}$, then this formula is the only element of $C(s,n)$ and has at most $s-1$ distinct nodes. For the nodes $x_{\nu,n}$ of every formula $Q_n \in C(s,n)$ which is not E-extremal in $C(s,n)$, it follows that

$$(5.9) \qquad x_{n,n}^{\min s} > x_{n,n} > x_{n,n}^{\max s}, \qquad (-1)^s x_{1,n}^{\min s} > (-1)^s x_{1,n} > (-1)^s x_{1,n}^{\max s}.$$

These inequalities may be helpful for the investigation of the question of whether all nodes of a formula $Q_n \in C(s,n)$ are contained in the interval $[-1,1]$. The E-extremal formulas are defined by the extremal property (5.7) resp. (5.8) of their remainder with respect to only one function, the monomial p_{s+1}. This extremal property remains valid for every function $f \in C^{s+1}$ with $f^{(s+1)} \geq 0$:

$$(5.10) \qquad R_n^{\min s}[f] \leq R_n[f] \leq R_n^{\max s}[f] \qquad \text{for every } Q_n \in C(s,n).$$

5.2.2. Optimal Chebyshev-type formulas. As mentioned above, there may exist infinitely many Chebyshev-type formulas Q_n of maximal possible degree d_n if $d_n < n$. Several authors have proposed to select so-called *E-optimal Chebyshev-type formulas*, where a (s,n) Chebyshev-type formula Q_n^E is E-optimal, if $s = d_n$ and $|R_n^E[p_{s+1}]| = \min\{|R_n[p_{s+1}]| \mid Q_n \in C(s,n)\}$. Rabinowitz & Richter (1970) have proved that such E-optimal Chebyshev-type formulas are optimal Chebyshev-type formulas for certain limit cases of two families of Hilbert spaces. E-optimal formulas Q_n^E exist for every $n \in \mathbb{N}$ (cf. Gautschi & Yanagiwara 1974). In particular, Anderson & Gautschi (1975) have proved that an E-optimal Chebyshev-type formula Q_n^E has at most d_n pairwise distinct nodes. The uniqueness of E-optimal Chebyshev-type formulas Q_n^E has been shown by Förster & Ostermeyer (1986). For small n, (d_n,n) Chebyshev-type formulas, in particular E-optimal formulas, have been computed by Kahaner (1970), Yanagiwara & Shibata (1973), Gautschi & Yanagiwara (1974), Yanagiwara & Fukutake & Shibata (1974), Anderson & Gautschi (1975), Gautschi & Monegato (1977) and Förster & Ostermeyer (1986). Some of these results are calculated for such Chebyshev-type formulas introduced by Barnhill et al. (1969), which minimize $\sum_{\nu=s+1}^{q}(R_n[p_\nu])^2$ in $C(s,n)$ for given s and q, $0 \leq s < q \leq \infty$. Results on multiple nodes of this latter type of Chebyshev formulas have been obtained by Salkauskas (1973a, 1973b, 1976). Chebyshev-type formulas Q_n which are optimal in the sense of Sard in $C(s,n)$, have been investigated by Krylov (1962, p. 138 ff), Franke (1971), Gautschi & Monegato (1977) and Förster (1982b). If the E-optimal Chebyshev-type formula Q_n^E has a definite remainder term R_n^E, then it follows from (5.10) that Q_n^E is also optimal in the sense of Sard in $C(d_n,n)$. For the weight function $w \equiv 1$, the midpoint rule is optimal in several classes of functions (cf. Brass 1977, p. 242 ff). For nonnegative weight functions, different Chebyshev-type extensions of the midpoint rule have been proposed by Jagermann (1966) and Stetter (1968). Using Peano kernel methods, it is easy to prove that for the class $\{f \mid V(f) \leq 1,\ V(f)\ :\ \text{total variation over } [-1,1]\}$ the optimal formula among *all* quadrature formulas Q_n is of Chebyshev-type.

5.3. Lower bounds for the maximal degree of Chebyshev-type formulas. By an important result of Seymor & Zaslavsky (1984) we have that

$$(5.11) \qquad \lim_{n\to\infty} 1/d_n = 0.$$

Recently, for w satisfying $c_2(1-|x|)^\beta \leq w(x) \leq c_1$ $(\beta \geq 0,\ 0 < c_2 \leq 1 \leq c_1)$ on $(-1,1)$, Wagner (1991) has stated lower bounds for d_n. For weight functions w with $w > 0$ a.e. on $(-1,1)$, interesting lower bounds for d_n have been proved by Rabau & Bajnok (1991).

5.3.1. Ultraspherical and Jacobian weight functions. Specialising their results to Jacobian weight functions $w_{\alpha,\beta}(x) = (1-x)^\alpha\,(1+x)^\beta$, Rabau & Bajnok (1991) obtained the lower bound $d_n > c(\alpha,\beta)\; n^{1/(2\rho+3)}$ for $\rho \geq -1/2$ and $d_n > c(\alpha,\beta)\sqrt{n}$ for $\rho < -1/2$, where $\rho = \max\{\alpha,\beta\}$ and $c(\alpha,\beta)$ is a positive constant depending on α and β only. For the weight function $w \equiv 1$, Bernstein (1937b) has proved the estimate $d_{2n} \geq 2l - 1$ for $n > 2\sqrt{2}(l+1)(l+4)$ and therefore that $d_{2n} \geq 2^{-1/4}\sqrt{2n}(1+\mathcal{O}(\frac{1}{n}))$. For the Jacobian weight function $w_{\alpha,\beta}$ with $\alpha \geq 0$, $\beta \geq 0$, Kuijlaars (1992a, 1992b) has shown by an extension of Bernstein's method the existence of a positive constant $\tilde{c}(\alpha, \beta)$ such that, for every $s \in \mathbb{N}$, there exists a (s, n) Chebyshev-type formula with $n < \tilde{c}(\alpha, \beta)\ s^{2+2\max\{\alpha,\beta\}}$. For nonnegative α and β, note that the order $\mathcal{O}(s^{2+2\max\{\alpha,\beta\}})$ is unimprovable, if s tends to infinity (see Section 5.4.2).

5.4. Upper bounds for the maximal degree of Chebyshev-type formulas. Let Q_n be positive with $deg(R_n) \geq 2m-1$ and let Q_n have pairwise distinct nodes. If Q_n is not the Gaussian formula Q_m^G, then by results of Bernstein (1936, 1937a) it is well-known that each interval

$$(5.12) \qquad (x_{\nu,m}^G, x_{\nu+1,m}^G) \qquad \nu = 0, 1, \ldots, m\,, \qquad x_{0,m}^G := -\infty, \quad x_{m,m+1}^G = \infty\,,$$

contains at least one node of Q_n. Furthermore, it follows from a method given by Bernstein (1937b) that

$$(5.13) \qquad \sum_{x_{k,m}^G \leq x_{\nu,n} \leq x_{l,m}^G} a_{\nu,n} \quad < \quad \sum_{\mu=k}^{l} a_{\mu,m}^G\,, \qquad 0 \leq k \leq l \leq m$$

(cf., e.g., [40] and Locher 1980). Relations (5.12) and (5.13) are a powerful tool to prove nonexistence results for (s, n) Chebyshev-type formulas. In particular, with $\nu = 0$ in (5.12) and $k = 0$, $l = 1$ in (5.13) we have for nonempty $C(2m-1, n)$ the necessary condition

$$(5.14) \qquad n \quad \geq \quad I[p_0]/a_{1,m}^G\,.$$

This inequality has been used by several authors to prove upper bounds for d_n. A detailed study of this method with several applications to different weight functions has been given by Geronimus (1969, 1970). For w being positive and continuous in $(-1, 1)$ Kahaner (1969) has shown that the following inequality is a necessary condition on w in order to admit a Chebyshev-type quadrature method $(Q_{n_i})_{i=1}^{\infty}$ with $\lim_{i\to\infty} \sup\ deg(R_{n_i})/n_i = d > 0$:

$$(5.15) \qquad w(x) \quad \geq \quad d\,\frac{I[p_0]}{2\pi}\,(1-x^2)^{-1/2} \qquad \text{for } x \in (-1, 1)\,.$$

For extensions of this result, see Section 4.4 and (Peherstorfer 1990).

5.4.1. Weight function $w \equiv 1$. The existence of (n, n) Chebyshev formulas for $n = 1, 2, \ldots, 7$ has been proved by Chebyshev (1874) and for $n = 9$ by Radau (1880b). If $n = 8$ or $n \geq 10$ then, using (5.14), Bernstein (1936, 1937a) has shown that (n, n) Chebyshev formulas do not exist. (At an earlier stage Bernstein 1932 has obtained the non-existence for $n \geq 15000$.) In particular, he has shown that $d_n < (2\pi\sqrt{6n})^{1/2}$ for every $n \in \mathbb{N}$. Recently Korevaar (1992) has proved the sharper result $d_n < (2/|J_0'(j_1)|)\sqrt{n} \approx 3.8525\sqrt{n}$, where j_1 is the first positive zero of the Bessel function J_0 (cf. [171]).

5.4.2. Ultraspherical and Jacobian weight functions. For the ultraspherical weight function w_λ and $\lambda > 0$ a result of Gatteschi (1963/64, p. 85) proves that $d_n \leq \bar{c}(\lambda)\ n^{1/(2\lambda+1)}$ with a constant $\bar{c}(\lambda)$ depending on λ only; see also the results of Gatteschi & Vinardi (1978), where very sharp estimates are given. For $\lambda < -0.30056...$ it has been shown in [43] that there exists no Chebyshev formula in the strict sense which has more than two nodes. For the remaining interval $(-0.30056..., 0)$ the question of whether there exists a Chebyshev method in the strict sense is still open. For the Jacobian weight function $w_{\alpha,\beta}$, by results of Achieser (1937) and Ossicini (1966) we have the estimate $d_n \leq \bar{c}(\alpha, \beta)\ n^{1/(2+2\max\{\alpha,\beta\})}$ for $\max\{\alpha, \beta\} > -1/2$. For similar results see Costabile (1974). Numerical examples for two special choices of α and β and small n have been given by Greenwood & Danford (1949) and Greenwood et al. (1959).

5.5. Weight functions admitting Chebyshev-type formulas of high algebraic degree. Until 1966, the only known weight function admitting Chebyshev-type methods in the strict sense was the Chebyshev weight function w_0. In 1966 Ullmann obtained a further weight function which admits a (n,n) Chebyshev-type formula for every $n \in \mathbb{N}$. Important progress has subsequently been made by Geronimus (1969), Gautschi (1984), Peherstorfer (1982, 1988, 1990) and others. In the following we give a short review on recent advances in this field.

5.5.1. Gaussian quadrature formulas of Chebyshev-type. Posse (1875) has proved that w_0 is the only weight function (except for a linear transformation) which admits $(2n-1,n)$ Chebyshev-type formulas for each $n \in \mathbb{N}$ (for further proofs see Sonin 1887, Krawtchouk 1935, Bailey 1936 and Geronimus 1944, 1946). Using recursion formula (5.6), Peherstorfer (1988b) has obtained the important result that w_0 is also the only weight function admitting $(1,1)$ and $(n,n+1)$ Chebyshev-type formulas for each $n > 1$. Applying Newton's identities, an extension of this result has been obtained in [49]. The question of whether there exist other weight functions admitting $(2n_i - 1, n_i)$ Chebyshev-type formulas for an increasing sequence $(n_i)_{i=1}^{\infty}$ was settled positive by Gautschi (1984). He has shown that the weight functions

$$(5.16) \qquad w_G(x) = \begin{cases} |x|\,(x^2-\xi^2)^{-1/2}\,(1-x^2)^{-1/2} & , \quad |x| \in (\xi,1) \\ 0 & , \quad \text{elsewhere} \end{cases} , \qquad \xi \in (0,1)$$

admit $(4n-1,2n)$ Chebyshev-type formulas for each $n \in \mathbb{N}$. Let us remark that for these weight functions the situation changes drastically as one proceeds from $2n$ to $2n+1$: $d_{2n+1} \leq 2\ln(4n+1)/\ln\frac{1+\xi}{1-\xi} + 3$ (see [42]), while $\lim_{n\to\infty} n\ Var(Q^G_{2n+1}) = (I[p_0])^2$ (see [47]). An explicite description of weight functions admitting $(2n_i - 1, n_i)$ Chebyshev-type methods has recently been given by Peherstorfer (1988a, 1990).

5.5.2. Weight functions admitting strict Chebyshev formulas for all numbers of nodes. Weight functions w admitting Chebyshev formulas Q_n in the strict sense for each $n \in \mathbb{N}$ are rare in a certain statistical sense, see (Salkauskas 1975). Most of those up to now known weight functions are of type

$$(5.17) \qquad w_U(x) = w_0(x)\,\frac{(1+ab)+(a+b)x}{1+a^2+2ax},$$

where, as above, w_0 is the Chebyshev weight function. For $b=0$ and $|a| < 1/2$ this is the result of Ullman (1966). For $b=-a$ and $|a| < 3-2\sqrt{2}$ this has been proven by Byrd & Stalla (1984). For $a=0$ and $|b| < 0.27846...$ as well as for several further choices of a and b this has recently been obtained by Xu (1983). Various extensions of Ullman's result have been given by Geronimus (1969), Geronimus & Medvedeva (1975) and in particular by Peherstorfer (1982). Considering a weight function w as a product $w(x) = w_0(x)\ v(x)$, for the above results v is a positive continuous function on $[-1,1]$. A quite different behavior have the following two weight functions, for which it has been proved in [44] that they also admit Chebyshev formulas Q_n in the strict sense for each $n \in \mathbb{N}$:

$$(5.18) \qquad w_{F1}(x) = w_0(x)\ |x|^{-1/2}\ (1+|x|)^{1/2},$$

$$(5.19) \qquad w_{F2}(x) = w_0(x)\ (1-x^2)^{-1/4}\ (1+(1-x^2)^{1/2})^{1/2};$$

see [48] for further weight functions of such type admitting Chebyshev-type methods of high algebraic degree in the sense of (1.6). Peherstofer (1982, 1988b) has described a general approach to construct weight functions for Chebyshev quadrature.

6. Further results and miscellaneous extensions.

6.1. Quadrature formulas having equal weights for part of the nodes. A further extension of Chebyshev-type formulas are such quadrature formulas Q_n, which have $n-k$ equal weights and k variable weights. For fixed k and increasing n nonexistence results for $deg(R_n) \geq n$ and upper bounds for $deg(R_n)$ have been proved by Erdös & Sharma (1965) for $w \equiv 1$ and by Meir & Sharma (1967) for ultraspherical weight functions w_λ. The question of the minimal order of increasing $k = k(n)$ such that $deg(R_n) \geq n-1$, seems to be open (by addition of nodes with weight zero to the Gaussian formula, an elementary upper bound is given by $k(n) \leq [n/2]$). Several authors have investigated Chebyshev-type quadrature formulas of closed or semiclosed type. These are quadrature formulas of type $Q_n[f] = A_n f(1) + B_n f(-1) + C_n \sum_{\nu=1}^{n-1} f(x_{\nu,n})$, where for the semiclosed type $B_n = 0$ is required. Nonexistence results for $deg(R_n) \geq n$ and upper bounds for $deg(R_n)$ for $w \equiv 1$ or ultraspherical weight functions w_λ have been proved by Gatteschi (1975, 1977), Gatteschi et al. (1976), Baratella (1977, 1978) and Gatteschi & Vinardi (1978). Notaris (1992) has shown that the Chebyshev weight function w_0 is the only symmetric weight function, for which the Gauss-Kronrod formula Q_{2n+1}^{GK} is a Chebyshev-type formula of closed type with $A_{2n+1} = B_{2n+1}$ for all $n \geq 2$. He has also proved that the Gauss-Kronrod formulas Q_{2n+1}^{GK} can not be Chebyshev-type formulas for all $n \in \mathbb{N}$. For functionals I of type (1.1) but with $I[p_0] = 0$, Achieser & Krein (1935) have studied quadrature formulas Q_{2n} of type $Q_{2n}[f] = A_n\{\sum_{\nu=1}^{n} f(x_\nu) - \sum_{\nu=1}^{n} f(y_\nu)\}$ with $deg(R_{2n}) \geq 2n+1$. Similar quadrature formulas for weight functions on infinite interval have been investigated by Bernstein (1938c).

6.2. Extensions of the functional *I*. Several results described in this paper remain valid if one considers in definition (1.1) a positive measure $d\mu(x)$ on $(-1,1)$ instead of $w(x)dx$. Some of the results even hold under the only assumption that there exists a Gaussian formula Q_n^G for one or for all $n \in \mathbb{N}$ for a given linear functional I on $C[-1,1]$, while for other special results an extension to arbitrary linear functionals on $C[-1,1]$ is possible. For the sake of brevity we refer to the references cited in the above sections.

6.3. Quadrature on infinite interval. A further extension of the functional I considered here is that to infinite intervals. Results on Chebyshev-type formulas on infinite intervals have been obtained by Salzer (1955, 1957), Bernstein (1937c), Krylov (1958), Wilf (1961), Tureckiĭ (1962), Ullman (1962, 1963), Burgoyne (1963), Gatteschi (1964/65), Kahaner & Ullman (1971), Janovič (1971), Nutfullin & Janovič (1972), Anderson (1974), Gautschi (1975) and Anderson & Gautschi (1975), see Gautschi (1976) for a detailed discussion. Later published papers are those of Nutfullin (1978), where weight functions of type $w(x) = x \sin \beta x/(\alpha^2 + x^2)^p$ and $w(x) = e^{-x^2}|x|^q$ are considered, and of Soliev & Solieva (1981).

6.4. Quadrature formulas involving derivative values. Such extended quadrature formulas having equal weights for all derivative terms of same order have been considered by Ghizzetti (1954/55), Ghizzetti & Ossicini (1970, p. 43 ff), Gatteschi et al. (1976), Vinardi (1977) and Guerra & Vincenti (1985).

6.5. Quadrature formulas being exact for other function spaces. Results on Chebyshev-type formulas being exact for trigonometric polynomials, have been proved by Keda (1962) and Rosati (1968). Ionescu-Pallas (1972) has obtained Chebyshev-type formulas for $I[f] = \int_{-\infty}^{\infty} \exp(-x^2) f(x)dx$ which are exact for polynomials of a given degree in $1/(1+x^2)$. To obtain lower bounds for d_n, a general approach has been given by Seymor & Zaslowsky (1984), Wagner (1991) and Rabau & Bajnok (1991) (cf. Section 5.3 above).

6.6. Chebyshev-type formulas in higher dimensions. For approximate integration over the s-dimensional cube several results have been obtained for lattice rules, which are Chebyshev-type formulas using all the points of a integration lattice. For recent advances see Lyness (1989), Niederreiter (1992) and Sloan (1992). Other Chebyshev-type formulas for multiple integrals have been investigated by Georgiev (1953), Coman (1970), Engels (1971), Dobrodeev (1978), Engels (1980b, p. 294 ff), Kuzjutin (1980), Shi (1981) and Yanovich (1983). Chebyshev-type formulas for path integrals have been considered by Marnevskaya & Yanovich (1988).

References

1. ACHIESER, N.I. (1937): *On the theorem of S.N. Bernstein concerning the Chebyshev quadrature formula.* (Ukrainian) Ž. Inst. Mat. Akad. Nauk USSR 3, 75–82. {Zbl. 18, 208}.
2. ACHIESER, N.I.; KREIN, M. (1935): *Sur une formule de quadrature de Tchebicheff.* C.R. Acad. Sci. Paris 200, 890–892. {Zbl. 11, 109}.
3. ALLASIA, G. (1976/77): *Proprieta statistiche delle formule di quadratura.* Rend. Sem. Mat. Univ. Politecn. Torino Vol. 35, 339–348. {MR 58 # 3330}.
4. ALLASIA, G. (1982/83): *Statistical properties of numerical integration formulas.* Atti Accad. Sci. Lett. Arti Palermo V(II), 149–160. {MR 88h : 65047}.
5. ANDERSON, L.A. (1974): *Optimal Chebyshev-type quadrature formulas for various weight functions.* Ph.D. Thesis, Purdue University.
6. ANDERSON, L.A.; GAUTSCHI, W. (1975): *Optimal weighted Chebyshev-type quadrature formulas.* Calcolo 12, 211–248. {MR 54 # 11738}.
7. BAILEY, R.P. (1936): *Convergence of sequences of positive linear functional operators.* Duke Math. J. 2, 287–303. {Zbl. 14, 312}.
8. BARATELLA, P. (1977): *Valutazione dell'errore per una formula di quadratura alla Tchebycheff di tipo chiuso.* Calcolo 14, 237–242. {MR 58 # 24866}.
9. BARATELLA, P. (1978): *Čebyšev quadrature formulas of semiclosed type.* Atti Accad. Sci. Torino Cl. Sci. Fis. Mat. Natur. 112, 179–184. {MR 80e : 65033}.
10. BARNHILL, R.E.; DENNIS, J.E., Jr.; NIELSON, G.M. (1969): *A new type of Chebyshev quadrature.* Math. Comp. 23, 437–441. {MR 39 # 3698}.
11. BEREZIN, I.S.; ZHIDKOV, N.P. (1965): *Computing Methods.* Vol. I, Pergamon Press, Oxford. {MR 30 # 4372}.
12. BERNSTEIN, S.N. (1932): *On a Chebyshev approximate integration formula.* (Russian) Izv. AN USSR, OMEN 9, 1219–1227.
13. BERNSTEIN, S.N. (1936): *Sur la formule de quadrature approchée de Tchebycheff.* C. R. Acad. Sci. Paris 203, 1305–1306. [Reprinted in Bernstein (1952), pp. 198–199 (Russian)]
14. BERNSTEIN, S.N. (1937a): *Sur les formules de quadrature de Cotes et Tchebycheff.* C. R. Acad. Sci. URSS 14, 323–326. [Reprinted in Bernstein (1952), pp. 200–204 (Russian)]
15. BERNSTEIN, S.N. (1937b): *On quadrature formulas with positive coefficients.* (Russian) Izv. AN USSR, Math. Series 4, 479–503. [Reprinted in Bernstein (1952), pp. 205–227 (Russian)]
16. BERNSTEIN, S.N. (1937c): *Modifications de la formule de quadrature de Tchebycheff.* C. R. Paris 204, 1526–1529.
17. BERNSTEIN, S.N. (1938a): *Examples of quadrature formulas similar to the Chebyshev formula.* (Russian) Trudy Leningrad. industr. in-ta, 5, Phys.-math. 1, 3–7. [Reprinted in Bernstein (1952), pp. 231–235 (Russian)]
18. BERNSTEIN, S.N. (1938b): *Sur un système d'équations indéterminées.* J. Math. Pures Appl. (9) 17, 179–186. [Reprinted in Bernstein (1952), pp. 236–242 (Russian)]
19. BERNSTEIN, S.N. (1938c): *Certain applications of the parametric method to study quadrature formulas.* (Russian) Zap. n.-d. inst. mat. i. mek. 4 (15) 1, 1–19. [Reprinted in Bernstein (1952), pp. 243–261 (Russian)]
20. BERNSTEIN, S.N. (1952): *Collected works.* Vol. II, Izdat. Akad. Nauk SSR, Moscow. {MR 16, 433}.
21. BRASS, H. (1977): *Quadraturverfahren.* Vandenhoeck & Ruprecht, Göttingen. {MR 56 # 1675}.
22. BRASS, H.; HÄMMERLIN, G. (Eds.) (1988): *Numerical integration III.* ISNM 85, Birkhäuser, Basel. {MR 90e : 65003}.
23. BURGOYNE, F.D. (1963): *The non-existence of certain Laguerre-Chebyshev quadrature formulas.* Math. Comp. 17, 196–197. {MR 28 # 2634}.
24. BYRD, P.F.; STALLA, L. (1984): *Chebyshev quadrature rules for a new class of weight functions.* Math. Comp. 173–181. {MR 85i : 65030}.

25. CHAWLA, M.M.; RAMAKRISHNAN, T.R. (1976): *Minimum variance approximate formulas.* SIAM J. Numer. Anal. 13, 113–128. {MR 55 # 11574}.
26. CHEBYSHEV, P.L. (1874): *Sur les quadratures.* J. Math. Pures Appl. (2) 19, 19–34. [Reprinted in: "Oeuvres", Vol. II, Chelsea, New York, 1962, 165–180] {MR 26 # 4870}.
27. COMAN, Gh. (1970): *Nouvelles formules de quadrature à coefficients égaux.* Mathematica (Cluj) 12 (35), 253–264. {MR 48 # 12781}.
28. COSTABILE, F. (1974): *Sulle formule di quadratura di Tschebyscheff.* Calcolo 11, 191–200. {MR 56 # 9916}.
29. DAVIS, P.J.; RABINOWITZ, P. (1984): *Methods of numerical integration.* Academic Press, London. {MR 86d : 65004}.
30. DOBRODEEV, L.N. (1978): *Cubature rules with equal coefficients for the integration of functions over symmetric regions.* (Russian) Ž. Vyčisl. Mat. i Mat. Fiz. 18, 846–852. {MR 80a : 65042}.
31. DURIS, C.S. (1971): *Optimal quadrature formulas using generalized inverses. Part I: General theory and minimum variance formulas.* Math. Comp. 25, 495–504. {MR 45 # 4633}.
32. ENGELS, H. (1971): *Gleichgewichtete Kubaturformeln für ein dreieckiges ebenes Integrationsgebiet.* ZAMM 51, T 49–51.
33. ENGELS, H. (1980a): *Eine Klasse von angenähert gleichgewichteten Gauß-Typ-Quadraturformeln.* ZAMM 60, T 285–286. {MR 83g : 65030}.
34. ENGELS, H. (1980b): *Numerical quadrature and cubature.* Academic Press, London. {MR 83g : 65002}.
35. ENGELS, H.; MERSCHEN, B.A. (1984): *New minimality properties of Gaussian quadratures.* Math. Z. 187, 549–558. {MR 85j : 65053}.
36. ENGELS, H.; MERSCHEN, B.A. (1987): *Some properties of dual quadrature formulae.* J. Comp. Appl. Math. 17, 69–77. {MR 88e : 65024}.
37. ERDÖS, P.; SHARMA, A. (1965): *On Tchebycheff quadrature.* Canad. J. Math. 17, 652–658. {MR 31 # 3774}.
38. ESPELID, T.O.; GENZ, A. (Eds.) (1992): *Numerical integration – recent developments, software and applications.* NATO ASI Series C: Math. and Phys. Sciences, Vol. 357, Kluwer Acad. Pub., Dordrecht.
39. FEJÉR, L. (1933): *Mechanische Quadraturen mit positiven Cotesschen Zahlen.* Math. Zeitschr. 37, 287–309.
40. FÖRSTER, K.-J. (1982a): *A comparison theorem for linear functionals and its application in quadrature.* In: Hämmerlin (1982), 66–76. {MR 83k : 65003}.
41. FÖRSTER, K.-J. (1982b): *Bemerkungen zur optimalen Tschebyscheff-Typ Quadratur.* Numer. Math. 38, 421–425. {MR 83d : 65065}.
42. FÖRSTER, K.-J. (1986a): *On Chebyshev quadrature for a special class of weight functions.* BIT 26, 327–332. {MR 88h : 65057}.
43. FÖRSTER, K.-J. (1986b): *On Chebyshev quadrature for ultraspherical weight functions.* Calcolo 23, 355–381. {MR 89g : 65021}.
44. FÖRSTER, K.-J. (1987a): *On weight functions admitting Chebyshev quadrature.* Math. Comp. 49, 251–258. {MR 89b : 65061}.
45. FÖRSTER, K.-J. (1987b): *Schranken für die Varianz und die Gewichte von Quadraturformeln.* Habilitationsschrift, TU Braunschweig.
46. FÖRSTER, K.-J. (1988a): *On the minimal variance of quadrature formulas.* In: Brass & Hämmerlin (1988), 37–48. {MR 91e : 65040}.
47. FÖRSTER, K.-J. (1988b): *On Chebyshev-quadrature and variance of quadrature formulas.* BIT 28, 360–363. {MR 89j : 65025}.
48. FÖRSTER, K.-J. (1989): *Über gleichgewichtete Quadraturformeln.* ZAMM 69, T 77–79. {MR 90i : 65044}.
49. FÖRSTER, K.-J. (1991): *On a theorem of C. Posse concerning Gaussian quadrature of Chebyshev-type.* to appear in Math. Comp.
50. FÖRSTER, K.-J. (1992a): *On quadrature formulae near Gaussian quadrature.* In: Espelid & Genz (1992), 45–54.
51. FÖRSTER, K.-J. (1992b): *Estimates for the variance of positive quadrature formulas.* to appear in J. Comp. Appl. Math.
52. FÖRSTER, K.-J. (1992c): *Comparison theorems for the variance of quadrature formulas.* Hildesheimer Informatik-Berichte 8/92, submitted.
53. FÖRSTER, K.-J.; OSTERMEYER, G.-P. (1986): *On weighted Chebyshev-type quadrature formulas.* Math. Comp. 46, 591–599 and supp. 21–27. {MR 87h : 65042}.
54. FÖRSTER, K.-J.; PETRAS, K. (1990): *On estimates for the weights in Gaussian quadrature in the ultraspherical case.* Math. Comp. 55, 243–264. {MR 91d : 65043}.
55. FÖRSTER, K.-J.; PETRAS, K. (1992): *On the variance of Gaussian quadrature formulae in the ultraspherical case.* Hildesheimer Informatik-Berichte 19/92, submitted.
56. FRANKE, R. (1971): *Best Chebyshev quadratures.* Rocky Mountain J. Math. 1, 499–508. {MR 43 # 6641}.
57. FREUD, G. (1969): *Orthogonale Polynome.* Birkhäuser Verlag, Basel. {MR 58 # 1982}.

58. GATTESCHI, L. (1963/64): *Su di un problema connesso alle formule di quadratura di Tschebyscheff.* Univ. e Politec. Torino Rend. Sem. Mat. 23, 75–87. {MR 30 # 4386}.
59. GATTESCHI, L. (1964/65): *Sulla non esistenza di certe formule di quadratura.* Univ. e Politec. Torino Rend. Sem. Mat. 24, 157–172. {MR 32 # 4846}.
60. GATTESCHI, L. (1975): *Il problema di Tchebycheff per le formule di quadratura di tipo chiuso.* Boll. Un. Mat. Ital. 11 (3), 641–643. {MR 53 # 3565}.
61. GATTESCHI, L. (1977): *Alcuni risultati sulle formule di quadrature del tipo di Tchebycheff.* Rend. Mat (6) 10, 523–533. {MR 58 # 1904}.
62. GATTESCHI, L.; MONEGATO, G.; VINARDI, G. (1976): *Alcuni problemi relativi alle formule di quadratura del tipo di Tchebycheff.* Calcolo 13, 79–104. {MR 54 # 6463}.
63. GATTESCHI, L.; VINARDI, G. (1978): *Sul grado di precisione di formule di quadratura del tipo di Tchebycheff.* Calcolo 15, 59–85. {MR 80b : 65029}.
64. GAUTSCHI, W. (1975): *Nonexistence of Chebyshev-type quadratures on infinite intervals.* Math. Comp. 29, 93–99. {MR 51 # 4633}.
65. GAUTSCHI, W. (1976): *Advances in Chebyshev quadrature.* In: Watson, G.A. (Ed.), Numerical Analysis. Proc. Dundee Conf. Num. Anal. 1975, Springer, Lect. Notes Math. 506, Berlin, 100–121. {MR 57 # 7956}.
66. GAUTSCHI, W. (1976/77): *Qualche contributo recente sul problema di Chebyshev nella teoria dell'integrazione numerica.* Rend. Sem. Mat. Univ. e Politec. Torino 35, 39–44. {MR 58 # 3333}.
67. GAUTSCHI, W. (1981): *A survey of Gauss-Christoffel quadrature formulae.* In: Butzer, P.L.; Fehér, F. (Ed.): E.B. Christoffel, The influence of his work on mathematics and the physical sciences. Birkhäuser Verlag Basel. {MR 83g : 65031}.
68. GAUTSCHI, W. (1984): *On some orthogonal polynomials of interest in theoretical chemistry.* BIT 24, 473–483. {MR 86d : 65030}.
69. GAUTSCHI, W.; MONEGATO, G. (1977): *On optimal Chebyshev-type quadratures.* Numer. Math. 28, 59–67. {MR 56 # 7121}.
70. GAUTSCHI, W.; YANAGIWARA, H. (1974): *On Chebyshev-type quadratures.* Math. Comp. 28, 125–134. {MR 48 # 10063}.
71. GEORGIEV, G. (1953): *Formulas of mechanical quadratures with equal coefficients for multiple integrals.* (Russian) Dokl. Akad. Nauk SSSR 89, 389–392. {MR 14, 852}.
72. GERONIMUS, Ja.L. (1944): *On Gauss' and Tchebycheff's quadrature formulas.* Bull. Amer. Math. Soc. 50, 217–221. {MR 6, 63}.
73. GERONIMUS, Ja.L. (1946): *On Gauss' and Tchebycheff's quadrature formulae.* C.R. (Doklady) Acad. Sci. URSS (N.S.) 51, 655–658. {MR 10, 37}.
74. GERONIMUS, Ja.L. (1949a): *On some quadrature formulas.* (Russian) Dokl. Akad. Nauk SSSR 65, 437–440. {MR 10, 703}.
75. GERONIMUS, Ja.L. (1949b): *On the degree of precision of quadrature formulas.* (Russian) Dokl. Akad. Nauk SSSR 68. {MR 11, 236}.
76. GERONIMUS, Ja.L. (1969): *On the Čebyšev quadrature formula.* (Russian) Izv. Akad. Nauk SSSR Ser. Mat. 33, 1182–1207. [English translation in: Math. USSR-Izv. 3, 1115–1138] {MR 41 # 4092}.
77. GERONIMUS, Ja.L. (1970): *The order of the degree of precision of Čebyšev's quadrature formula.* (Russian) Dokl. Akad. Nauk SSSR 190, 263–265. [English translation in: Soviet Math. Dokl. 11, 70–72] {MR 41 # 7843}.
78. GERONIMUS, Ja.L.; MEDVEDEVA, A.K. (1975): *The validity of "Čebyšev's principle" for a two-parameter family of weight functions.* (Russian) Dokl. Akad. Nauk SSR 224. [English translation in: Soviet Math. Dokl. 16 (1975) 1231–1233] {MR 53 # 6186}.
79. GHIZZETTI, A. (1954/55): *Sulle formule di quadratura.* Rend. Sem. Mat. Fis. Milano 26, 1–16. {MR 18, 391}.
80. GHIZZETTI, A.; OSSICINI, A. (1970): *Quadrature Formulae.* Academic Press, New York. {MR 42 # 4012}.
81. GREENWOOD, R.E.; CARNAHAN, P.D.M.; NOLLEY, J.W. (1959): *Numerical integration formulas for use with weight functions x^2 and $x/\sqrt{1-x^2}$.* Math. Tables Aids Comput. 13, 37–40. {MR 21 # 968}.
82. GREENWOOD, R.E.; DANFORD, M.B. (1949): *Numerical integration with a weight function x.* J. Math. and Phys. 28, 99–106. {MR 11, 57}.
83. GUERRA, S.; VINCENTI, G. (1985): *Sul problema di Tschebyscheff per formule di quadratura alle derivate dell'integrando.* Calcolo 22, 335–349. {MR 87j : 65036}.
84. GUSTAFSON, S.-Å. (1970): *Control and estimation of computational errors in the evaluation of interpolation formulae and quadrature rules.* Math. Comp. 24, 847–854. {MR 43 # 4248}.
85. HÄMMERLIN, G. (Ed.) (1979): *Numerische Integration.* ISNM 45, Birkhäuser, Basel. {MR 81b : 65006}.
86. HÄMMERLIN, G. (Ed.) (1982): *Numerical integration.* ISNM 57, Birkhäuser, Basel. {MR 83k : 65003}.
87. IONESCU-PALLAS, N. (1972): *Čebyšev formulas for Lorentz and Gauss convolutions.* (Romanian) Stud. Cerc. Mat. 24, 1413–1420. {MR 51 # 2254}.

88. JAGERMAN, D. (1966): *Investigation of a modified mid-point quadrature formula.* Math. Comp. 20, 79–89. {MR 32 # 8499}.
89. JANOVIČ, L.A. (1971): *A quadrature formula with equal coefficients for a certain form of the integral.* (Russian) Dokl. Akad. Nauk BSSR 15, 873–876. {MR 44 # 6150}.
90. KAHANER, D.K. (1968): *Equal and almost equal quadrature formulas.* Ph.D. Thesis, Stevens Institute of Technology.
91. KAHANER, D.K. (1969): *On equal and almost equal weight quadrature formulas.* SIAM J. Numer. Anal. 6, 551–556. {MR 44 # 3492}.
92. KAHANER, D.K. (1970): *Chebyshev-type quadrature formulas.* Math. Comp. 24, 571–574. {MR 42 # 8694}.
93. KAHANER, D.K. (1971): *Some polynomials for complex quadrature.* Math. Comp. 25, 827–830. {MR 45 # 7990}.
94. KAHANER, D.K.; ULLMAN, J.L. (1971): *Equal weight quadrature on infinite intervals.* SIAM J. Numer. Anal. 8, 75–79. {MR 44 # 4902}.
95. KARLIN, S.; STUDDEN, W.J. (1966): *Tchebycheff systems: with applications in analysis and statistics.* Interscience, New York.
96. KATZ, D. (1984): *Optimal quadrature points for approximating integrals when function values are observed with error.* Math. Mag. 57, 284–290. {MR 86c : 65023}.
97. KEDA, N. P. (1962): *Chebyshev-type quadratures for periodic functions.* (Russian) Vesci Akad. Navuk BSSR Ser. Fiz.-Tèhn. Navuk 1962 1, 19–23. {MR 26 # 2010}.
98. KOREVAAR, J. (1992): *Behavior of Cotes numbers and other constants, with an application to Chebyshev-type quadrature.* Indag. Math. N. S. 3, 391–402.
99. KOZLOVSKIĬ, N. Ja. (1971): *On the question of estimation of the remainder term of the Chebyshev formula.* (Russian) Dokl. Akad. Nauk BSSR 15, 965–967. {MR 45 # 2919}.
100. KRAWTCHOUK, M. (1935): *On an algebraic question in the moment problem.* (Ukrainian) J. Inst. Math. Acad. Sci. Ukraine 2, 87–92. {Zbl. 12, 294}.
101. KRYLOV, V.I. (1957): *On the proof of impossibility of constructing quadrature formulas with equal coefficients and more than nine nodes.* (Russian) Trudy Inst. Fiz. i Mat. Akad. Nauk BSSR 2, 249–254.
102. KRYLOV, V.I. (1958): *Mechanical quadratures with equal coefficients for the integrals* $\int_0^\infty e^{-x} f(x)\, dx$ *and* $\int_{-\infty}^\infty e^{-x^2} f(x)\, dx$. (Russian) Dokl. Akad. Nauk BSSR 2, 187–192. {MR 22 # 861}.
103. KRYLOV, V.I. (1962): *Approximate Calculation of Integrals.* Transl. from Russian by A.H. Stroud, MacMillan, New York – London. [2nd edition (Russian), Izdat, "Nauka", Moscow, 1967] {MR 26 # 2008, MR 36 # 1104}.
104. KUIJLAARS, A. (1992a): *The minimal number of nodes in Chebyshev type quadrature formulas.* Univ. of Amsterdam, math. preprint ser. 92–14, to appear in Indag. Mat.
105. KUIJLAARS, A. (1992b): *Chebyshev-type quadrature for Jacobi weight functions.* Preprint, presented on 4th Intern. Symp. on Orthogonal Polynomials and their Applications, Evian, France, 1992.
106. KUZJUTIN, V.F. (1980): *Error of a cubature formula with equal coefficients and an equidistant grid of nodes on some classes of functions.* (Russian) Izv. Akad. Nauk Kazah. SSR Ser. Fiz. Mat. 73–75, 90. {MR 81m : 65038}.
107. KUZMIN, R.O. (1935): *Sur la méthode de Tschebicheff pour l'évaluation approchée des intégrales.* C.R. Acad. Sci. Paris 201, 1094–1995. {Zbl. 12, 398}.
108. KUZMIN, R.O. (1936): *Sur la méthode de quadrature de Tschebycheff.* C.R. Acad. Sci. Paris 202, 272–273. {Zbl. 13, 107}.
109. KUZMIN, R.O. (1938): *On the distribution of roots of polynomials connected with quadratures of Chebyshev.* (Russian) Izv. Akad. Nauk SSSR Ser. Mat. 2, 427–444. {Zbl. 19, 405}.
110. LOCHER, F. (1980): *Dividierte Differenzen und Monotonie von Quadraturformeln.* Numer. Math. 34, 99–109. {MR 81h : 65016}.
111. LYNESS, J.N. (1989): *An introduction to lattice rules and their generator matrices.* IMA J. Numer. Anal. 9, 405–419. {MR 91b : 65029}.
112. MARNEVSKAYA, L.A.; YANOVICH, L.A. (1988): *Construction of Chebyshev quadrature formulas for path integrals of a certain type.* (Russian) Dokl. Akad. Nauk BSSR 32, 591–594. {MR 89g : 65022}.
113. MAYOT, M. (1950): *Sur la méthode d'intégration approchée de Tchebychef.* C.R. Acad. Sci. Paris 230, 429–430. {MR 11, 464}.
114. MEIR, A.; SHARMA, A. (1967): *A variation of the Tchebycheff quadrature problem.* Illinois J. Math. 11, 535–546. {MR 35 # 7058}.
115. MENCKE, B. (1990): *Quadraturformeln mit minimaler Varianz: Existenz und numerische Berechnung.* Diplomarbeit, Institut für Angewandte Mathematik, TU Braunschweig.
116. MERSCHEN, B.A. (1983): *Duale Quadraturformeln und ihre Eigenschaften.* Ph.D. Thesis, RWTH Aachen.
117. NATANSON, I. P. (1965): *Constructive Function Theory.* Vol III, Interpolation and Approximation Quadratures. Frederick Ungar Publ. Co., New York. {MR 33 # 4529c}.

118. NEVAI, P. (1979): *Orthogonal polynomials*. Memoirs AMS 18, Rhode Island.
119. NEVAI, P. (1986): *Géza Freud, orthogonal polynomials and Christoffel functions, a case study*. J. Approx. Th. 48, 3–167. {MR 88b : 42032}.
120. NIEDERREITER, H. (1992): *Random number generation and quasi-Monte Carlo methods*. SIAM, Philadelphia.
121. NOTARIS, S.E. (1992): *On Gauss-Kronrod quadrature formulae of Chebyshev-type*. Math. Comp. 58, 745–753.
122. NUTFULLIN, Š.N. (1978): *Equally weighted quadratures with partially coinciding nodes*. (Russian) Voprosy Vycisl. i Prikl. Mat. (Tashkent), 97–103. {MR 80c : 65060}.
123. NUTFULLIN, Š.N.; JANOVIČ, L.A. (1972): *Čebyšev quadrature formulae with certain weight functions that depend on parameters*. (Russian) Vesci Akad. Navuk BSSR Ser. Fiz.-Mat. Navuk 1972, 24–30. {MR 48 # 9203}.
124. OSSICINI, A. (1966): *Sulle formule di quadratura di Tschebyscheff*. Pubbl. Ist. Naz. Appl. Calcolo, No. 660, quad. 7, 43–59.
125. OSTROWSKI, A.M. (1959): *On trends and problems in numerical approximation*. In: On Numerical Approximation (R.E. Langer, ed.), pp. 3–10. The Univ. of Wisconsin Press, Madison. {MR 20 # 7381}.
126. PASZKOWSKI, S. (1975): *Zastosowania numeryczne wielomianów i szeregów Czebyszewa*. (Polish) Państwowe Wydawnictwo Naukowe, Warszawa. {MR 56 # 13534}.
127. PECKA, F. (1974): *On equally weighted quadratures*. Numer. Math. 23, 19–22. {MR 50 # 8926}.
128. PEHERSTORFER, F. (1982): *Weight functions which admit Tchebycheff quadrature*. Bull. Austral. Math. Soc. 26, 29–38. {MR 84k : 65025}.
129. PEHERSTORFER, F. (1988a): *On Gauss quadrature formulas with equal weights*. Numer. Math. 52, 317–327. {MR 89c : 65025}.
130. PEHERSTORFER, F. (1988b): *On Chebyshev quadrature formulas*. In [Brass, Hämmerlin], 172–186. {MR 91c : 65021}.
131. PEHERSTORFER, F. (1990): *Gauss-Tchebycheff quadrature formulas*. Numer. Math. 58, 273–286. {MR 91j : 65052}.
132. PERRON, O. (1932): *Algebra*. Walter de Gruyter & Co, Berlin.
133. PIESSENS, R.; DE DONCKER-KAPENGA, E.; ÜBERHUBER, C.W.; KAHANER, D.K. (1983): *QUADPACK – a subroutine package for automatic integration*. Springer Ser. Comp. Math. 1, Springer, Berlin. {MR 85b : 65022}.
134. PONOMARENKO, A.K. (1983): *Estimates of the error functional for quadrature formulas with Chebyshev weight*. (Russian) Metody Vychislenij 13, 116–121. {MR 85a : 65046}.
135. POPOVICIU, T. (1964): *La simplicité du reste dans certaines formules de quadrature*. Mathematica (Cluj) 6 (29), 157–184. {MR 32 # 4848}.
136. POSSE, C. (1875): *Sur les quadratures*. Nouv. Ann. de Math. (2) 14, 49–62.
137. RABAU, R.; BAJNOK, B. (1991): *Bounds for the number of nodes in Chebyshev-type quadrature formulas*. J. Appr. Th. 67, 199–214.
138. RABINOWITZ, P.; RICHTER, N. (1970): *Chebyshev-type integration rules of minimum norm*. Math. Comp. 24, 831–845. {MR 45 # 7996}.
139. RADAU, R. (1880a): *Sur les formules de quadrature à coefficients égaux*. C.R. Acad. Sci. Paris 90, 520–523.
140. RADAU, R. (1880b): *Étude sur les formules d'approximation qui servent à calculer la valeur numérique d'une intégrale définie*. J. Math. Pures Appl. (3) 6, 283–336.
141. RADAU, R. (1880c): *Étude sur les formules de quadrature à coefficients égaux*. C.R. Acad. Sci. Paris 90, 520–523.
142. RIVLIN, T.J. (1990): *Chebyshev polynomials*. John Wiley & Sons, Inc., New York. {MR 92a : 6506}.
143. ROSATI, F. (1968): *Problemi di Gauss e Tchebychef relativi a formule di quadratura esatte per polinomi trigonometrici*. Matematiche (Catania) 23, 31–49. {MR 41 # 1223}.
144. SALKAUSKAS, K. (1971): *Existence of quadrature formulae with almost equal weights*. Math. Comp. 25, 105–109. {MR 44 # 7750}.
145. SALKAUSKAS, K. (1973a): *Almost-interpolatory Chebyshev quadrature*. Math. Comp. 27, 645–654. {MR 49 # 5658}.
146. SALKAUSKAS, K. (1973b): *Almost-interpolatory Čebyšev quadrature on* $[-1, 1]$. Proc. 2nd Manitoba Conf. on Num. Math. Congr. Numeration VII, Utilitas Math., Winnipeg, Man., 333–334. {MR 50 # 6122}.
147. SALKAUSKAS, K. (1975): *On weight functions for Chebyshev quadrature*. Numer. Math. 24, 13–18. {MR 51 # 14538}.
148. SALKAUSKAS, K. (1976): *Almost-interpolatory Chebyshev quadrature II*. Proc. 6th Manitoba Conf. on Num. Anal. {MR 80h : 65016}.
149. SALZER, H.E. (1947): *Tables for facilitating the use of Chebyshev's quadrature formula*. J. Math. and Phys. 26, 191–194. {MR 9, 251}.
150. SALZER, H.E. (1955): *Equally weighted quadrature formulas over semi-infinite and infinite intervals*. J. Math. and Phys. 34, 54–63. {MR 16, 1055}.

151. SALZER, H.E. (1957): *Equally-weighted quadrature formulas for inversion integrals.* Math. Tables Aids Comput. 11, 197–200. {MR 19, 771}.

152. SARD, A. (1949): *Smoothest approximation formulas.* Ann. Math. Statistics 20, 612–615. {MR 12, 84}.

153. SEATZU, S. (1981): *Best quadrature formulas in Sard's sense with almost all the coefficients equal.* (Italian) Rend. Mat. (7) 1, 159–176. {MR 82k : 65022}.

154. SEYMOUR, P.D.; ZASLAVSKY, T. (1984): *Averaging sets: a generalisation of mean values and spherical designs.* Adv. in Math. 52, 213–240. {MR 85m : 05031}.

155. SHI, S.Z. (1981): *Estimate of error for quadrature of a multidimensional continuous function.* (Chinese) Math. Numer. Sinica 3, 360–364. {MR 83d : 65069}.

156. SHOHAT, J. (1937): *On mechanical quadratures, in particular with positive coefficients.* Trans. Amer. Math. Soc. 42, 461–496.

157. SLOAN, I.H. (1992): *Numerical integration in high dimensions – the lattice rule approach.* In: Espelid & Genz (1992), 55–69.

158. SLOAN, I.H.; SMITH, W.E. (1978): *Product-integration with the Clenshaw-Curtis and related points.* Numer. Math. 30, 415–428. {MR 58 # 13646}.

159. SOLIEV, Ju.; SOLIEVA, M.A. (1981): *Quadrature formulas of Chebyshev-type.* (Russian) Izv. Akad. Nauk Tadžik. SSR Otdel. Fiz. – Mat. i. Geolog.-Him. Nauk 2 (80), 74–76. {MR 82m : 65023}.

160. SONIN, N.JA. (1887): *On the approximate evaluation of definite integrals and on the related integral functions.* (Russian) Warshawskia Univ. Izv. 1, 1–76.

161. STETTER, F. (1968): *On a generalization of the midpoint rule.* Math. Comp. 22, 661–663. {MR 37 # 2449}.

162. SZEGÖ, G. (1939): *Orthogonal polynomials.* AMS Providence, Rhode Island, 4th edition 1975. {MR 46 # 9631}.

163. TURECKIĬ, A.H. (1962): *On the existence of Chebyshev quadrature formulas for an infinite interval.* (Russian) Vesci Akad. Nauk BSSR Ser. Fiz.-Tehn. Navuk 1962, No. 2, 119–120. {Zbl. 178, 516}.

164. ULLMAN, J.L. (1960): *Studies in Faber polynomials.* Trans. Amer. Math. Soc. 94, 515–528. {MR 22 # 3800}.

165. ULLMAN, J.L. (1962): *Tchebycheff quadrature is possible on the infinite interval.* Bull. Amer. Math. Soc. 68, 574–576. {MR 26 # 526}.

166. ULLMAN, J.L. (1963): *Tchebycheff quadrature on the infinite interval.* Trans. Amer. Math. Soc. 107, 291–299. {MR 26 # 5330}.

167. ULLMAN, J.L. (1966a): *A class of weight functions for which Tchebycheff quadrature is possible.* Bull. Amer. Math. Soc. 72, 1073–1075. {MR 34 # 4766}.

168. ULLMAN, J.L. (1966b): *A class of weight functions that admit Tchebycheff quadrature.* Michigan Math. J. 13, 417–423. {MR 34 # 5290}.

169. VINARDI, G. (1977): *Sul calcolo numerico di nodi e pesi per una formula di quadratura alla Tchebycheff di tipo chiuso.* Atti Accad. Sci. Torino Cl. Sci. Fis. Mat. Natur. 111, 253–259. {MR 57 # 18040}.

170. WAGNER, G. (1991): *On averaging sets.* Mh. Math. 111, 69–78. {MR 92e : 65031}.

171. WATSON, G.N. (1962): *A treatise on the theory of Bessel functions.* Univ. Press, Cambridge.

172. WILF, H.S. (1961): *The possibility of Tschebycheff quadrature on infinite intervals.* Proc. Nat. Acad. Sci. USA 47, 209–213. {MR 23 # A2683}.

173. WILF, H.S. (1967): *Advances in numerical quadrature.* In: Mathematical Methods for Digital Computers, Vol. II (A. Ralston and H.S. Wilf, eds.), pp. 133–144. Wiley, New York – London – Sydney. {MR 35 # 2516}.

174. WILSON, M.W. (1970a): *Discrete least squares and quadrature formulas.* Math. Comp. 24, 271–282. {MR 43 # 1430}.

175. WILSON, M.W. (1970b): *Neccessary and sufficient conditions for equidistant quadrature formulas.* SIAM J. Numer. Anal. 7, 134–141. {MR 43 # 8241}.

176. XU, Y. (1989): *Weight functions for Chebyshev quadrature.* Math. Comp. 53, 297–302. {MR 90b : 65038}.

177. YANAGIWARA, H. (1981): *A Chebyshev-type quadrature formula.* Bull. Fukuoka Univ. Ed. III 31, 45–48. {MR 83d : 65071}.

178. YANAGIWARA, H.; FUKUTAKE, T.; SHIBATA, K. (1974): *Chebyshev-type quadrature formulas for fourteen – seventeen nodes.* Bull. Fukuoka Univ. Ed. III 24, 41–48. {MR 51 # 2260}.

179. YANAGIWARA, H.; SHIBATA, K. (1973): *A Chebyshev quadrature formula for twelve nodes.* Bull. Fukuoka Univ. Ed. III 23, 45–51. {MR 49 # 10130}.

180. YANOVICH, L.A. (1983): *Chebyshev's quadrature formulas for a class of Wiener double integrals.* (Russian) Dokl. Akad. Nauk BSSR 27, 4, 293–296. {MR 84c : 65053}.

Institut für Mathematik, Universität Hildesheim,
Marienburger Platz 22, D-31141 Hildesheim, Germany

Gauss-type Quadrature Rules for Rational Functions*

Walter Gautschi

Abstract. When integrating functions that have poles outside the interval of integration, but are regular otherwise, it is suggested that the quadrature rule in question ought to integrate exactly not only polynomials (if any), but also suitable rational functions. The latter are to be chosen so as to match the most important poles of the integrand. We describe two methods for generating such quadrature rules numerically and report on computational experience with them.

Introduction

Traditionally, Gauss quadrature rules are designed to integrate exactly polynomials of maximum possible degree. This is meaningful for integrand functions that are "polynomial-like". For integrands having poles (outside the interval of integration) it would be more natural to include also rational functions among the functions to be exactly integrated. In this paper we consider n-point quadrature rules that exactly integrate m rational functions (with prescribed location and multiplicity of the poles) as well as polynomials of degree $2n-m-1$, where $0 \leq m \leq 2n$. The limit case $m = 2n$, in which only rational functions are being integrated exactly, is a rational counterpart of the classical Gauss formula; the latter corresponds to the other limit case $m = 0$.

In §1 we characterize these new quadrature rules in terms of classical (polynomial) Gauss formulae with modified weight functions. We also identify special choices of poles that are of interest in applications. The computation of the quadrature rules is discussed in §2, and numerical examples are given in §3.

*Work supported in part by the National Science Foundation under grant DMS–9023403.

1. Gauss quadrature for rational functions

Let $d\lambda$ be a measure on the real line having finite moments of all orders. Let $\zeta_\mu \in \mathbb{C}$, $\mu = 1, 2, \ldots, M$, be distinct real or complex numbers such that

$$\zeta_\mu \neq 0 \text{ and } 1 + \zeta_\mu t \neq 0 \text{ for } t \in \overline{\text{supp}(d\lambda)}, \quad \mu = 1, 2, \ldots, M . \tag{1.1}$$

For given integers m, n with $1 \leq m \leq 2n$, we wish to find an n-point quadrature rule that integrates exactly (against the measure $d\lambda$) polynomials of degree $2n - m - 1$ as well as the m rational functions

$$(1 + \zeta_\mu t)^{-s}, \quad \mu = 1, 2, \ldots, M, \quad s = 1, 2, \ldots, s_\mu, \tag{1.2}$$

where $s_\mu \geq 1$ and

$$\sum_{\mu=1}^{M} s_\mu = m. \tag{1.3}$$

In the extreme case $m = 2n$ (where polynomials of degree –1 are understood to be identically zero) the formula integrates exactly $2n$ rational functions (with poles of multiplicities s_μ at $-1/\zeta_\mu$), but no nontrivial polynomials. The formula, therefore, can be thought of as the rational analogue of the classical Gauss formula; the latter corresponds to the other limit case $m = M = 0$.

The solution of our problem is given by the following theorem.

THEOREM 1.1. *Define*

$$\omega_m(t) = \prod_{\mu=1}^{M} (1 + \zeta_\mu t)^{s_\mu}, \tag{1.4}$$

a polynomial of degree m. Assume that the measure $d\lambda/\omega_m$ admits a (polynomial) *n-point Gaussian quadrature formula*

$$\int_{\mathbb{R}} f(t) \, \frac{d\lambda(t)}{\omega_m(t)} = \sum_{\nu=1}^{n} w_\nu^G f(t_\nu^G) + R_n^G(f), \quad R_n^G(\mathbb{P}_{2n-1}) = 0, \tag{1.5}$$

with nodes t_ν^G contained in the support of $d\lambda$,

$$t_\nu^G \in \text{supp}(d\lambda). \tag{1.6}$$

Define

$$t_\nu = t_\nu^G, \quad \lambda_\nu = w_\nu^G \omega_m(t_\nu^G), \quad \nu = 1, 2, \ldots, n. \tag{1.7}$$

Then

$$\int_{\mathbb{R}} g(t) d\lambda(t) = \sum_{\nu=1}^{n} \lambda_\nu g(t_\nu) + R_n(g), \tag{1.8}$$

where

$$R_n(g) = 0 \quad if \quad \begin{cases} g(t) = (1+\zeta_\mu t)^{-s}, & \mu = 1,2,\ldots,M;\ s = 1,2,\ldots,s_\mu, \\ g \in \mathbb{P}_{2n-m-1}\,. \end{cases} \tag{1.9}$$

Conversely, (1.8) *with* $t_\nu \in \mathrm{supp}(d\lambda)$ *and* (1.9) *imply* (1.5), (1.6) *with* t_ν^G, w_ν^G *as defined in* (1.7).

Remark. Theorem 1.1, for real ζ_μ and either all $s_\mu = 1$ and $m = 2n$, or all but one $s_\mu = 2$ and $m = 2n-1$, is due to Van Assche and Vanherwegen [13]. The quadrature rule (1.5), especially its convergence properties for analytic functions f, has previously been studied by López and Illán [9, 10].

Proof of Theorem 1.1. Assume first (1.5), (1.6). For $\mu = 1,2,\ldots,M; s = 1,2,\ldots,s_\mu$, define

$$q_{\mu,s}(t) = \frac{\omega_m(t)}{(1+\zeta_\mu t)^s}\,. \tag{1.10}$$

Since $m \le 2n$ and $s \ge 1$, we have $q_{\mu,s} \in \mathbb{P}_{m-s} \subset \mathbb{P}_{2n-1}$, and therefore, by (1.5),

$$\begin{aligned} \int_{\mathbb{R}} \frac{d\lambda(t)}{(1+\zeta_\mu t)^s} &= \int_{\mathbb{R}} q_{\mu,s}(t)\frac{d\lambda(t)}{\omega_m(t)} = \sum_{\nu=1}^{n} w_\nu^G q_{\mu,s}(t_\nu^G) \\ &= \sum_{\nu=1}^{n} w_\nu^G \frac{\omega_m(t_\nu^G)}{(1+\zeta_\mu t_\nu^G)^s} = \sum_{\nu=1}^{n} \frac{\lambda_\nu}{(1+\zeta_\mu t_\nu)^s}\,, \end{aligned}$$

where (1.7) has been used in the last step and none of the denominators on the far right vanishes by (1.6) and (1.1). This proves the assertion in the top line of (1.9). The bottom part of (1.9) follows similarly: Let p be an arbitrary polynomial in $\mathbb{P}_{2n-m-1}$. Then, since $p\omega_m \in \mathbb{P}_{2n-1}$, again by (1.5) and (1.7),

$$\begin{aligned} \int_{\mathbb{R}} p(t)d\lambda(t) &= \int_{\mathbb{R}} p(t)\omega_m(t)\frac{d\lambda(t)}{\omega_m(t)} \\ &= \sum_{\nu=1}^{n} w_\nu^G p(t_\nu^G)\omega_m(t_\nu^G) = \sum_{\nu=1}^{n} \lambda_\nu p(t_\nu). \end{aligned}$$

To prove the converse, we first note that w_ν^G is well defined by (1.7), since $\omega_m(t_\nu) \neq 0$ by the assumption on t_ν and (1.1). One then easily verifies that (1.5) holds for all polynomials (1.10) (of degree $< m$) and all polynomials of the form $p\omega_m$ where $p \in \mathbb{P}_{2n-1-m}$. The collection of these polynomials, however, spans $\mathbb{P}_{2n-1}$. □

We will concentrate on six special choices of the parameters ζ_μ that are of interest in applications.

Case 1 (Simple real poles). All $s_\mu = 1$ in (1.2) (hence $M = m$), and all ζ_μ are real, distinct, and nonzero,

$$\zeta_\nu = \xi_\nu \in \mathbb{R}, \qquad \xi_\nu \neq 0, \qquad \nu = 1, 2, \dots, m. \tag{1.11a}$$

In this case the polynomial ω_m has the form

$$\omega_m(t) = \prod_{\nu=1}^{m} (1 + \xi_\nu t), \qquad \xi_\nu \in \mathbb{R}. \tag{1.11b}$$

If the support of $d\lambda$ is an interval, ω_m does not change sign on it because of (1.1).

Case 2 (Simple conjugate complex poles). All $s_\mu = 1$ (hence $M = m$), m even, and the ζ_μ occur in $m/2$ (distinct) pairs of conjugate complex numbers (cf. [9]),

$$\zeta_\nu = \xi_\nu + i\eta_\nu, \qquad \zeta_{\nu+m/2} = \xi_\nu - i\eta_\nu, \qquad \nu = 1, 2, \dots, m/2, \tag{1.12a}$$

where $\xi_\nu \in \mathbb{R}$ and $\eta_\nu > 0$. Here,

$$\omega_m(t) = \prod_{\nu=1}^{m/2} [(1 + \xi_\nu t)^2 + \eta_\nu^2 t^2], \tag{1.12b}$$

which is strictly positive for real t.

Case 2′ (Simple conjugate complex poles plus a simple real pole). All $s_\mu = 1$ (hence $M = m$), m (odd) ≥ 3, and, slightly changing the indexing of the ζ's,

$$\zeta_0 \in \mathbb{R}, \quad \zeta_\nu = \xi_\nu + i\eta_\nu, \quad \zeta_{\nu+(m-1)/2} = \xi_\nu - i\eta_\nu, \quad \nu = 1, 2, \dots, (m-1)/2, \tag{1.13a}$$

where $\zeta_0 = \xi_0 \neq 0$ and $\xi_\nu \in \mathbb{R}$, $\eta_\nu > 0$ for $1 \leq \nu \leq (m-1)/2$. Then

$$\omega_m(t) = (1 + \xi_0 t) \prod_{\nu=1}^{(m-1)/2} [(1 + \xi_\nu t)^2 + \eta_\nu^2 t^2]. \tag{1.13b}$$

Case 3 (Real poles of order 2). All $s_\mu = 2$ in (1.2) (hence $2M = m$), and all ζ_μ are nonzero, real and distinct,

$$\zeta_\nu = \xi_\nu \in \mathbb{R}, \qquad \xi_\nu \neq 0, \qquad s_\nu = 2, \qquad \nu = 1, 2, \dots, m/2. \tag{1.14a}$$

The polynomial ω_m now has the form

$$\omega_m(t) = \prod_{\nu=1}^{m/2} (1 + \xi_\nu t)^2 \tag{1.14b}$$

and is nonnegative for real t, and positive on the support of $d\lambda$.

Case 3′ (Real poles of order 2 plus a simple real pole). Here, all $\zeta_\mu = \xi_\mu$ are nonzero, real and distinct, $s_\nu = 2$ for $\nu = 1, 2, \ldots, M-1$ and $s_M = 1$. Thus, $m = 2M - 1$, and

$$\omega_m(t) = (1 + \xi_M t) \prod_{\nu=1}^{M-1} (1 + \xi_\nu t)^2, \quad M = \frac{m+1}{2}, \quad m(\text{odd}) \geq 3. \tag{1.15}$$

If $M = n$, i.e., $m = 2n - 1$, the quadrature rule (1.5) is then identical with the "orthogonal quadrature rule" of [13], having as nodes the zeros of the rational function $(1 + \zeta_n t)^{-1} + \sum_{\nu=1}^{n-1} a_\nu (1 + \zeta_\nu t)^{-1}$ which is orthogonal (relative to the measure $d\lambda$) to 1 and to $(1 + \zeta_\mu t)^{-1}$, $\mu = 1, 2, \ldots, n-1$. As in Case 1, the polynomial ω_m preserves its sign on the interval on which $d\lambda$ is supported.

Case 4 (Conjugate complex poles of order 2). All $s_\mu = 2$ (hence $2M = m$), $m = 0$ (mod 4), and the ζ_μ, $\mu = 1, 2, \ldots, m/2$, occur in $m/4$ conjugate complex pairs, similarly as in Case 2. Thus,

$$\omega_m(t) = \prod_{\nu=1}^{m/4} [(1 + \xi_\nu t)^2 + \eta_\nu^2 t^2]^2. \tag{1.16}$$

In all six cases, the measure $d\lambda/\omega_m$ admits a Gaussian n-point formula for each $n = 1, 2, 3, \ldots$, so that the assumption of Theorem 1.1 is fulfilled for each n.

Putting $f(t) = \omega_m(t) g(t)$ in (1.5) and using (1.7), we get

$$\int_{\mathbb{R}} g(t) d\lambda(t) = \sum_{\nu=1}^{n} \lambda_\nu g(t_\nu) + R_n^G(\omega_m g), \tag{1.17}$$

where from the well-known expression for the remainder term of Gaussian quadrature rules, one has

$$R_n^G(\omega_m g) = \gamma_n (\omega_m g)^{(2n)}(\tau), \qquad \gamma_n = \frac{\hat{\beta}_0 \hat{\beta}_1 \cdots \hat{\beta}_n}{(2n)!}. \tag{1.18}$$

Here, τ is some number in the smallest interval containing the support of $d\lambda$, and $\hat{\beta}_k = \beta_k(d\lambda/\omega_m)$ are the β-recursion coefficients for the measure $d\lambda/\omega_m$ (cf. (2.1) below). The latter are computed as part of the algorithms to be described in the next section.

2. Computation of the quadrature rule (1.5)

We propose essentially two methods for generating the basic quadrature rule (1.5), the first being most appropriate if the support of $d\lambda$ is a finite interval, the other more effective, though possibly slower, when the support interval of $d\lambda$ is unbounded.

2.1. Method based on partial fraction decomposition and modification algorithms. To compute the n-point formula (1.5), it suffices to compute the nth-degree orthogonal polynomial $\hat{\pi}_n(\,\cdot\,) = \pi_n(\,\cdot\,; d\hat{\lambda})$ relative to the measure $d\hat{\lambda} = d\lambda/\omega_m$, or, more precisely, the recursion coefficients $\hat{\alpha}_k = \alpha_k(d\hat{\lambda})$, $\hat{\beta}_k = \beta_k(d\hat{\lambda})$, $k = 0, 1, \dots, n-1$, in the three-term recurrence relation satisfied by these (monic) polynomials:

$$\begin{gathered} \hat{\pi}_{k+1}(t) = (t - \hat{\alpha}_k)\hat{\pi}_k(t) - \hat{\beta}_k\hat{\pi}_{k-1}(t), \\ k = 0, 1, \dots, n-1, \\ \hat{\pi}_0(t) = 1, \quad \hat{\pi}_{-1}(t) = 0. \end{gathered} \tag{2.1}$$

The nodes t_ν^G and weights w_ν^G in (1.5) can then be obtained by standard techniques via an eigensystem problem for the (symmetric, tridiagonal) Jacobi matrix of order n having the $\hat{\alpha}_k$, $k = 0, 1, \dots, n-1$, on the diagonal, and $\sqrt{\hat{\beta}_k}$, $k = 1, 2, \dots, n-1$, on the side diagonals (see, e.g., [7], [4, §6]). The coefficients $\hat{\alpha}_k$, $\hat{\beta}_k$ in turn are expressible in terms of the orthogonal polynomials $\hat{\pi}_k$ as

$$\begin{aligned} \hat{\alpha}_k &= \frac{(t\hat{\pi}_k, \hat{\pi}_k)}{(\hat{\pi}_k, \hat{\pi}_k)}, \quad 0 \le k \le n-1, \\ \hat{\beta}_0 = (\hat{\pi}_0, \hat{\pi}_0), \; \hat{\beta}_k &= \frac{(\hat{\pi}_k, \hat{\pi}_k)}{(\hat{\pi}_{k-1}, \hat{\pi}_{k-1})}, \quad 1 \le k \le n-1, \end{aligned} \tag{2.2}$$

where $(\,\cdot\,, \,\cdot\,)$ denotes the inner product

$$(u, \nu) = \int_{\mathbb{R}} u(t)\nu(t)\,d\hat{\lambda}(t). \tag{2.3}$$

(If the error constant γ_n in (1.18) is desired, one needs to compute, in addition, $\hat{\beta}_n$.)

The basic idea of computing the coefficients in (2.2) is as follows. Suppose we can construct an N-point quadrature rule for $d\hat{\lambda} = d\lambda/\omega_m$, where $N > n$, which is exact for polynomials of degree $\le 2n-1$:

$$\int_{\mathbb{R}} p(t)\,d\hat{\lambda}(t) = \sum_{k=1}^{N} W_k p(T_k), \qquad p \in \mathbb{P}_{2n-1}. \tag{2.4}$$

Here the weights W_k are not necessarily all positive. Denote the discrete measure implied by the sum on the right by $d\Lambda_N$:

$$\int_{\mathbb{R}} p(t)\,d\Lambda_N(t) = \sum_{k=1}^{N} W_k p(T_k). \tag{2.5}$$

Then from the formulae in (2.2) one easily sees by induction that

$$\begin{aligned} \alpha_k(d\hat{\lambda}) &= \alpha_k(d\Lambda_N), \\ \beta_k(d\hat{\lambda}) &= \beta_k(d\Lambda_N), \end{aligned} \qquad k = 0, 1, \ldots, n-1. \tag{2.6}$$

Thus, the desired recursion coefficients are the first n of the α- and β-coefficients belonging to the discrete measure $d\Lambda_N$. These can be generated by Stieltjes's procedure, which is implemented in the routine `sti` of [4]. (The faster routine `lancz` of [4], implementing the Lanczos method, would also be applicable here, even though $d\Lambda_N$ is not necessarily a positive measure.)

We next show how a quadrature rule of type (2.4), with $N = O(mn)$, can be constructed by means of partial fraction decomposition and suitable modification algorithms. For this, we consider separately Cases 1–3′ identified in §1. The analysis of Case 4 becomes so tedious that we will not pursue it any further in this context; see, however, §2.2.

2.1.1. *Simple real poles.* We set up the partial fraction decomposition of $1/\omega_m$ in the form

$$\frac{1}{\omega_m(t)} = \frac{1}{\prod_{\nu=1}^{m}(1+\xi_\nu t)} = \sum_{\nu=1}^{m} \frac{c_\nu}{t + (1/\xi_\nu)}, \tag{2.7}$$

where

$$c_\nu = \frac{\xi_\nu^{m-2}}{\prod_{\substack{\mu=1 \\ \mu\neq\nu}}^{m} (\xi_\nu - \xi_\mu)}, \quad \nu = 1, 2, \ldots, m, \tag{2.8}$$

and an empty product in (2.8) (when $m = 1$) is to be taken as 1. Then, with $d\hat{\lambda} = d\lambda/\omega_m$,

$$\int_{\mathbb{R}} p(t)d\hat{\lambda}(t) = \sum_{\nu=1}^{m} \int_{\mathbb{R}} p(t)\, \frac{c_\nu d\lambda(t)}{t + (1/\xi_\nu)}\,.$$

The integrals on the right involve measures $c_\nu d\lambda$ modified by linear divisors. For such measures, the associated recursion coefficients can be obtained from those of $c_\nu d\lambda$ (assumed known) by a suitable modification algorithm (cf. [4, §5]). Unless $x_\nu = -1/\xi_\nu$ is very close to the support interval of $d\lambda$, the most appropriate algorithm is the one embodied in the routine `gchri` of [4] with `iopt` = 1. Otherwise, the routine `chri` of [4] (again with `iopt` = 1) is preferable. A basic ingredient of the routine `gchri` is the modified Chebyshev algorithm (cf. [2, §2.4]) using modified moments $\int_{\mathbb{R}} \pi_k(t; d\lambda) c_\nu d\lambda(t)/(t - x_\nu)$, $k = 0, 1, 2, \ldots, 2n-1$. These in turn are generated by backward recurrence as minimal solution of the three-term recurrence relation for the measure $d\lambda$; cf. [1, §5].

Having obtained, in whichever way, the first n of the α- and β-coefficients for the modified measure $c_\nu d\lambda(t)/(t-x_\nu)$, and hence the Gaussian quadrature formula[1]

$$\int_{\mathbb{R}} p(t)\,\frac{c_\nu d\lambda(t)}{t+(1/\xi_\nu)} = \sum_{r=1}^{n} w_r^{(\nu)} p(t_r^{(\nu)}), \qquad p \in \mathbb{P}_{2n-1}, \tag{2.9}$$

via eigensystem techniques, we then get

$$\begin{aligned}\int_{\mathbb{R}} p(t)\,\frac{d\lambda(t)}{\omega_m(t)} &= \sum_{\nu=1}^{m} \int_{\mathbb{R}} p(t)\,\frac{c_\nu d\lambda(t)}{t+(1/\xi_\nu)} \\ &= \sum_{\nu=1}^{m}\sum_{r=1}^{n} w_r^{(\nu)} p(t_r^{(\nu)}), \qquad p \in \mathbb{P}_{2n-1},\end{aligned}$$

hence the desired quadrature rule (2.4), with $N = mn$ and

$$\begin{aligned} T_{(\nu-1)n+r} &= t_r^{(\nu)}, \\ W_{(\nu-1)n+r} &= w_r^{(\nu)}, \end{aligned} \qquad \nu = 1,2,\ldots,m; \quad r = 1,2,\ldots,n. \tag{2.10}$$

The procedure described works best if the support of $d\lambda$ is a finite interval. Otherwise, the modified Chebyshev algorithm underlying the procedure is likely to suffer from ill-conditioning; cf. Example 3.4. Another difficulty that may adversely affect the accuracy of the results, in particular if $m = 2n$, is the possibility that the constants $c_\nu\ \mathrm{sgn}_{t\in \mathrm{supp}(d\lambda)}(t+1/\xi_\nu)$ become very large and alternate in sign; cf. Example 3.2. This will cause serious cancellation errors in evaluating inner products relative to the measure $d\Lambda_N$ (there being blocks of weights W_k which are very large positive alternating with blocks of weights which are very large negative). In such cases, either m has to be lowered, perhaps down to $m = 1$, or else the method discussed in §2.2 invoked, which will be more effective (but possibly more expensive).

2.1.2. *Simple conjugate complex poles.* We now consider Case 2 of §1, i.e., conjugate complex parameters $\zeta_\nu = \xi_\nu + i\eta_\nu$, $\zeta_{\nu+m/2} = \bar{\zeta}_\nu$, where $\xi_\nu \in \mathbb{R}$, $\eta_\nu > 0$ and m is even. In this case, an elementary computation yields the partial fraction decomposition

$$\frac{1}{\omega_m(t)} = \sum_{\nu=1}^{m/2} \frac{c_\nu + d_\nu t}{\left(t+\frac{\xi_\nu}{\xi_\nu^2+\eta_\nu^2}\right)^2 + \left(\frac{\eta_\nu}{\xi_\nu^2+\eta_\nu^2}\right)^2}, \qquad t \in \mathbb{R}, \tag{2.11}$$

[1] In order to produce positive β-coefficients, as required in the routine for Gauss quadrature formulae, one inputs the measure $|c_\nu/(t-x_\nu)|d\lambda(t)$ and, if this entails a change of sign, reverses the sign of all Gauss weights after exiting from the Gauss quadrature routine.

where

$$c_\nu = \frac{1}{\eta_\nu}\left(\frac{\xi_\nu}{\xi_\nu^2+\eta_\nu^2}\ \mathrm{Im}\ p_\nu + \frac{\eta_\nu}{\xi_\nu^2+\eta_\nu^2}\ \mathrm{Re}\ p_\nu\right), \qquad d_\nu = \frac{1}{\eta_\nu}\mathrm{Im}\ p_\nu \tag{2.12}$$

and

$$p_\nu = \prod_{\substack{\mu=1\\ \mu\neq\nu}}^{m/2} \frac{(\xi_\nu + i\eta_\nu)^2}{(\xi_\nu-\xi_\mu)^2-(\eta_\nu^2-\eta_\mu^2)+2i\eta_\nu(\xi_\nu-\xi_\mu)} \tag{2.13}$$

with $p_1 = 1$ if $m = 2$. One can then proceed as in §2.1.1, except that the modification of the measure $d\lambda$ now involves multiplication by a nonconstant linear function (if $d_\nu \neq 0$) in addition to division by a quadratic. The former modification is handled by the routine `chri` of [4] with `iopt` = 1, the latter by the routine `gchri` with `iopt` = 2 (or, if more appropriate, by `chri` with `iopt` = 5). The quadrature rule (2.4) so obtained has $N = mn/2$.

If the poles $-1/\zeta_\mu$ are located in conjugate pairs on a line parallel to the imaginary axis, then by an elementary calculation one can show that all p_ν are real, hence $d_\nu = 0$, and there is no need to call `chri`.

2.1.2′. *Simple conjugate complex poles plus a simple real pole.* We are now in Case 2′ of §1, with m odd, $\zeta_0 = \xi_0 \in \mathbb{R}$ and the remaining ζ_μ conjugate complex as in Case 2. This yields

$$\frac{1}{\omega_m(t)} = \frac{c_0'}{t+(1/\xi_0)} + \sum_{\nu=1}^{(m-1)/2} \frac{c_\nu' + d_\nu' t}{\left(t+\frac{\xi_\nu}{\xi_\nu^2+\eta_\nu^2}\right)^2 + \left(\frac{\eta_\nu}{\xi_\nu^2+\eta_\nu^2}\right)^2}, \quad t\in\mathbb{R}, \tag{2.14}$$

where

$$c_0' = \frac{\xi_0^{m-2}}{\prod_{\nu=1}^{(m-1)/2}[(\xi_0-\xi_\nu)^2+\eta_\nu^2]},$$
$$c_\nu' = \frac{1}{\eta_\nu}\left(\frac{\xi_\nu}{\xi_\nu^2+\eta_\nu^2}\ \mathrm{Im}\ p_\nu' + \frac{\eta_\nu}{\xi_\nu^2+\eta_\nu^2}\ \mathrm{Re}\ p_\nu'\right), \tag{2.15}$$
$$d_\nu' = \frac{1}{\eta_\nu}\ \mathrm{Im}\ p_\nu'$$

and

$$p_\nu' = \frac{\xi_\nu + i\eta_\nu}{\xi_\nu - \xi_0 + i\eta_\nu}\, p_\nu, \tag{2.16}$$

with p_ν the same as in (2.13) with m replaced by $m-1$. The technique called for is a combination of the one in §2.1.1, to deal with the first term in (2.14), and the one in §2.1.2, to deal with the remaining terms, and yields a quadrature rule (2.4) with $N = (m+1)n/2$.

2.1.3. *Real poles of order 2.* This is Case 3 of §1, and leads to the partial fraction decomposition

$$\frac{1}{\omega_m(t)} = \sum_{\nu=1}^{m/2}\left(\frac{c_\nu}{t+1/\xi_\nu} + \frac{d_\nu}{(t+1/\xi_\nu)^2}\right), \tag{2.17}$$

$$c_\nu = -\frac{2\xi_\nu^{m-3}\sum_{\substack{\mu=1\\ \mu\neq\nu}}^{m/2}\frac{\xi_\mu}{\xi_\nu-\xi_\mu}}{\prod_{\substack{\mu=1\\ \mu\neq\nu}}^{m/2}(\xi_\nu-\xi_\mu)^2}, \tag{2.18}$$

$$d_\nu = \frac{\xi_\nu^{m-4}}{\prod_{\substack{\mu=1\\ \mu\neq\nu}}^{m/2}(\xi_\nu-\xi_\mu)^2}, \tag{2.19}$$

where $c_1 = 0$, $d_1 = \xi_1^{-2}$ when $m = 2$. Here, $N = mn$ in (2.4).

2.1.3′. *Real poles of order 2 plus a simple real pole.* Similarly as in §2.1.3, the partial fraction decomposition has now the form

$$\frac{1}{\omega_m(t)} = \frac{c'_M}{t+1/\xi_M} + \sum_{\nu=1}^{M-1}\left(\frac{c'_\nu}{t+1/\xi_\nu} + \frac{d'_\nu}{(t+1/\xi_\nu)^2}\right), \quad M=(m+1)/2, \quad m \text{ odd}, \tag{2.20}$$

$$c'_M = \frac{\xi_M^{m-2}}{\prod_{\nu=1}^{M-1}(\xi_M-\xi_\nu)^2},$$

$$c'_\nu = -\frac{\xi_\nu^{m-3}\left(\xi_M + 2(\xi_\nu-\xi_M)\sum_{\substack{\mu=1\\ \mu\neq\nu}}^{M-1}\frac{\xi_\mu}{\xi_\nu-\xi_\mu}\right)}{(\xi_\nu-\xi_M)^2\prod_{\substack{\mu=1\\ \mu\neq\nu}}^{M-1}(\xi_\nu-\xi_\mu)^2},$$

$$d'_\nu = \frac{\xi_\nu^{m-4}}{(\xi_\nu-\xi_M)\prod_{\substack{\mu=1\\ \mu\neq\nu}}^{M-1}(\xi_\nu-\xi_\mu)^2}.$$

Empty sums and products (when $M = 2$) have their conventional values 0 and 1, respectively. Again, $N = mn$ in (2.4).

The presence of two terms in the summations of (2.17) and (2.20) complicates matters considerably, as they call for two applications of the routine `gchri`: First, we must generate sufficiently many of the recursion coefficients for the measure $d\lambda(t)/(t-x_\nu)$, $x_\nu = -1/\xi_\nu$, in order next to generate the desired recursion coefficients for $d\lambda(t)/(t-x_\nu)^2$ by backward recursion – a recursion based on the recurrence relation generated in the first application of `gchri` (which in turn requires backward recursion!). The procedure nevertheless works well if the x_ν are not too close to the support interval of $d\lambda$; see Example 3.3.

2.2. Discretization method. In this method, the inner product (2.3) is approximated by a discrete (positive) inner product,

$$(u,v) = \int_{\mathbb{R}} u(t)v(t)\,\frac{d\lambda(t)}{\omega_m(t)} \approx \sum_{k=1}^{N} \omega_k^{(N)} u(\tau_k^{(N)}) v(\tau_k^{(N)}) =: (u,v)_N, \quad N > n, \tag{2.21}$$

whereupon the formulae (2.2) are applied with the inner product $(\,\cdot\,,\cdot\,)$ replaced by $(\,\cdot\,,\cdot\,)_N$ throughout. This yields approximations

$$\hat{\alpha}_{k,N} \approx \hat{\alpha}_k, \qquad \hat{\beta}_{k,N} \approx \hat{\beta}_k, \qquad k = 0,1,\ldots,n-1. \tag{2.22}$$

In effect we are generating the polynomials orthogonal with respect to the discrete inner product $(\,\cdot\,,\cdot\,)_N$ in order to approximate the desired orthogonal polynomials.

The computation of the approximate coefficients (2.22) can be done by either Stieltjes's procedure or Lanczos's algorithm (cf., e.g., [3, §§6–7]). Both are implemented in the routine `mcdis` of [4].

With any reasonable choice of the discretization (2.21), it will be true that the procedure converges as $N \to \infty$,

$$\lim_{N\to\infty} \hat{\alpha}_{k,N} = \hat{\alpha}_k, \quad \lim_{N\to\infty} \hat{\beta}_{k,N} = \hat{\beta}_k, \qquad 0 \le k \le n-1. \tag{2.23}$$

A natural choice, indeed, is given by

$$\tau_k^{(N)} = t_k^{(N)}(d\lambda), \qquad \omega_k^{(N)} = \frac{w_k^{(N)}(d\lambda)}{\omega_m(\tau_k^{(N)})}, \qquad k = 1,2,\ldots,N, \tag{2.24}$$

where $t_k^{(N)}(d\lambda)$ are the zeros of the orthogonal polynomial $\pi_N(\,\cdot\,;d\lambda)$, and $w_k^{(N)}(d\lambda)$ the respective Christoffel numbers.

The discretization method is conceptually simpler, and sometimes more stable, than the methods of §2.1, but may become significantly more expensive, regardless of the choice of m, if poles are close to the interval of integration, or if high accuracy is desired; cf. Examples 3.1 and 3.5. Note also that Case 4 that was skipped in §2.1 can easily be handled by the present method; see Example 3.6.

3. Numerical Examples

All examples in this section were computed on the Cyber 205 in both single and double precision. The respective machine precisions are 7.11×10^{-15} and 5.05×10^{-29}.

Example 3.1. $I_1(\omega) = \int_{-1}^{1} \frac{\pi t/\omega}{\sin(\pi t/\omega)}\,dt, \quad \omega > 1.$

Here, $d\lambda(t) = dt$, and the poles of the integrand are located at the integer multiples of ω. It is natural, then, to make our quadrature rule (1.8) exact for m elementary rational functions matching the m poles closest to the origin, say those at

$-(m/2)\omega, \dots, -\omega, \omega, \dots, (m/2)\omega$ when m is even. This suggests to identify $-1/\zeta_\mu$ in (1.2) with these poles, i.e., in (1.11a) to set

$$\xi_\nu = (-1)^\nu/(\omega\lfloor(\nu+1)/2\rfloor), \quad \nu = 1,2,\dots,m. \tag{3.1}$$

Best accuracy is expected when $m = 2n$, in which case the method described in §2.1.1 was found to work rather well, the only difficulty being the relatively slow convergence of the backward recurrence algorithm for computing the $2n$ modified (Legendre) moments associated with the measure $dt/(t \pm \omega)$ when ω is very close to 1. For single-precision accuracy $\epsilon = \frac{1}{2} \times 10^{-10}$ and double-precision accuracy $\epsilon^d = \frac{1}{2} \times 10^{-25}$, and $n = 20$, the respective starting indices k_0 and k_0^d in the backward recursion yielding the desired accuracy are shown in Table 3.1 for selected values of ω.

ω	k_0	k_0^d
2.0	50	63
1.5	53	71
1.1	67	106
1.01	124	247

TABLE 3.1. *Starting indices for backward recurrence when* $n = 20$

Other than that, the method appears to be very stable and produces quadrature rules that are rapidly converging. In Table 3.2, the results of the n-point rule (1.17)

ω	n	n-point rational Gauss	γ_n	err. Gauss
2.0	1	2.1	3.94(−1)	1.43(−1)
	4	2.33248722	3.50(−7)	7.18(−5)
	7	2.332487232246550235	2.61(−15)	2.73(−8)
	10	2.332487232246550241107076	1.48(−24)	1.02(−11)
1.1	2	4.43	1.73(−2)	2.60(−1)
	5	4.467773637	2.00(−9)	2.09(−2)
	8	4.46777364638776571	5.61(−18)	1.53(−3)
	11	4.467773646387765789236123	1.66(−27)	1.09(−4)
1.01	3	8.429	2.53(−4)	4.20(−1)
	6	8.4301845803	6.27(−12)	1.85(−1)
	9	8.4301845804708420582	7.52(−21)	8.37(−2)
	12	8.430184580470842058971264	1.23(−30)	3.75(−2)

TABLE 3.2. *Numerical results for* $I_1(\omega)$, *error constants, and comparison with Gauss quadrature*

applied to $g(t) = (\pi t/\omega)/\sin(\pi t/\omega)$ in double precision are shown for $\omega = 2$, 1.1 and 1.01, along with the error constants γ_n of (1.18). Also shown in the last column are the relative errors of the n-point Gauss-Legendre rule. For $\omega = 2$, the exact answer is known to be $8C/\pi$, where C is Catalan's constant (cf. [8, Eq. 3.747(2)]). The value shown in Table 3.2 for $n = 10$ agrees with it to all 25 decimal digits given. Ordinary Gauss-Legendre quadrature is seen to converge rather slowly, as ω approaches 1. In contrast, convergence of the rational Gauss quadrature rule is fast even for ω very close to 1. The extra effort required in this case is expended, as illustrated in Table 3.1, at the time when the rule is generated.

		method of §2.1		method of §2.2	
ω	n	SP	DP	SP	DP
2.0	1	.001	.004	.007	.178
	4	.008	.036	.010	.224
	7	.027	.125	.028	.220
	10	.065	.296	.016	.423
1.1	2	.002	.014	.060	1.554
	5	.013	.063	.068	1.649
	8	.036	.177	.098	1.274
	11	.080	.376	.104	1.461
1.01	3	.005	.037	.460	20.10
	6	.020	.104	.446	41.51
	9	.050	.244	.458	10.44
	12	.102	.487	.525	12.65

TABLE 3.3. *Timings* (in seconds) *of the methods in* §§2.1 *and* 2.2 *applied to Example* 3.1

Identical results were obtained by the discretization method of §2.2, but with substantially greater effort, particularly for higher accuracies and for ω close to 1. Respective timings are shown in Table 3.3, both for single-precision (SP) and double-precision (DP) accuracy requirements of $\frac{1}{2} \times 10^{-10}$ and $\frac{1}{2} \times 10^{-25}$, respectively.

While the choice $m = 2n$ indeed gives best accuracy, other choices of m may be preferable if the effort and time to generate the quadrature rule is of any importance. It turns out that with the method of partial fractions, $m = 2\lfloor (n+1)/2 \rfloor$ gives almost the same accuracy at about half the effort, whereas $m = 2$ gives considerably less accuracy but requires only about one-tenth the effort. The discretization method of §2.2, on the other hand, requires essentially the same effort regardless of the choice of m. Some timings required to generate the quadrature rules for various m and n, and the relative errors achieved, are shown in Table 3.4

			method of §2.1		method of §2.2		
ω	n	m	SP	DP	SP	DP	err
2.0	10	2	.008	.037	.014	.399	1.10(−17)
		10	.033	.153	.015	.409	1.47(−25)
		20	.065	.296	.016	.423	1.58(−25)
1.1	11	2	.009	.047	.095	1.402	2.20(−13)
		12	.045	.213	.100	1.426	2.80(−23)
		22	.080	.376	.104	1.461	6.68(−26)
1.01	12	2	.011	.065	.526	12.65	1.15(−13)
		12	.052	.258	.511	12.46	9.10(−25)
		24	.102	.487	.526	12.65	3.55(−27)

TABLE 3.4. *Timings and errors for selected* $m \le 2n$

If $m = 2$, the values of n for which full accuracy of about 10^{-25} is attained are 15, 21 and 22 for $\omega = 2.0$, 1.1 and 1.01, respectively. Interestingly, the timings involved are only about half those for $m = 2\lfloor (n+1)/2 \rfloor$ shown in Table 3.4.

Example 3.2. $I_2(\omega) = \int_0^1 \frac{t^{-1/2}\Gamma(1+t)}{t+\omega}\,dt, \quad 0 < \omega < 1.$

Here we take $d\lambda(t) = t^{-1/2}\,dt$ on [0,1]. If we wish to match the first $2n-1$ poles of the gamma function at the negative integers as well as the pole at $-\omega$, we set $m = 2n$ in

$$\begin{aligned} \xi_1 &= \tfrac{1}{\omega}, \\ \xi_\nu &= \tfrac{1}{\nu-1}, \qquad \nu = 2, 3, \ldots, m. \end{aligned} \tag{3.2}$$

The rational n-point Gauss rule (1.7), (1.17), generated by the method of §2.1.1, then produces (in double precision) results as shown in Table 3.5, where $\omega = \frac{1}{2}$. In the last column we list the absolute value of the difference between double-precision and single-precision results. In contrast to Example 3.1, we now see a case in which the accuracy reaches a limit (at about $n = 10$) and deteriorates, rather than improves, as n is further increased. (When $n = 20$, the calculation even breaks down in single precision!). The last column in Table 3.5 provides a clear hint as to what is happening: a steady growth in numerical instability. Closer examination reveals the true cause of this instability. The constants c_ν in the partial fraction decomposition (2.7) become very large and alternate in sign. Thus, for example, $c_{18} = -2.3375\ldots \times 10^9$ and $c_{19} = 2.3336\ldots \times 10^9$ when $n = 18$. This produces blocks of large coefficients W_k in (2.10) that alternate in sign from block to block, causing severe cancellation errors in summations such as

(2.5) (which are abundant in Stieltjes's algorithm). The phenomenon evidently is a manifestation of the asymmetric distribution of the poles of the gamma function.

n	n-point rational Gauss	$\|DP - SP\|$
2	1.746	2.24(−13)
6	1.75012059121	7.05(−11)
10	1.750120591261335415386	3.36(−8)
14	1.7501205912613354159	1.52(−5)
18	1.7501205912613356	4.31(−3)

TABLE 3.5. *Numerical results for* $I_2(\omega)$, $\omega = .5$

The method of §2.2, in contrast, does not suffer from any numerical instability and produces for $n = 12$, with comparable effort, the value

$$I_2(.5) = 1.750120591261335415394610, \tag{3.3}$$

believed to be correct to all 25 digits shown.

Matching only n poles, and thus taking $m = n$ in (3.2), stabilizes the procedure of §2.1.1 considerably, and as a consequence produces the correct result (3.3) (except for a discrepancy of 1 unit in the last decimal place) for $n = 11$. An even more stable procedure results from taking $m = 2$ and, amazingly, yields the correct answer (to all digits shown!) already for $n = 13$.

Example 3.3. $I_3(\omega) = \int_{-1}^{1} \left(\frac{\pi t/\omega}{\sin(\pi t/\omega)}\right)^2 dt, \quad \omega > 1.$

Similarly as in Example 3.1, we take

$$\xi_\nu = (-1)^\nu/(\omega\lfloor(\nu+1)/2\rfloor), \qquad \nu = 1, 2, \ldots, m/2. \tag{3.4}$$

We applied the procedure described in §2.1.3 for $\omega = 2$, 1.5, 1.1 and 1.01, both in single and double precision, requesting accuracies of $\epsilon = \frac{1}{2} \times 10^{-10}$ and $\epsilon^d = \frac{1}{2} \times 10^{-25}$,

ω	k_1	k_1^d	k_2	k_2^d
2.0	130	163	50	63
1.5	133	191	53	71
1.1	177	306	67	106
1.01	384	727	124	247

TABLE 3.6. *Starting indices for two backward recurrences when* $n = 20$

respectively. When $m = 2n$ and $n = 20$, starting indices k_1, k_1^d in the first application of the backward recursion that were found to meet the accuracy requirements for the poles closest to [−1,1], and the analogous starting indices k_2, k_2^d in the second application, are shown in Table 3.6 for the four values of ω. As expected, the procedure becomes laborious as ω approaches 1. Selected double-precision results produced by the n-point rational Gauss rule, along with error constants, are shown in Table 3.7. The last column shows the relative error of results generated by the n-point Gauss-Legendre rule. For $\omega = 2$, the exact answer is known to be $I_3(2) = 4 \ln 2$ ([8, Eq. 3.837(2)]) and is correctly reproduced to 25 digits when $n = 11$. Note again the fast convergence of the rational Gauss quadrature rule, even for ω very close to 1, in contrast to the relatively slow convergence of the ordinary Gauss rule, especially for ω close to 1.

ω	n	n-point rational Gauss	γ_n	err. Gauss
2.0	2	2.75	9.90(−3)	4.36(−2)
	5	2.77258868	1.16(−9)	4.70(−5)
	8	2.7725887222397811	3.28(−18)	2.92(−8)
	11	2.7725887222397812376689 28	9.72(−28)	1.52(−11)
1.1	2	15.5	3.00(−2)	6.69(−1)
	6	16.5328175	8.54(−12)	6.79(−2)
	10	16.5328177384604181	7.21(−24)	3.42(−3)
	14	16.532817738460418301558 98	2.35(−37)	1.40(−4)
1.01	2	184.	6.34(−2)	9.64(−1)
	6	188.674782	2.20(−11)	7.06(−1)
	10	188.674784224994172	1.88(−23)	4.00(−1)
	14	188.674784224994174270832 5	6.15(−37)	1.92(−1)

TABLE 3.7. *Numerical results for $I_3(\omega)$, error constants, and comparison with Gauss quadrature*

While the choice $m = 2\lfloor (n+1)/2 \rfloor$ produced similar advantages as in Example 3.1 — an increase of speed by a factor of about 2 at only a slight loss of accuracy — the choice $m = 2$ offered no significant gains in accuracy over the Gauss-Legendre rule, unlike $m = 4$, which did (since a symmetric *pair* of double poles is now accounted for).

We also applied the discretization method of §2.2 and obtained identical results with somewhat less effort in the case $\omega = 2$, and about the same effort in the case $\omega = 1.1$. For $\omega = 1.01$, however, we were unable to attain the requested double-precision accuracy with a discretization parameter $N \leq 800$ (in (2.21)).

Example 3.4. $I_4 = \int_0^\infty \frac{t}{e^t - 1} e^{-t} dt$.

The appropriate measure here is $d\lambda(t) = e^{-t}dt$ on $[0,\infty]$. Since the integrand has poles at the integer multiples of $2\pi i$, we let $\zeta_\nu = -1/(2\nu\pi i) = i/(2\nu\pi)$, and thus in

(1.12) take

$$\xi_\nu = 0, \qquad \eta_\nu = \frac{1}{2\nu\pi}, \qquad \nu = 1, 2, \ldots, m/2. \tag{3.5}$$

The quantity p_ν in (2.13) being real, and thus $d_\nu = 0$ in (2.12), there is no nonconstant linear factor in the numerators of (2.11). This simplifies somewhat the procedure in §2.1.2, as it obviates the need to apply the routine `chri`.

In Table 3.8 we compare the performance (in double precision and for $m = 2n$) of our rational quadrature routine with Gauss-Laguerre quadrature (applied to $f(t) = t/(e^t - 1)$) and the Gaussian quadrature rule (applied to $f(t) = e^{-t}$) associated with "Einstein's weight function" $t/(e^t - 1)$; for the latter see [6]. The respective relative errors are shown in the last two columns. It can be seen that the Gauss-Laguerre and Gauss-Einstein quadratures are comparable in accuracy, the former being somewhat more accurate for small values of n, the latter for larger values of n. Both quadrature rules, however, are incomparably inferior to the rational Gauss formula, which for $n = 15$ produces the true value of the integral, $\zeta(2) - 1 = (\pi^2/6) - 1$, to 25 correct decimal digits. (Actually, the last digit is off by one unit.) The results become even slightly more accurate when we choose $m = 2\lfloor (n+1)/2 \rfloor$, and are still better, by several orders of magnitude, than those for Gauss-Laguerre and Gauss-Einstein quadrature when $m = 2$.

n	n-point rational Gauss	err GL	err GE
1	.59	9.76(−2)	4.09(−1)
5	.644934055	1.50(−5)	2.97(−4)
10	.644934066848226428	2.22(−8)	1.15(−8)
15	.6449340668482264364724151	1.59(−11)	3.25(−13)

TABLE 3.8. *Numerical results for I_4 and comparison with Gauss-Laguerre and Gauss-Einstein quadrature*

The high accuracy of our rational quadrature rules in this example is all the more remarkable as the routine `gchri`, used in their construction (by the methods of §2.1.2), is subject to ill-conditioning, causing the recursion coefficients for the relevant orthogonal polynomials to gradually lose accuracy (by as much as 10 decimals, when $n = 15$ and $m = 2n$).

This weakness is accentuated when one tries to deal with more difficult integrals, for example,

$$I(\theta) = \int_0^\infty \frac{t}{e^t - 1} \sqrt{1 + \tfrac{1}{2}\theta t}\, dt = \int_0^\infty \frac{t}{1 - e^{-t}} \sqrt{1 + \tfrac{1}{2}\theta t} \cdot e^{-t} dt, \quad \theta > 0, \tag{3.6}$$

which has an additional branch point singularity at $t = -2/\theta$. Here, when $\theta = .75$, the n-point rational Gauss formula (in double precision and for $m = 2n$) gives only about

13 correct decimal places for $n = 15$, and 18 for $n = 30$. By the time n reaches 33, the ill-conditioning in the routine gchri has built up to such a level that the method fails (by producing a negative β-recursion coefficient). To get higher accuracy, one needs to apply the discretization method of §2.2, which is more stable, but becomes fairly expensive if pushed much beyond $n = 30$. Using $d\lambda(t) = e^{-t}dt$, and hence the N-point Gauss-Laguerre formula, to effect the discretization in (2.21), and requesting an accuracy of $\frac{1}{2} \times 10^{-25}$ for the desired recursion coefficients, we have observed timings of the order 12–16 seconds, and discretization parameters N as large as $N = 370$, for $33 \le n \le 40$. The rational Gauss formula so produced then yields relative errors of 4.26×10^{-21} for $n = 35$, and 9.65×10^{-23} for $n = 40$. This is still better, by about 4 decimal orders of accuracy, than Gauss-Laguerre quadrature applied to the second form of the integral in (3.6), and Gauss-Einstein quadrature applied to the first form.

Generalized Fermi-Dirac integrals (cf. [12]) are similar to $I(\theta)$ except that $t/(e^t - 1)$ is replaced by $t^k/(e^{-\eta+t} + 1)$, where η is a real parameter and $k = 1/2$, $3/2$ or $5/2$. The poles are now located at $\eta \pm (2\nu - 1)i\pi$, $\nu = 1, 2, 3, \ldots$. The use of rational Gauss quadrature to compute such integrals is dealt with elsewhere [5].

Example 3.5. $I_5(\eta) = \int_0^\infty \frac{t}{e^{-\eta+t} - 1} e^{-t} dt, \quad \eta < 0.$

Again, we take $d\lambda(t) = e^{-t}dt$ and note that the poles are now at $\eta \pm 2\nu\pi i$, $\nu = 0, 1, 2, \ldots$. Accordingly, in (1.13) we take

$$\xi_0 = -\frac{1}{\eta}, \quad \xi_\nu = -\frac{\eta}{\eta^2 + 4\nu^2\pi^2}, \quad \eta_\nu = \frac{2\nu\pi}{\eta^2 + 4\nu^2\pi^2}, \quad \nu = 1, 2, \ldots, (m-1)/2, \qquad (3.7)$$

and use the procedure of §2.1.2′. Selected results (for $m = 2n - 1$), comparing rational Gauss formulae with Gauss-Laguerre formulae, are shown in Table 3.9. In the case

η	n	n-point rational Gauss	err GL
−.1	3	.4503	1.16(−1)
	6	.4501936153	5.13(−2)
	9	.450193614441350	2.70(−2)
	12	.450193614441347840 96	1.55(−2)
−1.0	2	.113	2.14(−1)
	6	.1111093520	5.07(−3)
	11	.1111093516052317322	1.81(−4)
	16	.1111093516052317320105065	1.26(−5)
−10.0	2	.122(−4)	1.79(−1)
	6	.113502121(−4)	1.57(−4)
	11	.1135021146353905 78(−4)	7.31(−9)
	16	.113502114635390570187 0968(−4)	1.20(−12)

TABLE 3.9. *Numerical results for I_5 and comparison with Gauss-Laguerre quadrature*

$\eta = -.1$, we were able to go only up to $n = 13$; when $n = 14$, our procedure failed by producing a negative β-coefficient in (2.2). The difficulty is caused by the ill-conditioning (mentioned after (2.10)) affecting the modified Chebyshev procedure. Even though our procedure was successful for $n = 13$, it had to work hard to take care of the pole at $\eta = -.1$: Backward recursion to compute modified moments had to start at $\nu = 584$ to get single-precision accuracy $\frac{1}{2} \times 10^{-10}$, and at $\nu = 2650$ to get double-precision accuracy $\frac{1}{2} \times 10^{-25}$.

Replacing the numerator te^{-t} in the integrand by t^k, where $k = 1/2, 3/2$ or $5/2$, and adding a factor $\sqrt{1 + \frac{1}{2}\theta t}$ as in (3.6), produces the Bose-Einstein integral whose computation by rational Gauss quadrature is discussed in [5].

Example 3.6. $I_6 = \int_0^\infty \left(\frac{t}{e^t - 1}\right)^2 e^{-t} dt$.

Here, as in Example 3.4, we take $d\lambda(t) = e^{-t}dt$ and parameters ξ_ν, η_ν as in (3.5), except that there are only $m/4$ of them, m being divisible by 4. In Table 3.10 we give

n	n-point rational Gauss	err GL	err GE
2	.47	3.71(−2)	1.61(−2)
8	.4816405209	1.16(−6)	5.99(−10)
14	.48164052105807573311	4.36(−9)	7.26(−18)
20	.48164052105807573134587777	2.80(−11)	1.09(−25)

TABLE 3.10. *Numerical results for I_5 and comparison with Gauss-Laguerre and Gauss-Einstein quadrature*

the results for $m = 2n$ (n even) obtained by the discretization method of §2.2, analogous to those of Table 3.8 but using the square of the Einstein function as weight function in GE. (The method of §2.1, as mentioned earlier, was not implemented.) What is remarkable in this example is the competitiveness of the Gauss-Einstein quadrature rule vis-à-vis the rational Gauss rule.

References

1. GAUTSCHI, W., Minimal solutions of three-term recurrence relations and orthogonal polynomials, *Math. Comp. 36* (1981), 547–554.

2. GAUTSCHI, W., On generating orthogonal polynomials, *SIAM J. Sci. Stat. Comput. 3* (1982), 289–317.

3. GAUTSCHI, W., Computational problems and applications of orthogonal polynomials, in *Orthogonal Polynomials and Their Applications* (C. Brezinski et al., eds.), IMACS Annals Comput. Appl. Math., Vol. 9, Baltzer, Basel, 1991, pp. 61–71.

4. GAUTSCHI, W., Algorithm xxx — ORTHPOL: A package of routines for generating orthogonal polynomials and Gauss-type quadrature rules, *ACM Trans. Math. Software*, submitted.

5. GAUTSCHI, W., On the computation of generalized Fermi-Dirac and Bose-Einstein integrals, *Comput. Phys. Comm.*, to appear.

6. GAUTSCHI, W. and MILOVANOVIĆ, G.V., Gaussian quadrature involving Einstein and Fermi functions with an application to summation of series, *Math. Comp.* *44* (1985), 177–190.

7. GOLUB, G.H. and WELSCH, J.H., Calculation of Gauss quadrature rules, *Math. Comp.* *23* (1969), 221–230.

8. GRADSHTEYN, I.S. and RYZHIK, I.M., *Table of Integrals, Series, and Products*, Academic Press, Orlando, 1980.

9. LÓPEZ LAGOMASINO, G., and ILLÁN, J., A note on generalized quadrature formulas of Gauss-Jacobi type, in *Constructive Theory of Functions*, Publ. House Bulgarian Acad. Sci., Sofia, 1984, pp. 513–518.

10. LÓPEZ LAGOMASINO, G. and ILLÁN GONZÁLEZ, J., Sobre los métodos interpolatorios de integración numérica y su conexión con la aproximación racional, *Rev. Ciencias Matém.* *8* (1987), no. 2, 31–44.

11. PICHON, B., Numerical calculation of the generalized Fermi-Dirac integrals, *Comput. Phys. Comm.* *55* (1989), 127–136.

12. SAGAR, R.P., A Gaussian quadrature for the calculation of generalized Fermi-Dirac integrals, *Comput. Phys. Comm.* *66* (1991), 271–275.

13. VAN ASSCHE, W. and VANHERWEGEN, I., Quadrature formulas based on rational interpolation, *Math. Comp.*, to appear.

Professor Walter Gautschi
Department of Computer Sciences
Purdue University
West Lafayette, Indiana 47907, U.S.A.
wxg@cs.purdue.edu

International Series of Numerical Mathematics, Vol. 112, © 1993 Birkhäuser Verlag Basel

Subdivision Methods for Adaptive Integration over Hyperspheres

Alan Genz

Abstract. An adaptive algorithm for numerical multiple integration over an n-dimensional hypersphere is considered. Two subdivision methods are analyzed and formulas are derived that permit the efficient application of the associated subregion transformations.

1. Introduction

The problem considered in this paper is the numerical evaluation of integrals

$$I(f) = \int_S f(\mathbf{x})d\mathbf{x},$$

where $\mathbf{x} = (x_1, x_2, ..., x_n)^t$ and $S = \{\mathbf{x} \mid 0 \leq ||\mathbf{x}||_2 \leq R^2\}$, the n-sphere with radius R. Although there has been extensive work done on the development of integration rules for the whole n-sphere (see the books by Engels 1980, Davis and Rabinowitz 1984 and Stroud 1971, for numerous references), there has been only limited work done on adaptive algorithms, which could be useful for some of the difficult practical problems that arise in application areas like statistics.

Subregion adaptive algorithms for numerical integration dynamically subdivide the integration region S and try to concentrate the computational work in the subregions of S where the integrand $f(\mathbf{x})$ is less smooth. A *globally* adaptive algorithm, which will be used as background for this paper, has the following general structure.

- Input: S, $f(\mathbf{x})$, an absolute error tolerance ϵ and a work limit W.
- Algorithm:
    1. Begin with an initial subdivision of S, and estimates for $I(f)$ and the error.
    2. **While** (error $> \epsilon$) and (total work $<$ W) **Do**
        (a) Determine a new subdivision by subdividing a largest error subregion.
        (b) Apply a local integration rule to $f(\mathbf{x})$ in the new subregions.
        (c) Update the subregion list and estimates for $I(f)$ and the error.
        Endwhile
- Output: Estimates for $I(f)$ and the absolute error.

A similar algorithm has been used for hyperrectangles and simplices; see the papers by van Dooren and de Ridder (1976), Genz and Malik (1980), Berntsen, Espelid and Genz (1991) and Genz (1991). The specific components that are needed (in steps a and b) by this algorithm when it is used for hyperspherical regions are, a clearly defined method for subdividing S and, local integration and error estimation rules for the resulting subregions.

A major problem with the implementation of adaptive algorithms for hyperspheres is that there is no regular way (for $n > 2$) to subdivide a hypersphere into arbitrarily smaller and smaller subregions which are suitable for the application of standard integration rules. This is in contrast to the situation for integration regions like hyperrectangles and simplices, for which there are straightforward methods for subdivision that produce subregions of the same type as the original region. A subdivision algorithm for hyperspheres must therefore generate irregular subregions where there is no natural choice for the integration rules that are needed for estimation of the integrals over the subregions. A transformation from some other type of region where good integration rules are available is therefore needed. But the use of such a transformation may increase both the cost of applying the integration rule, and the difficulty of the resulting transformed integrand.

This primary aim of this paper is to consider two methods for subdivision of S and some transformations from more standard subregions to the resulting subregions. The paper also considers how to make the transformation process efficient and derives formulas for the efficient evaluation of the transformation Jacobians.

2. Subregions

In this section two methods are described for subdividing S into successively smaller subregions. The first subdivision method, which will be called the *radial* subdivision method, uses subregions that are a product of a radial interval and a spherical (n-1)-simplex. A spherical (n-1)-simplex $\tilde{T}$ is defined from n vertices $\{\mathbf{v}_j\}_{j=1}^n$, with $||\mathbf{v}_j||_2 = 1$ for all j, using

$$\tilde{T} = \{ \mathbf{x} \mid \mathbf{x} = s\mathbf{q},\ \mathbf{q} = \sum_{j=1}^{n} \alpha_j \mathbf{v}_j,\ s = ||\mathbf{q}||_2^{-1},\ \sum_{j=1}^{n} \alpha_j = 1 \text{ with } \alpha_j \geq 0 \text{ for all } j \ \}.$$

A radial subregion has the form $S_k = [r_{k_1}, r_{k_2}] \times \tilde{T}_k$, where $\tilde{T}_k$ is a spherical simplex generated from the spherical vertices $\{\mathbf{v}_j^k\}_{j=1}^n$ and $r_{k_i} \epsilon [0, R]$. The initial subdivision of S that is used with the radial subdivision has n+1 subregions $\{S_k\}_{k=1}^{n+1}$. Each subregion has the form $S_k = [0, R] \times \tilde{T}_k$, where $\tilde{T}_k$ has a set of generating vertices that is obtained by deleting the kth vertex from the set of n+1 vertices for a regular n-simplex circumscribed by S. An example (Mysovskikh, 1980) of such a simplex has vertices $\{\mathbf{u}_j\}_{j=0}^n$ with components

$$u_{i,j} = \begin{cases} -\sqrt{(n+1)/((n-i+1)n(n-i+2))} & \text{if } 0 < i \leq j \\ \sqrt{(n+1)(n-j)/(n(n-j+1))} & \text{if } i = j+1 \\ 0 & \text{otherwise.} \end{cases}$$

The radial subregions are divided into two pieces using one of two procedures. Assuming a priority subregion S_k has been selected by the adaptive algorithm, the first procedure cuts the radial interval at the point $\hat{r} = ((r_{k_1}^n + r_{k_2}^n)/2)^{1/n}$, to produce two new subregions $S_k^{(1)} = [r_{k_1}, \hat{r}] \times \tilde{T}_k$ and $S_k^{(2)} = [\hat{r}, r_{k_2}] \times \tilde{T}_k$. The second procedure cuts one of the $n(n-1)/2$ spherical simplex edges to produce two new spherical simplices. An edge from vertex $\mathbf{v}_i^k$ to vertex $\mathbf{v}_l^k$ has midpoint $\hat{\mathbf{v}}$ on the surface of S at $\hat{\mathbf{v}} = \beta(\mathbf{v}_i^k + \mathbf{v}_l^k)$, if $\beta = ||\mathbf{v}_i^k + \mathbf{v}_l^k||_2^{-1}$. In this case the two new subregions are $S_k^{(1)} = [r_{k_1}, r_{k_2}] \times (\tilde{T}_k \backslash \mathbf{v}_i^k \cup \{\hat{\mathbf{v}}\})$ and $S_k^{(2)} = [r_{k_1}, r_{k_2}] \times (\tilde{T}_k \backslash \mathbf{v}_l^k \cup \{\hat{\mathbf{v}}\})$. The choice of the radial or edge direction to cut can be made by computing some heuristic measure of smoothness (an approximate partial derivative has often been used with adaptive algorithms for other types of region) along each of the possible directions and choosing the direction with minimal smoothness for the cut. This procedure divides the chosen subregion S_k into two subregions with approximately equal volume.

The second subdivision method, which is called the *simplex* subdivision method, uses subregions that are determined by n+1 vertices. A subregion S_k is defined by specifying $m_k + 1$ vertices $\{\mathbf{v}_j^k\}_{j=0}^{m_k}$ inside of S and $n - m_k$ vertices $\{\mathbf{v}_j^k\}_{j=m_k+1}^{n}$ on the surface of S, with at least one vertex inside of S $(0 \le m_k < n+1)$. S_k is then defined by

$$S_k = \{\, \mathbf{x} \mid \mathbf{x} = s\sum_{j=0}^{n} \alpha_j \mathbf{v}_j^k,\ s = (1 - \sum_{j>m_k} \alpha_j + ||\sum_{j>m_k} \alpha_j \mathbf{v}_j^k||_2)^{-1},\ \sum_{j=0}^{n} \alpha_j = 1,\ \alpha_j \ge 0\}.$$

The initial subdivision of S that is used with the simplex subdivision method also has the same n+1 subregions that were used with the radial subdivision method. Each initial subregion S_k has $m_k = 0$ and vertices given by the set $\{\{\mathbf{0}\} \cup \{R\mathbf{u}_j\}_{j=0}^{n+1} \setminus R\mathbf{u}_k\}$.

The simplex subregions are divided into two pieces by first selecting a priority edge. If this edge is between vertex $\mathbf{v}_i^k$ to vertex $\mathbf{v}_l^k$ (with $i < l$), and if $i \le m_k$, then the midpoint is $\hat{\mathbf{v}} = (\mathbf{v}_i^k + \mathbf{v}_l^k)/2$. If $i > m_k$, then the midpoint is $\hat{\mathbf{v}} = \beta(\mathbf{v}_i^k + \mathbf{v}_l^k)$, with $\beta = ||\mathbf{v}_i^k + \mathbf{v}_l^k||_2^{-1}$. The vertices for two new subregions $S_k^{(1)}$ and $S_k^{(2)}$ are $\{\{\mathbf{v}_j^k\}_{j=0}^{n+1} \backslash \mathbf{v}_i \cup \{\hat{\mathbf{v}}\}\}$ and $\{\{\mathbf{v}_j^k\}_{j=0}^{n+1} \setminus \mathbf{v}_l \cup \{\hat{\mathbf{v}}\}\}$, respectively. If $l < m_k$ or $m_k < i$, then $m_k^{(1)} = m_k^{(2)} = m_k$; otherwise $m_k^{(1)} = m_k + 1$ and $m_k^{(2)} = m_k$. This simplex subdivsion method also divides the chosen subregion S_k into two approximately equal volume subregions.

3. Integration Rules and Jacobians

There are no known integration rules of polynomial degree greater than one for $n > 2$ for either of the general types of subregions described in the previous section. Such formulas are difficult to construct because the are no general formulas for the integrals of monomials for these regions. There are no simple formulas even for the volume of the spherical n-simplices for $n > 3$ (see Coxeter, 1973, Chapter 7). Because of the nonlinear nature of the definition of these subregions, even if a formula of some polynomial degree was available for some prototypical region, there is unlikely to be any polynomial degree preserving transformation which could be used to transform the formula for use in other subregions. However, because of the paramaterizations used to define these subregions, it is natural to try to approximate integrals over these subregions with transformed integration rules derived from standard integration rules for simplices. Other parameterizations could motivate the use of other types of rules, but this section focuses on the use of standard simplex rules.

3.1 Radial Subregions

For a radial subregion S_k, let V_k be the $n \times n$ matrix with columns that are the vertices of the spherical simplex $\tilde{T}_k$. A point $\mathbf{x} \epsilon S_k$ can be written $\mathbf{x} = r\mathbf{q}s = r(\sum_{j=1}^{n-1} \alpha_j \mathbf{v}_j + (1 - \sum_{j=1}^{n-1} \alpha_j)\mathbf{v}_n)s = rsV_k\boldsymbol{\alpha}$ with $s = ||V_k\boldsymbol{\alpha}||_2^{-1}$, $\boldsymbol{\alpha} = (\alpha_1, \alpha_2, ..., \alpha_n)^t$ and $r\epsilon[r_{k_1}, r_{k_2}]$. Therefore

$$\int_{S_k} f(\mathbf{x})d\mathbf{x} = \int_{r_{k_1}}^{r_{k_2}} \int_{T_{n-1}} f(rsV_k\boldsymbol{\alpha})J(\boldsymbol{\alpha}, r)d\boldsymbol{\alpha}dr,$$

where $J(\boldsymbol{\alpha}, r)$ is the Jacobian of the transformation $\mathbf{x} = rV\boldsymbol{\alpha}s$ from $[r_{k_1}, r_{k_2}] \times T_{n-1}$ to S_k, and T_n is the standard n-simplex with vertices $\mathbf{0}, \mathbf{e}_1, \mathbf{e}_2, ..., \mathbf{e}_n$ (Stroud, 1971; the Euclidean basis vector $\mathbf{e}_j$, is defined to have 1 in position j and zeros elsewhere). Defining $s_j = \frac{\partial s}{\partial \alpha_j}$, and using standard rules for manipulating determinants

$$\begin{aligned}
J(\boldsymbol{\alpha}, r) &= |det([\frac{\partial \mathbf{x}}{\partial \alpha_1}, \frac{\partial \mathbf{x}}{\partial \alpha_2}, ..., \frac{\partial \mathbf{x}}{\partial \alpha_{n-1}}, \frac{\partial \mathbf{x}}{\partial r}])| \\
&= |det([rs(v_1^k - v_n^k) + r\mathbf{q}s_1, \; ..., \; rs(v_{n-1}^k - v_n^k) + r\mathbf{q}s_{n-1}, \; s\mathbf{q}])| \\
&= r^{n-1}s^n|det([v_1^k - v_n^k + \mathbf{q}s_1/s, \; ..., \; v_{n-1}^k - v_n^k + \mathbf{q}s_{n-1}/s, \; \mathbf{q}])| \\
&= r^{n-1}s^n|det([v_1^k - v_n^k, \; ..., \; v_{n-1}^k - v_n^k, \; \mathbf{q}])| \\
&= r^{n-1}s^n|det([v_1^k - v_n^k, \; ..., \; v_{n-1}^k - v_n^k, \; v_n^k])| \\
&= r^{n-1}s^n|det(V_k)|.
\end{aligned}$$

The transformed integral therefore becomes

$$\int_{S_k} f(\mathbf{x})d\mathbf{x} = |det(V_k)| \int_{r_{k_1}}^{r_{k_2}} r^{n-1} \int_{T_{n-1}} f(rsV_k\boldsymbol{\alpha})s^n d\boldsymbol{\alpha}dr.$$

These integrals can be approximated with the product of a radial integration rule for integrals of the form $\int_a^b r^{n-1}f(r)dr$, and an integration rule for the T_{n-1}. A different radial rule needs to be constructed whenever a new radial interval is produced by the subdivision process. These radial integrals can be reduced to integrals in the form $\int_c^1 r^{n-1}f(r)dr$ and standard methods can be used to construct Gauss rules for these integrals. The time for rule construction will not be long, compared to time taken for other work in an adaptive algorithm, as long as the polynomial degree of the rule is not too high.

The determinants $det(V_k)$ need to be computed for each new subregion. The value for each of the determinants for the initial subdivision of S into n+1 subregions is $det(V_k) = \sqrt{(n+1)^{n-1}/n^n}$. This formula can be derived from the formula for the volume of a regular simplex given in the book by Coxeter (1973, p. 159). A subdivision with the cut taken along the radial direction does not affect $det(V_k)$, so the two new subregions $S_k^{(1)}$ and $S_k^{(2)}$ have $det(V_k^{(1)}) = det(V_k^{(2)}) = det(V_k)$. If the cut is made along the edge from vertex $\mathbf{v}_i^k$ to vertex $\mathbf{v}_l^k$, with $\hat{\mathbf{v}} = \beta(\mathbf{v}_i^k + \mathbf{v}_l^k)$, then $V_k^{(1)}$ is obtained by replacing column i in V_k by $\hat{\mathbf{v}}$. Subtracting β times column l from column i in $V_k^{(1)}$ does not change $det(V_k^{(1)})$ and the result is $det(V_k^{(1)}) = \beta det(V_k)$. A similar argument applied to $det(V_k^{(2)})$ shows $det(V_k^{(2)}) = det(V_k^{(1)}) = \beta det(V_k)$, so the determinants that are required for the approximate evaluation of the integrals can be efficiently computed.

As the adaptive algorithm proceeds to refine the subregions, some of the subregions will become smaller, the spherical simplices $\tilde{T}_k$ will becomes closer to standard simplices and the subregion integrals should be easier to compute. In the early stages of the algorithm, when the spherical simplices are poorly approximated by standard simplices, it is natural to consider other parameterizations to define the subregions S_k. One possibility is to define the spherical simplices $\tilde{T}_k$ by

$$\tilde{T}_k = \{\ \mathbf{x} \mid \mathbf{x} = s\mathbf{q},\ \mathbf{q} = \sum_{j=1}^{n} \alpha_j \mathbf{v}_j^k,\ s = ||\mathbf{q}||^{-1},\ \sum_{j=1}^{n} \alpha_j^2 = 1 \text{ with } \alpha_j \geq 0 \text{ for all } j\ \}.$$

An analysis similar to the one given for the original paramaterization shows the transformed integrals now take the form

$$\int_{S_k} f(\mathbf{x})d\mathbf{x} = |det(V_k)| \int_{r_{k_1}}^{r_{k_2}} r^{n-1} \int_{\sum_{j=1}^{n}\alpha_j^2=1} \frac{f(rsV_k\boldsymbol{\alpha})s^n}{\sqrt{1-\sum_{j=1}^{n-1}\alpha_j^2}} d\boldsymbol{\alpha} dr.$$

These integrals could be approximated by rules that are a product of a radial rule and a rule for the hyperspherical orthant surface defined by $\sum_{j=1}^{n} \alpha_j^2 = 1$ and $\alpha_j \geq 0$ for all j, with weight function $1/\sqrt{1-\sum_{j=1}^{n-1}\alpha_j^2}$. Rules of this type have not been developed, and so this approach cannot at present be used for algorithm implementation and testing.

A further transformation that is based on a more standard transformation from angular coordinates to spherical coordinates uses $\alpha_j = cos(\theta_1)cos(\theta_2)...cos(\theta_{j-1})sin(\theta_j)$, for $j < n$ and $\alpha_n = cos(\theta_1)...cos(\theta_{n-1})$, with $0 \leq \theta_i \leq \pi/2$. After this additional transformation, the integrals become

$$\int_{S_k} f(\mathbf{x})d\mathbf{x} = |det(V_k)| \int_{r_{k_1}}^{r_{k_2}} r^{n-1} \int_0^{\frac{\pi}{2}} ... \int_0^{\frac{\pi}{2}} f(rsV_k\boldsymbol{\alpha})s^n cos(\theta_1)^{n-2} ... cos(\theta_{n-2}) d\boldsymbol{\theta} dr.$$

Rules for these integrals could be constructed as products of radial rules and rules for integrals of the form $\int_0^{\frac{\pi}{2}} cos(\theta)^j f(\theta)d\theta$, but this approach will be inefficient. A more direct approach could just use good rules for hypercubes.

3.2 Simplex Subregions

Now consider the simplex subdivision method, and let V_k be the $n \times n$ matrix with columns $\{\mathbf{v}_j^k\}_{j=1}^n$. These column vectors are assumed to be linearly independent so $\mathbf{v}_0^k = V_k\delta^k$ for some unique n-vector δ^k. Defining $\mathbf{w} = (1, 1, ..., 1)^t$, a point $\mathbf{x} \epsilon S_k$ can be written in the form $\mathbf{x} = V_k(\boldsymbol{\alpha} + \delta^k(1-\mathbf{w}^t\boldsymbol{\alpha}))s$, and an integral over S_k takes the form

$$\int_{S_k} f(\mathbf{x})d\mathbf{x} = \int_{T_n} f(V_k(\boldsymbol{\alpha} + \delta^k(1-\mathbf{w}^t\boldsymbol{\alpha}))s)J(\boldsymbol{\alpha})d\boldsymbol{\alpha}.$$

Using $s_j = \frac{\partial s}{\partial \alpha_j}$ again, and defining $\mathbf{s} = (s_1, s_2, ..., s_n)^t$, the Jacobian becomes

$$\begin{aligned} J(\alpha) &= |det([\frac{\partial \mathbf{x}}{\partial \alpha_1}, \frac{\partial \mathbf{x}}{\partial \alpha_2}, ..., \frac{\partial \mathbf{x}}{\partial \alpha_n}])| \\ &= |det(V_k((I - \delta^k \mathbf{w}^t)s + (\alpha + \delta^k(1 - \mathbf{w}^t\alpha))\mathbf{s}^t))|, \end{aligned}$$

where I is the $n \times n$ identity matrix. Next, using $(I - \delta^k\mathbf{w}^t)^{-1} = (I + \frac{\delta^k \mathbf{w}^t}{1 - \mathbf{w}^t\delta^k})$ and some algebra,

$$\begin{aligned} J(\alpha) &= |det(V_k(I - \delta^k\mathbf{w}^t)(I + \frac{\delta^k\mathbf{w}^t}{1-\mathbf{w}^t\delta^k})((I - \delta^k\mathbf{w}^t)s + (\alpha + \delta^k(1-\mathbf{w}^t\alpha))\mathbf{s}^t))| \\ &= |det(V_k(I - \delta^k\mathbf{w}^t)(sI + (\alpha + \delta^k(1-\mathbf{w}^t\alpha) + \frac{\delta^k\mathbf{w}^t(\alpha + \delta^k(1-\mathbf{w}^t\alpha))}{1 - \mathbf{w}^t\delta^k})\mathbf{s}^t))| \\ &= |det(V_k(I - \delta^k\mathbf{w}^t)(sI + (\alpha + \frac{\delta^k\mathbf{w}^t\alpha + \delta^k(1-\mathbf{w}^t\alpha)}{1 - \mathbf{w}^t\delta^k})\mathbf{s}^t))| \\ &= |det(V_k(I - \delta^k\mathbf{w}^t)(sI + (\alpha + \frac{\delta^k}{1-\mathbf{w}^t\delta^k})\mathbf{s}^t))| \\ &= s^n|det(V_k(I - \delta^k\mathbf{w}^t)(I + (\alpha + \frac{\delta^k}{1-\mathbf{w}^t\delta^k})\frac{\mathbf{s}^t}{s}))|. \end{aligned}$$

Then, using the result $det(I - \mathbf{x}\mathbf{y}^t) = 1 - \mathbf{x}^t\mathbf{y}$, $J(\alpha)$ becomes

$$J(\alpha) = s^n|det(V_k)|\ |(1-\mathbf{w}^t\delta^k)(1 + (\alpha^t + \frac{(\delta^k)^t}{1-\mathbf{w}^t\delta^k})\frac{\mathbf{s}}{s}))|.$$

$J(\alpha)$ can be further simplified if the partial derivatives s_j are determined. Because $s = (1 - \sum_{j>m_k}\alpha_j + ||\mathbf{q}||)^{-1}$, with $\mathbf{q} = \sum_{j>m_k}\alpha_j\mathbf{v}_j^k$,

$$s_j = \begin{cases} s^2(1-\mathbf{q}^t\mathbf{v}_j^k/||\mathbf{q}||) & \text{if } j > m_k \\ 0 & \text{otherwise.} \end{cases}$$

By defining $\mu = 1 - \mathbf{w}^t\delta^k = 1 - \sum_{i=1}^n \delta_i^k$, substituting the s_j's, expanding the sums and using more algebra, $J(\alpha)$ now becomes

$$\begin{aligned} J(\alpha) &= s^n|det(V_k)|\ |\mu + \sum_{j=1}^n(\mu\alpha_j + \delta_j^k)\frac{s_j}{s}| \\ &= s^n|det(V_k)|\ |\mu + s\sum_{j>m_k}(\mu\alpha_j + \delta_j^k)(1-\mathbf{q}^t\mathbf{v}_j^k/||\mathbf{q}||)| \\ &= s^n|det(V_k)|\ |\mu + s\sum_{j>m_k}(\mu\alpha_j + \delta_j^k) - \frac{s\mathbf{q}^t}{||\mathbf{q}||}\sum_{j>m_k}(\mu\alpha_j + \delta_j^k)\mathbf{v}_j^k| \\ &= s^n|det(V_k)|\ |\mu + s\sum_{j>m_k}(\mu\alpha_j + \delta_j^k) - s\mu||\mathbf{q}|| - \frac{s\mathbf{q}^t}{||\mathbf{q}||}\sum_{j>m_k}\delta_j^k\mathbf{v}_j^k| \\ &= s^n|det(V_k)|\ |\mu(1 + s(\sum_{j>m_k}\alpha_j - ||\mathbf{q}||)) + s\sum_{j>m_k}\delta_j^k(1 - \frac{\mathbf{q}^t\mathbf{v}_j^k}{||\mathbf{q}||})| \end{aligned}$$

$$= s^n|det(V_k)|\ |\mu(1+s(1-1+\sum_{j>m_k}\alpha_j-||\mathbf{q}||))+s\sum_{j>m_k}\delta_j^k(1-\frac{\mathbf{q}^t\mathbf{v}_j^k}{||\mathbf{q}||})|$$

$$= s^n|det(V_k)|\ |s(\mu+\sum_{j>m_k}\delta_j^k(1-\frac{\mathbf{q}^t\mathbf{v}_j^k}{||\mathbf{q}||}))|$$

$$= s^{n+1}|det(V_k)|\ |1-\sum_{j\leq m_k}\delta_j^k-\mathbf{q}^t\sum_{j>m_k}\delta_j^k\frac{\mathbf{v}_j^k}{||\mathbf{q}||}|.$$

Defining $\mathbf{p}=\sum_{j>m_k}\delta_j^k\mathbf{v}_j^k$ and $\hat{\mathbf{q}}=\mathbf{q}/||\mathbf{q}||$, the transformed integral finally becomes

$$\int_{S_k} f(\mathbf{x})d\mathbf{x} = |det(V_k)|\int_{T_n} f(sV_k(\alpha+\delta^k(1-\mathbf{w}^t\alpha)))s^{n+1}|1-\sum_{j\leq m_k}\delta_j^k-\hat{\mathbf{q}}^t\mathbf{p}|d\alpha.$$

These integrals can be approximated with standard integration rules for the n-simplex. As with the radial transformations, the efficient application of the simplex rules may require efficient methods for evaluation of the determinants $det(V_k)$ and maintenance of the δ^k vectors that are needed for each subregion S_k. Initially, $det(V_k)=\sqrt{(n+1)^{n-1}/n^n}$ and $\delta^k=\mathbf{0}$ for all of the n+1 subregions in the original subdivision of S. If $det(V_k)$ and δ^k are given for a particular subregion, then the cost of determining each integrand evaluation point $\mathbf{x}$, and the factors inside the integral associated with $J(\alpha)$, is $O(n^2)$. A simplex rule of polynomial degree $d\geq 3$ requires at least $O(n)$ integrand evaluation points, so the cost of applying such a rule (given $det(V_k)$ and δ^k) is at least $O(n^3)$. It would therefore be tempting to use standard $O(n^3)$ methods from linear algebra to compute $det(V_k)$ and determine δ^k for each new subregion. However, as the adaptive algorithm proceeds to produce successively smaller subregions, the columns of successive V_k's become closer together, making the computation of $det(V_k)$ and determination of δ^k successively more illconditioned. Therefore, some methods for directly updating $det(V_k)$ and δ^k are presented.

Suppose an edge from vertex $\mathbf{v}_i^k$ to vertex $\mathbf{v}_l^k$, with $i<l$, is selected for subdivision at some stage in the algorithm. The updating of $det(V_k)$ and δ^k for the two new subregions requires the consideration of several cases that depend on the positions of i and l relative to 0 and m_k. The simplest cases occur when $i>0$. Then $\hat{\mathbf{v}}=\beta(\mathbf{v}_i^k+\mathbf{v}_l^k)$, for some β (with $\beta=\frac{1}{2}$ when $i\leq m_k$), and $\hat{\mathbf{v}}$ replaces $\mathbf{v}_i^k$ and $\mathbf{v}_l^k$ in V_k in the formation of $V_k^{(1)}$ and $V_k^{(2)}$, respectively. Therefore $det(V_k^{(1)})=det(V_k^{(2)})=\beta det(V_k)$. In both cases the $\mathbf{v}_0$ vector is unchanged, with $\mathbf{v}_0^k=\sum\delta_j^k\mathbf{v}_j^k$ before the subdivision. After subdivision, $\mathbf{v}_0^k$ needs be rewritten in terms of the columns of both of the matrices $V_k^{(1)}$ and $V_k^{(2)}$. In the first case, $\mathbf{v}_0^k$ can be rewritten as

$$\mathbf{v}_0^k=\frac{\delta_i^k}{\beta}(\beta(\mathbf{v}_i^k+\mathbf{v}_l^k))+(\delta_l^k-\delta_i^k)\mathbf{v}_l^k+\sum_{j\neq i\neq l}\delta_j^k\mathbf{v}_j^k,$$

and this determines δ^{k1} for the subregion $S_k^{(1)}$. A similar analysis can be done to determine δ^{k2} for the subregion $S_k^{(2)}$.

If $i=0$ then $\hat{\mathbf{v}}=(\mathbf{v}_0^k+\mathbf{v}_l^k)/2$. For the subregion $S_k^{(1)}$, $\hat{\mathbf{v}}$ has replaced $\mathbf{v}_0^k$, so $det(V_k^{(1)})=det(V_k)$. The vector $\mathbf{v}_0^{k1}=\mathbf{v}_l^k/2+\sum_{j=1}^n\delta_j^k\mathbf{v}_j^k/2$, so δ^{k1} is determined for the subregion $S_k^{(1)}$. The subregion $S_k^{(2)}$ has had $\mathbf{v}_l^k$ replaced by $\hat{\mathbf{v}}$. This can be expressed in matrix form by $V_k^{(2)} = V_k + (\hat{\mathbf{v}} - \mathbf{v}_l^k)\mathbf{e}_l^t = V_k + (\mathbf{v}_0^k - \mathbf{v}_l^k)\mathbf{e}_l^t/2 = V_k(I + (\boldsymbol{\delta}^k - \mathbf{e}_l)\mathbf{e}_l^t/2)$. Therefore $det(V_k^{(2)}) = det(V_k)(1 + (\delta_l^k - 1)/2) = det(V_k)(1 + \delta_l^k)/2$. The vector $\mathbf{v}_0^{k2} = \mathbf{v}_0^k$, but it needs to be rewritten in terms of the new basis. This can be done using straightforward matrix algebra.

$$\begin{aligned}
\mathbf{v}_0^{k2} &= V^k\boldsymbol{\delta}^k = V_k^{(2)}(I + (\boldsymbol{\delta}^k - \mathbf{e}_l)\mathbf{e}_l^t/2)^{-1}\boldsymbol{\delta}^k = V_k^{(2)}(I - \frac{(\boldsymbol{\delta}^k - \mathbf{e}_l)\mathbf{e}_l^t}{2 + \mathbf{e}_l^t(\boldsymbol{\delta}^k - \mathbf{e}_l)})\boldsymbol{\delta}^k \\
&= V_k^{(2)}(I - \frac{(\boldsymbol{\delta}^k - \mathbf{e}_l)\mathbf{e}_l^t}{1 + \delta_l^k})\boldsymbol{\delta}^k = V_k^{(2)}(\boldsymbol{\delta}^k - \frac{(\boldsymbol{\delta}^k - \mathbf{e}_l)\delta_l^k}{1 + \delta_l^k}) = V_k^{(2)}(\boldsymbol{\delta}^k + \delta_l^k\mathbf{e}_l)/(1 + \delta_l^k),
\end{aligned}$$

so $\boldsymbol{\delta}^{k2}$ is determined for the subregion $S_k^{(2)}$.

The updating formulas for $det(V_k)$ and $\boldsymbol{\delta}^k$ are now summarized.

$$det(V_k^{(\nu)}) = det(V_k) \times \begin{cases} 1 & \text{if } i = 0 \text{ and } \nu = 1 \\ (1 + \delta_l^k)/2 & \text{if } i = 0 \text{ and } \nu = 2 \\ \beta & \text{otherwise.} \end{cases}$$

$$\delta_j^{k1} = \begin{cases} \left.\begin{cases} (1 + \delta_l^k)/2 & \text{if } j = l \\ \delta_j^k/2 & \text{if } j \neq l \end{cases}\right\} & \text{and } i = 0, \\ \left.\begin{cases} \delta_i^k/\beta & \text{if } j = i \\ \delta_l^k - \delta_i^k & \text{if } j = l \\ \delta_j^k & \text{if } j \neq l \neq i \end{cases}\right\} & \text{and } i > 0. \end{cases}$$

$$\delta_j^{k2} = \begin{cases} \left.\begin{cases} 2\delta_l^k/(1 + \delta_l^k) & \text{if } j = l \\ \delta_j^k/(1 + \delta_l^k) & \text{if } j \neq l \end{cases}\right\} & \text{and } i = 0, \\ \left.\begin{cases} \delta_l^k/\beta & \text{if } j = l \\ \delta_i^k - \delta_l^k & \text{if } j = i \\ \delta_j^k & \text{if } j \neq l \neq i \end{cases}\right\} & \text{and } i > 0. \end{cases}$$

This completes the derivation of the formulas for updating the quantities $det(V_k)$ and $\boldsymbol{\delta}^k$. These formulas allow the updating process to be completed in $O(n)$ time.

Two desirable stability properties of the $\boldsymbol{\delta}^k$ vectors are $\delta_j^k \geq 0$ and $\sum_{j=1}^n \delta_j^k \leq 1$. Initially, $\boldsymbol{\delta}^k = \mathbf{0}$, for the n+1 original subregions of S, so both properties are satisfied. However, the differences used in some of the updating formulas for $\boldsymbol{\delta}^k$ indicate that negative components are possible, and as the subdivision process proceeds, it is easy to see that subregions with negative components could be generated. An example of such a subregion is generated from one of the original subregions, say S_1, by first dividing edge $(0, 1)$ and then dividing edge $(1, 2)$ of $S_1^{(1)}$. The first of the resulting subregions will have at least one negative component in $\boldsymbol{\delta}^k$. It is possible that restrictions could be placed on the subdivision algorithm in order to maintain the stability conditions, but such restrictions might also reduce the overall efficiency of the adaptive integration algorithm.

4. Implementation Progress and Concluding Remarks

The two subdivision methods described in the previous sections have been separately implemented as part of two different FORTRAN subroutines for adaptive integration over hyperspheres. The overall structure of these subroutines is similar to the structure used in the adaptive integration algorithms implemented for hyperrectangles and simplices (Berntsen, Espelid and Genz 1991 and Genz 1991). Fourth differences are used to determine priority edge directions for subdivision in the hypercube and simplex algorithms, and fourth differences are also used in the both spherical subdivision implementations. The simplex integration rules that are used by the subroutines are the rules developed by Grundmann and Möller (1978). These rules are implemented with a choice of degrees 5, 7 and 9. Differences of two successive degree rules are used for error estimation. The radial subdivision method also requires rules for integrals of the form $\int_c^1 r^{n-1} f(r) dr$. Formulas for Gauss rules of degrees 3 and 5 were derived and an attempt was made to construct higher degree Kronrod rules from the Gauss rules, so that error estimates could be efficiently computed. It was found, however, that the Kronrod rules with real nodes do not exist for some values of c and n, so the Kronrod rules are not implemented. At present, only degree 3 and 5 Gauss rules are implemented. It was also found that the determination of the weights and nodes for these rules becomes somewhat unstable as c approaches 1.

The number of integrand evaluations needed for a degree $2s+1$ Grundmann-Möller rule for the n-simplex is $(n+s+1)!/((n+1)!s!)$. A degree 5 formula, for example, requires $(n+3)(n+2)/2$ integrand values for the simplex subdivision method, but $3(n+2)(n+1)/2$ total integrand values are needed for the radial subdivision method because the product of an (n-1)-simplex rule and 3 point Gauss rule are used. There are similar differences between the two subdivision methods in the integrand evaluation numbers for other choices of degree, and in general, the rule application cost per subregion will be higher for the radial subdivision method.

However, there are two disadvantages with the simplex subdivision method. One disadvantage is that the Jacobian of the transformation for the simplex subdivision method is more complicated than the Jacobian for the radial subdivision method. This makes the complete integrand, which the integration rule is actually applied to, more difficult to integrate approximately, and could reduce the rate of the convergence for the adaptive algorithm. A second disadvantage is that the subregions that are produced are more complicated to maintain and, although direct methods were developed for updating the associated determinants and basis vectors, it is not known how stable these methods are.

Some testing has been performed with both implementations using fairly simple selected test integrands. Preliminary results indicate that neither subdivision method works substantially better than the other, and more extensive testing needs to be done before properly substantiated conclusions can be drawn about the relative strengths of the methods.

Acknowledgments

This work was supported in part by the United States National Science Foundation Grant DMS-9211640.

5. References

J. Berntsen, T. O. Espelid and A. Genz (1991), An Adaptive Algorithm for the Approximate Calculation of Multiple Integrals, *ACM Trans. Math. Softw.* 17, pp. 437-451.

H. S. M. Coxeter (1973), *Regular Polytopes*, Dover Publications, Inc., New York.

P. J. Davis and P. Rabinowitz (1984), *Methods of Numerical Integration*, Academic Press, New York.

P. vanDooren and L. deRidder (1976), An Adaptive Algorithm for Numerical Integration over an N-Dimensional Rectangular Region, *J. Comp. Appl. Math.* 2, pp. 207-217.

H. Engels (1980), *Numerical Quadrature and Cubature*, Academic Press, New York.

A. Genz (1991), An Adaptive Numerical Integration Algorithm for Simplices, in *Computing in the 90s, Proceedings of the First Great Lakes Computer Science Conference*, N. A. Sherwani, E. de Doncker and J. A. Kapenga (Eds.), Lecture Notes in Computer Science Volume 507, Springer-Verlag, New York, pp. 279-292.

A. Genz and A. A. Malik (1980), An Adaptive Algorithm for Numerical Integration over an N-Dimensional Rectangular Region, *J. Comp. Appl. Math.*, 6, pp. 295-302.

A. Grundmann and H. M. Möller (1978), Invariant Integration Formulas for the N-Simplex by Combinatorial Methods, *SIAM J. Numer. Anal.* 15, pp. 282-290.

I. P. Mysovskikh (1980), The Approximation of Multiple Integrals by Using Interpolatory Cubature Formula, in *Quantitative Approximation*, R. A. DeVore and K. Scherer (Eds.), Academic Press, New York, pp. 217-243.

A. H. Stroud (1971), *Approximate Calculation of Multiple Integrals*, Prentice-Hall, Englewood Cliffs, New Jersey.

Author's address:

Alan Genz
Department of Pure and Applied Mathematics
Washington State University
Pullman, WA 99164-3113 USA
acg@eecs.wsu.edu

Formules de quadrature dans R^2 avec "réseau" minimal de droites

Allal Guessab

Abstract

We will consider the problem of approximating a double integral on a convex compact set K as a minimal linear combination of integrals on the real line. We obtain Cubature formulae, which are exact on the space $Q_{2k+1}(K)$ of all polynomials of degree $2k+1$ respectively to each variable $x_i, i = 1, 2$.

1 Introduction

La théorie sur les formules de quadrature numérique pour les fonctions d'une variable est bien développée : citons par exemple Davis et Rabinowitz [2], Krylov [5], Stroud et Secrest [6] et Gautschi [3] (voir aussi Brass [1]) pour les formules de quadrature de type Gauss.

Il est clair qu'en dimension 1, le domaine d'intégration est beaucoup plus simple géométriquement qu'en dimension quelconque ($n \geq 2$), et que, par ailleurs, la théorie sur les polynômes orthogonaux y est plus facilement utilisable pour la construction de telles formules.

Ici, dans ce travail, le point de vue abordé consiste à approcher une intégrale double par une combinaison linéaire d'intégrales simples. Ainsi, cet article est consacré à la recherche des formules de quadrature avec "réseau" minimal de droites de la forme :

$$I_{K,w}(f) = \iint_K w(x,y)f(x,y)\,dxdy \approx A_{K,w}(f), \tag{1}$$

où

$$A_{K,w}(f) = \sum_{i=1}^{m} \lambda_i \int_{\Gamma_i} w(x_1,x_2)f(x_1,x_2)\,d\sigma, \tag{2}$$

et

$$\lambda_i > 0, \forall i = 1, \ldots, m, \tag{3}$$

$$\Gamma_i = \left\{ (x_1,x_2) \in K; x_1 \sin(\psi_i) - x_2 \cos(\psi_i) + a_i = 0 \right\} \tag{4}$$

avec

$$\psi_i \in [0,\pi], a_i \in \mathrm{R} \tag{5}$$

et où K est un compact convexe de R^2, à frontière de mesure nulle, et $mesK > 0$, w étant continue et strictement positive sur K. De plus (1) est exacte sur $Q_{2k+1}(K)$, i.e.

$$I_{K,w}(R) = A_{K,w}(R), \ \forall R \in Q_{2k+1}(K), \tag{6}$$

où $k \in \mathbb{N}$, et $Q_{2k+1}(K)$ désigne l'espace vectoriel des polynômes de degré $2k+1$ par rapport à chacune des 2 variables x_1, x_2.

Etant donné K et w, le problème est donc de déterminer le plus petit entier naturel m, les coefficients λ_i et les droites Γ_i de sorte que la formule de quadrature (1) soit exacte sur $Q_{2k+1}(K)$.

2 Une caractérisation de formules de quadrature de type (1)

Dans toute la suite on définit

$$d_i(x_1, x_2) = x_1 \sin(\psi_i) - x_2 \cos(\psi_i) + a_i. \tag{7}$$

D'autre part, on note par $\mathcal{O}_{k+1}(K, w)$ l'espace vectoriel des polynômes appartenant à $Q_{k+1}(K)$ et orthogonaux à $Q_k(K)$ relativement à K et la fonction w, i.e.

$$R \in \mathcal{O}_{k+1}(K, w) \iff I_{K,w}(RQ) = 0, \forall Q \in Q_k(K). \tag{8}$$

On rappelle que $\mathcal{O}_{k+1}(K, w)$ est de dimension $2k+3$, et admet une base de la forme (cf. Guessab [4])

$$\mathcal{B}_{k+1}(K, w) = \mathcal{B}^{(1)}_{k+1}(K, w) \cup \mathcal{B}^{(2)}_{k+1}(K, w) \tag{9}$$

où

$$\mathcal{B}^{(1)}_{k+1}(K, w) = \left\{ B^{(1)}_{(k+1,i)} = x_1^{k+1} x_2^i + Q^{(1)}_i, i = 0, 1, \cdots, k+1 \right\}, \tag{10}$$

$$\mathcal{B}^{(2)}_{k+1}(K, w) = \left\{ B^{(2)}_{(i,k+1)} = x_1^i x_2^{k+1} + Q^{(2)}_i, i = 0, 1, \cdots, k \right\}, \tag{11}$$

où $Q^{(1)}_i, Q^{(2)}_i$ sont déterminés par la relation d'orthogonalité.

Dans la proposition qui suit, on détermine le plus petit nombre m de droites de la formule de quadrature (1), qui est supposée exacte sur $Q_{2k+1}(K)$.

Proposition 2.1 *On suppose donnée une formule de quadrature de type (1) exacte sur $Q_{2k+1}(K)$. Alors $m \geq k+1$.*

Preuve Supposons que $m \leq k$, et considérons le polynôme

$$V = \prod_{i=1}^{m} d_i, \tag{12}$$

il est clair que le polynôme V^2 appartient à $Q_{2k+1}(K)$, donc

$$0 < I_{K,w}(V^2) = A_{K,w}(f) = 0, \tag{13}$$

puisque la formule de quadrature est exacte sur $Q_{2k+1}(K)$. D'où la contradiction, donc $m \geq k+1$. □

Définition 2.1 *Toute formule de quadrature de type (1) exacte sur* $Q_{2k+1}(K)$ *avec* $m = k + 1$ *est dite à "réseau" minimal de droites.*

On a alors la

Proposition 2.2 *On suppose donnée une formule de quadrature de type (1) exacte sur* $Q_{2k+1}(K)$*, avec "réseau" minimal de droites* ($m = k + 1$). *Alors tous les coefficients* λ_i *sont strictement positifs.*

Preuve Pour tout i fixé dans $\{1, 2, ..., m\}$, posons:

$$R_i = \prod_{j=1, j\neq i}^{k+1} d_j \tag{14}$$

où d_j est la droite définie par (7), alors le polynôme R_i^2 appartient à $Q_{2k+1}(K)$, ainsi compte tenu de l'exactitude de la formule de quadrature sur $Q_{2k+1}(K)$, on obtient :

$$0 < I_{K,w}(R_i^2) = \lambda_i \int_{\Gamma_i} w(x_1, x_2) R_i^2(x_1, x_2)\, d\sigma \tag{15}$$

par suite $\lambda_i > 0$, pour tout i. □

Proposition 2.3 *On suppose donnée une formule de quadrature de type (1) exacte sur* $Q_{2k+1}(K)$*, avec "réseau" minimal de droites* ($m = k + 1$). *Alors*

1) Le polynôme $V = \prod_{i=1}^{m} d_i$ *appartient à* $\mathcal{O}_{k+1}(K, w)$.

2) Tout polynôme R *appartenant à* $\mathcal{O}_{k+1}(K, w)$ *vérifie*

$$J_i = \int_{\Gamma_i} w(x_1, x_2) R(x_1, x_2) R_i(x_1, x_2)\, d\sigma = 0 \tag{16}$$

où

$$R_i(x_1, x_2) = \prod_{j=1, j\neq i}^{m} d_j. \tag{17}$$

Preuve 1) Pour tout polynôme Q appartenant à $Q_k(K)$, il est clair que QV appartient à $Q_{2k+1}(K)$, ainsi

$$I_{K,w}(QV) = A_{K,w}(QV) = 0, \forall Q \in Q_k(K), \tag{18}$$

donc V appartient à $\mathcal{O}_{k+1}(K, w)$.

2) Soit R un polynôme de $\mathcal{O}_{k+1}(K, w)$, alors pour tout i le polynôme RR_i appartient à $Q_{2k+1}(K)$, donc

$$0 = I_{K,w}(RR_i) = A_{K,w}(RR_i) = \lambda_i J_i, \tag{19}$$

or λ_i est strictement positif (cf. Proposition 2.2), donc $J_i = 0$. □

Remarque En utilisant (7), on vérifie que le polynôme $\prod_{i=1}^{k+1} d_i$ peut s'écrire sous la forme

$$\prod_{i=1}^{k+1} d_i = c_1 x_1^{k+1} + c_2 x_2^{k+2} + Q,$$

où $Q \in Q_k(K)$. Il vient alors de la proposition 2.3 que s'il n'existe pas deux constantes réelles c_1 et c_2 telles que

$$\prod_{i=1}^{k+1} d_i = c_1 B^{(1)}_{(k+1,0)} + c_2 B^{(2)}_{(0,k+1)},$$

où les polynômes $B^{(1)}_{(k+1,0)}, B^{(2)}_{(0,k+1)}$ sont définis dans (10) et (11). Alors il n'existe pas de formule de quadrature de type (1) à "réseau" minimal de droites.

3 Conditions d'existence d'une formule de quadrature à réseau minimal de droites

Il vient alors le résultat fondamental suivant :

Théorème 3.1 *On suppose donné k, tel que*

1) Il existe deux constantes réelles c_1 et c_2, et $k+1$ droites distinctes de la forme (7) telles que

$$\prod_{i=1}^{k+1} d_i = c_1 B^{(1)}_{(k+1,0)} + c_2 B^{(2)}_{(0,k+1)}. \tag{20}$$

2) Pour tout i, j les polynômes $B^{(1)}_{(k+1,i)}, B^{(2)}_{(i,k+1)}$ définis dans (10) et (11) vérifient pour tout polynôme Q appartenant à $Q_k(K)$

$$J^{(1)}_{i,j} = \int_{\Gamma_j} w(x_1,x_2) B^{(1)}_{(k+1,i)}(x_1,x_2) Q(x_1,x_2)\, d\sigma = 0 \tag{21}$$

$$J^{(2)}_{i,j} = \int_{\Gamma_j} w(x_1,x_2) B^{(2)}_{(i,k+1)}(x_1,x_2) Q(x_1,x_2)\, d\sigma = 0 \tag{22}$$

3) La formule de quadrature (1) est exacte sur $Q_k(K)$.

Alors la formule de quadrature (1) est exacte sur $Q_{2k+1}(K)$.

Preuve En effet, supposons donnée une formule de quadrature de type (1) exacte sur $Q_k(K)$, et telle que les polynômes $B^{(1)}_{(k+1,i)}, B^{(2)}_{(i,k+1)}$ vérifient les relations (21) et (22).

Soit maintenant Q un polynôme appartenant à $Q_{2k+1}(K)$, alors Q peut s'écrire sous la forme

$$Q = \sum_{i=0}^{k+1} Q_i^{(1)} B^{(1)}_{(i,k+1)} + \sum_{i=0}^{k} Q_i^{(2)} B^{(2)}_{(i,k+1)} + R \tag{23}$$

où les polynômes $Q_i^{(1)}, Q_i^{(2)}, R$ appartiennent à $Q_k(K)$, ainsi, compte tenu de l'orthogonalité des polynômes $B^{(1)}_{(k+1,i)}, B^{(2)}_{(i,k+1)}$ à $Q_k(K)$, relativement à K et w, il vient

$$I_{K,w}(Q) = I_{K,w}(R) \tag{24}$$

or d'après 3)

$$I_{K,w}(R) = A_{K,w}(R), \tag{25}$$

puisque R appartient à $Q_k(K)$ et la formule de quadrature est exacte sur $Q_k(K)$, il vient alors

$$I_{K,w}(Q) = A_{K,w}(R). \tag{26}$$

Enfin, puisque

$$A_{K,w}\left(\sum_{i=0}^{k+1} Q_i^{(1)} B^{(1)}_{(k+1,i)} + \sum_{i=0}^{k} Q_i^{(2)} B^{(2)}_{(i,k+1)}\right) = 0, \tag{27}$$

on obtient

$$I_{K,w}(Q) = A_{K,w}(Q), \tag{28}$$

d'où le résultat. □

A partir des résultats connus en dimension 1 pour l'obtention de formule de quadrature de Gauss-Christoffel exacte sur $P_{2k+1}([-1,1])$ (cf. [2] page 74)

Il vient le

Corollaire 3.1 *On suppose que* $K = [-1,1] \times [-1,1], w = 1$. *Alors pour tout* $k \in \mathbb{N}$, *il existe une formule de quadrature à "réseau" minimal de droites, exacte sur* $Q_{2k+1}(K)$ *de la forme*

$$A_{K,w}(f) = \sum_{i=1}^{k+1} \lambda_i \int_{-1}^{1} f(x_1, x_{i,2})\, dx_1, \tag{29}$$

où $x_{i,2}$ *sont les zéros du polynôme orthogonal de degré* $k+1$ *relativement à* $[-1,1]$ *et à la fonction* $w = 1$, *et où* λ_i *est le coefficient d'intégration correspondant à* $x_{i,2}$ *dans la formule de quadrature de Gauss-Christoffel.*

Preuve Le résultat se déduit de l'application du théorème 3.1. En effet dans ce cas, les polynômes $B^{(1)}_{(k+1,i)}, B^{(2)}_{(i,k+1)}$ peuvent s'écrire sous la forme

$$B^{(1)}_{(k+1,i)}(x_1, x_2) = P_{k+1}(x_1) P_i(x_2) \tag{30}$$

$$B^{(2)}_{(i,k+1)}(x_1, x_2) = P_i(x_1) P_{k+1}(x_2) \tag{31}$$

où P_i désigne le polynôme orthogonal normalisé de degré i relativement à $[-1,1]$ et à la fonction $w = 1$, et où par convention $P_0 = 1$.

D'autre part, le polynôme $B^{(2)}_{(0,k+1)}$ peut s'écrire sous la forme

$$B^{(2)}_{(0,k+1)}(x_1, x_2) = \prod_{i=1}^{k+1} (x_2 - x_{i,2}), \tag{32}$$

où les $x_{i,2}$ sont les zéros du polynôme de degré $k+1$ de Legendre. Ainsi pour tout i, les droites d_i peuvent être choisies de la forme (avec le choix particulier $c_1 = 0$ et $c_2 = 1$ dans le théorème 3.1 1))

$$d_i(x_1, x_2) = x_2 - x_{i,2}. \tag{33}$$

Pour le point 2), on a dans ce cas pour tout $Q \in Q_k(K)$

$$J_{i,j}^{(1)} = \int_{-1}^{1} P_{k+1}(x_1) P_i(x_{j,2}) Q(x_1, x_{j,2})\, dx_1 = 0, \tag{34}$$

car le polynôme P_{k+1} est orthogonal à $P_k([-1,1])$, et

$$J_{i,j}^{(2)} = \int_{-1}^{1} P_i(x_1) P_{k+1}(x_{j,2}) Q(x_1, x_{j,2})\, dx_1 = 0 \tag{35}$$

car $P_{k+1}(x_{j,2}) = 0$.

Pour le point 3), on a pour tout x_1 fixé dans $[-1,1]$, et pour tout Q appartenant à $Q_{2k+1}(K)$

$$\int_{-1}^{1} Q(x_1, x_2)\, dx_2 = \sum_{i=1}^{k+1} \lambda_i Q(x_1, x_{i,2}), \tag{36}$$

où pour tout $i = 1, \ldots, k+1$, λ_i désigne le coefficient d'intégration correspondant à $x_{i,2}$ dans la formule de Gauss-Christoffel de degré $2k+1$ (cf. [2], page 74). Il résulte alors que

$$I_{K,w}(Q) = A_{K,w}(Q), \forall Q \in Q_{2k+1}(K) \tag{37}$$

et le résultat suit. □

Remarque Si on applique la formule de quadrature de Gauss-Christoffel de degré $2k+1$

$$A_{[-1,1],1}(g) = \sum_{i=1}^{k+1} \lambda_i g(x_{i,1}), \tag{38}$$

aux intégrales $\int_{-1}^{1} f(x_1, x_{j,2}) dx_1, \forall j = 1, \ldots k+1$. On obtient alors la formule de quadrature

$$\int_{-1}^{1} \int_{-1}^{1} f(x_1, x_2)\, dx_1 dx_2 \approx \Pi_{K,1}(f) = \sum_{i=1, j=1}^{k+1} \lambda_i \lambda_j f(x_{i,1}, x_{i,2}), \tag{39}$$

qui est exacte sur $Q_{2k+1}([-1,1] \times [-1,1])$. On retrouve la formule produit de Gauss-Christoffel, qui est exacte sur $Q_{2k+1}([-1,1] \times [-1,1])$ et qui est à nombre de noeuds minimal (cf. [4]).

De la même manière on a le

Corollaire 3.2 *On suppose que $K = [-1,1] \times [-1,1]$, et $w = 1$. Alors pour tout $k \in \mathbb{N}$, il existe une formule de quadrature à "réseau" minimal de droites, exacte sur $Q_{2k+1}(K)$ de la forme*

$$A_{K,w}(f) = \sum_{i=1}^{k+1} \lambda_i \int_{-1}^{1} f(x_{i,1}, x_2)\, dx_2 \tag{40}$$

où $x_{i,1}$ sont les zéros du polynôme orthogonal de degré $k+1$ relativement à $[-1,1]$ et à la fonction $w=1$, et où λ_i est le coefficient d'intégration correspondant à $x_{i,1}$ dans la formule de quadrature de Gauss-Christoffel.

Preuve Analogue à celle du corollaire 3.1. □

Remarque Dans le cas où $K=[-1,1]\times[-1,1]$ et $w=1$. Les formules de quadrature du corollaire 3.1 et 3.2 ont été établies, avec $c_1=1$ et $c_2=0$, où $c_1=0$ et $c_2=1$ (cf. théorème 3.1 1)). On montre que si $k\geq 2$, alors le point 1) du théorème 3.1 n'est pas vérifié si $c_1\neq 0$ et $c_2\neq 0$. (cf. [4]).

4 Exemples de formule de quadrature de type (1) dans le cas où K n'est pas un carré

Dans ce paragraphe, on suppose que K et w vérifient de plus les conditions suivantes, pour tout $(x_1,x_2)\in K$, on a

$$\begin{cases} (-x_1,-x_2)\in K, (x_2,x_1)\in K, \\ w(x_1,x_2)=w(x_2,x_1) \end{cases} \tag{41}$$

D'autre part on note pour tout $(\alpha_1,\alpha_2)\in \mathbb{N}\times\mathbb{N}$ par

$$I_{\alpha_1,\alpha_2}=\int_K w(x_1,x_2)x_1^{\alpha_1}x_2^{\alpha_2}\,dx_1dx_2 \tag{42}$$

On construit alors les formules de quadrature à "réseau" minimal de droites. Tout d'abord on a la:

Proposition 4.1 *Soient K et w vérifiant (39), on suppose que $I_{2,0}^2=I_{0,0}I_{2,2}$. Si de plus le réel r supposé non nul défini par*

$$r=\sup\left\{x_1,(x_1,\sqrt{I_{2,0}/I_{0,0}})\in K\right\} \tag{43}$$

vérifie

$$\int_{-r}^{r} w\left(x_1,\sqrt{I_{2,0}/I_{0,0}}\right)(x_1^2-I_{2,0}/I_{0,0})dx_1=0. \tag{44}$$

Alors il existe une formule de quadrature à "réseau" minimal de droites, exacte sur $Q_3(K)$ de la forme

$$\begin{aligned} A_{K,w}(f)= {} & \lambda_1\int_{-r}^{r} w(x_1,x_{1,2})f(x_1,x_{1,2})dx_1 \\ & +\lambda_2\int_{-r}^{r} w(x_1,-x_{1,2})f(x_1,-x_{1,2})dx_1 \end{aligned} \tag{45}$$

et

$$x_{1,2}=\sqrt{I_{2,0}/I_{0,0}}. \tag{46}$$

Preuve Le résultat se déduit de l'application du théorème 3.1. En effet, le polynôme $B_{(2,0)}^{(2)}$ (cf. (1)) peut s'écrire sous la forme

$$\begin{aligned} B_{(2,0)}^{(2)}(x_1, x_2) &= x_2^2 - I_{2,0}/I_{0,0} && (47)\\ &= (x_2 - x_{1,2})(x_2 + x_{1,2}) && (48)\end{aligned}$$

avec $x_{1,2} = \sqrt{I_{2,0}/I_{0,0}}$ ainsi (en choisissant $c_1 = 0$ et $c_2 = 1$ dans (20)) on a

$$\begin{aligned} d_1(x_1, x_2) &= x_2 - I_{2,0}/I_{0,0}, && (49)\\ d_2(x_1, x_2) &= x_2 + I_{2,0}/I_{0,0}, && (50)\end{aligned}$$

et on note que si r vérifie la relation (41), alors le point 2) du théorème 3.1 suit. Enfin les expressions de λ_1 et λ_2 sont obtenues en résolvant le système suivant

$$I_{K,w}(x_1^i x_2^j) = A_{K,w}(x_1^i x_2^j), \forall i = 0, 1, j = 0, 1. \qquad (51)$$

□

De même on a la

Proposition 4.2 *On suppose que K et w vérifient (39), et*

$$I_{2,2}I_{4,0} = I_{2,0}I_{4,2}, I_{2,2}I_{4,4} = I_{4,2}.$$

Si de plus les réels r, r' supposés non nul définis par

$$r = \sup\left\{x_1; (x_1, 0) \in K\right\}, \qquad (52)$$

$$r' = \sup\left\{x_1; (x_1, \sqrt{I_{4,2}/I_{2,2}}) \in K\right\} \qquad (53)$$

vérifient

$$\int_{-r}^{r} w(x_1, 0)\left(1 - \frac{I_{0,0}I_{4,2} - I_{2,0}I_{2,2}}{I_{2,0}I_{4,2} - I_{2,2}^2}\right)dx_1 = 0, \qquad (54)$$

$$\int_{-r}^{r} w(x_1, 0)x_1^2\left(x_1^2 + \frac{I_{2,2}I_{4,2} - I_{4,0}I_{4,2}}{I_{2,0}I_{4,2} - I_{2,2}^2}\right)dx_1 = 0, \qquad (55)$$

$$\int_{-r'}^{r'} w(x_1, x_{1,2})\left(x_1^2 - \frac{I_{2,2}}{I - 2,0}\right)dx_1 = 0, \qquad (56)$$

$$\int_{-r'}^{r'} w(x_1, x_{1,2})x_1^2\left(x_1^2 + \frac{I_{2,2}I_{4,2} - I_{4,0}I_{4,2}}{I_{2,0}I_{4,2} - I_{2,2}^2}\right)dx_1 = 0, \qquad (57)$$

avec $x_{1,2} = \sqrt{I_{4,2}/I_{2,2}}$. Alors il existe une formule de quadrature à "réseau" minimal de droites, exacte sur $Q_5(K)$ de la forme

$$\begin{aligned} A_{K,w}(f) = {} & \lambda_1 \int_{-r}^{r} w(x_1, 0)f(x_1, 0)dx_1 \\ & + \lambda_2 \int_{-r'}^{r'} w(x_1, x_{1,2})f(x_1, x_{1,2})dx_1 \\ & + \lambda_3 \int_{-r'}^{r'} w(x_1, -x_{1,2})f(x_1, -x_{1,2})dx_1, && (58)\end{aligned}$$

avec

$$\lambda_1 = \frac{I_{2,0} - I_{2,2}/x_{1,2}}{\int_{-r}^{r} w(x_1, 0)dx_1} \tag{59}$$

$$\lambda_2 = \lambda_3 = \frac{I_{2,2}/2x_{1,2}^2}{\int_{-r'}^{r'} w(x_1, x_{1,2})x_1^2 dx_1} \tag{60}$$

Preuve Analogue à celle de la proposition 4.1 □

References

1. Brass H. (1977) "Quadraturverfahren", Vandenhoech and Ruprecht, Göttingen.
2. Davis P. and J. Rabinowitz P. (1984) "Methods of numerical integration". Academic Press, New York.
3. Gautschi W. (1981): "A survey of Gauss-Christoffel quadrature formulae." Ed. by P. L. Butzer and F. Féher, Aachen, Birkhauser Verlag, Basel.
4. Guessab A. (1986): "Cubature formulae which are exact on spaces P, intermediate between P_k and $Q_{k'}$. Numer. Math., Vol. 49, pp. 561-576.
5. Krylov H.I. (1962): " Approximate calculation of integrals. MacMillan New York, London.
6. Stroud A.H. and Secrest D. (1966): "Gaussian quadrature formulas". Prentice Hall, Englewood Cliffs, N.J.

Allal Guessab
University of Pau,
I.P.R.A. Department of Applied Mathematics,
URA 1204-CNRS,
Avenue de l'Université, 64000 Pau, France.
E-mail: guessab @ iprvs1.univ-pau.fr

International Series of Numerical Mathematics, Vol. 112, © 1993 Birkhäuser Verlag Basel

Quadrature rules derived from linear convergence accelerations schemes

Sven-Åke Gustafson

Abstract. We shall develop efficient schemes for approximating the integral over the entire real line with a finite linear combination of functional values. To do this we first approximate the integral with a trapezoidal sum, then apply a linear convergence acceleration scheme to approximate this infinite sum with a linear combination of a finite number of terms. We will study two important cases in detail, namely when the integrand is oscillating or decays exponentially. If we also require the integrand to be analytic on a strip along the real axis, the trapezoidal approximation can be shown to be very accurate, provided some simple regularity assumptions are satisfied.

1 Introduction

We shall describe methods for deriving quadrature rules of the form

$$\int_{-\infty}^{\infty} f(s)\, ds = \sum_{r=-N_1}^{N_2} A_r f(rh) + R_T, \qquad (1)$$

where the step-size h is a positive number, N_1 and N_2 are integers. The coefficients A_r depend on N_1, N_2 and h and may be either negative or positive. We shall describe conditions, which need to be imposed in order to guarantee that the truncation error satisfies a bound of the form

$$|R_T| \le c_1 e^{-c_2\sqrt{N}}, \; N = \max(N_1, N_2), \qquad (2)$$

where c_1 and c_2 are positive constants. The general idea is, first to approximate the integral in (1) with a trapezoidal sum, then to evaluate the latter with a suitable linear convergence acceleration scheme. Thus

$$T(h) = h \sum_{r=-\infty}^{\infty} f(rh), \tag{3}$$

and

$$T(h) \approx \sum_{r=-N_1}^{N_2} A_r f(rh). \tag{4}$$

If now f is analytic on the strip

$$S = \{|\Im(s)| \leq D\} \tag{5}$$

then, under certain regularity assumptions given in Stenger (1981), (1993) and Lund and Bowers (1992), $T(h)$ approximates the integral very accurately. In fact, we may show that

$$|\int_{-\infty}^{\infty} f(s)\, ds - T(h)| \leq d_1 e^{-d_2/h}, \tag{6}$$

where d_1, d_2 are positive numbers, which do not depend on the step-size h. This holds for:

Example 1

$$f(s) = \frac{1}{1+s^2}, \tag{7}$$

for which function we may take any number $D < 1$ in (5). In this case we get

$$T(h) = h \sum_{r=-\infty}^{\infty} \frac{1}{1+r^2h^2}. \tag{8}$$

This example illustrates a situation where the trapezoidal sum $T(h)$ gives an accurate approximation of the integral in (1) but $T(h)$ itself converges slowly if summed up term by term. A different situation occurs for:

Example 2

$$f(s) = e^{-s^2}, \tag{9}$$

where the sum $T(h)$ converges rapidly. Hence a formula (1) is obtained by taking $A_r = h$ and truncating the sum (3), retaining a suitable number of terms. Due to (2) we may estimate the integral in (1) by selecting a suitable number h and then evaluating the sequence of trapezoidal approximations

$$T(h),\ T(h/2),\ T(h/4), \ldots$$

which converges rapidly towards the true value of the integral.

Definition 1 *The function f is said to belong to the class $E(\alpha_1, \alpha_2)$, if there are positive numbers α_1, α_2 and a constant C such that*

$$\begin{aligned} |f(x)| &\leq Ce^{-\alpha_1 |x|},\ x \in (-\infty, 0) \\ |f(x)| &\leq Ce^{-\alpha_2 |x|},\ x \in [0, \infty) \end{aligned}$$

Example 3

$$f(s) = \frac{1}{e^{-0.5s} + e^s}$$

belongs to $E(0.5, 1)$.

We write

$$T(h) = h[f(0) + \sum_{r=1}^{\infty} f(rh) + \sum_{r=1}^{\infty} f(-rh)] \tag{10}$$

If $f \in E(\alpha_1, \alpha_2)$, then the first sum in (10) converges like a geometric series with terms $e^{-\alpha_1 rh}$ the second like a geometric series with terms $e^{-\alpha_2 rh}$. We may approximate $T(h)$ with the finite sum

$$T_{N_1,N_2}(h) = h[f(0) + \sum_{r=1}^{N_1} f(rh) + \sum_{r=1}^{N_2} f(-rh)].$$

It has $N_1 + N_2 + 1$ terms. The error occurring when the integral in (1) is approximated with this sum consists of the discretization error caused by replacing the integral with the infinite trapezoidal sum and the truncation error, which arises when this sum is truncated, retaining $N_1 + N_2 + 1$ terms.

The discretization error decreases when h is reduced but then N_1, N_2 must be chosen larger to control the truncation error. A suitable strategy for choosing the parameters N_1, N_2 and h is described in Stenger (1993) and Lund and Bowers (1992).

The case of oscillating integrands is illustrated by:

Example 4

$$f(s) = e^{ist} \frac{1}{\sqrt{1+s^2}}$$

where t is a real number.

Here the trapezoidal sum converges slowly and direct summation is not feasible in practice. Instead, convergence acceleration is used to approximate the sum with a finite number of terms, and for a large and important class of functions the truncation error of the accelerated series decreases exponentially in the number of terms used for approximating the sum. This fact enables us to modify and extend the strategies in Stenger (1993) and Lund and Bowers (1992) for selecting N_1, N_2 and h to the case of oscillating integrands. This will be discussed elsewhere.

The use of convergence acceleration is essential in the oscillating case and it results in significant savings in computational effort in the exponentially decaying case. Both of these cases are included if write

$$f(s) = w(s)g(s)$$

where w may have one of the forms:

$$w(s) = e^{ist},\ w(s) = e^{-\alpha_1|s|},\ w(s) = e^{-\alpha_2|s|}$$

and g is bounded on the real line and analytic on the strip S in (5). Then the sums in (10) may be written

$$\sum_{r=0}^{\infty} z^r g((r+1)h) \tag{11}$$

where z is given by one of the formulas:

$$z = e^{iht},\ z = e^{-\alpha_1 h},\ z = e^{-\alpha_2 h}.$$

In Section 2 we will describe some linear methods for evaluating the power series (11).

The fact that the trapezoidal rule, under certain conditions gives a very accurate approximation to integrals over the entire real line has been known for a long time. See e.g. Davis and Rabinowitz (1984), Bowers and Lund (1992) and Stenger (1993) and the references given there. Sidi (1988) and (1992) considers oscillating functions and advocates replacing the integral with an integral over a finite interval and using extrapolation for calculating the infinite integral. In Sidi (1982) we find efficient quadrature rules similar to those developed here for the case of exponentially decaying integrands. The schemes given here are applicable for both decaying and oscillating integrands. They have the advantage that no precomputed weights and abscissas need to be stored. Also the calculations may be arranged such that no functional values are discarded when one seeks to improve the accuracy by halving the step-size h or increasing the parameter n to be introduced in Section 2. This is similar to the Romberg scheme but in contrast to Gaussian rules when an increase in the order means that all functional values which have been calculated earlier must be discarded. Another strategy to deal with slow convergence of the trapezoidal sum is to use some variable transformation which exploits the specific properties of the problem at hand. In addition to earlier quoted sources we mention here the papers by Takahasi and Mori (1973) and Talbot (1979). Numerous convergence acceleration and extrapolation schemes have been developed in recent years. See e.g. Wimp (1981) and Brezinski and Zaglia (1991) who list hundreds of papers including many contributions by Sidi and Wynn. Nonlinear methods are often believed to be most accurate but their stability properties are often uncertain and computational break-downs are difficult to avoid in a general context. In contrast, linear methods may be applied to any real or complex-valued function f and the influence of uncertainties in input data is easy to study. In the case of (1) the sum of the absolute values of the weights determines the condition of the problem. In order to bound the truncation error in the calculated sum we need to impose general assumptions on f as described in Section 2.

2 Linear convergence acceleration schemes derived from quadrature and interpolation

Consider the general power series

$$\mathcal{F}(z) = \sum_{r=0}^{\infty} c_r z^r, \tag{12}$$

where the function $\mathcal{F}$ is defined for those z which are such that the series (12) converges. We next introduce:

Definition 2 *Let n be a positive integer, $\eta_0(z), \ldots, \eta_{n-1}(z)$ n numbers, which as indicated may depend on z. We call*

$$\mathcal{F}_n(z) = \sum_{r=0}^{n-1} \eta_r(z) c_r, \tag{13}$$

a linear transformation of the power series (12).

Note that if we use (13) to approximate the sums in (10) $\eta_r(z)$ define the weights in the quadrature rule (1). It is desirable that $\mathcal{F}_n(z)$ should give an approximation to $\mathcal{F}(z)$ and often one seeks to construct sequences of transformations $\mathcal{F}_1(z)$, $\mathcal{F}_2(z), \ldots$ such that

$$\lim_{n\to\infty} \mathcal{F}_n(z) = \mathcal{F}(z).$$

Definition 3 *Associated with the transformation $\mathcal{F}_n(z)$ we introduce its generating polynomial $Q_n(z;\cdot)$ with argument t given by*

$$Q_n(z;t) = \sum_{r=0}^{n-1} \eta_r(z) t^r. \tag{14}$$

We now show how to derive rational expressions approximating the sum (12) by fitting linear combinations of geometric series to this power series. Let namely

$$t_1, t_2, \ldots, t_n, \tag{15}$$

be n fixed numbers. Next determine the unique solution

$$x_1, x_2, \ldots, x_n, \tag{16}$$

to the linear system

$$\sum_{i=1}^{n} x_i t_i^r = c_r, \quad r = 0, 1, \ldots, n-1. \tag{17}$$

Next put

$$\hat{c}_r = \sum_{i=1}^{n} x_i t_i^r, \quad r = 0, 1, \ldots, \tag{18}$$

and set

$$\mathcal{F}_n(z) = \sum_{r=0}^{\infty} \hat{c}_r z^r. \tag{19}$$

We note that

$$\hat{c}_r = c_r, \quad r = 0, 1, \ldots, n-1. \tag{20}$$

We show now that $\mathcal{F}_n(z)$ can be expressed equivalently as a rational expression with z as variable or as a linear transformation of the power series (12). We find namely

$$\mathcal{F}_n(z) = \sum_{r=0}^{\infty} \hat{c}_r z^r = \sum_{r=0}^{\infty} \sum_{i=1}^{n} x_i t_i^r z^r = \sum_{i=1}^{n} x_i \sum_{r=0}^{\infty} z^r t_i^r = \sum_{i=1}^{n} \frac{x_i}{1 - z t_i}. \tag{21}$$

Next we demonstrate that $\mathcal{F}_n(z)$ may be expressed as a linear transformation of the form (12).

Theorem 4 *Let $t_1, \ldots, t_n$ be as before and determine $\eta_0(z), \ldots, \eta_{n-1}(z)$ as the solution of the linear system*

$$\sum_{r=0}^{n-1} \eta_r(z) t_i^r = \frac{1}{1 - t_i z}, \quad i = 1, 2, \ldots, n. \tag{22}$$

Then if $x_1, \ldots, x_n$ solve (17),

$$\mathcal{F}_n(z) = \sum_{i=1}^{n} \frac{x_i}{1 - z t_i} = \sum_{r=0}^{n-1} \eta_r(z) c_r. \tag{23}$$

Proof
Combining (17), (21) and (22) we find

$$\mathcal{F}_n(z) = \sum_{i=1}^{n} \frac{x_i}{1 - z t_i} = \sum_{i=1}^{n} x_i \sum_{r=0}^{n-1} \eta_r(z) t_i^r = \sum_{r=0}^{n-1} \eta_r(z) c_r. \tag{24}$$

Remark 1 *Using (24) we may study how the calculated value $F_n(z)$ is influenced by errors in the given numbers c_r and conclude that the stability of the calculated sum is governed by the number*

$$\sum_{r=0}^{n-1} |\eta_r(z)|$$

which is determined by the choice of $t_1, \ldots, t_n$ and the value of z.

Remark 2 *We note that we determined the numbers x_i by solving a linear system of the same type as that encountered when one seeks to determine the weights of a mechanical quadrature rule while $\eta_r(z)$ were obtained after determining an interpolating polynomial. Since these systems are of Vandermonde type they may be treated by using e.g. the codes in Gustafson (1971). The results obtained up to now in the present section are valid for any sequence. In order to obtain bounds on $|\hat{\mathcal{F}}(z) - \mathcal{F}((z)|$ we need to introduce assumptions on the coefficients $c_0, c_1, \ldots$.*

We discuss here the case when the function g occurring in (11) is completely monotonic on the interval $[0, \infty]$ or can be written as the difference of two functions belonging to this class. This case is illustrated of Examples 5 and 6 in Section 3. Using a theorem by Bernstein (See e.g. Widder (1971) page 155) we verify directly

Lemma 5 *Let g be completely monotonic on $[0, \infty]$, i. e. be such that*

$$(-1)^n g^{(n)}(s) \geq 0,\ 0 \leq s \leq \infty,\ n = 0, 1, \ldots$$

and let h be a fixed positive number. Then there is a nondecreasing Stieltjes integrator α of bounded variation such that

$$g(rh) = \int_0^1 t^r \, d\alpha(t)\, dt,\ r = 0, 1, \ldots \tag{25}$$

Examples of completely monotonic functions are $g(s) = \exp(-s)$ and $g(s) = 1/(1+s)$. If g is the difference between two completely monotonic functions then the representation (25) still holds if we remove the demand that the integrator α is nondecreasing. In many situations $d\alpha(t)$ may be represented as $\rho(t)\, dt$ where ρ is a suitable weighting function. We establish

Lemma 6 *Let the sequence* $c_0, c_1, \ldots$ *admit the representation*

$$g(rh) = \int_0^1 t^r d\alpha(t)\, dt, \; r = 0, 1, \ldots \tag{26}$$

where the Stieltjes integrator α *is of bounded variation. Then*

$$\mathcal{F}(z) = \sum_{r=0}^{\infty} c_r z^r = \int_0^1 \frac{1}{1-zt}\, d\alpha(t), \tag{27}$$

$$\mathcal{F}_n(z) = \sum_{r=0}^{n-1} c_r \eta_r(z) = \int_0^1 Q_n(z;t)\, d\alpha(t). \tag{28}$$

We note that the integral (27) defines the analytic continuation of the function defined by (12) to the complex plane with exception for z real and $z \geq 1$. Our task is to evaluate this integral numerically. This may be done by approximating the integrand by $Q_n(z;t)$ and then use (13) or constructing an equivalent quadrature rule from (16) and (17) to arrive at the rational approximation (21). We now prove

Theorem 7 *Let* $t_1, t_2, \ldots, t_n$ *be* n *distinct points in* $[0,1]$ *and let* $Q_n(z;t)$ *be the uniquely determined polynomial of degree less than* n *which interpolates* $1/(1-tz)$ *at these points. Put*

$$P_n(t) = \prod_{i=1}^{n} (t - t_i)$$

Then

$$Q_n(z;t) = \frac{P_n(1/z) - P_n(t)}{(1-zt)P_n(1/z)} \tag{29}$$

and

$$F(z) - F_n(z) = \frac{1}{P_n(1/z)} \int_0^1 \frac{P_n(t)}{(1-zt)}\, d\alpha(t). \tag{30}$$

Proof:
The denominator of the expression at the right hand side of (29) is 0 *only for* $t = 1/z$ *but then the numerator is* 0 *as well and we may establish that the expression is bounded at a neighbourhood of this point. Hence it is a polynomial in* t *and its degree is less than* n. *Since* P_n *is* 0 *at* t_i, $i = 1, 2, \ldots, n$ *the desired interpolation property follows. The second statement follows upon integration with respect to* $d\alpha(t)$ *and using (27), (28).*

We specialise to the case of Chebyshev acceleration and establish

Theorem 8 *Use the assumptions and notations of the preceding Theorem and put*

$$P_n(t) = T_n^*(t),$$

the shifted Chebyshev polynomials, which are defined by the recurrence relations

$$T_0^*(t) = 1,\ T_1^*(t) = 2t - 1, \tag{31}$$
$$T_n^*(t) = 2(2t-1)T_{n-1}^*(t) - T_{n-2}^*(t). \tag{32}$$

Then

$$|F(z) - F_n(z)| = \mathcal{O}(\lambda^n(z)), \tag{33}$$

where

$$\lambda(z) = \min(|\lambda_1(z)|, |\lambda_2(z)|), \tag{34}$$
$$\lambda_1(z) = \frac{z}{(\sqrt{1-z}+1)^2}, \tag{35}$$
$$\lambda_2(z) = \frac{z}{(\sqrt{1-z}-1)^2}. \tag{36}$$

Proof.
We have $|T_n^*(t)| \leq 1,\ 0 \leq t \leq 1$. *Using (31) and (32) we verify by direct calculation*

$$T_n^*(t) = \frac{1}{2}(\lambda_1^n(z) + \lambda_2^n(z))$$

where $\lambda_1(z), \lambda_2(z)$ *are defined in (34). Combining these facts with (30) we arrive at the desired conclusion.*

If Chebyshev transformation is used and the assumptions of Theorem 8 are satisfied, then the transformed series converges like a geometric series, whose quotient depends on the value of z. Thus for $z = -1$, $z = i$ and $z = 1/e$ the absolute value of the quotient is 0.1716, 0.2168 and 0.1142 respectively. When z approaches 1 the quotient also approaches this value and hence the task of summing the series (12) becomes difficult. Also the stability of the calculated sum as measured by (23) deteriorates. These difficulties may be overcome by using batching as described in Gustafson (1991a) We consider the first sum of (10) which may be written

$$\sum_{r=0}^{\infty} f((r+1)h)$$

and let (12) be the corresponding power series. Let k be a positive integer and consider the k different series

$$\sum_{r=0}^{\infty} f((kr+1+\ell)h), \ \ell = 0, 1, \dots, k-1$$

Each of these corresponds to a power series with argument z^k and for a suitable choice of k the acceleration scheme may be more efficient than for the original series with argument z. A good choice is to determine k such that $z^k \approx i$ in the oscillating case and $z^k \approx e^{-1}$ in the decaying case.

3 Numerical examples

Example 5

$$\int_{-\infty}^{\infty} e^{is} \frac{1}{1+is}\, ds = \frac{2\pi}{e}$$

The integrand is oscillating and the absolute value decays slowly when $|s| \to \infty$. We use (10) and (11) employing Chebyshev acceleration. It is easily verified that (11) satisfies (26) with

$$g(s) = \frac{1}{1+is}$$

We have namely

$$g(s) = \frac{1}{1+is} = \frac{1-is}{1+s^2}$$

and

$$\frac{1}{1+s^2} = \int_0^{\infty} e^{-su} \sin u\, du, \quad \frac{s}{1+s^2} = \int_0^{\infty} e^{-su} \cos u\, du.$$

Next make the substitution $e^{-hu} = t$ to arrive at the desired conclusion. The step-zise h corresponds to $z = e^{ih}$. The results in Table 1 emerged. We note

step-size	# of func.-values	absolute error
$\pi/2$	9	$4.3 \cdot 10^{-2}$
$\pi/4$	10	$7.7 \cdot 10^{-4}$
$\pi/8$	10	$1.5 \cdot 10^{-4}$
$\pi/16$	10	$1.3 \cdot 10^{-3}$

Table 1: Results of evaluating Example 5 without batching

step-size	# of func.-values	absolute error
$\pi/2$	9	$4.3 \cdot 10^{-2}$
$\pi/4$	20	$7.8 \cdot 10^{-4}$
$\pi/8$	48	$4.8 \cdot 10^{-7}$
$\pi/16$	88	0

Table 2: Results of Example 5 with batching

that the observed accuracy deteriorates for $h < \pi/8$, since the convergence acceleration is influenced by round-offs. This effect is counter-acted efficiently by means of batching when k is determined such that $z^k = i$. These results are given in Table 2. All calculations were carried out in single precision, i.e. with a working relative accuracy of $2^{-23} \approx 1.2 \cdot 10^{-7}$.

Example 6

$$\int_0^\infty \sqrt{x}e^{-x}\,dx = \frac{\sqrt{\pi}}{2}$$

Here we put $x = e^s$ and get

$$\int_{-\infty}^\infty e^{1.5s-exp(s)}\,ds = I_R + I_S$$

where

$$I_R = \int_0^\infty e^{1.5s-\exp s}\,ds \qquad (37)$$

$$I_S = \int_0^\infty e^{-1.5s-\exp(-s)}\,ds. \qquad (38)$$

The integrand of I_S decays exponentially while that of I_R decays more rapidly. We discretize both integrals using a common step-size h. Then the trapezoidal sum from I_R converges rapidly and is not treated by means of any acceleration method. The sum arising from I_S converges like a geometric series with quotient $e^{-1.5h}$ and it could be summed up term by term. This was compared to using Chebyshev acceleration with batching where k is selected such that $hk = 1$ and the corresponding z-value hence becomes $e^{-1.5}$. We verify that (26) is satisfied with

$$g(s) = e^{-\exp(-s)}.$$

We have namely that

$$g(s) = \sum_{n=0}^{\infty} \frac{(-1)^n}{n!} e^{-sn}.$$

Thus

$$g(s) = g_1(s) - g_2(s),$$

where

$$g_1(s) = \sum_{n=0}^{\infty} \frac{e^{-2ns}}{(2n)!}, \quad g_2(s) = \sum_{n=0}^{\infty} \frac{e^{-(2n+1)s}}{(2n+1)!}.$$

Using the definition of complete monotonicity we check directly that g_1, g_2 are completely monotonic and hence we conclude that g satisfies (26). All calculations were carried out in double precision, i.e with computer accuracy $2^{-52} \approx 2.2 \cdot 10^{-16}$. We note that the number of terms required for calculating both sums were inversely proportional to h and that some gains were achieved by using Chebyshev acceleration. The observed error was the same with or without acceleration of the slow sum. The error decreases rapidly when the step-size h is decreased. The results are given in Table 3.

step-size	# of terms			observed error
h	I_R	I_S	I_S acc.	
1	4	26	2	$1.4 \cdot 10^{-3}$
0.5	8	50	14	$5.8 \cdot 10^{-8}$
0.25	15	98	60	$4.4 \cdot 10^{-16}$

Table 3: Results of Example 6

References

Brezinski, C and M. R. Redivo Zaglia, (1991) *Extrapolation Methods- Theory and Practice*, North-Holland, Amsterdam, London, New York, Tokyo

Davis, P. J. and P. Rabinowitz, (1984) *Methods of Numerical Integration,*, Second Edition, Academic Press, San Diego, USA

Gustafson, S.-Å. (1971) Rapid computation of general interpolation formulas and mechanical quadrature rules, *CACM* v. 14, pp. 797–801.

Gustafson, S.-Å. (1978a) Convergence acceleration on a general class of power series, *Computing*, v. 21, pp. 53–69.

Gustafson, S.-Å. (1978b) Two computer codes for convergence acceleration, *Computing*, v.21. pp. 87–91.

Gustafson, S.-Å. (1991a) Computing inverse Laplace transforms using convergence acceleration, Bowers, K. and J. R. Lund (Eds) *Computation and Control II*, Birkhäuser Verlag, Basel pp. 151–163

Gustafson, S.-Å. (1991b), Numerical inversion of Laplace transforms using integration and convergence acceleration, *Techn. Rep.*91-18 Swedish Nuclear Fuel and Waste Management Co., Stockholm, Sweden, 1991.

Lund, J. R. and K. L. Bowers (1992), *Sinc Methods for Differential Equations*, SIAM, Philadelphia

Sidi, A. (1982) Numerical quadrature rules for some infinite range integrals, *Math. Comp.* v. 38, pp. 127–142.

Sidi, A. (1988) A user-friendly extrapolation method for oscillatory infinite integrals, *Math. Comp.* v. 51, pp. 249-266.

Sidi, A. (1992) Computation of oscillatory infinite integrals by extrapolation methods, T. O. Espelid and A. Genz (Eds) *Numerical Integration* Kluwer Academic Publishers, Amsterdam.

Stenger, F. (1981), Numerical Methods Based on Whittaker Cardinal, or Sinc Functions, *SIAM Rev.*, v. 23, pp. 165–224.
Stenger, F. (1993), *Numerical Methods Based on Sinc and Analytic Functions*, Springer-Verlag, to appear
Talbot, A. (1979), The accurate inversion of Laplace transforms, *J. Inst. Maths. Applics.* v. 23., pp. 97–120
Takahasi, M. and M. Mori (1973), Quadrature formulas obtained by variable transformation, *Numer. Math.* v. 21, pp. 206–219
Widder, D. V. (1971), *An Introduction to Transform Theory*, Academic Press, New York, London, Toronto, Sydney
Wimp, J. (1981), *Sequence Transformations and Their Applications*, Academic Press, New York, London, Toronto, Sydney

Author's address:

Sven-Åke Gustafson
HSR
Box 2557 Ullandhaug
N-4004 Stavanger
Norway
gustafson@hsr.no

Construction of fully symmetrical cubature rules of very high degree for the square

Ann Haegemans and Pierre Verlinden

Abstract. A new method for the construction of fully symmetrical cubature formulae for the square is proposed. Even for high degrees the formulae are surprisingly easy to compute: the hardest computation to perform is the solution of a system of two polynomial equations of relatively low degree in two unknowns.

1 Introduction

In this paper we search for cubature formulae of the form :

$$Q[f] = \sum_{i=1}^{N_1} w_i^{(1)} Q_i^{(1)}[f] + \sum_{i=1}^{N_2} w_i^{(2)} Q_i^{(2)}[f] + \sum_{i=1}^{N_3} w_i^{(3)} Q_i^{(3)}[f] \tag{1}$$

where

$$\begin{aligned} Q_i^{(1)}[f] &= f(x_i^{(1)},0) + f(-x_i^{(1)},0) + f(0,x_i^{(1)}) + f(0,-x_i^{(1)}) \\ Q_i^{(2)}[f] &= f(x_i^{(2)},x_i^{(2)}) + f(x_i^{(2)},-x_i^{(2)}) + f(-x_i^{(2)},x_i^{(2)}) + f(-x_i^{(2)},-x_i^{(2)}) \\ Q_i^{(3)}[f] &= f(x_i^{(3)},y_i^{(3)}) + f(x_i^{(3)},-y_i^{(3)}) + f(-x_i^{(3)},y_i^{(3)}) + f(-x_i^{(3)},-y_i^{(3)}) + \\ & \quad f(y_i^{(3)},x_i^{(3)}) + f(y_i^{(3)},-x_i^{(3)}) + f(-y_i^{(3)},x_i^{(3)}) + f(-y_i^{(3)},-x_i^{(3)}) \end{aligned} \tag{2}$$

as an approximation of the integral :

$$I[f] = \int_{-1}^{1}\int_{-1}^{1} f(x,y)\,dx\,dy. \tag{3}$$

The number of points in the cubature formula is $4N_1 + 4N_2 + 8N_3$ and we want the cubature formula to be of degree $d = 4k - 1$, i.e. $I[f] = Q[f]$ for all polynomials of degree $\leq d$; but not for at least one polynomial of degree $d + 1$.
Since the region and the formula are both D_4-invariant Sobolev's (1962) theorem says that a sufficient condition for a formula of the form (1) to be a cubature formula of degree d is:

$$I(g) = Q(g) \tag{4}$$

for all D_4-invariant polynomials g of degree $\leq d$. We denote the vectorspace of all D_4-invariant polynomials of degree $\leq d$ by $P_d(D_4)$.
Let $s(x,y) = x^2 + y^2$ and $u(x,y) = x^2y^2$. Then, every D_4-invariant polynomial f is a polynomial in s and u:

$$f(x,y) = g(s(x,y), u(x,y))$$

and thus

$$Q[f] = \sum_{i=1}^{N_1} 4w_i^{(1)} g(s_i^{(1)}, 0) + \sum_{i=1}^{N_2} 4w_i^{(2)} g(s_i^{(2)}, \frac{(s_i^{(2)})^2}{4}) + \sum_{i=1}^{N_3} 8w_i^{(3)} g(s_i^{(3)}, u_i^{(3)}) \tag{5}$$

where

$$\begin{array}{lll} s_i^{(1)} = s(x_i^{(1)}, 0), & & i = 1, \ldots, N_1 \\ s_i^{(2)} = s(x_i^{(2)}, x_i^{(2)}), & & i = 1, \ldots, N_2 \\ s_i^{(3)} = s(x_i^{(3)}, y_i^{(3)}), & u_i^{(3)} = u(x_i^{(3)}, y_i^{(3)}), & i = 1, \ldots, N_3. \end{array}$$

2 Construction of a basis

According to Sobolev's (1962) theorem we have to solve a system

$$Q[\phi_i] = I[\phi_i], \qquad i = 1, \ldots, \dim P_d(D_4) \tag{6}$$

where $\{\phi_i | i = 1, \ldots, \dim P_d(D_4)\}$ form a basis for the vectorspace $P_d(D_4)$.
The most easy choice for a basis is :

$$s^i u^j, \qquad 2i + 4j \leq d, \quad i, j \in \mathbb{N}$$

but then the system (6) is difficult to solve.
We propose here the following basis :

$$\begin{array}{ll} \zeta_l(s) = s^l, & l = 0, 1, \ldots, 2k-1 \\ \eta_l(s,u) = (s^2 - 4u)s^{l-2}, & l = 2, 3, \ldots, 2k-1 \\ \theta_{lj}(s,u) = (s^2 - 4u)s^{l-2j}u^{j-1}, & l = 4, 5, \ldots, 2k-1, \quad j = 2, 3, \ldots, \lfloor \frac{l}{2} \rfloor. \end{array} \tag{7}$$

For these polynomials (7) we have :

$$\begin{array}{llll} \forall l, \forall j : & & \theta_{lj}(s_i^{(1)}, 0) = 0, & i = 1, \ldots, N_1 \\ \forall l, \forall j : & \eta_l(s_i^{(2)}, \frac{(s_i^{(2)})^2}{4}) = 0, & \theta_{lj}(s_i^{(2)}, \frac{(s_i^{(2)})^2}{4}) = 0, & i = 1, \ldots, N_2. \end{array} \tag{8}$$

3 Computation of the points and weights of type 3

Let for k even ($k = 2m$) :

$$\begin{array}{lll} P_1^{(3)}(s,u) & = & s^{2m-2} + \sum_{i=1}^{m-2} \lambda_i s^{2m-2i-2} u^i + \sum_{i=0}^{2m-3} \sum_{j=0}^{\lfloor \frac{i}{2} \rfloor} \alpha_{ij} s^i u^j \\ P_2^{(3)}(s,u) & = & u^{m-1} + \sum_{i=1}^{m-2} \mu_i s^{2m-2i-2} u^i + \sum_{i=0}^{2m-3} \sum_{j=0}^{\lfloor \frac{i}{2} \rfloor} \beta_{ij} s^i u^j \end{array} \tag{9}$$

and for k odd ($k = 2m+1$) :

$$\begin{aligned} P_1^{(3)}(s,u) &= s^{2m-1} + \textstyle\sum_{i=1}^{m-2} \lambda_i s^{2m-2i-1} u^i + \sum_{i=0}^{2m-2} \sum_{j=0}^{\lfloor \frac{i}{2} \rfloor} \alpha_{ij} s^i u^j \\ P_2^{(3)}(s,u) &= s u^{m-1} + \textstyle\sum_{i=1}^{m-2} \mu_i s^{2m-2i-2} u^i + \sum_{i=0}^{2m-2} \sum_{j=0}^{\lfloor \frac{i}{2} \rfloor} \beta_{ij} s^i u^j \end{aligned} \tag{10}$$

be such that :

$$\begin{aligned} P_1^{(3)}(s_i^{(3)}, u_i^{(3)}) = 0 \\ P_2^{(3)}(s_i^{(3)}, u_i^{(3)}) = 0 \end{aligned} \tag{11}$$

with $\lambda_1, \ldots, \lambda_{m-2}, \mu_1, \ldots, \mu_{m-2}$ free parameters and α_{ij} and β_{ij} coefficients that have to be defined later on.
For k even ($k = 2m$), $P_1^{(3)}$ and $P_2^{(3)}$ are of degree $4m-4 = 2k-4$ in x and y, there are $M = 2m-4 = k-4$ parameters λ_i and μ_i and $2m(m-1) = k(k-2)/2$ coefficients α_{ij} and β_{ij}.

Example:
$m = 3, k = 6, d = 23$

$$\begin{aligned} P_1^{(3)}(s,u) &= \underbrace{\begin{matrix} s^4 + \lambda_1 s^2 u + \\ u^2 + \mu_1 s^2 u + \end{matrix}}_{\text{terms of degree 8}} \underbrace{\begin{matrix} \alpha_{30}s^3 + \alpha_{31}su + \\ \beta_{30}s^3 + \beta_{31}su + \end{matrix}}_{\text{terms of degree 6}} \underbrace{\begin{matrix} \alpha_{20}s^2 + \alpha_{21}u + \\ \beta_{20}s^2 + \beta_{21}u + \end{matrix}}_{\text{terms of degree 4}} \underbrace{\begin{matrix} \alpha_{10}s + \\ \beta_{10}s + \end{matrix}}_{\text{terms of degree 2}} \underbrace{\begin{matrix} \alpha_{00} \\ \beta_{00} \end{matrix}}_{\text{terms of degree 0}} \\ P_2^{(3)}(s,u) &= \end{aligned}$$

For k odd ($k = 2m+1$), $P_1^{(3)}$ and $P_2^{(3)}$ are of degree $4m-2 = 2k-4$ in x and y, there are $M = 2m-4 = k-5$ parameters λ_i and μ_i and $2m^2 = (k-1)^2/2$ coefficients α_{ij} and β_{ij}.

Example:
$m = 3, k = 7, d = 27$

$$\begin{aligned} P_1^{(3)}(s,u) &= \underbrace{\begin{matrix} s^5 + \lambda_1 s^3 u + \\ su^2 + \mu_1 s^3 u + \end{matrix}}_{\text{terms of degree 10}} \underbrace{\begin{matrix} \alpha_{40}s^4 + \alpha_{41}s^2u + \alpha_{42}u^2 + \\ \beta_{40}s^4 + \beta_{41}s^2u + \beta_{42}u^2 + \end{matrix}}_{\text{terms of degree 8}} \underbrace{\begin{matrix} \alpha_{30}s^3 + \alpha_{31}su + \\ \beta_{30}s^3 + \beta_{31}su + \end{matrix}}_{\text{terms of degree 6}} \\ P_2^{(3)}(s,u) &= \end{aligned}$$

$$\underbrace{\begin{matrix} \alpha_{20}s^2 + \alpha_{21}u + \\ \beta_{20}s^2 + \beta_{21}u + \end{matrix}}_{\text{terms of degree 4}} \underbrace{\begin{matrix} \alpha_{10}s + \\ \beta_{10}s + \end{matrix}}_{\text{terms of degree 2}} \underbrace{\begin{matrix} \alpha_{10} \\ \beta_{10} \end{matrix}}_{\text{terms of degree 0}}$$

$P_1^{(3)}(s,u)\theta_{lj}(s,u)$ is of degree $2k-4+2l$ and must be integrated exactly if $l \leq k+1$. Taking into account (8) and (11) we have :

$$\begin{aligned} I[P_1^{(3)}(s,u)\theta_{lj}(s,u)] = Q[P_1^{(3)}(s,u)\theta_{lj}(s,u)] = 0 \\ l = 4, \ldots, k+1, \quad j = 2, \ldots, \lfloor \tfrac{l}{2} \rfloor \end{aligned} \tag{12}$$

Equations (12) are linear in the unknows α_{ij}. If k is even there are $k(k-2)/4$ equations for the $k(k-2)/4$ unknows and if k is odd there are $(k-1)^2/4$ equations for the $(k-1)^2/4$ unknowns α_{ij}.
Integrating

$$P_2^{(3)}(s,u)\theta_{lj}(s,u), \qquad l=4,\ldots,k+1, \qquad j=2,\ldots,\lfloor\frac{l}{2}\rfloor \tag{13}$$

gives sufficient linear equations for the β_{ij}.
For most choices of the parameters, the number of common zeros of $P_1^{(3)}$ and $P_2^{(3)}$ is $N_3 = (k-2)^2/2$ if k is even and $N_3 = (k-1)(k-3)/2$ if k is odd. This can be seen by a careful examination of the degree of the eliminant when one eliminates u between $P_1^{(3)}(s,u)$ and $P_2^{(3)}(s,u)$.
Once the points $(s_i^{(3)}, u_i^{(3)})$, $i=1,\ldots,N_3$ are found we can compute the weights $w_i^{(3)}$ as the solution of a linear system :

- for k even :

$$\begin{gathered} I[\theta_{lj}(s,u)] = Q[\theta_{lj}(s,u)] = \textstyle\sum_{i=1}^{N_3} 8w_i^{(3)}\theta_{lj}(s_i^{(3)},u_i^{(3)})], \\ j=2,\ldots,k/2, \quad l=2j,\ldots,k+2j-3 \end{gathered} \tag{14}$$

- for k odd :

$$\begin{gathered} I[\theta_{lj}(s,u)] = Q[\theta_{lj}(s,u)] = \textstyle\sum_{i=1}^{N_3} 8w_i^{(3)}\theta_{lj}(s_i^{(3)},u_i^{(3)})], \\ j=2,\ldots,(k-1)/2, \quad l=2j,\ldots,k+2j-3 \end{gathered} \tag{15}$$

and

$$\begin{gathered} I[\theta_{lj}(s,u)] = Q[\theta_{lj}(s,u)] = \textstyle\sum_{i=1}^{N_3} 8w_i^{(3)}\theta_{lj}(s_i^{(3)},u_i^{(3)})], \\ j=(k+1)/2,\ldots,k-2, \quad l=k+1+2j. \end{gathered} \tag{16}$$

Remark that the number of polynomials that we have integrated exactly until now is :

- for k even:

$k(k-2)/2$	for the determination of the α_{ij} and the β_{ij}
$+(k-2)^2/2$	for the determination of the $w_i^{(3)}$
$=(k-2)(k-1)$	

- for k odd:

$(k-1)^2/2$	for the determination of the α_{ij} and the β_{ij}
$+(k-1)(k-3)/2$	for the determination of the $w_i^{(3)}$
$=(k-2)(k-1)$	

and this is just the number of elements $\theta_{lj}(s,u)$ in the basis (7). For almost all choices of the parameters the polynomials integrated exactly by equations (12), (13), (14) for k even and (12), (13), (15), (16) for k odd, form a basis for the vectorspace spanned by the polynomials

$$\theta_{lj}(s,u) = (s^2-4u)s^{l-2j}u^{j-1}, \quad l=4,5,\ldots,2k-1, \quad j=2,3,\ldots,\lfloor\frac{l}{2}\rfloor. \tag{17}$$

4 Computation of the points and weights of type 1

Let

$$P^{(1)}(s) = s^{k-1} + \sum_{i=0}^{k-2} \gamma_i s^i \tag{18}$$

be such that

$$P^{(1)}(s_i^{(1)}) = 0. \tag{19}$$

$P^{(1)}(s)\eta_l(s,u)$ is of degree $2k-2+2l$ and must be integrated exactly if $l \leq k$. Taking into account (8) and (19) we have :

$$\begin{aligned} I[P^{(1)}(s)\eta_l(s,u)] &= Q[P^{(1)}(s)\eta_l(s,u)] \\ &= \textstyle\sum_{i=1}^{N_3} 8w_i^{(3)} P^{(1)}(s_i^{(3)})\eta_l(s_i^{(3)},u_i^{(3)}), \quad l=2,\ldots,k. \end{aligned} \tag{20}$$

Equations (20) are linear in the unknows γ_i. There are $k-1$ equations for the $k-1$ unknows.
The $s_i^{(1)}$ are the zeros of $P^{(1)}(s)$, so $N_1 = k-1$.
The $k-1$ weights $w_i^{(1)}$ are found as the solution of the linear system :

$$\begin{aligned} I[\eta_l(s,u)] &= Q[\eta_l(s,u)] \\ &= \textstyle\sum_{i=1}^{N_3} 8w_i^{(3)}\eta_l(s_i^{(3)},u_i^{(3)}) + \sum_{i=1}^{N_1} 4w_i^{(1)}\eta_l(s_i^{(1)},0), \quad l=2,\ldots,k. \end{aligned} \tag{21}$$

5 Computation of the points and weights of type 2

Let

$$P^{(2)}(s) = s^k + \sum_{i=0}^{k-1} \delta_i s^i \tag{22}$$

be such that

$$P^{(2)}(s_i^{(2)}) = 0. \tag{23}$$

$P^{(2)}(s)\zeta_l(s)$ is of degree $2k+2l$ and must be integrated exactly if $l \leq k-1$. Taking into account (8) and (23) we have :

$$\begin{aligned} I[P^{(2)}(s)\zeta_l(s)] &= Q[P^{(2)}(s)\zeta_l(s)] \\ &= \textstyle\sum_{i=1}^{N_3} 8w_i^{(3)} P^{(2)}(s_i^{(3)})\zeta_l(s_i^{(3)}) + \\ &\quad \textstyle+ \sum_{i=1}^{N_1} 4w_i^{(1)} P^{(2)}(s_i^{(1)})\zeta_l(s_i^{(1)}), \quad l=0,\ldots,k-1. \end{aligned} \tag{24}$$

Equations (24) are linear in the unknows δ_i. There are k equations for the k unknows.
The $s_i^{(2)}$ are the zeros of $P^{(2)}(s)$, so $N_2 = k$.
The k weights $w_i^{(2)}$ are found as the solution of the linear system :

$$\begin{aligned} I[\zeta_l(s)] &= Q[\zeta_l(s)] \\ &= \textstyle\sum_{i=1}^{N_3} 8w_i^{(3)}\zeta_l(s_i^{(3)}) + \sum_{i=1}^{N_1} 4w_i^{(1)}\zeta_l(s_i^{(1)}) + \\ &\quad \textstyle+ \sum_{i=1}^{N_2} 4w_i^{(2)}\zeta_l(s_i^{(2)}), \quad l=0,\ldots,k-1. \end{aligned} \tag{25}$$

6 Number of points in the cubature formula

The number of points in the cubature formula is :

$$\begin{aligned} N &= 8(k-2)^2/2 + 4(k-1) + 4k &= 4(k^2 - 2k + 3) \quad \text{if } k \text{ is even} \\ N &= 8(k-1)(k-3)/2 + 4(k-1) + 4k &= 4(k^2 - 2k + 2) \quad \text{if } k \text{ is odd.} \end{aligned}$$

The number of free parameters λ_i and μ_i is :

$$\begin{aligned} M &= k - 4 \quad \text{if } k \text{ is even} \\ M &= k - 5 \quad \text{if } k \text{ is odd.} \end{aligned}$$

The method that is explained in the preceding pages only works for $k \geq 4$. However for low values of k ($k \leq 3$) we can adapt the method :

- $k = 1, d = 3$:
  There are no orbits of type 3 and 1. The polynomial (22) has degree 1 ($= N_2$) and thus $N = 4$. There are no free parameters in the formula.

- $k = 2, d = 7$:
  There are no orbits of type 3. The polynomial (18) has degree 1 ($= N_1$) and the polynomial (22) has degree 2 ($= N_2$) and thus $N = 12$. There are no free parameters in the formula.

- $k = 3, d = 11$) :
  The two polynomials (10) are identical and equal $s + \alpha_{00}$. So the value of $u_1^{(3)}$ is arbitrary and can be seen as a parameter. There are 2 orbits of type 1 and 3 orbits of type 2 and thus $N = 28$.

7 Results

In the following table we give for the different degrees (d) the number of points (N) and free parameters (M) in the formula. Until now for high degrees only Gauss-product formulae are known. In the table we also give the number of points of the corresponding Gauss-product formula (N_G) and Möller's (1976) lower bound (N_M). For low degrees ($d = 3, 7, 11, 15$) the method constructs existing formulae that are listed in Stroud (1971). The formula of degree 19 was computed by Verlinden and Cools (1991), (Cools 1992).

k	d	N_3	N_2	N_1	N	N_G	N_M	M
1	3	0	1	0	4	4	4	0
2	7	0	2	1	12	16	12	0
3	11	1	3	2	28	36	24	1
4	15	2	4	3	44	64	40	0
5	19	4	5	4	68	100	60	0
6	23	8	6	5	108	144	84	2
7	27	12	7	6	148	196	112	2
8	31	18	8	7	204	256	144	4
9	35	24	9	8	260	324	180	4
10	39	32	10	9	332	400	220	6

For degree 11 we have one free parameter, so we have an infinite number of 28-point formulae, one of them being C2:11-2 (Stroud 1971). For two values of the parameter one of the $s_i^{(2)}$ becomes zero. The origin is then a point of the cubature formula and the number of points is 25. In this way we found the formulae C2:11-1 (Stroud 1971) and the 25-point formula of degree 11 of Haegemans and Piessens (1977).
For the formulae with more than one free parameter we can try to choose them such that the formula have all points inside the region, and/or that the weights are positive and/or that some weights become zero or some points of type 2 or 1 go to the origin.

8 Conclusions

The proposed method for constructing fully symmetrical cubature formulae for the square has the big advantage that one does not have to solve big nonlinear systems. The most difficult computation are the solution of a linear system and the solution of a nonlinear system of two polynomial equations of low degree with two unknows.
The fact that we have a lot of free parameters is both an advantage and a disadvantage. On the one hand one can use them to construct formulae with a good structure but on the other hand it asks for a bit of luck to tune them good.

References

R. Cools (1992), A survey of methods for constructing cubature formulae. In A. Genz and T.O. Espelid, editors , *Numerical Integration*, volume 357 of *Nato ASI Series*, pages 1–24. Academic Publishers, Dordrecht.

A. Haegemans and R. Piessens (1977), Construction of cubature formulas of degree eleven for symmetric planar regions, using orthogonal polynomials. *SIAM J. Numer. Anal.*, 14:139–148.

H.M. Möller (1976), Kubaturformeln mit minimaler Knotenzahl. *Numer. Math.*, 25:185–200.

S.L. Sobolev (1962), Cubature formulas on the sphere invariant under finite groups of rotation. *Dokl. Akad. Nauk SSSR*, 146:310–313 (Russian). *Soviet Math. Dokl.* 3: 1307–1310 (English).

A.H. Stroud (1971), *Approximate calculation of multiple integrals.* Prentice Hall, Englewood Cliffs, N.J.

P. Verlinden and R. Cools (1991), A cubature formula of degree 19 with 68 nodes for the integration over the square. *Report TW 149, Dept. of Computer Science, K.U.Leuven.*

Authors' address:

Ann Haegemans
Dept. of Computer Science
Katholieke Universiteit Leuven
Celestijnenlaan 200 A
B-3001 Leuven, Belgium
Ann.Haegemans@cs.kuleuven.ac.be

Pierre Verlinden
Dept. of Computer Science
Katholieke Universiteit Leuven
Celestijnenlaan 200 A
B-3001 Leuven, Belgium
Pierre.Verlinden@cs.kuleuven.ac.be

Numerical Integration of Nearly Singular Functions

Takemitsu Hasegawa Tatsuo Torii

Abstract. An automatic quadrature is presented for approximating the integral of functions having poles (single pole on the real axis or complex conjugate pair of poles) near the interval of integration. After the smooth part of the integrand is expanded in terms of the Chebyshev polynomials, the integral is approximated and evaluated by using recurrence relations and some extrapolation method. The present scheme can efficiently approximate, to a required tolerance, integrals with different poles with the same number of function evaluations. Numerical examples are also included to illustrate the performance of the method.

1. Introduction

An automatic quadrature is presented for approximating the integral

$$I(f;K) \equiv \int_{-1}^{1} f(x)K(x)\,dx, \tag{1}$$

where $f(x)$ is assumed to be a given smooth function on $[-1,1]$ and $K(x)$ is a nearly singular function. In particular, we construct quadrature formulas of interpolatory type for the weight function $K(x)$ having a single pole on the real axis near the interval of integration $[-1,1]$ (the discussion below holds for the more general case of a single pole in the complex plane except for $[-1,1]$) or a complex conjugate pair of poles, *i.e.*, K is of the form

$$K_1(x) = 1/(x-c) \quad \text{or} \quad K_2(x) = 1/(x^2+\delta^2), \quad -1 \le x \le 1, \tag{2}$$

where $|\delta| << 1$ and c is outside but very close to $[-1,1]$; for avoiding the loss of significant figures in the computation described in this paper we write $c = \delta \pm 1$ (we take $\delta > 0$ for plus sign and $\delta < 0$ for minus sign). If $|\delta|$ is close to zero, the weights $K_1(x)$ and $K_2(x)$ given in (2) are badly–behaved functions on the interval $[-1,1]$.

Only a small amount of literature (Lether 1977a, 1977b, Monegato 1986 and Bialecki 1989) is concerned with the evaluation of the integral (1) with the kernel $K_1(x)$ or $K_2(x)$

(2), although the numerical integration of integrals having several types of typical singular kernel functions has been treated by many authors, see for example Davis and Rabinowitz (1984). Furthermore, there appears to exist no automatic quadrature method for the integral we are concerned with in this paper.

The present scheme is an extension of the Clenshaw–Curtis method (Clenshaw and Curtis 1960) for the integral $I(f; K = 1)$ (1). The function $f(x)$ is approximated by a sum of Chebyshev polynomials of the first kind $T_k(x)$:

$$p_N(x) = \sum_{k=0}^{N}{}'' a_k^N \, T_k(x), \qquad -1 \le x \le 1, \tag{3}$$

interpolating $f(x)$ at the $N+1$ abscissae $\cos(\pi j/N)$ $(0 \le j \le N)$. In (3) the double prime denotes the summation where the first and last terms are halved. The coefficients a_k^N are efficiently computed by using the Fast Fourier Transform (FFT) (Gentleman 1972, Branders and Piessens 1975, and Hasegawa et al. 1990). If $f(x)$ is a smooth function, p_N converges rapidly to f as N increases.

The use of p_N given by (3) instead of f in the integral (1) yields an approximation $Q_N(f; K)$,

$$Q_N(f; K) = \int_{-1}^{1} p_N(x) K(x)\, dx = I(p_N; K). \tag{4}$$

To evaluate the approximate integral $Q_N(f; K_i)$ with $i = 1, 2$, write, respectively for each $K_i(x)$ $(i = 1, 2)$,

$$p_N(x) \;=\; 2K_1(x)^{-1} \sum_{k=0}^{N-1}{}' b_k^{(1)} T_k(x) + \tau \, T_N(x), \tag{5}$$

$$p_N(x) \;=\; 4K_2(x)^{-1} \sum_{k=0}^{N-2}{}' b_k^{(2)} T_k(x) + \tau_1 \, T_{N-1}(x) + \tau_2 \, T_N(x), \tag{6}$$

where the prime denotes the summation whose first term is halved. Then we have the quadrature rules $Q_N(f; K_1)$ and $Q_N(f; K_2)$:

$$Q_N(f; K_1) \;=\; 4 \sum_{k=0}^{N/2-1}{}' b_{2k}^{(1)}/(1-4k^2) + \tau \int_{-1}^{1} T_N(x) K_1(x)\, dx, \tag{7}$$

$$Q_N(f; K_2) \;=\; 8 \sum_{k=0}^{N/2-1}{}' b_{2k}^{(2)}/(1-4k^2) + \tau_2 \int_{-1}^{1} T_N(x) K_2(x)\, dx, \tag{8}$$

respectively, where here and henceforth we assume that N is even (note that $\int_{-1}^{1} T_{N-1}(x)$ $K_2(x)\, dx = 0$ because the integrand is an odd function).

In section 2 we show that the coefficients $b_k^{(1)}$ and $b_k^{(2)}$ satisfy three term recurrence relations and present an algorithm to stably compute all the coefficients in (5) and (6) from the solution of a linear system of equations having a tridiagonal coefficient matrix, derived from the recurrence relation. Section 3 shows that the integrals $\int_{-1}^{1} T_N(x) K_i(x)\,dx$ $(i = 1, 2)$, in (7) and (8) are efficiently evaluated partly by using an extrapolation method of Richardson type (Levin and Sidi 1981). Section 4 discusses error estimates for the approximations $Q_N(f; K_i)$. In section 5, numerical examples are given.

2. Computation of the coefficients

We now show an algorithm to compute the coefficients $b_k^{(i)}$ $(i = 1, 2)$ in (5) and (6).

2.1 The coefficients for $K_1(x)$. Using the relation

$$2\,x\,T_k(x) = T_{k+1}(x) + T_{k-1}(x), \qquad k = 1, 2, \ldots, \tag{9}$$

and comparing the coefficients of both sides in (5), we have the three term recurrence relation

$$b_{k+1}^{(1)} - 2\,c\,b_k^{(1)} + b_{k-1}^{(1)} = a_k^N \quad (k = 0, 1, \ldots, N-1), \qquad b_{N-1}^{(1)} + \tau = a_N^N/2, \tag{10}$$

where we have assumed $b_{-1}^{(1)} = b_1^{(1)}$ and $b_N^{(1)} = 0$ for convenience.

We can stably compute the values $b_k^{(1)}$ by solving the linear system given by (10). Let $\mathbf{b}^{(1)} = [b_0^{(1)}, b_1^{(1)}, \ldots, b_{N-1}^{(1)}]^T$, $\mathbf{a}^{(1)} = [a_0^N/2, a_1^N, \ldots, a_{N-1}^N]^T$ and let $A^{(1)}$ denote a tridiagonal matrix such that $A_{0,0}^{(1)} = -c$, $A_{i,i}^{(1)} = -2c$ $(i = 1, \ldots, N-1)$, $A_{i,i+1}^{(1)} = A_{i+1,i}^{(1)} = 1$ $(i = 0, \ldots, N-2)$. Then the coefficients $b_k^{(1)}$ in (5) are obtained from the solution of the linear equations $A^{(1)}\,\mathbf{b}^{(1)} = \mathbf{a}^{(1)}$ without pivoting, because $A^{(1)}$ is a diagonally dominant matrix when $|c| > 1$. The use of $b_{N-1}^{(1)}$ obtained above in the last relation of (10) gives the value of the coefficient τ.

2.2 The coefficients for $K_2(x)$. Using the relation (9) twice yields

$$4\,x^2\,T_k(x) = T_{k+2}(x) + 2\,T_k(x) + T_{k-2}(x), \qquad k \geq 2. \tag{11}$$

Applying (11) to the right in (6) and comparing the coefficients of the both sides, we have, defining $d = 1 + 2\delta^2$,

$$\begin{aligned} &b_{k+2}^{(2)} + 2\,d\,b_k^{(2)} + b_{k-2}^{(2)} = a_k^N \qquad (k = 0, 1, \ldots, N-2), \\ &b_{N-3}^{(2)} + \tau_1 = a_{N-1}^N, \qquad b_{N-2}^{(2)} + \tau_2 = a_N^N/2, \end{aligned} \tag{12}$$

where we have defined $b_{-2}^{(2)} = b_2^{(2)}$, $b_{-1}^{(2)} = b_1^{(2)}$ and $b_i^{(2)} = 0$ $(i \geq N-1)$, for convenience. Similarly to the previous subsection, we can stably solve the linear system of equations $A^{(2)}\,\mathbf{b}^{(2)} = \mathbf{a}^{(2)}$ to obtain the coefficients $b_{2k}^{(2)}$ $(k = 0, 1, \ldots, N/2-1)$, required for the quadrature rule (8), where we define $\mathbf{b}^{(2)} = [b_0^{(2)}, b_2^{(2)}, \ldots, b_{N-2}^{(2)}]^T$, $\mathbf{a}^{(2)} = [a_0^N/2, a_2^N, \ldots, a_{N-2}^N]^T$

and let $A^{(2)}$ denote the tridiagonal matrix such that $A^{(2)}_{0,0} = d$, $A^{(2)}_{i,i} = 2d$ $(i = 1, \ldots, N/2-1)$, and $A^{(2)}_{i,i+1} = A^{(2)}_{i+1,i} = 1$ $(i = 0, \ldots, N/2-2)$.

3. Use of extrapolation

We show how to evaluate the integrals $\int_{-1}^{1} T_N(x)K_i(x)dx$ $(i = 1, 2)$ in (7) and (8) partly by using an extrapolation technique due to Levin and Sidi (1981).

We begin with the evaluation of the integral $\int_{-1}^{1} T_N(x)/(x-c)dx$, for which the following relation (Hasegawa et al. 1983 p.552) plays an important role:

$$1/(x-c) = -4/(\alpha^{-1}-\alpha)\sum_{n=0}^{\infty}{}' \alpha^n T_n(x), \quad |x| \le 1, \quad |c| > 1, \tag{13}$$

where we set $c = (\alpha + \alpha^{-1})/2$ and choose $\alpha = c \pm \sqrt{c^2-1}$ so that $|\alpha| < 1$, *i.e.*,

$$\begin{aligned} \alpha &= 1/(c+\sqrt{c^2-1}), && \text{if } c = \delta + 1 > 1, \\ \alpha &= c+\sqrt{c^2-1}, && \text{if } c = \delta - 1 < -1. \end{aligned} \tag{14}$$

Theorem 1 *Let N be an even integer and let α be defined by (14). Further let $\phi(\alpha, N)$ and $\psi(\alpha, N)$ be defined by*

$$\phi(\alpha, N) = \sum_{n=0}^{N/2-1} \frac{2\alpha^{2n+1}}{N-2n-1} - \alpha^N \ln\left(\frac{1+\alpha}{1-\alpha}\right), \tag{15}$$

$$\psi(\alpha, N) = \frac{1}{\alpha}\ln\left(\frac{1+\alpha}{1-\alpha}\right) - \sum_{n=0}^{N/2-1}\frac{2\alpha^{2n}}{2n+1}, \tag{16}$$

respectively. Then we have

$$\int_{-1}^{1}\frac{T_N(x)}{x-c}dx = \phi(\alpha, N) - \alpha^{1-N}\psi(\alpha, N). \tag{17}$$

Proof: Using the relation

$$2T_n(x)T_m(x) = T_{n+m}(x) + T_{|n-m|}(x), \qquad n \ge 0, \quad m \ge 0, \tag{18}$$

gives

$$2\sum_{n=0}^{\infty}{}'\alpha^n T_n(x)T_m(x) = \sum_{n=-\infty}^{\infty} \alpha^{|n|} T_{|n+m|}(x). \tag{19}$$

Now noting that $\int_{-1}^{1} T_{|n|}(x)dx = 2/(1-n^2)$ if n is even and vanishes otherwise, we have, from (13) and (19),

$$\begin{aligned} \frac{\alpha^{-1}-\alpha}{-2}\int_{-1}^{1}\frac{T_N(x)}{x-c}dx &= \sum_{n=-\infty}^{\infty}\alpha^{|n-N|}\int_{-1}^{1}T_{|n|}(x)dx = \sum_{n=-\infty}^{\infty}\frac{\alpha^{|2n-N|}-\alpha^{|2n+2-N|}}{2n+1} \\ &= (1-\alpha^2)\sum_{n=0}^{\infty}\left\{\frac{\alpha^{2n}}{2n+1-N}+\frac{\alpha^{2n}}{2n+N+1}\right\}. \end{aligned} \tag{20}$$

The Theorem 1 follows if we use in the far right of (20)

$$\sum_{n=0}^{\infty} \frac{\alpha^{2n}}{2n+1-N} = \sum_{n=0}^{N/2-1} \frac{\alpha^{2n}}{2n+1-N} + \sum_{n=0}^{\infty} \frac{\alpha^{2n+N}}{2n+1} = -\phi(\alpha, N)/(2\alpha),$$

and

$$\sum_{n=0}^{\infty} \frac{\alpha^{2n}}{2n+1+N} = \alpha^{-N} \left\{ \sum_{n=0}^{\infty} \frac{\alpha^{2n}}{2n+1} - \sum_{n=0}^{N/2-1} \frac{\alpha^{2n}}{2n+1} \right\} = \alpha^{-N} \psi(\alpha, N)/2. \quad \square \tag{21}$$

Remark 1: Some significant digits might be lost in the computation of $\ln\{(1+\alpha)/(1-\alpha)\}$ in $\phi(\alpha, N)$ and $\psi(\alpha, N)$ defined by (15) and (16), respectively if $|c|$ is very close to unity. To avoid this difficulty, it is helpful to use the expression

$$2\ln\left(\frac{1+\alpha}{1-\alpha}\right) = \ln\left(\frac{c+1}{c-1}\right) = \begin{cases} \ln\{\delta/(c-1)\} & \text{if } \delta < 0 \\ \ln\{(c+1)/\delta\} & \text{if } \delta > 0. \end{cases} \tag{22}$$

Remark 2: Another numerical difficulty may happen in the computation of $\psi(\alpha, N)$ (16) when $|\alpha|^{-N}$ is large. Let $M = [-N \log_{10} \alpha]$, then more than M significant digits might be lost in the subtraction in the right hand side of (16). The reason can be simply seen as follows. Since for every even $N \geq 2$

$$0 \leq \sum_{n=0}^{\infty} \frac{\alpha^{2n}}{2n+N+1} < \sum_{n=0}^{\infty} \frac{\alpha^{2n}}{2n+1} = \frac{1}{2\alpha} \ln\left(\frac{1+\alpha}{1-\alpha}\right),$$

it follows from (21) that

$$0 \leq \psi(\alpha, N) \big/ [(1/\alpha) \ln\{(1+\alpha)/(1-\alpha)\}] < \alpha^N \leq 10^{-M}.$$

Suppose that we want to evaluate $\psi(\alpha, N)$ as accurately as possible, say, to within the accuracy at most by one digit worse than the roundoff error level of computer. If $|\alpha|^{-N} \leq s$, where s is a small number such as $s = 10$ at most, then $\psi(\alpha, N)$ can be evaluated in the numerically stable manner from the right hand side of (16). However, when the condition $|\alpha|^{-N} \leq s$ is not satisfied, which may happen when $|\alpha|$ is not very close to 1, the computation of the right hand side of (16) becomes very susceptible to the loss of significance as shown in *Remark 2* above. In this case we may choose to work directly with the infinite series $\sum_{n=0}^{\infty} \alpha^{2n}/(2n+1+N)$; see (21). This is achieved by employing the d-transformation of Levin and Sidi (1981) by picking $m = 1$ for this transformation. With this value of m, the transformation can be implemented by the W-algorithm of Sidi (1982). In our computations we used the computer program given in Ford and Sidi (1987) by letting $m = 1$ there and setting the integer parameter INCR in this program so that the required infinite sum is computed to very high accuracy. In particular, the value of INCR is taken to satisfy $\alpha^{\text{INCR}} \approx 0.66$, where the constant 0.66 was empirically determined. (The program in Ford and Sidi 1987 actually implements the so-called $W^{(m)}$-algorithm of these authors and reduces to the W-algorithm of Sidi 1982 when $m = 1$. We have also verified that the program due to Ford and Sidi can be easily modified to work for the input α of complex value to cope with a complex-valued c as well as real c.)

Now we proceed to the evaluation of $\int_{-1}^{1} T_N(x)/(x^2 + \delta^2)dx$, which is accomplished similarly to that above.

Lemma 2 *Let* $|\delta| = (\alpha - \alpha^{-1})/2$ *and choose* $\alpha = |\delta| - \sqrt{\delta^2 + 1}$. *Then we have*

$$\frac{1}{x^2 + \delta^2} = \frac{-4}{|\delta|(\alpha + \alpha^{-1})} \sum_{n=0}^{\infty}{}' (-\alpha^2)^n T_{2n}(x). \tag{23}$$

Proof: Use (13) with c replaced by $i|\delta| = \{i\alpha + 1/(i\alpha)\}/2$, then we have

$$\frac{1}{x - i|\delta|} = \frac{-4}{(i\alpha)^{-1} - i\alpha} \sum_{n=0}^{\infty}{}' (i\alpha)^n T_n(x). \tag{24}$$

The lemma follows if we use (24) in

$$\frac{1}{x^2 + \delta^2} = \frac{1}{|\delta|} \Im \frac{1}{x - i|\delta|}. \quad \square$$

Theorem 3 *Let* N *be an even integer and let* α *be defined in Lemma 2. Then we have*

$$\int_{-1}^{1} \frac{T_N(x)}{x^2 + \delta^2} dx = \frac{1}{|\delta|} \left\{ 2\alpha \sum_{n=0}^{N/2-1} \frac{(-\alpha^2)^n}{N - 2n - 1} - 2\alpha \sum_{n=0}^{\infty} \frac{(-\alpha^2)^n}{N + 2n + 1} + (-\alpha^2)^{N/2} \tan^{-1} \frac{1}{|\delta|} \right\}. \tag{25}$$

Proof: Similarly to the proof of Theorem 1, using the relations (19) and (23) we have

$$\begin{aligned} \frac{|\delta|(\alpha + \alpha^{-1})}{-2} \int_{-1}^{1} \frac{T_N(x)}{x^2 + \delta^2} dx &= \sum_{n=-\infty}^{\infty} (-\alpha^2)^{|n-N/2|} \int_{-1}^{1} T_{|2n|}(x) dx \\ &= \sum_{n=-\infty}^{\infty} (-\alpha^2)^{|n-N/2|} \left\{ \frac{1}{2n+1} - \frac{1}{2n-1} \right\} \\ &= (1 + \alpha^2) \sum_{n=0}^{\infty} \left\{ \frac{(-\alpha^2)^n}{2n + 1 + N} + \frac{(-\alpha^2)^n}{2n + 1 - N} \right\}. \end{aligned} \tag{26}$$

The second term of the far right in (26) can be rewritten as follows:

$$\sum_{n=0}^{\infty} \frac{(-\alpha^2)^n}{2n + 1 - N} = - \sum_{n=0}^{N/2-1} \frac{(-\alpha^2)^n}{N - 2n - 1} + \sum_{n=0}^{\infty} \frac{(-\alpha^2)^{n+N/2}}{2n + 1}. \tag{27}$$

It is easy to verify (25) if we use (26) and (27) along with the relation

$$-2\alpha \sum_{n=0}^{\infty} (-\alpha^2)^n/(2n + 1) = -2 \tan^{-1} \alpha = \tan^{-1}(1/|\delta|),$$

the last equality of which can be seen from the fact that $-1/|\delta| = 2\alpha/(1-\alpha^2)$. □

In our numerical experiments we observed that the W–algorithm due to Sidi (1982) (we set INCR = 1, *i.e.* Levin–transformation, Levin 1973) can make the alternating series, the second term of the right in (25) converge so quickly that at most twenty terms are sufficient to achieve the accuracy of the roundoff error in the double precision for any positive N and $|\alpha| < 1$.

4. Error estimates

Here we describe error estimates for the quadrature rules $Q_N(f;K_i) = I(p_N;K_i)$ $(i=1,2)$ based on the polynomial $p_N(x)$ interpolating $f(x)$ at $\cos(\pi j/N)$ $(0 \le j \le N)$, the zeros of a polynomial $\omega_{N+1}(t)$ defined by

$$\omega_{N+1}(t) = T_{N+1}(t) - T_{N-1}(t) = 2\,(t^2-1)\,U_{N-1}(t), \tag{28}$$

where $U_k(t)$ is the Chebyshev polynomial of the second kind defined by $U_k(t) = \sin(k+1)\theta/\sin\theta$, $t = \cos\theta$.

Let ε_ρ denote the ellipse in the complex plane $z = x + iy$ with foci $(-1,0)$, $(1,0)$ and semimajor axis $a = (\rho+\rho^{-1})/2$ and semiminor axis $b = (\rho-\rho^{-1})/2$ for a constant $\rho > 1$. Assume that $f(z)$ is single–valued and analytic inside and on ε_ρ. Then, the error of $p_N(x)$ can be expressed in terms of a contour integral (Elliott 1965 and Hasegawa and Torii 1991a), which is also expanded in a Chebyshev series (Hasegawa et al. 1983):

$$f(x) - p_N(x) = \frac{1}{2\pi i}\oint_{\varepsilon_\rho} \frac{\omega_{N+1}(x)\,f(z)\,dz}{(z-x)\,\omega_{N+1}(z)} = \omega_{N+1}(x)\sum_{k=0}^{\infty}{}' V_k^N(f)\,T_k(x). \tag{29}$$

In (29) the coefficients $V_k^N(f)$ are given by

$$V_k^N(f) = \frac{1}{\pi^2 i}\oint_{\varepsilon_\rho} \frac{\tilde{U}_k(z)\,f(z)\,dz}{\omega_{N+1}(z)}, \qquad k \ge 0. \tag{30}$$

The Chebyshev function of the second kind $\tilde{U}_k(z)$ is defined by

$$\tilde{U}_k(z) = \int_{-1}^{1} \frac{T_k(t)\,dt}{(z-t)\sqrt{1-t^2}} = \frac{\pi}{\sqrt{z^2-1}\,w^k} = \frac{2\pi}{(w-w^{-1})\,w^k}, \tag{31}$$

where $w = z + \sqrt{z^2-1}$ and $|w| > 1$ for $z \notin [-1,1]$ (Gautschi and Varga 1983 and Hasegawa et al. 1983).

Using (29) in (1) and (4) yields the error for the approximate integral $Q_N(f;K)$:

$$I(f;K) - Q_N(f;K) = I(f-p_N;K) = \sum_{k=0}^{\infty}{}' V_k^N(f)\,\Omega_k^N(K), \tag{32}$$

where $\Omega_k^N(K)$ is defined by

$$\Omega_k^N(K) = \int_{-1}^{1} K(x)\,\omega_{N+1}(x)\,T_k(x)dx. \tag{33}$$

The estimation of the error in (32) requires a bound for $\Omega_k^N(K)$ as well as one for $V_k(f)$. For $K_1(x) = 1/(x-c)$, the following lemma is proven in Appendix A.

Lemma 4 *Let $K_1(x)$ be defined by (2) and further let $\Omega_k^N(K)$ be defined by (33). Then $\Omega_k^N(K_1)$ is bounded independently of c by*

$$|\Omega_k^N(K_1)| \le 8. \tag{34}$$

Remark. The proof of Lemma 4 given in Appendix A reveals that the relation (34) holds for any value of c in the complex plane as well as on the real line, provided α is chosen so that $|\alpha| < 1$ for complex $c \notin [-1,1]$.

In Appendix B we prove the following lemma for $K_2(x) = 1/(x^2+\delta^2)$.

Lemma 5 *Let $K_2(x)$ be defined by (2) and further let $\Omega_k^N(K)$ be defined by (33). Then for even N, $\Omega_k^N(K_2)$ is bounded independently of δ by*

$$|\Omega_k^N(K_2)| \le 8k/\sqrt{\delta^2+1} \le 8k, \tag{35}$$

if k is odd, and vanishes otherwise.

Suppose that $f(z)$ is a meromorphic function which has M simple poles at the points z_m $(m = 1,2,\ldots,M)$ outsides ε_ρ with residues $\mathrm{Res} f(z_m)$. Then, performing the contour integral of (30) yields

$$V_k^N(f) = -\frac{2}{\pi}\sum_{m=1}^{M} \mathrm{Res} f(z_m)\,\tilde{U}_k(z_m)/\omega_{N+1}(z_m), \quad k \ge 0. \tag{36}$$

Now, noting that $T_k(z) = (w^k + w^{-k})/2$ for complex $z = (w + w^{-1})/2 \notin [-1,1]$ where $|w| > 1$, we have from (28) $\omega_{N+1}(z) = \sqrt{z^2-1}(w^N - w^{-N})$, which is combined with (31) to yield

$$\frac{\tilde{U}_k(z)}{\omega_{N+1}(z)} = \frac{\pi}{z^2-1}\,\frac{1}{w^k\,(w^N - w^{-N})}.$$

The most dominant term in the right of (36) is obtained for the poles z_j for which $\left|z_j + \sqrt{z_j^2-1}\right| = \min_{1\le m\le M}\left|z_m + \sqrt{z_m^2-1}\right| \equiv r > 1$. If we assume that there is only one such z_j, we have $V_k^N(f) \sim V_0^N(f) w_j^{-k}$ for sufficiently large N, where $w_j = z_j + \sqrt{z_j^2-1}$. This fact and (32) along with (34) and (35) permit us to estimate the errors

$$\begin{aligned} &|I(f;K_1) - Q_N(f;K_1)| \\ &\le 8\sum_{k=0}^{\infty}{}' \left|V_k^N(f)\right| \sim 8|V_0^N(f)| \sum_{k=0}^{\infty}{}' r^{-k} = 8\left|V_0^N(f)\right| \frac{r+1}{2(r-1)}, \end{aligned} \tag{37}$$

$$
\begin{aligned}
&|I(f;K_2) - Q_N(f;K_2)| \\
&\leq \sum_{k=0}^{\infty} 8(2k+1)\left|V_{2k+1}^N(f)\right| \sim 8\left|V_0^N(f)\right| \sum_{k=0}^{\infty} 8(2k+1)\,r^{-2k-1} \\
&= 8\left|V_0^N(f)\right|\left\{\frac{r}{r^2-1} + \frac{2r}{(r^2-1)^2}\right\} < 4\left|V_0^N(f)\right|\frac{r^2}{(r-1)^2}, \qquad (38)
\end{aligned}
$$

In the last inequality of (38) we have used the fact that $r + 1 > 2$.

Now, we wish to estimate $|V_0^N(f)|$ in terms of the available coefficients a_k^N of $p_N(x)$. Elliott (1965) gives

$$
a_k^N = \frac{2}{\pi i}\oint_{\varepsilon_\rho} \frac{T_{N-k}(z)\,f(z)}{\omega_{N+1}(z)}\,dz, \qquad 0 \leq k \leq N.
$$

Performing the contour integral and comparing the result with (36) gives the relations $|V_0^N| \sim |a_N^N| r/(r^2-1)$ and $|a_k^N| \sim r|a_{k+1}^N|$ for sufficiently large N. From these relations and (37) we have the error estimate for $Q_N(f;K_1)$

$$
E_N(f;K_1) = \left(\left|a_N^N\right|/2\right)\frac{8r}{(r-1)^2}. \qquad (39)
$$

The constant r may be estimated from the asymptotic behavior of $\{a_k^N\}$ (Hasegawa et al. 1983). Similarly, for $K_2(x)$ we have, from (38),

$$
E_N(f;K_2) = \left(\left|a_N^N\right|/2\right)\frac{2\,r^3}{(r-1)^3}. \qquad (40)
$$

We remark that the error estimates (39) and (40) for the approximations $Q_N(f;K_1)$ and $Q_N(f;K_2)$, respectively, are independent of the value of δ. This indicates that the approximate polynomial $p_N(x)$ can be used common to a set of the integrals $I(f;K_1)$ (or $I(f;K_2)$) with different values of δ if a stopping criterion (39) (or (40)) is satisfied.

Incidentally, an automatic quadrature of non–adaptive type is generally constructed from the sequence of the approximations $\{Q_N\}$ converging to the integral $I(f;K)$, until a stopping criterion is satisfied. It is an usual and simple way to double the degree N of $p_N(x)$ (3) for generating the sequence $\{Q_N(f;K)\}$ (4) (Gentleman 1972 and Branders and Piessens 1975). In order to make an automatic quadrature efficient, however, it is advantageous to have more chances of checking the stopping criterion than doubling N. To this end, as shown in Hasegawa et al. (1990) we may generate the sequence of $\{p_N\}$, increasing the degree N more slowly as follows:

$$
N = 6, 8, 10, \ldots, 3\times 2^n, 4\times 2^n, 5\times 2^n, \ldots, \qquad (n = 1, 2, 3, \ldots),
$$

and by using the FFT; see also Hasegawa and Torii (1991b) for the application of the FFT to the problem of product type integration.

5. Numerical examples

We compute the integrals

$$\text{(P1)} \quad \int_{-1}^{1} f(x)/(x-c)\, dx, \qquad c = -1+\delta, \quad \delta < 0,$$

$$\text{(P2)} \quad \int_{-1}^{1} f(x)/(x^2+\delta^2)\, dx,$$

where we take $f(x) = (1-a^2)/(1-2ax+a^2)$, which has a Chebyshev series expansion $f(x) = 2\sum_{n=0}^{\infty}{}' a^n T_n(x)$ for $|a| < 1$. If the parameter a is close to unity in magnitude, the Chebyshev series for $f(x)$ converges slowly and the numerical integration of the integrals (P1) and (P2) by the present method is difficult. Here we choose a close to unity such that $(a+a^{-1})/2 = 1+5^{-3}$ and $a < 1$. Table 1 lists the number of function evaluations $N+1$ (= number of sample points used for the polynomial $p_N(x)$ given by (3) to interpolate $f(x)$) required to satisfy the requested relative tolerances $\epsilon_r = 10^{-6}$ and 10^{-10} and the actual relative accuracies achieved.

Table 1: Performance of the present method for the integrals (P1) $\int_{-1}^{1} f(x)/(x-c)\, dx$, $c = -1+\delta$, and (P2) $\int_{-1}^{1} f(x)/(x^2+\delta^2)\, dx$, where $f(x) = (1-a^2)/(1-2ax+a^2)$ with a satisfying $(a+a^{-1})/2 = 1+5^{-3}$ and $0 < a < 1$. The number of function evaluations $N+1$ required to satisfy the requested relative tolerances $\epsilon_r = 10^{-6}$ and 10^{-10} are listed in the forth and sixth columns.

			$\epsilon_r = 10^{-6}$		$\epsilon_r = 10^{-10}$	
Kernel	δ	Integral	$N+1$	error	$N+1$	error
(P1)	-1×10^{-1}	0.5152711361031522	161	7.9×10^{-13}	257	6.4×10^{-16}
	-1×10^{-3}	0.8281473339333300	161	4.0×10^{-12}	257	8.0×10^{-15}
	-1×10^{-5}	1.119198712640058	161	2.2×10^{-11}	257	3.1×10^{-13}
	-1×10^{-7}	1.409879668356110	161	2.6×10^{-11}	257	1.0×10^{-12}
	-1×10^{-9}	1.700555499268794	161	2.2×10^{-11}	257	2.2×10^{-12}
(P2)	1×10^{-1}	4.346003191058892	129	5.7×10^{-12}	257	1.4×10^{-15}
	1×10^{-2}	$3.993558372439038\times10^{1}$	129	4.0×10^{-9}	193	4.7×10^{-12}
	1×10^{-3}	$3.954547456719288\times10^{2}$	97	2.6×10^{-7}	161	2.9×10^{-11}
	1×10^{-4}	$3.950611386192320\times10^{3}$	65	4.8×10^{-7}	161	3.8×10^{-12}
	1×10^{-5}	$3.950217432400829\times10^{4}$	65	4.8×10^{-8}	129	1.5×10^{-11}

Appendix A

Here we prove Lemma 4. Using (13), (18) and (19) we have

$$(\alpha - \alpha^{-1})\frac{T_{N+1}(x)T_k(x)}{x-c} = 2\sum_{n=-\infty}^{\infty} \alpha^{|n|} T_{|n+N+1|}(x)\, T_k(x)$$

$$= \sum_{n=-\infty}^{\infty} \alpha^{|n|} \left\{ T_{|n+N+1|+k}(x) + T_{||n+N+1|-k|}(x) \right\}$$

$$= \sum_{n=-\infty}^{\infty} \alpha^{|n|} \left\{ T_{|n+N+1+k|}(x) + T_{|n+N+1-k|}(x) \right\}. \quad (41)$$

Similarly, we have

$$(\alpha - \alpha^{-1}) \frac{T_{N-1}(x) T_k(x)}{x - c} = \sum_{n=-\infty}^{\infty} \alpha^{|n+2|} \left\{ T_{|n+N+1+k|}(x) + T_{|n+N+1-k|}(x) \right\}. \quad (42)$$

Using (28), (41) and (42) in (33) yields

$$(\alpha - \alpha^{-1}) \Omega_k^N(K_1) = \sum_{n=-\infty}^{\infty} (\alpha^{|n|} - \alpha^{|n+2|}) \int_{-1}^{1} \left\{ T_{|n+N+1+k|}(x) + T_{|n+N+1-k|}(x) \right\} dx.$$

Since $|\alpha^{|n|} - \alpha^{|n+2|}| = |\alpha - \alpha^{-1}|\, |\alpha|^{|n+1|}$ if $n + 1 \neq 0$ and vanishes otherwise, we have

$$|\Omega_k^N(K_1)| \leq \sum_{n=-\infty}^{\infty} |\alpha|^{|n+1|} \left| \int_{-1}^{1} \left\{ T_{|n+N+1+k|}(x) + T_{|n+N+1-k|}(x) \right\} dx \right|. \quad (43)$$

Using in (43) the facts that $|\alpha| < 1$ and that $\int_{-1}^{1} T_{|n|}(x)\, dx = 2/(1 - n^2)$ if n is even and vanishes otherwise, gives

$$|\Omega_k^N(K_1)| \leq 2 \sum_{n=-\infty}^{\infty} \frac{2}{|1 - 4n^2|} = 8,$$

which establishes the Lemma 4.

Appendix B

Here we prove Lemma 5. It is trivial that $\Omega_k^N(K_2)$ vanishes if k is even since then the integrand is an odd function. The proof of the case where k is odd requires some manipulation. From (28) and (33) we have

$$|\Omega_k^N(K_2)| \leq |F_k^{N+1}(K_2)| + |F_k^{N-1}(K_2)|,$$

where we define

$$F_k^{N\pm1}(K_2) = \int_{-1}^{1} T_{N\pm1}(x)\, T_k(x)\, K_2(x)\, dx. \quad (44)$$

To see (35), it is sufficient to show that

$$|F_k^{N\pm1}(K_2)| \leq 4k/\sqrt{\delta^2 + 1}. \quad (45)$$

Here we verify (45) only for the case $F_k^{N+1}(K_2)$. The case $F_k^{N-1}(K_2)$ is similarly verified.

Similarly to the derivation to (41), using (18), (19) and (23) we have

$$
\begin{aligned}
-|\delta|(\alpha+\alpha^{-1})\frac{T_{N+1}(x)T_k(x)}{x^2+\delta^2} &= 2\sum_{n=-\infty}^{\infty}(-\alpha^2)^{|n|}\,T_{|2n+N+1|}(x)\,T_k(x)\\
&= \sum_{n=-\infty}^{\infty}(-\alpha^2)^{|n|}\left\{T_{|2n+N+1+k|}(x)+T_{|2n+N+1-k|}(x)\right\}\\
&= \sum_{n=-\infty}^{\infty}\left\{(-\alpha^2)^{|n-k|}+(-\alpha^2)^{|n|}\right\}T_{|2n+N+1-k|}(x). \quad (46)
\end{aligned}
$$

Noting that $\int_{-1}^{1} T_{|2n|}(x)\,dx = 2/(1-4n^2)$, from (44) and (46) we have, for odd k and even N,

$$
-|\delta|\,(\alpha+\alpha^{-1})F_k^{N+1}(K_2) = 2\sum_{n=-\infty}^{\infty}\frac{(-\alpha^2)^{|n-k|}+(-\alpha^2)^{|n|}}{1-(2n+N-k+1)^2}. \quad (47)
$$

To prove (45) for the case of $N+1$, it remains to verify in (47) that

$$
\left|(-\alpha^2)^{|n-k|}+(-\alpha^2)^{|n|}\right| \le (1-\alpha^2)k = -\alpha(\alpha-\alpha^{-1})k = -2\alpha|\delta|k, \quad (48)
$$

because

$$
\sum_{n=-\infty}^{\infty} 2/\left|1-(2n+N-k+1)^2\right| = \sum_{n=-\infty}^{\infty} 2/\left|1-4n^2\right| = 4,
$$

and $\alpha+\alpha^{-1} = -2\sqrt{1+\delta^2}$. We use the relations $-1\le\alpha\le 0$, and for odd integer k,

$$
1+(-\alpha^2)^k = 1-\alpha^{2k} = (1-\alpha^2)(1+\alpha^2+\cdots+\alpha^{(2k-2)}) \le -2\,\alpha\,|\delta|\,k,
$$

to verify (48). It is easy to see (48) for the case $0<k\le n$, where $\left|(-\alpha^2)^{|n-k|}+(-\alpha^2)^{|n|}\right| = \left|(-\alpha^2)^{n-k}\right|(1-\alpha^{2k}) \le 1-\alpha^{2k}$. Similarly for $n\le 0$ we have that $\left|(-\alpha^2)^{|n-k|}+(-\alpha^2)^{|n|}\right| = |\alpha^{-2n}|(1-\alpha^{2k})$. For $0<n<k/2$ it can be seen that

$$
\left|(-\alpha^2)^{|n-k|}+(-\alpha^2)^{|n|}\right| = \left|(-\alpha^2)^n\right|\left(1-\alpha^{2(k-2n)}\right) \le (1-\alpha^2)(k-2n) < (1-\alpha^2)k,
$$

while $\left|(-\alpha^2)^{|n-k|}+(-\alpha^2)^{|n|}\right| = \left|\alpha^{2(k-n)}\right|\left(1-\alpha^{2(2n-k)}\right)$ for $k/2<n<k$.

Acknowledgements

We are indebted to H. Sugiura for helpful comments. We are grateful to the referee for useful comments for improving the presentation. The first author wishes to thank Professor Avram Sidi for valuable comments during the Oberwolfach conference.

References

Bialecki B. (1989), A modified Sinc quadrature rule for functions with poles near the arc of integration, *BIT* **29** 464–476.

Branders M. and Piessens R. (1975), An extension of Clenshaw–Curtis quadrature, *J. Comp. Appl. Math.* **1** 55–65.

Clenshaw C. W. and Curtis A. R. (1960), A method for numerical integration on an automatic computer, *Numer. Math.* **2** 197–205.

Davis P. J. and Rabinowitz P. (1984), *Methods of Numerical Integration, 2nd ed.* Academic Press, Orlando.

Elliott D. (1965), Truncation errors in two Chebyshev series approximations, *Math. Comp.* **19** 234–248.

Ford W. F. and Sidi A. (1987), An algorithm for a generalization of the Richardson extrapolation process, *SIAM J. Numer. Anal.* **24** 1212–1232.

Gentleman W. M. (1972), Implementing Clenshaw–Curtis quadrature II. Computing the cosine transformation, *Comm. ACM* **15** 343–346.

Gautschi W. and Varga R. S. (1983), Error bounds for Gaussian quadrature of analytic functions *SIAM J. Numer. Anal.* **20** 1170–1186.

Hasegawa T. and Torii T. (1991a), An automatic quadrature for Cauchy principal value integrals, *Math. Comp.* **56** 741–754.

Hasegawa T. and Torii T. (1991b), Application of a modified FFT to product type integration *J. Comp. Appl. Math.* **38** 157–168.

Hasegawa T. et al. (1983), Generalized Chebyshev interpolation and its application to automatic quadrature, *Math. Comp.* **41** 537–553.

Hasegawa T. et al. (1990), An algorithm based on the FFT for a generalized Chebyshev interpolation, *Math. Comp.* **54** 195–210.

Lether F. G. (1977a), Subtracting out complex singularities in numerical integration, *Math. Comp.* **31** 223–229.

Lether F. G. (1977b), Modified quadrature formulas for functions with nearby poles, *J. Comp. Appl. Math.* **3** 3–9.

Levin D. (1973), Development of non–linear transformations for improving convergence of sequences, *Internat. J. Comut. Math.* **B3** 371–388.

Levin D. and Sidi A. (1981), Two new classes of nonlinear transformations for accelerating the convergence of infinite integrals and series, *Appl. Math. Comp.* **9** 204–215.

Monegato G. (1986), Quadrature formulas for functions with poles near the interval of integration, *Math. Comp.* **47** 301–312.

Sidi A. (1982) An algorithm for a special case of a generalization of the Richardson extrapolation process, *Numer. Math.* **39** pp.299–307.

Takemitsu Hasegawa
Department of Information Science
Faculty of Engineering
Fukui University, Fukui, 910, Japan
hasegawa@fuis.fuis.fukui-u.ac.jp

Tatsuo Torii
Department of Information Engineering
Faculty of Engineering
Nagoya University, Nagoya, 464–01, Japan

International Series of Numerical Mathematics, Vol. 112, © 1993 Birkhäuser Verlag Basel

Some Problems involving Orthogonal Polynomials

David B. Hunter
Harry V. Smith

Abstract Suppose a weight-function w is continuous, positive and integrable over (-1,1). In the first part of this paper we generalize an expression due to Stenger (1966) for the error in the associated Gaussian quadrature formula. In the later part, we consider the Rodrigue's function U_n associated with the polynomial p_n in the orthogonal sequence over (-1,1) with respect to w. Some properties of U_n are proved, and a number of conjectures, with some supporting evidence, are made.

1. Errors in Gaussian quadrature

Suppose w is a weight-function which is positive and continuous on the open interval (-1,1) and is such that the integral

$$\int_{-1}^{1} w(x)dx$$

exists. Let p_n be the polynomial of degree n in the orthogonal sequence associated with w over (-1,1), scaled so that the coefficient of x^n in the expansion of $p_n(x)$ is positive. If x_{n1}, x_{n2}, ..., x_{nn} are the zeros of p_n, the Gaussian quadrature formula for

$$I(f) = \int_{-1}^{1} w(x)f(x)dx \tag{1}$$

can be written in the form

$$I(f) = \sum_{k=1}^{n} w_{nk}\ f(x_{nk}) + E_n(f). \tag{2}$$

Suppose now that f is analytic in some circle with centre 0, radius $R > 1$ in the complex plane, and that its Taylor series about 0 is

$$\sum_{k=0}^{\infty} a_k z^k. \tag{3}$$

Stenger (1966) showed that the error $E_n(f)$ in (2) can be expressed in the form

$$E_n(f) = \sum_{k=0}^{\infty} a_{2n+k}\, e_{nk} \tag{4}$$

where the coefficients e_{nk} are independent of f. He then proved the following result

Theorem 1 If w is an even function, then

$$\left.\begin{aligned} e_{n,2k} &> 0, \\ e_{n,2k+1} &= 0. \end{aligned}\right\} \quad (k = 0, 1, 2, \ldots.)$$

From this, he derived a number of results on monotonicity of convergence of Gaussian quadrature formulae, and obtained some error bounds.

If w is not an even function, equation (4) holds, but the sign properties of the coefficients e_{nk} are less predictable, in general. We shall now derive a condition which guarantees positivity of these coefficients. Stenger's analysis involves two other series:

$$1/p_n(z) = \sum_{j=0}^{\infty} b_{nj}\, z^{-n-j} \tag{5}$$

and

$$q_n(z) = \sum_{j=0}^{\infty} c_{nj}\, z^{-n-j-1} \tag{6}$$

where

$$q_n(z) = \int_{-1}^{1} \frac{w(x)\, p_n(x)}{z - x}\, dx, \quad (z \notin [-1, 1]) \tag{7}$$

In terms of the coefficients in (5) and (6) the coefficients e_{nk} of (4) are given by

$$e_{nk} = \sum_{j=0}^{k} b_{nj}\, c_{n,k-j}. \tag{8}$$

In Hunter (1975) the following two results are proved:

Lemma 1. $c_{n,2j} > 0$, $(j = 0, 1, 2, \ldots..)$.

Lemma 2. If $w(x)/w(-x)$ is strictly increasing then

$$b_{nj} > 0, \quad (j = 0, 1, 2, \ldots.)$$

These suggest two further results:

Lemma 3. $b_{n,2j} > 0$ $(j = 0, 1, 2, \ldots.)$

Lemma 4. If $w(x)/w(-x)$ is strictly increasing, then

$$c_{nj} > 0 \quad (j = 0, 1, 2, \ldots..)$$

Lemma 3 is proved, in the following sharper form, in Hunter (1977).

Theorem 2. Suppose

$$p(z) = z^n + \sum_{j=0}^{n-1} a_j z^j$$

is a monic polynomial of degree n, where the coefficients a_0, a_1, ..., a_{n-1} are real and are not all zero. Let

$$1/p(z) = \sum_{k=0}^{\infty} b_k z^{-n-k}.$$

Then, provided the zeros of p are all real,

$$b_{2j} \geq (a_{n-1}^2 - 2a_{n-2})^j/2^j j! > 0, \tag{9}$$

with equality in the first inequality if and only if (i) $j = 0$, or (ii) $j = 1$ and $a_{n-1} = 0$.

Note that in (9)

$$a_{n-1}^2 - 2a_{n-2} = \sum_{j=1}^{n} x_j^2$$

where x_1, x_2, ..., x_n are the zeros of p. This theorem was further generalized in Hunter and McDonald (1989).

Lemma 4, finally, was proved by Gautschi (1983). This now enables us to extend Stenger's theorem as follows:

Theorem 3. Provided $w(x)/w(-x)$ is strictly increasing, the coefficients e_{nk} in (4) are all positive.

Proof. The theorem follows immediately from (8), on applying Lemmas 2 and 4.

The above theorem is closely related to Theorem 2.1 of Gautschi and Varga (1983). It should enable us to extend the monotonicity results of Stenger (1966). Note, too, that there are analogues of Lemmas 2 and 4 if $w(x)/w(-x)$ is strictly decreasing (see Hunter 1975 and Gautschi 1983). They lead to the following corollary to Theorem 3.

Corollary. If $w(x)/w(-x)$ is strictly decreasing the coefficients e_{nk} in (4) satisfy

$$(-1)^k e_{nk} > 0 \qquad (k = 0, 1, 2, \ldots).$$

2. Properties of the Rodrigues function

It is well-known that under the conditions mentioned at the beginning of this paper there is a function U_n with the properties

$$\begin{aligned} &U_n^{(r)}(1) = U_n^{(r)}(-1) = 0, \qquad (r = 0, 1, \ldots, n-1), \\ \text{and} \quad &U_n^{(n)}(x) = w(x)\, p_n(x), \quad (-1 < x < 1) \end{aligned} \tag{10}$$

For example, if $w(x)$ is the Jacobi weight-function $(1 - x)^\alpha (1 + x)^\beta$, then

$$U_n(x) = \frac{(-1)^n}{2^n\, n!} (1 - x)^{n+\alpha} (1 + x)^{n+\beta}; \tag{11}$$

here eqns (10) are equivalent to Rodrigues' formula for the Jacobi polynomials. Otherwise, not much is known about the function U_n. We wish to argue that if $w(x)/w(-x)$ is strictly increasing then U_n is biased towards the right-hand half of the interval $(-1,1)$. Some evidence for this is given by the following result.

Theorem 4 If $w(x)/w(-x)$ is strictly increasing then

$$\int_{-1}^{1} (-1)^n U_n(x)\, x^j \, dx > 0, \qquad (j = 0, 1, 2, \ldots\ldots)$$

Proof From Hunter (1975), eqn (11),

$$c_{nj} = \frac{(n + j)!}{j!} \int_{-1}^{1} (-1)^n U_n(x) x^j \, dx.$$

So, as $c_{nj} > 0$ from Lemma 4,

$$\int_{-1}^{1} (-1)^n U_n(x)x^j \, dx > 0 \qquad (j = 0, 1, 2, \ldots).$$

Note that $(-1)^n U_n(x) > 0$ for $n = 0, 1, 2, \ldots$. It follows from Theorem 4 that

$$\int_{0}^{1} (-1)^n [U_n(x) - U_n(-x)] \, x^{2j+1} \, dx > 0 \quad (j = 0, 1, 2, \ldots).$$

This shows that, in a sense, U_n is biased towards the right-hand half of $(-1, 1)$. We suspect that the bias is, in fact, stronger than that shown by Theorem 4. It is shown by Smith (1990) that, for $1 \leqslant r \leqslant n$, $U_n^{(r)}$ has exactly r real, distinct zeros in $(-1, 1)$. Suppose these are

$$x_{n1}^{(r)} > x_{n2}^{(r)} > \ldots\ldots > x_{nr}^{(r)} .$$

Then, setting

$$p_{nr}(x) = \gamma_{nr} \prod_{j=1}^{r} (x - x_{nj}^{(r)}),$$

where γ_{nr} is a suitable positive scale-factor, we can write

$$U_n^{(r)}(x) = (-1)^{n-r} (1 - x^2)^{n-r} w_r(x) \, p_{nr}(x) \tag{12}$$

where w_r is some weight-function. Smith showed that $w_r(x) > 0$ for $-1 < x < 1$, and derived a sequence of quadrature formulae of the form

$$\int_{-1}^{1} w_r(x) \, f(x)dx = \sum_{j=1}^{r} w_{nj}^{(r)} \, f(x_{nj}^{(r)}) + \sum_{j=0}^{n-r-1} [K_{nj}^{'(r)} f^{(j)}(-1) + K_{nj}^{(r)} \, f^{(j)}(1)] \tag{13}$$
$$+ E_n^{(r)}(f), \qquad (0 \leqslant r \leqslant n)$$

As with Gaussian quadrature, these formulae are exact for all polynomials of degree less than 2n, and the weights $w_{nj}^{(r)}$ are positive for $0 < r \leqslant n$, $1 \leqslant j \leqslant r$. In general $w_r(x)$ differs from $w(x)$; but for the case $w(x) = (1 - x)^{\alpha} (1 + x)^{\beta}$ we can choose the value of γ_{nr} to ensure that $w_r(x) = w(x)$. In any case, the formulae have some theoretical interest, and they may provide a means of proving Conjecture 1 below. This conjecture generalizes a theorem of Markoff (1886).

Conjecture 1. Let $W(x,\tau)$ be a weight-function, depending on a parameter τ, which is positive and continuous for $-1 < x < 1$, $\tau_1 < \tau < \tau_2$. Suppose also that the partial derivative $W_\tau(x,\tau)$ exists and is continuous for $-1 < x < 1$, $\tau_1 < \tau < \tau_2$, and that the integrals

$$\int_{-1}^{1} W_\tau(x, \tau)x^j \, dx, \qquad (0 \leqslant j \leqslant 2n-1)$$

converge uniformly in every closed subinterval of (τ_1, τ_2). Let $p_n(x,\tau)$ be the polynomial of degree n (in x) in the orthogonal sequence associated with $W(x,\tau)$, and let $U_n(x,\tau)$ be such that

$$U^{(r)}(-1,\tau) = U^{(r)}(1,\tau) = 0, \qquad (r = 0, 1, 2, \ldots, n-1)$$

and

$$U^{(n)}(x,\tau) = W(x,\tau)\, p_n(x,\tau), \qquad (-1 < x < 1)$$

where the superscripts denote differentiation with respect to x. Let the zeros of $U_n^{(r)}(x,\tau)$ for a given value of τ be

$$x_{n1}^{(r)}(\tau) > x_{n2}^{(r)}(\tau) > \ldots . > x_{nr}^{(r)}(\tau).$$

Then for each $r(1 \leqslant r \leqslant n)$ and each $\upsilon(1 \leqslant \upsilon \leqslant r)$ $x_{n\upsilon}^{(r)}$ is an increasing function of τ, provided $W_\tau(x, \tau)/W(x, \tau)$ is an increasing function of x in $(-1, 1)$.

Markoff's Theorem (1886) is the special case $r = n$ of the above conjecture. A simpler proof than Markoff's is given in Szegö (1967), Theorem 6.12.1. It may be possible to prove the general result by Szegö's method, using the quadrature rule (13) instead of Gaussian quadrature but some of the details are troublesome. We are grateful to Professor K. Jetter for suggesting an alternative approach, which may well be useful. Meanwhile, we suggest two pieces of evidence in support of the conjecture:

(i) The zeros of p_{nr} interlace with those of $p_{n,r-1}$ (if $r > 1$) and with those of $p_{n,r+1}$ (if $r < n$). So it is feasible that, since the zeros of p_{nn} are increasing functions of τ, by Markoff's Theorem, they will tend to "drag" the other zeros to the right.

(ii) The conjecture is true for the Jacobi weight-function $W(x,\beta) = (1 - x)^\alpha (1 + x)^\beta$ for fixed α, if we regard β as a parameter.

In this case

$$W_\beta(x, \beta)/W(x,\beta) = \ln(1 + x),$$

which is strictly increasing. Also, from (11),

$$U_n^{(r)}(x,\beta) = \frac{(-1)^{n-r}\, r!}{2^{n-r}\, n!} (1 - x)^{n-r+\alpha} (1 + x)^{n-r+\beta} P_r^{(n-r+\alpha,n-r+\beta)}(x).$$

So the zeros $x_{n1}^{(r)}, x_{n2}^{(r)}, x_{nr}^{(r)}$ are simply those of $P_r^{(n-r+\alpha,n-r+\beta)}$, which are increasing functions of β, by Markoff's Theorem.

Assuming the truth of Conjecture 1, we can now deduce the following corollary.

Corollary. If $w(x)/w(-x)$ is strictly increasing, the zeros $x_{nj}^{(r)}$ of $U_n^{(r)}$ satisfy

$$x_{nj}^{(r)} + x_{n,r-j+1}^{(r)} > 0, \quad (j = 1, 2, \ldots, r)$$

Proof. Let $W(x, \tau) = \tau\, w(x) + (1 - \tau)\, w(-x)$, where $0 < \tau \leq 1$.

Then

$$W_\tau(x, \tau)/W(x,\tau) = \frac{w(x) - w(-x)}{\tau w(x) + (1 - \tau)\, w(-x)}$$

$$= \tau^{-1} - \frac{\tau^{-1}}{w(x)/w(-x) - 1 + \tau^{-1}}$$

This is strictly increasing with respect to x. So the zeros $x_{nj}^{(r)}(\tau)$ are strictly increasing with respect to τ. When $\tau = \frac{1}{2}$, $W(x,\tau)$ is an even function of x. So

$$x_{nj}^{(r)}(\tfrac{1}{2}) + x_{n,r-j+1}^{(r)}(\tfrac{1}{2}) = 0.$$

Hence if $\tau > \frac{1}{2}$,

$$x_{nj}^{(r)}(\tau) + x_{n,r-j+1}^{(r)}(\tau) > 0.$$

The required result follows on setting $\tau = 1$.

The above corollary enables us to generalize Lemma 2 as follows.

Lemma 5. If

$$1/p_{nr}(z) = \sum_{j=0}^{\infty} b_{nj}^{(r)} z^{-r-j}$$

then, provided $w(x)/w(-x)$ is strictly increasing, $b_{nj}^{(r)} > 0$.

The proof is similar to that of Hunter (1975) and Gautschi (1983); and, of course, still depends on the truth of Conjecture 1.

As a special case of the above corollary, we have the result $x_{n1}^{(1)} > 0$. That is, the single maximiser of $(-1)^n U_n$ is positive. Thus the graph of U_n is biased to the right-hand half of $(-1,1)$, as indicated below:

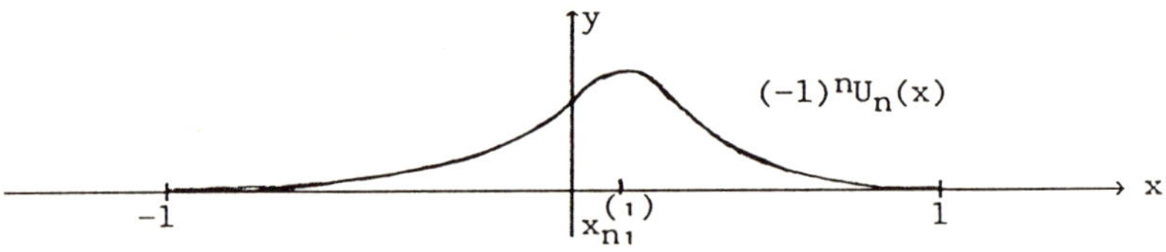

Finally, we suggest the following further conjecture.

Conjecture 2. If $w(x)/w(-x)$ is strictly increasing then, for $0 < r < n$, $w_r(x)/w_r(-x)$ is also strictly increasing. In particular, in the case $r = 0$, $U_n(x)/U_n(-x)$ is strictly increasing.

Our only evidence for the truth of this conjecture is that it holds if $w(x) = (1 - x)^{\alpha} (1 + x)^{\beta}$, with $\alpha < \beta$, since then $w_r(x)$ is simply a constant multiple of $w(x)$. Incidentally, Conjecture 1 would follow immediately from Conjecture 2.

References

Gautschi W. (1983), On Padé approximants associated with Hamburger series. Calcolo, v.20, 111-127.

Gautschi, W. and Varga R.S. (1983), Error bounds for Gaussian quadrature of analytic functions. SIAM J. Numer. Anal. v.20, 1170-1186.

Hunter, D.B. (1975), Some properties of orthogonal polynomials. Math. Comp., v.29, 559-565.

Hunter, D.B. (1977), The positive-definiteness of the complete symmetric functions of even order. Math. Proc. Camb Phil. Soc., v.82, 255-258.

Hunter, D.B. and McDonald I.G. (1989), Some sign properties of symmetric functions. Math. Proc. Camb. Phil. Soc., v.105, 193-196.

Markoff, A. (1886), Sur les racines de certaines équations (seconde note). Mathematische Annalen, v.27, 177-182.

Smith, H.V. (1990), A class of quadrature formulae. J. Inst. Math. and Comp. Sci. (Math. Ser.), v.3, 181-183.

Stenger, F. (1966), Bounds on the error of Gauss-type quadratures. Numer. Math., v.8, 150 - 160.

Szegö G. (1967), Orthogonal Polynomials, 3rd ed., Amer. Math. Soc. Colloq. Publ., v.23, Amer. Math. Soc., Providence, RI.

Authors' addresses

Dr. D.B. Hunter,
Department of Mathematics,
University of Bradford,
Bradford,
West Yorkshire,
England.

Dr. H.V. Smith,
Faculty of Information and Engineering Systems,
Leeds Metropolitan University,
Calverley Street,
Leeds,
West Yorkshire,
England.
email H.SMITH@UK.AC.LMU

Intermediate Error Estimates for Quadrature Formulas

Peter Köhler

Abstract: Let R be the error of a quadrature formula which is exact for polynomials of degree less r. We consider the constants e_j in the estimate $|R[f]| \le e_j \|f^{(j)}\|_\infty$, with the aim to obtain estimates for $e_1, \ldots, e_{r-1}$ in terms of e_0 and e_r or related expressions.

1. Introduction

Let $R[f] = I_w[f] - Q[f]$ be the error of the quadrature formula $Q[f] = \sum_{i=1}^n a_i f(x_i)$ for the integral $I_w[f] = \int_a^b f(x)w(x)\,dx$, where w is an integrable weight function (i.e., $w \in L_1[a,b]$), and $-\infty < a \le x_1 < \ldots < x_n \le b < \infty$. Further, let $R[\mathcal{P}_{r-1}] = 0$, where $\mathcal{P}_{r-1}$ denotes the polynomials of degree less r. With

$$\|R\|_j = \sup\{\frac{|R[f]|}{\|f^{(j)}\|_\infty} \mid f \in C^j[a,b],\ \|f^{(j)}\|_\infty \ne 0\}$$

for $j = 0, \ldots, r$, the error estimates $|R[f]| \le \|R\|_j \|f^{(j)}\|_\infty$ hold. In this paper, estimates for the intermediate error constants $\|R\|_1, \ldots, \|R\|_{r-1}$ in terms like $\|R\|_0$ and $\|R\|_r$ are considered. If such estimates are available, it is not necessary to compute all those error constants separately (unless one wants to have more precise results than are obtainable in this way).

In Köhler (1992), it was proven that an estimate of Ligun (1973) for the periodic case holds in the nonperiodic case (and for more general functionals than considered here), too, and we immediately obtain

$$\frac{\|R\|_j}{c_j} \le \|R\|_0^{\,1-\frac{j}{r}} \left(\frac{\|R\|_r}{c_r}\right)^{\frac{j}{r}} \qquad \text{for } j = 0, \ldots, r, \tag{1.1}$$

where

$$c_j = \inf_{c \in \mathbb{R}} \int_0^1 |B_j(x) - c|\,dx \qquad \text{for } j \ge 1, \tag{1.2}$$

and $c_0 = 1$ ($c_1 = 1/4$, $c_2 = 1/32$, $c_3 = 1/192$, $c_4 = 5/6144$). Here, B_j is the Bernoulli polynomial of degree j, with main coefficient $1/j!$. Instead of the c_j, also Favard's constants K_j can be used, which are related to the c_j by $K_j = (2\pi)^j c_j$. It was also shown that (1.1) can be improved for quadrature formulas with weight function $w \equiv 1$. But these improved estimates are not adequate for arbitrary weight functions, and it is the aim of this paper to obtain improvements also for arbitrary w. As it is well known,

$$\|Q\|_0 = \sum_{i=1}^{n} |a_i| \quad \text{and} \quad \|R\|_0 = \|w\|_1 + \|Q\|_0,$$

where $\|w\|_1 = \int_a^b |w(x)|\,dx$; further, there exist Peano kernels G_j with

$$R[f] = \int_a^b f^{(j)}(t) G_j(t)\,dt \qquad \text{and} \qquad \|R\|_j = \|G_j\|_1$$

for $j = 1, \ldots, r$ and $f \in W_\infty^j[a,b]$, where $W_\infty^j[a,b]$ denotes the functions with $f^{(j-1)}$ absolutely continuous and $\|f^{(j)}\|_\infty < \infty$. It will be shown that, in (1.1), $\|R\|_0$ can be replaced by $\|Q\|_0$ and $\|R\|_r$ by $\|G_r - v\|_1$, respectively, where $v \in L_1[a,b]$ can be chosen suitably to improve the estimate; however, this makes some correction terms necessary.

In section 2, the results and applications are given. In section 3, some identities for the c_i are proven, and section 4 contains the proofs for section 2.

2. The results

We will need the following expressions for the correction terms. For $h \in [0, b-a]$, let

$$\Omega_0(w,h) = \int_a^{a+h} |w(x) - w(x+b-a-h)|\,dx + \int_{a+h}^b |w(x) - w(x-h)|\,dx,$$

and

$$\Omega_1(w,h) = \sup_{0 \le t \le h} \Omega_0(w,t).$$

Obviously,

$$\Omega_j(w,h) = 0 \qquad \text{if } w = constant,$$

$$\Omega_j(w,h) \le 2\|w\|_1,$$

and

$$\Omega_j(w,h) \le \Big(\int_a^{a+h} + \int_{b-h}^b\Big) |w(x)|\,dx + \omega(w,h)_1,$$

for $j = 0, 1$, where $\omega(w,h)_1 = \sup_{0 \le t \le h} \int_{a+t}^b |w(x) - w(x-t)|\,dx$ is the L_1-modulus of smoothness. Especially, $\lim_{h \to 0} \Omega_j(w,h) = 0$ for all $w \in L_1[a,b]$. Examples:

– Let w be even, i.e., $w(x) = w(b+a-x)$, and monotonic on $[a,(a+b)/2]$, and let $t \in [0,(b-a)/2]$. Then

$$\Omega_0(w,t) = 4\Big|\int_{b-t/2}^{b} w(x)\,dx - \int_{(a+b)/2}^{(a+b+t)/2} w(x)\,dx\Big|.$$

– Let $w(x) = 1/\sqrt{1-x^2}$ on $[-1,1]$ and $t \in [0,1]$. Then

$$\Omega_0(w,t) = 4\arcsin\frac{1}{4}\big(\sqrt{4t-t^2}\sqrt{4-t^2} - t(2-t)\big) \le 4\arcsin\sqrt{t} \le 2\pi\sqrt{t}. \tag{2.1}$$

– Let $w(x) = (1-x)^\alpha(1+x)^\beta$ on $[-1,1]$, where $\alpha,\beta > -1$, and let $t \in [0,1]$. Then

$$\Omega_0(w,t) \le \frac{t^{\min(1,\alpha+1,\beta+1)}}{\min(\alpha+1,\beta+1)}\Big((|\alpha|+1)2^{|\alpha|} + (|\beta|+1)2^{|\beta|} + 2^{|\alpha|+|\beta|+2}\Big).$$

The following two theorems (one for odd and one for even j) are the main result of this paper.

THEOREM 2.1. *Let $1 \le j \le r-1$, with odd j, and let an arbitrary function $v \in L_1[a,b]$ be given ($v \ne G_r$). Then*

$$\frac{\|R\|_j}{c_j} \le \|Q\|_0^{\,1-\frac{j}{r}}\Big(\frac{\|G_r - v\|_1}{c_r}\Big)^{\frac{j}{r}}\Big(1 + \frac{j}{2r}\frac{\Omega_0(v,\tau)}{\|G_r - v\|_1} + \frac{r-j}{2r}\frac{\Omega_1(w,\tau)}{\|Q\|_0}\Big),$$

where

$$\tau = \frac{1}{2}\Big(\frac{1}{\|Q\|_0}\frac{\|G_r - v\|_1}{c_r}\Big)^{\frac{1}{r}}. \tag{2.2}$$

THEOREM 2.2. *Let $1 \le j \le r-1$, with even j, and let an arbitrary function $v \in L_1[a,b]$ be given ($v \ne G_r$). Then*

$$\frac{\|R\|_j}{c_j} \le \|R\|_0^{\frac{1}{j+1}}\|Q\|_0^{\frac{j}{j+1}-\frac{j}{r}}\Big(\frac{\|G_r - v\|_1}{c_r}\Big)^{\frac{j}{r}}\Big(1 + \frac{j}{2r}\frac{\Omega_0(v,\tau)}{\|G_r - v\|_1} + \frac{r-j-1}{2r}\frac{\Omega_1(w,\tau)}{\|Q\|_0}\Big),$$

with τ as in (2.2) (for $j = r-1$, the term $\|Q\|_0$ in the denominator of τ can be replaced by $\|R\|_0$).

Remarks. 1. In the form stated, these theorems hold if the parameter τ defined by (2.2) satisfies $\tau \in (0,b-a]$. Typically, one has $\|Q\|_0 = \|w\|_1 + o(1)$, and $\|G_r\|_1 = O(n^{-r})$, so that $\tau = O(n^{-1})$ (for $v \equiv 0$ or another convenient choice of v), and therefore $\tau \in (0,b-a]$ for a sufficiently large number of nodes. If $\Omega_0(v,\tau)$ is replaced by $\Omega_0(v,\tau - [\tau/(b-a)](b-a))$, and correspondingly for $\Omega_1(w,\tau)$, where $[x]$ is the largest integer less or equal x, then the theorems remain valid for arbitrary $\tau > 0$.

2. For odd j, $K_j \le \pi/2$, while for even $j \ge 2$ and $\|R\|_0 \le 2\|Q\|_0$ (i.e., for $\|Q\|_0 \ge \|w\|_1$) $K_j(\|R\|_0/\|Q\|_0)^{1/(j+1)} \le K_j 2^{1/(j+1)} < \pi/2$ (remember that $c_j = K_j/(2\pi)^j$). This enables to unify the estimates of Theorem 2.1 and 2.2, in case that the condition on $\|Q\|_0$ is satisfied.

Applications of Theorem 2.1 and 2.2:

1. *Asymptotically best quadrature formulas.* Consider a sequence of quadrature formulas Q_n with equidistant nodes, i.e., $x_i = a + ih$ for $i = 0, \ldots, n$, where $h = (b-a)/n$, satisfying

$$\lim_{n\to\infty} h^{-r}\|R_n\|_r = c_r\|w\|_1.$$

Since this is the least possible value for this limit, the formulas are called asymptotically best for the class $W_\infty^r[a,b]$ (see Braß 1974 for $w \equiv 1$, and Köhler 1991b for arbitrary $w \in L_1[a,b]$). Further, assume that

$$\lim_{n\to\infty} \|Q_n\|_0 = \|I_w\|_0 = \|w\|_1 \tag{2.3}$$

(which holds for all "reasonable" quadrature formulas). For $1 \le j \le r-1$, we obtain from Theorem 2.1 (with $v \equiv 0$)

$$\|R_n\|_j \le c_j h^j \|w\|_1(1+o(1)) \qquad \text{for odd } j,$$

and from Theorem 2.2

$$\|R_n\|_j \le 2^{\frac{1}{j+1}} c_j h^j \|w\|_1(1+o(1)) \qquad \text{for even } j.$$

This means that a sequence of quadrature formulas which is asymptotically best for some $r \ge 2$, and satisfies (2.3), is also asymptotically best for any odd $j < r$ (i.e., asymptotically best for the class $W_\infty^j[a,b]$), and that it is not far from being asymptotically best for even $j < r$.

2. *Compound quadrature formulas.* Let nodes t_i with $0 \le t_1 < \ldots < t_r \le 1$ be given, and partition points $z_i = a + ih$ for $i = 0, \ldots, N$, where $h = (b-a)/N$. Let $p[f; t_1, \ldots, t_r] \in \mathcal{P}_{r-1}$ denote the polynomial interpolating f at the t_i, and define the "elementary" quadrature formula Q^{el} by

$$Q^{el}[f] = \int_0^1 p[f; t_1, \ldots, t_r](x)\,dx,$$

so that Q^{el} is the interpolatory quadrature formula for $\int_0^1 f(x)\,dx$ with nodes t_i. The corresponding compound quadrature formula for $\int_a^b f(x)w(x)\,dx$ is defined by

$$Q_N[f] = \sum_{i=1}^N \int_{z_{i-1}}^{z_i} p[f; z_{i-1}+t_1h, \ldots, z_{i-1}+t_rh](x)w(x)\,dx.$$

Using the methods of Köhler (1991a), yields $\|Q_N\|_0 = \|Q^{el}\|_0 \|w\|_1 + O(\omega(w,h)_1)$ and $\|G_{r,N} - ch^r w\|_1 = h^r \|G_r^{el} - c\|_1 \|w\|_1 + h^r O(\omega(w,h)_1)$, where G_r^{el} and $G_{r,N}$ are the Peano kernels of order r of Q^{el} and Q_N, respectively (explicit estimates for the $O(\omega(w,h)_1)$-terms can be obtained). Here, the function $v = ch^r w$ has been chosen, and c can be used to minimize $\|G_r^{el} - c\|_1$. We obtain

$$\|R_N\|_j \le c_j h^j \|w\|_1 \|Q^{el}\|_0^{1-\frac{j}{r}} \left(\frac{\|G_r^{el} - c\|_1}{c_r} \right)^{\frac{j}{r}} \left(1 + O(\Omega_1(w,h))\right)$$

for odd j, while for even j, a factor $2^{1/(j+1)}$ has to be added at the right-hand side.

3. *Gaussian quadrature formulas.* Let $w \in L_1[-1,1]$, $w \ge 0$, $\int_{-1}^{1} w(x)\,dx > 0$, and let Q_n be the corresponding n-point Gauss formula. Then $\|Q_n\|_0 = \|w\|_1$ and $\|R_n\|_0 = 2\|w\|_1$. Further, by Lebesgue's inequality,

$$\|R_n\|_{2n} \le \|R_n\|_0 \|E_{2n-1}\|_{2n} = \frac{2\|w\|_1}{2^{2n-1}(2n)!},$$

where E_n denotes the error functional of best approximation by polynomials of degree n in the sup-norm. From Theorem 2.1 and 2.2 (with $v = 0$ and $r = 2n$) and Stirling's formula, we obtain

$$\|R_n\|_j < \frac{\pi}{2} \|w\|_1 \left(\frac{e}{4n}\right)^j \left(1 + \frac{2n-j}{4n} \frac{\Omega_1(w, \frac{\pi e}{4n})}{\|w\|_1}\right)$$

for $1 \le j \le 2n$. For $j < n$, better estimates can be obtained as follows. Using the Corollary of Braß and Förster (1987) yields

$$\|R_n\|_n \le \frac{1}{2^{n-1}} \left(\frac{(2n)!}{n!(3n)!} \frac{3n}{2n-1} \right)^{\frac{1}{2}} \sup_{\mu \ge n} |R_n[T_{\mu+n}]|,$$

where T_μ denotes the Tschebyscheff polynomial of degree μ. Using the rough estimate $|R_n[T_{\mu+n}]| \le \|R_n\|_0 \|T_{\mu+n}\|_\infty = \|R_n\|_0$, Stirling's formula for the factorials, and Theorem 2.1 and 2.2 with $r = n$, we arrive at

$$\|R_n\|_j < \frac{\pi}{2} \|w\|_1 \left(\frac{e}{3\sqrt{3}n}\right)^j \left(1 + \frac{n-j}{2n} \frac{\Omega_1(w, \frac{\pi e}{3\sqrt{3}n})}{\|w\|_1}\right) \tag{2.4}$$

for $1 \le j \le n$ and $n \ge 25$. The restriction $n \ge 25$ can be removed at least in some cases, if $|R_n[T_{\mu+n}]|$ is estimated more carefully. E.g., for Gauss-Legendre formulas Q_n^{GL} ($w \equiv 1$) and Gauss-Tschebyscheff formulas Q_n^{GT} ($w(x) = 1/\sqrt{1-x^2}$), the estimate holds for $n \ge 1$. Further, $\Omega_1 = 0$ for Q_n^{GL}, while (2.1) can be used for Q_n^{GT}. For comparison, (2.4) yields $\|R_n^{GL}\|_1 < \pi e/(3\sqrt{3}n) = 1.643.../n$ and $\|R_n^{GL}\|_2 < \pi e^2/(27n^2) = 0.8597.../n^2$, whereas the best possible estimates are $\|R_n^{GL}\|_1 \le \pi^2/(8n) = 1.233.../n$ and $\|R_n^{GL}\|_2 \le 2\pi^2/(27\sqrt{3}n^2) = 0.4220.../n^2$, respectively (Petras 1989 and 1992),

and $\|R_n^{GT}\|_1 < (1+1.282/\sqrt{n})2.582/n$ instead of $\|R_n^{GT}\|_1 = 2\tan(\pi/(4n)) \le 2/n$ (Ponomarenko 1983).
Using (1.1), gives the simpler estimate

$$\|R_n\|_j < \pi\|w\|_1\Big(\frac{e}{3\sqrt{3}n}\Big)^j$$

for $0 \le j \le n$ and $n \ge 3$. Finally let us note that the terms $e/4$ and $e/(3\sqrt{3})$ in estimates of the type above cannot be improved, because they are best possible for $\|R_n^{GL}\|_{2n}$ and $\|R_n^{GL}\|_n$, respectively.

3. Some identities for the c_j

The aim of this section is to prove some identities for the numbers c_j defined in (1.2). These identities are needed for the proof of Theorem 2.1. From Braß (1974), we have

$$c_j = \begin{cases} \int_0^1 |B_j(x) - B_j(\frac{1}{4})|\,dx = 4|B_{j+1}(\frac{1}{4})| & \text{for even } j, \text{ and} \\ \int_0^1 |B_j(x)|\,dx = 4(1-2^{-1-j})|B_{j+1}(0)| & \text{for odd } j. \end{cases}$$

By the second relation,

$$\frac{c_{j-1}}{2(2^j-1)} = \frac{|B_j(0)|}{2^{j-1}} = \frac{|B_j(\frac{1}{2})|}{2^{j-1}-1} \qquad \text{for even } j. \tag{3.1}$$

The following lemma will be used with $\alpha_j = (j-1)(m-j-1)$ and $\alpha_j = (j-1)(j-2)$.

LEMMA 3.1. *For even $m \ge 4$ and arbitrary $\alpha_j \in \mathbb{R}$, there holds*

$$\sum_{\substack{j=2\\ j\,even}}^{m-2} \frac{\alpha_j}{2^j-1} c_{j-1}c_{m-j-1} = (-1)^{\frac{m}{2}} 8\Big((1-2^{1-m})\sum_{\substack{j=2\\ j\,even}}^{m-2} \alpha_j B_j(0)B_{m-j}(0) + \sum_{\substack{j=2\\ j\,even}}^{m-2} \alpha_j B_j(\frac{1}{2})B_{m-j}(0)\Big).$$

PROOF. Using (3.1), we obtain

$$\begin{aligned}\sum_{\substack{j=2\\ j\,even}}^{m-2} \frac{\alpha_j}{2^j-1} c_{j-1}c_{m-j-1} &= \sum_{\substack{j=2\\ j\,even}}^{m-2} \alpha_j \frac{2(2^{m-j}-1)}{2^{m-j-1}}\frac{c_{j-1}}{2^j-1}|B_{m-j}(0)| \\ &= 8\sum_{\substack{j=2\\ j\,even}}^{m-2} \alpha_j\Big((1-2^{1-m})2^{j-1} - (2^{j-1}-1)\Big)\frac{c_{j-1}}{2(2^j-1)}|B_{m-j}(0)| \\ &= 8\sum_{\substack{j=2\\ j\,even}}^{m-2} \alpha_j\Big((1-2^{1-m})|B_j(0)| - |B_j(\frac{1}{2})|\Big)|B_{m-j}(0)|,\end{aligned}$$

with $\operatorname{sign}B_j(0) = (-1)^{\frac{j}{2}+1}$ and $\operatorname{sign}B_j(\frac{1}{2}) = (-1)^{\frac{j}{2}}$ for $j \geq 2$ even. ■

Let us now consider the functions

$$A_{m,n}(x) = \sum_{\substack{j=2 \\ j\,even}}^{m-2} \binom{m}{j} B_j^{(1)}(x) B_{m-j}^{(n)}(1)$$

for $x \in \{0, \frac{1}{2}\}$ and $n \in \{1,2,3\}$, where $B_j^{(n)}$ denotes the Bernoulli polynomial of order n and degree j. The Bernoulli polynomials B_j used above are essentially the Bernoulli polynomials of order 1, i.e.,

$$B_j = \frac{1}{j!} B_j^{(1)}. \tag{3.2}$$

The following relations will be used in the sequel:

$$B_m^{(n+1)}(x+y) = \sum_{j=0}^{m} \binom{m}{j} B_j^{(1)}(x) B_{m-j}^{(n)}(y), \tag{3.3}$$

$$B_j^{(n+1)}(x) = (1 - \frac{j}{n}) B_j^{(n)}(x) + j(\frac{x}{n} - 1) B_{j-1}^{(n)}(x), \tag{3.4}$$

$$B_j^{(n+1)}(x+1) = B_j^{(n+1)}(x) + j B_{j-1}^{(n)}(x) \tag{3.5}$$

(see Nörlund 1954, p. 133 and p. 139, eq. (50) for (3.3), and p. 145, eq. (81) for (3.4), and Milne-Thomson 1981, p. 129, eq. (2) for (3.4), and p. 129 for (3.5)). Using (3.4), $B_j^{(n)}(1)$ can be computed recursively, which, together with $B_j^{(1)}(1) = B_j^{(1)}(0)$ (for $j \geq 2$) and (3.2), yields

$$B_j^{(n)}(1) = (-1)^{n-1} \binom{j-1}{n-1} j! B_j(0) \quad \text{for } j \geq 2 \text{ even and } n = 1,2,3.$$

Hence,

$$A_{m,n}(x) = (-1)^{n-1} m! \sum_{\substack{j=2 \\ j\,even}}^{m-2} \binom{m-j-1}{n-1} B_j(x) B_{m-j}(0),$$

and therefore

$$\sum_{\substack{j=2 \\ j\,even}}^{m-2} (j-1)(m-j-1) B_j(x) B_{m-j}(0)$$

$$= \sum_{\substack{j=2 \\ j\,even}}^{m-2} \left((m-3)\binom{m-j-1}{1} - 2\binom{m-j-1}{2} \right) B_j(x) B_{m-j}(0)$$

$$= -\frac{1}{m!}\Big((m-3) A_{m,2}(x) + 2A_{m,3}(x) \Big). \tag{3.6}$$

In the same way, we obtain

$$\sum_{\substack{j=2\\ j\,even}}^{m-2} (j-1)(j-2)B_j(x)B_{m-j}(0) = -\frac{1}{m!}\Big((m-3)A_{m,2}(x) + 2A_{m,3}(x)\Big) + \frac{m-3}{m!}\Big((m-2)A_{m,1}(x) + A_{m,2}(x)\Big) \quad (3.7)$$

for $n = 1, 2, 3$. These expressions can be inserted into the right-hand side of Lemma 3.1. We now have to calculate $A_{m,n}(x)$ for even m, $n = 1, 2, 3$ and $x \in \{0, \frac{1}{2}\}$. Using (3.3), $B_j^{(1)}(0) = 0$ for $j \geq 3$ odd, $B_j^{(1)}(\frac{1}{2}) = 0$ for $j \geq 1$ odd, $B_0^{(n)}(x) = 1$ for all n and x, and $B_1^{(1)}(x) = x - \frac{1}{2}$, we obtain

$$A_{m,n}(x) = B_m^{(n+1)}(x+1) - B_m^{(n)}(1) - m(x - \frac{1}{2})B_{m-1}^{(n)}(1) - B_m^{(1)}(x).$$

After some elementary calculations, making use of the properties of the Bernoulli polynomials noted above, we arrive at

$$\begin{aligned}
\frac{1}{m!}A_{m,1}(0) &= -(m+1)B_m(0),\\
\frac{1}{m!}A_{m,2}(0) &= \frac{1}{2}(m+1)(m-2)B_m(0),\\
\frac{1}{m!}A_{m,3}(0) &= -\Big(\frac{1}{6}m(m-1)(m-2)+1\Big)B_m(0) - \frac{m-3}{12}B_{m-2}(0),\\
\frac{1}{m!}A_{m,1}(\frac{1}{2}) &= -mB_m(\frac{1}{2}) - B_m(0),\\
\frac{1}{m!}A_{m,2}(\frac{1}{2}) &= \frac{1}{2}m(m-3)B_m(\frac{1}{2}) - \frac{1}{8}B_{m-2}(\frac{1}{2}) + (m-1)B_m(0),
\end{aligned}$$

and

$$\frac{1}{m!}A_{m,3}(\frac{1}{2}) = -\Big(\frac{1}{6}(m-1)(m-2)(m-3)+1\Big)B_m(\frac{1}{2}) + \frac{m-3}{24}B_{m-2}(\frac{1}{2}) - \frac{1}{2}(m-1)(m-2)B_m(0).$$

By inserting this into (3.6) and (3.7), and applying Lemma 3.1, we obtain, again after some elementary calculations, the following lemma, which is one of the main tools for the proof of Theorem 2.1.

LEMMA 3.2. *For even $m \geq 4$, there holds*

$$\sum_{\substack{j=2\\ j\,even}}^{m-2} \frac{(j-1)(m-j-1)}{2^j-1}c_{j-1}c_{m-j-1} = -4(m-1)c_{m-1} + \frac{(m-3)c_{m-3}}{4(1-2^{2-m})},$$

and

$$\sum_{\substack{j=2\\ j\,even}}^{m-2} \frac{(j-1)(j-2)}{2^j-1}c_{j-1}c_{m-j-1} = 8c_{m-1} - \frac{(m-3)c_{m-3}}{2(2^{m-2}-1)}.$$

LEMMA 3.3. *For all $m \geq 1$, there holds*

$$\sum_{j=0}^{m} |B_j(\frac{1}{2})B_{m-j}(\frac{1}{2})| = |(m-1)B_m(1)|.$$

PROOF. From (3.3) and (3.4), we obtain

$$\sum_{j=0}^{m} \binom{m}{j} B_j^{(1)}(x)B_{m-j}^{(1)}(y) = B_m^{(2)}(x+y)$$

$$= (1-m)B_m^{(1)}(x+y) + m(x+y-1)B_{m-1}^{(1)}(x+y).$$

With (3.2) and $x = y = \frac{1}{2}$, this yields

$$\sum_{j=0}^{m} B_j(\frac{1}{2})B_{m-j}(\frac{1}{2}) = (1-m)B_m(1),$$

where $B_j(\frac{1}{2}) = 0$ for odd r and $\mathrm{sign} B_j(\frac{1}{2}) = (-1)^{j/2}$ for even r. ∎

LEMMA 3.4. *For all $r \geq 1$, there holds*

$$\sum_{\mu=0}^{r-1} (\mu-1)\frac{|B_\mu(\frac{1}{2})|}{2^\mu} c_{r+1-\mu} = -\frac{rb_r}{2^{r+2}}$$

and

$$\sum_{\mu=0}^{r-1} (r+1-\mu)\frac{|B_\mu(\frac{1}{2})|}{2^\mu} c_{r+1-\mu} = \frac{rc_r}{4},$$

where $b_r = \int_0^1 |B_r(x) - B_r(\frac{1}{2})|\, dx$.

PROOF. If R is a linear functional or a seminorm defined on $C_{2\pi}$, then

$$\|R\|_r \leq 2\sum_{\mu=0}^{r-1} h^{\mu-1}|B_\mu(\frac{1}{2})|\|R\|_{r+1-\mu} + h^r b_r\|R\|_0$$

for $h > 0$, by formula (12) of Ligun (1973) (see also Köhler 1992, formula (2.2)). Especially, for best approximation by trigonometric polynomials of degree n, there holds $\|R\|_r = K_r/(n+1)^r = c_r(2\pi/(n+1))^r$ for all r, so that, with $t = h(n+1)/\pi$,

$$c_r \leq \phi(t) := 4\sum_{\mu=0}^{r-1} \frac{t^{\mu-1}}{2^\mu}|B_\mu(\frac{1}{2})|c_{r+1-\mu} + \frac{t^r}{2^r}b_r$$

holds for $t>0$ and $r \geq 1$. But

$$\phi(1) = 4\sum_{\mu=0}^{r-1} \frac{|B_\mu(\frac{1}{2})|}{2^\mu}c_{r+1-\mu} + \frac{b_r}{2^r} = c_r \tag{3.8}$$

(the latter equality follows from Lemma 2 of Ligun 1973), so that $\phi(1) = c_r \leq \phi(t)$ for all $t>0$, which implies $\phi'(1) = 0$. From this, we immediately obtain the first identity of the lemma, while the second is a linear combination of the first and of (3.8). ∎

4. The proofs of Theorem 2.1 and 2.2

We will first prove the following weaker versions of Theorem 2.1 and 2.2. As before, $\tau \in (0, b-a]$ should hold, but if not, only a minor modification is necessary.

THEOREM 4.1. *Let $1 \le j \le r-1$, with odd j. Then, for arbitrary $c \in \mathbb{R}$, there holds*

$$\frac{\|R\|_j}{c_j} \le \|Q\|_0^{1-\frac{j}{r}} \Big(\frac{\|G_r - c\|_1}{c_r}\Big)^{\frac{j}{r}} \Big(1 + \frac{r-j}{2r} \frac{\Omega_1(w,\tau)}{\|Q\|_0}\Big),$$

where

$$\tau = \frac{1}{2}\Big(\frac{1}{\|Q\|_0} \frac{\|G_r - c\|_1}{c_r}\Big)^{\frac{1}{r}}.$$

THEOREM 4.2. *Let $1 \le j \le r-1$, with even j. Then, for arbitrary $c \in \mathbb{R}$, there holds*

$$\frac{\|R\|_j}{c_j} \le \|R\|_0^{\frac{1}{j+1}} \|Q\|_0^{\frac{j}{j+1}-\frac{j}{r}} \Big(\frac{\|G_r - c\|_1}{c_r}\Big)^{\frac{j}{r}} \Big(1 + \frac{r-j-1}{2r} \frac{\Omega_1(w,\tau)}{\|Q\|_0}\Big),$$

with τ as in Theorem 4.1 for $j = 1, \ldots, r-2$ (for $j = r-1$, τ is meaningless).

In the following, let $e_j = \|R\|_j$ and

$$\tilde{e}_j = \sup\{\frac{|R[f]|}{\|f^{(j)}\|_\infty} \mid f \in C^j[a,b], \|f^{(j)}\|_\infty \ne 0, \int_a^b f^{(j)}(x)\,dx = 0\}$$

($e_j = \|G_j\|_1$ and $\tilde{e}_j = \inf_{c\in\mathbb{R}} \|G_j - c\|_1$ for $j = 1, \ldots, r$).

PROOF of Theorem 4.1. Let $f \in W_\infty^r[a,b]$ ($r \ge 2$) and $l = b-a$. For odd k with $0 < k < r$, let $\tilde{f}^{(k)} \in W_\infty^k(\mathbb{R})$ be the l-periodic extension of $f^{(k)}$. Define $\tilde{f}$ by

$$\tilde{f}(t) = \sum_{i=0}^{k-1} \frac{f^{(i)}(a)}{i!}(t-a)^i + \int_a^t \tilde{f}^{(k)}(u) \frac{(t-u)^{k-1}}{(k-1)!}\,du.$$

Then, by a special form of the Euler-MacLaurin formula (see Lemma 1 of Ligun 1973, or Lemma 2.1 of Köhler 1992), there holds

$$\tilde{f}(t) = \sum_{j=0}^{k-1} B_j(\frac{1}{2})\tau^j \tilde{f}_\tau^{(j)}(t) + \tilde{F}_k(t), \tag{4.1}$$

for arbitrary $\tau > 0$, where

$$\tilde{f}_\tau(t) = \frac{1}{\tau}\int_{-\frac{\tau}{2}}^{\frac{\tau}{2}} \tilde{f}(t-z)\,dz \quad \text{and} \quad \tilde{F}_k(t) = -\tau^{k-1}\int_{-\frac{\tau}{2}}^{\frac{\tau}{2}} \Psi_k(\frac{z}{\tau})\tilde{f}^{(k)}(t-z)\,dz.$$

Here, $\Psi_k(x) = B_k^*(x) - B_k(\frac{1}{2})$, and B_k^* is the 1-periodic extension of $B_k\big|_{[0,1)}$, i.e., the Bernoulli monospline. We obtain

$$|R[f]| = |R[\tilde{f}]| \leq \sum_{j=0}^{k-1} |B_j(\frac{1}{2})|\tau^j|R[\tilde{f}_\tau^{(j)}]| + |R[\tilde{F}_k]|. \tag{4.2}$$

$|R[\tilde{F}_k]|$ is estimated in the following way. From $\Psi_k(x) = (-1)^k\Psi_k(-x)$ and the l-periodicity of $\tilde{f}^{(k)}$, we obtain, with $w(x) = 0$ for $x \notin [a,b]$,

$$\begin{aligned} I_w[\tilde{F}_k] &= -\tau^{k-1}\int_{-\infty}^{\infty}\int_{-\frac{\tau}{2}}^{\frac{\tau}{2}} \Psi_k(\frac{z}{\tau})\tilde{f}^{(k)}(t-z)\,dz\,w(t)\,dt \\ &= -\tau^{k-1}\int_{-\frac{\tau}{2}}^{\frac{\tau}{2}} \Psi_k(\frac{z}{\tau})\int_{-\infty}^{\infty} \tilde{f}^{(k)}(u)w(u+z)\,du\,dz \\ &= -\tau^{k-1}\int_{0}^{\frac{\tau}{2}} \Psi_k(\frac{z}{\tau})\int_a^b f^{(k)}(u)\sum_{j=-\infty}^{\infty}(w(u+jl+z)+(-1)^k w(u+jl-z))\,du\,dz. \end{aligned}$$

This yields

$$|I_w[\tilde{F}_k]| \leq \tau^{k-1}\int_0^{\frac{\tau}{2}} |\Psi_k(\frac{z}{\tau})|W(z)\,dz\|f^{(k)}\|_\infty,$$

where

$$W(z) = \int_a^b \Big|\sum_{j=-\infty}^{\infty}(w(u+jl+z)+(-1)^k w(u+jl-z))\Big|\,du \leq \begin{cases} 2\|w\|_1 & \text{for all } k, \\ \Omega_0(w,2z) & \text{for odd } k. \end{cases}$$

Since we consider only odd k, we obtain

$$|I_w[\tilde{F}_k]| \leq \frac{\tau^k}{2} b_k\Omega_1(w,\tau)\|f^{(k)}\|_\infty$$

for $\tau \in (0, b-a]$, where $b_k = \int_0^1 |\Psi_k(x)|\,dx$. Further,

$$|Q[\tilde{F}_k]| \leq \|Q\|_0\|\tilde{F}_k\|_\infty \leq \|Q\|_0\tau^k b_k\|f^{(k)}\|_\infty,$$

so that

$$|R[\tilde{F}_k]| \leq \tau^k b_k\hat{e}_0(\tau)\|f^{(k)}\|_\infty, \tag{4.3}$$

where

$$\hat{e}_0(\tau) = \|Q\|_0 + \frac{1}{2}\Omega_1(w,\tau).$$

To estimate $R[\tilde{f}_\tau^{(j)}]$, let $g = \tilde{f}_\tau^{(j)}$ and apply (4.1) to g, i.e., $g = \sum_{\mu=0}^{k-j} B_\mu(1/2)\tau^\mu g_\tau^{(\mu)} + G_{k+1-j}$. Here,

$$g_\tau(t) = \frac{1}{\tau}\int_{-\frac{\tau}{2}}^{\frac{\tau}{2}} g(t-z)\,dz = \frac{1}{\tau^2}\int_{-\frac{\tau}{2}}^{\frac{\tau}{2}}\int_{-\frac{\tau}{2}}^{\frac{\tau}{2}} \tilde{f}^{(j)}(t-z-u)\,du\,dz = \frac{1}{\tau^2}\Delta_\tau^2\tilde{f}^{(j-2)}(t-\tau),$$

and therefore

$$g_\tau^{(k+2-j)}(t) = \frac{1}{\tau^2}\Delta_\tau^2 \tilde{f}^{(k)}(t-\tau).$$

As is easily verified, $g_\tau^{(k+2-j)}$ has mean value zero, so that

$$|R[g_\tau^{(\mu)}]| \le \tilde{e}_{k+2-\mu-j}\|g_\tau^{(k+2-j)}\|_\infty \le \frac{4}{\tau^2}\tilde{e}_{k+2-\mu-j}\|f^{(k)}\|_\infty. \tag{4.4}$$

Further,

$$G_{k+1-j}(t) = -\tau^{k-j}\int_{-\frac{\tau}{2}}^{\frac{\tau}{2}} \Psi_{k+1-j}(\frac{z}{\tau})g^{(k+1-j)}(t-z)\,dz$$

with

$$g^{(k+1-j)}(t) = \tilde{f}_\tau^{(k+1)}(t) = \frac{1}{\tau}\Delta_\tau \tilde{f}^{(k)}(t-\frac{\tau}{2}),$$

which implies

$$|Q[G_{k+1-j}]| \le \|Q\|_0\|G_{k+1-j}\|_\infty \le 2\tau^{k-j}b_{k+1-j}\|Q\|_0\|f^{(k)}\|_\infty$$

and

$$I_w[G_{k+1-j}] = -\tau^{k-j-1}\int_{-\frac{\tau}{2}}^{\frac{\tau}{2}} \Psi_{k+1-j}(\frac{z}{\tau})\phi(z)\,dz,$$

where

$$\phi(z) = \int_{-\infty}^{\infty}(\Delta_\tau \tilde{f}^{(k)}(t-z-\frac{\tau}{2}))w(t)\,dt = -\int_{-\infty}^{\infty}\tilde{f}^{(k)}(u)\Delta_\tau w(u+z-\frac{\tau}{2})\,du.$$

Because of the periodicity of $\tilde{f}^{(k)}$, we obtain, in the same way as for $I_w[\tilde{F}_k]$,

$$\|\phi\|_\infty \le \|f^{(k)}\|_\infty \Omega_0(w,\tau)$$

for $\tau \in (0, b-a]$, and therefore

$$|R[G_{k+1-j}]| \le 2\tau^{k-j}b_{k+1-j}\hat{e}_0(\tau)\|f^{(k)}\|_\infty.$$

This yields

$$\begin{aligned}|R[\tilde{f}_\tau^{(j)}]| =& |R[g]| \le \sum_{\mu=0}^{k-j}|B_\mu(\frac{1}{2})|\tau^\mu|R[g_\tau^{(\mu)}]| + |R[G_{k+1-j}]| \\ \le& \left(4\sum_{\mu=0}^{k-j}\tau^{\mu-2}|B_\mu(\frac{1}{2})|\tilde{e}_{k+2-\mu-j} + 2\tau^{k-j}b_{k+1-j}\hat{e}_0(\tau)\right)\|f^{(k)}\|_\infty \end{aligned} \tag{4.5}$$

for $0 \le j \le k$ and $k \ge 1$. Inserting (4.5) and (4.3) into (4.2), yields

$$\begin{aligned}|R[f]| \le& \Bigg(4\sum_{j=0}^{k-1}\sum_{\mu=0}^{k-j}|B_j(\frac{1}{2})B_\mu(\frac{1}{2})|\tau^{\mu+j-2}\tilde{e}_{k+2-\mu-j} \\ &+ \Big(2\sum_{j=0}^{k-1}|B_j(\frac{1}{2})|b_{k+1-j} + b_k\Big)\tau^k\hat{e}_0(\tau)\Bigg)\|f^{(k)}\|_\infty,\end{aligned}$$

and therefore, by rearranging the double sum, we obtain

$$e_k \leq 4\sum_{\nu=0}^{k} \tau^{\nu-2}\tilde{e}_{k+2-\nu} \sum_{j=0}^{\min(\nu,k-1)} |B_j(\frac{1}{2})B_{\nu-j}(\frac{1}{2})| + \Big(2\sum_{j=0}^{k-1} |B_j(\frac{1}{2})|b_{k+1-j} + b_k\Big)\tau^k\hat{e}_0(\tau).$$

From Lemma 3.3, $B_j(\frac{1}{2}) = 0$ for odd j, $b_j = |B_j(\frac{1}{2})|$ for even j, and

$$b_k = \int_0^1 |B_k(x)|\,dx = 2|\int_0^{\frac{1}{2}} B_k(x)\,dx| = 2|B_{k+1}(\frac{1}{2})| + 2|B_{k+1}(0)|$$

for odd k, we obtain

$$e_k \leq 4\sum_{\nu=0}^{k-1} \tau^{\nu-2}|(\nu-1)B_\nu(0)|\tilde{e}_{k+2-\nu} + 2\tau^k(k+1)|B_{k+1}(0)|\hat{e}_0(\tau). \tag{4.6}$$

With $T = 2\tau$ and (3.1), this yields

$$e_k \leq \frac{24}{T^2}\tilde{e}_{k+2} + 6\sum_{\substack{\nu=4\\ \nu\,\text{even}}}^{k-1} T^{\nu-2}\frac{\nu-1}{2^\nu-1}c_{\nu-1}\tilde{e}_{k+2-\nu} + \frac{3}{2}\frac{k+1}{2^{k+1}-1}T^k c_k\hat{e}_0(\tau) \tag{4.7}$$

for $k \geq 3$. We will now proceed by induction. From (4.6), we obtain

$$\frac{e_1}{c_1} \leq T^{-2}\frac{1}{3}\frac{\tilde{e}_3}{c_3} + T\frac{2}{3}\hat{e}_0(\tau).$$

Now assume that

$$\frac{e_j}{c_j} \leq T^{j-k}\frac{j}{k}\frac{\tilde{e}_k}{c_k} + T^j\frac{k-j}{k}\hat{e}_0(\tau) \tag{4.8}$$

for j and k odd and $j < k$ has already been proven. From this estimate with $j = k+2-\nu$ and $\tilde{e}_j \leq e_j$, we obtain, by inserting into (4.7),

$$e_k \leq \frac{24}{T^2}\tilde{e}_{k+2} + \frac{6}{k}\frac{e_k}{c_k}\sum_{\substack{\nu=4\\ \nu\,\text{even}}}^{k-1} \frac{(\nu-1)(k+2-\nu)}{2^\nu-1}c_{\nu-1}c_{k+2-\nu} + \Big(\frac{6}{k}\sum_{\substack{\nu=4\\ \nu\,\text{even}}}^{k-1} \frac{(\nu-1)(\nu-2)}{2^\nu-1}c_{\nu-1}c_{k+2-\nu} + \frac{3(k+1)}{2(2^{k+1}-1)}c_k\Big)T^k\hat{e}_0(\tau). \tag{4.9}$$

Applying Lemma 3.2 with $m = k+3$ to determine $\sum_{\nu=4}^{k-1}\ldots$, we obtain

$$\frac{e_k}{c_k} \leq T^{-2}\frac{k}{k+2}\frac{\tilde{e}_{k+2}}{c_{k+2}} + T^k\frac{2}{k+2}\hat{e}_0(\tau).$$

Inserting this into (4.8), yields

$$\frac{e_j}{c_j} \leq T^{j-k-2}\frac{j}{k+2}\frac{\tilde{e}_{k+2}}{c_{k+2}} + T^j\frac{k+2-j}{k+2}\hat{e}_0(\tau).$$

Therefore, we have proven the following lemma.

Lemma 4.1. *For odd j and odd r with $1 \le j \le r-1$, $\tau \in (0, b-a]$ and $T = 2\tau$, there holds*

$$\frac{e_j}{c_j} \le T^{j-r}\frac{j}{r}\frac{\tilde{e}_r}{c_r} + T^j\frac{r-j}{r}\hat{e}_0(\tau).$$

Inserting $T = 2\tau = (\tilde{e}_r/(\hat{e}_0(0)c_r))^{\frac{1}{r}}$ into this estimate, we finally obtain

$$\frac{e_j}{c_j} \le \hat{e}_0(0)^{\frac{r-j}{j}}\left(\frac{\tilde{e}_r}{c_r}\right)^{\frac{j}{r}}\left(1 + \frac{r-j}{2r}\frac{\Omega_1(w,\tau)}{\hat{e}_0(0)}\right).$$

This proves Theorem 4.1 for odd r. We now have to estimate e_j with j odd by $\tilde{e}_r$ with r even. Let k be odd, $k = r-1$, so that r is even. This time, we will apply (4.5) only for $j = 1, \ldots, k-1$, and estimate $R[\tilde{f}_\tau]$ by

$$|R[\tilde{f}_\tau]| \le \tilde{e}_{k+1}\|\tilde{f}_\tau^{(k+1)}\|_\infty \le \frac{2}{\tau}\tilde{e}_{k+1}\|f^{(k)}\|_\infty. \tag{4.10}$$

Proceeding in the same way as above, we first obtain

$$\begin{aligned} e_k \le &\frac{2}{\tau}\tilde{e}_{k+1} + 4\sum_{\nu=2}^{k-1}\tau^{\nu-2}\left((\nu-1)|B_\nu(0)| - |B_\nu(\tfrac{1}{2})|\right)\tilde{e}_{k+2-\nu} \\ &+2\left((k+1)|B_{k+1}(0)| - |B_{k+1}(\tfrac{1}{2})|\right)\tau^k\hat{e}_0(\tau), \end{aligned}$$

and from this, by applying Lemma 4.1 (with r replaced by k and j replaced by $k+2-\nu$) to estimate $\tilde{e}_{k+2-\nu}$ by e_k (with $k+2-\nu$ odd, since the summands with odd ν vanish),

$$\begin{aligned} e_k \le &\frac{4}{T}\tilde{e}_{k+1} + \frac{4}{k}\frac{e_k}{c_k}\sum_{\nu=2}^{k-1}\frac{k+2-\nu}{2^{\nu-2}}\left((\nu-1)|B_\nu(0)| - |B_\nu(\tfrac{1}{2})|\right)c_{k+2-\nu} \\ &+\left(\frac{4}{k}\sum_{\nu=2}^{k-1}\frac{\nu-2}{2^{\nu-2}}\left((\nu-1)|B_\nu(0)| - |B_\nu(\tfrac{1}{2})|\right)c_{k+2-\nu}\right. \\ &\left.+\frac{1}{2^{k-1}}\left((k+1)|B_{k+1}(0)| - |B_{k+1}(\tfrac{1}{2})|\right)\right)T^k\hat{e}_0(\tau). \end{aligned}$$

Applying Lemma 3.2 (together with (3.1)) and Lemma 3.4 to the sums involved above, yields

$$\frac{e_k}{c_k} \le \frac{1}{T}\frac{k}{k+1}\frac{\tilde{e}_{k+1}}{c_{k+1}} + T^k\frac{1}{k+1}\hat{e}_0(\tau) \tag{4.11}$$

for $k \ge 1$ odd, $k+1 = r$. Combining this with Lemma 4.1, yields

Lemma 4.2. *The inequality of Lemma 4.1 is also true for odd j and even r.*

In the same way as above, we now obtain Theorem 4.1 for even r. ■

PROOF of Theorem 4.2. Let $j < r$, j even, r arbitrary. By Theorem 3.1c) of Köhler (1992), there holds

$$\frac{e_j}{c_j} \leq e_0^{\frac{1}{j+1}} \left(\frac{\tilde{e}_{j+1}}{c_{j+1}}\right)^{\frac{j}{j+1}} \leq e_0^{\frac{1}{j+1}} \left(\frac{e_{j+1}}{c_{j+1}}\right)^{\frac{j}{j+1}}. \tag{4.12}$$

This proves Theorem 4.2 for $j = r - 1$. Otherwise, since $j + 1$ is odd, we can apply Theorem 4.1 to e_{j+1}, which completes the proof of Theorem 4.2. ■

The proofs of Theorem 2.1 and 2.2 are not essentially different from those of Theorem 4.1 and 4.2. We indicate the necessary modifications.

PROOF of Theorem 2.1. We first consider the case k odd, $k = r - 1$. Instead of (4.10), we start with

$$\begin{aligned} R[\tilde{f}_\tau] &= \int_a^b G_{k+1}(t)\tilde{f}_\tau^{(k+1)}(t)\,dt \\ &= \int_a^b (G_{k+1}(t) - v(t))\tilde{f}_\tau^{(k+1)}(t)\,dt + \int_a^b v(t)\tilde{f}_\tau^{(k+1)}(t)\,dt. \end{aligned}$$

From $\tilde{f}_\tau^{(k+1)}(t) = \frac{1}{\tau}(\tilde{f}^{(k)}(t+\frac{\tau}{2}) - \tilde{f}^{(k)}(t-\frac{\tau}{2}))$ and the l-periodicity of $\tilde{f}^{(k)}$, we obtain, with $v(x) = 0$ for $x \notin [a, b]$,

$$\int_a^b v(t)\tilde{f}_\tau^{(k+1)}(t)\,dt = \frac{1}{\tau}\int_a^b \tilde{f}^{(k)}(t-\frac{\tau}{2}) \sum_{j=-\infty}^{\infty} (v(t+jl-\tau) - v(t+jl))\,dt,$$

so that

$$|R[\tilde{f}_\tau]| \leq (\|G_{k+1} - v\|_1 + \frac{1}{2}\Omega_0(v,\tau))\frac{2}{\tau}\|f^{(k)}\|_\infty.$$

Instead of (4.11), we obtain

$$\frac{e_k}{c_k} \leq \frac{1}{T}\frac{k}{k+1}\frac{1}{c_{k+1}}(\|G_{k+1} - v\|_1 + \frac{1}{2}\Omega_0(v,\tau)) + T^k\frac{1}{k+1}\hat{e}_0(\tau).$$

Combining this with Lemma 4.1 (with r replaced by k), yields

$$\frac{e_j}{c_j} \leq T^{j-r}\frac{j}{r}\frac{1}{c_r}(\|G_r - v\|_1 + \frac{1}{2}\Omega_0(v,\tau)) + T^j\frac{r-j}{r}\hat{e}_0(\tau)$$

for odd j and even r with $1 \leq j \leq r-1$. The same estimate holds for odd r. To obtain this, we have to choose $k = r - 2$ (k odd), and to estimate $R[g_\tau]$ in the same way as $R[\tilde{f}_\tau]$ above (instead of using (4.4)), which gives us

$$|R[g_\tau]| \leq (\|G_{k+2} - v\|_1 + \frac{1}{2}\Omega_0(v,\tau))\frac{4}{\tau^2}\|f^{(k)}\|_\infty.$$

The other parts of the proof of Theorem 2.1 remain unchanged. ■

PROOF of Theorem 2.2. For $j = r - 1$, an appropriate modification of the proof of Theorem 2.1 of Köhler (1992) yields

$$\frac{e_j}{c_j} \leq \frac{1}{T}\frac{j}{j+1}\frac{1}{c_{j+1}}(\|G_{j+1} - v\|_1 + \frac{1}{2}\Omega_0(v,\tau)) + T^j\frac{1}{j+1}e_0,$$

which proves Theorem 2.2 for $j = r - 1$. For $j < r - 1$, we use again (4.12) and apply Theorem 2.1 to e_{j+1}. ■

References

H. Braß (1974), Asymptotisch optimale Quadraturverfahren. In: J. Albrecht, L. Collatz (eds.), Numerische Methoden bei Differentialgleichungen. ISNM 19, Birkhäuser Verlag, Basel, 47–65.

H. Braß/K.-J. Förster (1987), On the estimation of linear functionals, Analysis 7, 237–258.

P. Köhler (1991a), Error estimates for generalized compound quadrature formulas. *To appear in* IMA J. Numer. Anal.

P. Köhler (1991b), Construction of asymptotically best quadrature formulas. Institute für Mathematik der TU Braunschweig, Bericht 91/11.

P. Köhler (1992), Intermediate error estimates. To appear in J. Comp. Appl. Math.

A.A. Ligun (1973), Exact inequalities for the upper bounds of seminorms on a class of periodic functions. Math. Notes 13, 389–393.

L.M. Milne-Thomson (1981), The calculus of finite differences. 2nd ed., Chelsea, New York.

N.E. Nörlund (1954), Vorlesungen über Differenzenrechnung. Chelsea, New York.

K. Petras (1989), Normabschätzungen für die ersten Peanokerne der Gauß-Formeln. ZAMM 69, T81–T83.

K. Petras (1992), Gaussian quadrature formulae – second Peano kernels, nodes, weights and Bessel functions. Institute für Mathematik der TU Braunschweig, Bericht 92/15.

A.K. Ponomarenko (1983), Estimates of the error functional for quadrature formulas with Chebyshev weight (russian). Metody Vychislenij 13, 116–121.

Author's address:

Peter Köhler
Institut für Angewandte Mathematik
TU Braunschweig
Pockelsstr. 14
3300 Braunschweig
Germany

International Series of Numerical Mathematics, Vol. 112, © 1993 Birkhäuser Verlag Basel

Stability tests for linear difference forms

Franz Locher

Dedicated to Professor Dr. Manfred Reimer, University of Dortmund, on his sixtieth birthday (November 22, 1993)

Abstract. We show that Dahlquist's root condition for the zeros of the characteristic polynomial p_n, $p_n(z) = \sum_{\nu=0}^{n} a_\nu z^\nu$, of a linear difference form with constant coefficients is equivalent to the problem whether the polynomials $\varphi_n := \sum_{\nu=0}^{n} a_\nu T_\nu$ and $\psi_{n-1} := \sum_{\nu=0}^{n} a_\nu U_{\nu-1}$ have common simple zeros in the interval $(-1, 1)$ or whether their zeros interlace. The proper tool to attack the stability problem is then Euclidean division in connection with Sturm sequence arguments. According to results of F. Peherstorfer (1990) interlacing properties of the zeros of orthogonal polynomial expansions are also responsible for the existence of positive interpolatory quadrature formulas. In this context we give a proof of a Kakeya-Eneström type theorem based only on real arguments and therefore not restricted to Chebyshev expansions.

1. Introduction

According to the famous Dahlquist root condition (cf. e.g. Reimer 1982, p. 182 ff.) a linear difference form or a linear multistep method

$$(L) \qquad \sum_{\nu=0}^{n} a_\nu x_{k+\nu} = h \sum_{\nu=0}^{n} b_\nu f_{k+\nu}\,, \quad a_\nu, b_\nu \in \mathbb{R}\,, \quad a_n \neq 0\,,$$

is stable iff the characteristic polynomial p_n, $p_n(z) = \sum_{\nu=0}^{n} a_\nu z^\nu$, has all its zeros in the closed unit disk of the complex plane and if the unimodular zeros are all simple. Hence, one is interested either in a simple test applied to the coefficients a_ν, $\nu = 0, \ldots, n$, or in an easily implementable algorithm which guarantee the stability of the difference form (L). For a homogeneous linear differential equation with constant coefficients

$$(D) \qquad \sum_{\nu=0}^{n} \alpha_\nu x^{(\nu)} = 0\,, \quad \alpha_\nu \in \mathbb{R},\ \alpha_n \neq 0\,,$$

the corresponding problem is to decide whether all zeros of the associated characteristic polynomial q_n, $q_n(z) := \sum_{\nu=0}^{n} \alpha_\nu z^\nu$, lie in the left half plane. Here, the Routh-Hurwitz criterion (cf. Henrici 1974, p. 485 ff.), which is based on the principle of the argument of complex function theory, gives information about the location of the zeros of q_n. The stability problem for (L) has been attacked by the Schur-Cohn test (Henrici 1974, p. 491 ff.). But it seems to be remained unnoticed that it can also be decided with the aid of the principle of the argument and Sturm sequence techniques. We will show that (L) is stable iff the polynomials $\varphi_n := \sum_{\nu=0}^{n} a_\nu T_\nu$ and $\psi_{n-1} := \sum_{\nu=0}^{n} a_\nu U_{\nu-1}$ have only simple common zeros in the interval $(-1, 1)$ and if the other zeros of φ_n and ψ_{n-1} interlace there. So, the stability problem for a linear difference form – or in the notation of digital signal processing – for a discrete linear time invariant system is reduced to interlacing properties of the zeros of orthogonal polynomial expansions. This fact gives a bridge to the results of F. Peherstorfer (1990) on the existence of positive interpolatory quadrature formulas. He showed that such formulas exist iff the zeros of an orthogonal polynomial expansion and its associated one interlace. Hence, our stability test is also a test for the existence of certain positive quadrature formula. Finally, we give a proof of a Kakeya-Eneström type theorem which is based on interlacing properties of the zeros of orthogonal expansions and does not refer to complex function theoretic arguments. Therefore it is not restricted to expansions in Chebyshev polynomials which can be interpreted as the real resp. imaginary part of $\exp(int)$, $t = \arccos(x)$, but can be applied to other orthogonal expansions, e.g. of Gegenbauer type.

2. Stability of linear difference forms

Let Γ be the positively oriented unit circle with the usual parametric representation $\Gamma = \{\exp(it) \mid 0 \le t \le 2\pi\}$ and $p_n(\Gamma) := \{p_n(\exp(it)) \mid 0 \le t \le 2\pi\}$ be the image curve of Γ under p_n ("Ortskurve"), where p_n, $p_n(z) = \sum_{\nu=0}^{n} a_\nu z^\nu$, $a_\nu \in \mathbb{R}$, $a_n \ne 0$, is a real n-th degree polynomial. By splitting into its real and imaginary part we get

$$p_n(\exp(it)) = \sum_{\nu=0}^{n} a_\nu \cos(\nu t) + i \sum_{\nu=0}^{n} a_\nu \sin(\nu t) =: u_n(t) + i\tilde{v}_n(t) .$$

As p_n is real, the image curve $p_n(\Gamma)$ is symmetric with respect to the real axis. So, it suffices to consider the parameter interval $[0, \pi]$. But if $t \in [0, \pi]$, then $t = \arccos x$ is bijective for $x \in [-1, 1]$. Hence we get

$$\varphi_n(x) := u_n(\arccos(x)) = \sum_{\nu=0}^{n} a_\nu T_\nu(x) ,$$

where T_ν, $T_\nu(x) = \cos(\nu \arccos(x))$, is the Chebyshev polynomial of the first kind of degree ν. As $\tilde{v}_n(k\pi) = 0$ for $k \in \mathbb{Z}$, we consider

$$v_{n-1}(t) := \frac{\tilde{v}_n(t)}{\sin(t)} = \sum_{\nu=0}^{n} a_\nu \frac{\sin(\nu t)}{\sin(t)} .$$

Then we get

$$\psi_{n-1}(x) := v_{n-1}(\arccos(x)) = \sum_{\nu=0}^{n} a_\nu U_{\nu-1}(x) ,$$

where $U_\nu, U_\nu(x) = \frac{\sin((\nu+1)\arccos(x))}{\sin(\arccos(x))}$, is the Chebyshev polynomial of the second kind of degree ν ($U_{-1} := 0$). At all, we have for $t \in [0,\pi]$, $x \in [-1,1]$, $t = \arccos(x)$,

$$\begin{aligned} p_n(\exp(it)) &= u_n(t) + i\sin(t)v_{n-1}(t) \\ &= \sum_{\nu=0}^{n} a_\nu T_\nu(x) + i\sqrt{1-x^2}\sum_{\nu=0}^{n} a_\nu U_{\nu-1}(x) \\ &=: \varphi_n(x) + i\sqrt{1-x^2}\psi_{n-1}(x) . \end{aligned}$$

First we can characterize the unimodular zeros of p_n. They occur only at common zeros of the real part u_n and the imaginary part $\tilde{v}_n$. Hence they are either $z = 1$, namely if $p_n(1) = \sum_{\nu=0}^{n} a_\nu = 0$, or $z = -1$, namely if $p_n(-1) = \sum_{\nu=0}^{n}(-1)^\nu a_\nu = 0$, or they originate as a conjugate pair $z, \bar{z}$, $z = \exp(it)$, via $t = \arccos(x)$ from a common zero $x \in [-1,1]$ of φ_n and ψ_{n-1}. A zero of p_n at $x = 1$ or $x = -1$ can easily be detected by Horner's scheme, whereas a common zero of φ_n and ψ_{n-1} leads to a nontrivial common divisor which can be found by Euclidean division.

Now we focus on the zeros of p_n in the interior of the unit disk and assume that $p_n(z) \neq 0$ if $z \in \Gamma$. The principle of the argument (cf. Henrici 1974, p. 278 ff.) states that the number $N_\Gamma(p_n)$ of zeros of p_n in the interior of Γ equals the winding number $n(p_n(\Gamma), 0)$ of $p_n(\Gamma)$ with respect to 0:

$$N_\Gamma(p_n) = \frac{1}{2\pi i}\int_\Gamma \frac{p_n'(z)}{p_n(z)}dz = \frac{1}{2\pi i}\int_{p_n(\Gamma)} \frac{dw}{w} =: n(p_n(\Gamma), 0) .$$

As $p_n(\Gamma)$ is symmetric with respect to the real axis, one half of the winding number $n(p_n(\Gamma), 0)$ must occur on the upper half of the unit circle. Hence, we have $N_\Gamma(p_n) = n$ iff for $0 < t < \pi$ the image curve $p_n(\Gamma)$ crosses the imaginary axis n times in positive direction and no crossing in negative direction and no tangential crossing takes place. Such an intersection corresponds to a zero of φ_n in the interval $(-1,1)$. As φ_n is a polynomial of degree n, φ_n must have n simple zeros in $(-1,1)$. The intersections with the real axis correspond to zeros of ψ_{n-1}, which then must interlace with the zeros of φ_n. So we get

Theorem 2.1. *Let* p_n, $p_n(z) = \sum_{\nu=0}^{n} a_\nu z^\nu$, $a_\nu \in \mathrm{IR}$, $a_n \neq 0$, *be given and* $\varphi_n := \sum_{\nu=0}^{n} a_\nu T_\nu$, $\varphi_{n-1} := \sum_{\nu=0}^{n} a_\nu U_{\nu-1}$. *Then we have:*
(1) p_n *has all its zeros in the open unit disk iff* φ_n *has* n *simple zeros in the open interval* $(-1,1)$ *which are separated by the* $n-1$ *simple zeros of* ψ_{n-1}.
(2) p_n *has unimodular zeros iff*

(2.1) $p_n(1) = 0$ *or* $p_n(-1) = 0$ *or*

(2.2) φ_n *and* ψ_{n-1} *have a nontrivial common divisor with some zero in the open interval* $(-1,1)$.

This theorem is constructive and enables to formulate an algorithm, which is essentially based on Euclidean division. For it is well-known that starting with two polynomials φ_n and ψ_{n-1} of degree n and $n-1$ resp. and the same highest degree coefficients one gets by repeated Euclidean division either a nontrivial greatest common divisor or a Sturm sequence which has its maximally possible length n if additionally the zeros of φ_n are real and simple and separated by the $n-1$ zeros of ψ_{n-1}. So we start the Euclidean algorithm with $f_0 := \sum_{\nu=0}^{n} a_\nu^{(0)} T_\nu$ and $f_1 := \sum_{\nu=0}^{n} a_\nu^{(1)} U_{\nu-1}$, $a_\nu^{(0)} := a_\nu^{(1)} := a_\nu$, $\nu = 0,\ldots,n$. It is not necessary to rearrange these polynomials with respect to the monomial basis as one knows the identities

$$T_\nu(x) = xU_{\nu-1}(x) - U_{\nu-2}(x)\ , \quad \nu = 1,2,\ldots,$$
$$U_k(x) := 0 \quad \text{if} \quad k < 0\ .$$

So we get

$$\begin{aligned} f_0(x) &= \sum_{\nu=0}^{n} a_\nu T_\nu(x) = a_0 + \sum_{\nu=1}^{n} a_\nu (xU_{\nu-1}(x) - U_{\nu-2}(x)) \\ &= x\sum_{\nu=1}^{n} a_\nu U_{\nu-1}(x) - \left((a_2 - a_0) + \sum_{\nu=3}^{n} a_\nu U_{\nu-2}(x)\right) \\ &=: xf_1(x) - f_2(x)\ , \end{aligned}$$

$$f_2(x) := \sum_{\nu=2}^{n} a_\nu^{(2)} U_{\nu-2}(x)\ , \quad a_2^{(2)} := a_2 - a_0\ , \quad a_\nu^{(2)} := a_\nu\ , \quad \nu = 3,\ldots,n\ .$$

In the next and the following steps one has to divide expansions in Chebyshev polynomials of the second kind. This can be done with the aid of the recursion formula. For $k = 2,\ldots,n$ one uses the identity

$$\sum_{\mu=k-2}^{n} a_\mu^{(k-2)} U_{\mu-k+2}(x) = (\alpha_k U_1(x) + \beta_k) \sum_{\mu=k-1}^{n} a_\mu^{(k-1)} U_{\mu-k+1}(x) - \sum_{\mu=k}^{n} a_\mu^{(k)} U_{\mu-k}(x)$$

to get by rearrangement and by equating coefficients

$$\begin{aligned}
\alpha_k &= \frac{a_n^{(k-2)}}{a_n^{(k-1)}}, \quad \beta_k = \frac{a_{n-1}^{(k-2)} - \alpha_k a_{n-2}^{(k-1)}}{a_n^{(k-1)}}, \\
a_k^{(k)} &= \alpha_k a_k^{(k-1)} + \beta_k a_{k-1}^{(k-1)} - a_{k-2}^{(k-2)}, \\
a_\mu^{(k)} &= \alpha_k a_\mu^{(k-1)} + \beta_k a_{\mu-1}^{(k-1)} + \alpha_k a_{\mu-2}^{(k-1)} - a_{\mu-2}^{(k-2)}, \quad \mu = k+1, \dots, n.
\end{aligned}$$

This recursion can break down. If $a_\mu^{(k)} = 0$, $\mu = k, \dots, n$, then f_{k-1} is the greatest common divisor of f_0 and f_1 and if additionally f_{k-1} has some zeros in $(-1,1)$, then p_n has unimodular zeros. If $a_k^{(k)} = 0$ for $k \in \{2, \dots, n\}$, then the sequence $f_0, f_1, f_2, \dots$, has not its maximally possible length and therefore f_0 has fewer than n zeros in $(-1,1)$; hence, p_n is not stable.

Now we have the following stability test for a linear difference form (L).

Stability test 2.2. *(1) Check by repeated application of Horner's scheme whether* 1 *or* -1 *are simple zeros of* p_n.
(2) By Euclidean division find the greatest common divisor χ *of* φ_n *and* ψ_{n-1}. *Check wether* χ *has only simple zeros in* $(-1,1)$; *this can be done by Euclidean division of* χ *by* χ' *and real Sturm sequence arguments.*
(3) By Horner's scheme "purify" φ_n *and* ψ_{n-1} *from their common zeros in* $(-1,1)$ *which have been found in (2) getting new starting polynomials* φ_s *and* ψ_{s-1}. *By Euclidean division build up a Sturm sequence* $f_0, f_1, \dots, f_s$ *starting with* $f_0 := \varphi_s$, $f_1 := \psi_{s-1}$. *Check whether this sequence has its maximally possible length* $s+1$ *and whether the numerical sequence* $f_0(-1), f_1(-1), \dots, f_s(-1)$ *has its maximally possible number* s *of sign changes.*

This algorithm has been implemented, and the results, especially with respect to the speed, are very satisfactory. For polynomials up to degree 80 the stability test consumed only a few milliseconds of CPU time on a PC AT 386 (Locher 1993).

3. A stability test based on alternation properties.

For a *monic* orthogonal polynomial p_n the monic *associated* polynomial $p_n^{(1)}$ is defined by

$$p_n^{(1)}(x) := \int_I \frac{p_{n+1}(x) - p_{n+1}(t)}{x - t} d\mu(t)$$

(cf. Chihara 1978, p. 86 ff.). Here we have assumed the usual properties of the probability measure μ on the real line: infinitely many points of increase in I, all moments are finite. As the monic Chebyshev polynomial $\tilde{U}_n = 2^{-n}U_n$ of the second kind is associated to the monic Chebyshev polynomial $\tilde{T}_n = 2^{1-n}T_n$ of the first kind (cf. Chihara 1978, p. 89), we can interpret the results of Chapter 2 in the following way: The polynomial p_n, $p_n(z) = \sum_{\nu=0}^{n} a_\nu z^\nu$, $p_n(1) \neq 0 \neq p_n(-1)$, $a_\nu \in \mathbb{R}$, $a_n \neq 0$, has all its zeros in the interior of the unit circle Γ iff the n zeros of $\varphi_n := \sum_{\nu=0}^{n} a_\nu 2^{\nu-1}\tilde{T}_\nu$ lie in the interval $(-1,1)$ and are separated by the $n-1$ zeros of $\psi_{n-1} := \sum_{\nu=0}^{n} a_\nu 2^{\nu-1}\tilde{T}^{(1)}_{\nu-1}$; $\tilde{T}^{(1)}_{-1} := 0$. We point to the fact that there is an intimate connection between the polynomial expansions $\varphi_n := \sum_{\nu=0}^{n} \alpha_\nu p_\nu$ and $\psi_{n-1} := \sum_{\nu=0}^{n} \alpha_\nu p^{(1)}_{\nu-1}$, $p^{(1)}_{-1} := 0$. At the Oberwolfach conference on Numerical Integration IV I heard of the results of F. Peherstorfer (1990) on positive quadrature formulas. His basic result says that a positive interpolatory quadrature formula based on the zeros of $\varphi_n := \sum_{\nu=0}^{n} a_\nu p_\nu$ exists iff the n zeros of φ_n are separated by the $n-1$ zeros of $\psi_{n-1} := \sum_{\nu=0}^{n} a_\nu p^{(1)}_{\nu-1}$. The location of the zeros of orthogonal expansions was also treated by Iserles and Saff (1989). We state in this context a rather useful lemma which avoids any complex function theory and which is often used when orthogonal, especially Chebyshev expansions are considered. Hence, this argument can be applied for wider classes of orthogonal polynomials.

Lemma 3.1. *Let* $\varphi_n := \sum_{\nu=0}^{n} a_\nu p_\nu$, $\psi_{n-1} := \sum_{\nu=0}^{n} a_\nu p^{(1)}_{\nu-1}$. *If the polynomial*

$$\Phi := \varphi_n p_n^{(1)} - \psi_{n-1} p_{n+1}$$

is strictly positive in I, *then* φ_n *has* n *zeros in* I *which are separated by the* $n-1$ *zeros of* ψ_{n-1}.

Proof. (1) Let $\xi_{n+1} < \xi_n < \cdots < \xi_2 < \xi_1$ be the $n+1$ zeros of p_{n+1} and $\eta_n < \cdots < \eta_2 < \eta_1$ the n zeros of $p_n^{(1)}$. Then we have

$$\xi_{n+1} < \eta_n < \xi_n < \cdots < \xi_2 < \eta_2 < \xi_1$$

(Chihara 1978, p. 86). Hence

$$\Phi(\xi_k) = \varphi_n(\xi_k) p_n^{(1)}(\xi_k) > 0 \ , \quad k = 1, \ldots, n+1 \ ,$$

and as

$$\operatorname{sign} p_n^{(1)}(\xi_k) = (-1)^{k-1} \ ,$$

we have

$$\operatorname{sign} \varphi_n(\xi_k) = (-1)^{k-1} \ , \quad k = 1, \ldots, n+1 \ .$$

Consequently, there exist n zeros x_k, $k = 1, \ldots, n$, of φ_n with

$$\xi_{n+1} < x_n < \xi_n < \cdots < \xi_2 < x_1 < \xi_1 .$$

(2) Similarly, we have

$$\Phi(x_k) = -\psi_{n-1}(x_k) p_{n+1}(x_k) > 0 ,$$
$$\operatorname{sign} p_{n+1}(x_k) = (-1)^k , \quad k = 1, \ldots, n .$$

So we conclude, that

$$\operatorname{sign} \psi_{n-1}(x_k) = (-1)^{k+1} , \quad k = 1, \ldots, n .$$

Hence we get $n-1$ zeros y_k, $k = 1, \ldots, n-1$, of ψ_{n-1} where

$$x_n < y_{n-1} < x_{n-1} < \cdots < x_2 < y_1 < x_1 . \qquad \square$$

The argumentation of this lemma was used for $\varphi_n := p_n$ and $\psi_{n-1} := p_{n-1}^{(1)}$ to show the interlacing property of the zeros of an orthogonal polynomial p_n and its associated $p_{n-1}^{(1)}$. But it seems to be remained unnoticed that we can use it to get location and separation theorems for the zeros of orthogonal expansions. We demonstrate this fact in the special case of Chebyshev expansions: We consider the polynomials $\varphi_n := \sum_{\nu=0}^{n} a_\nu T_\nu$ and $\psi_{n-1} = \sum_{\nu=0}^{n} a_\nu U_{\nu-1}$ in the interval $[-1, 1]$. As $x = T_2(\tau)$, $x \in [-1, 1]$, $\tau \in [0, 1]$ is bijective and $T_\nu(T_2(\tau)) = T_{2\nu}(\tau)$, $U_{\nu-1}(T_2(\tau)) = \frac{U_{2\nu-1}(\tau)}{T_1(\tau)}$, we get

$$\begin{aligned} u_{2n}(\tau) &:= \varphi_n(T_2(\tau)) = \sum_{\nu=0}^{n} a_\nu T_{2\nu}(\tau) , \\ \tilde{v}_{2n-1}(\tau) &:= \psi_{n-1}(T_2(\tau)) = \frac{1}{T_1(\tau)} \sum_{\nu=0}^{n} a_\nu U_{2\nu-1}(\tau) =: \frac{1}{T_1(\tau)} v_{2n-1}(\tau) . \end{aligned}$$

Hence, if u_{2n} has n zeros in $(0, 1)$, which are separated by $n-1$ zeros of v_{2n-1} there, then φ_n has n zeros in $(-1, 1)$, which are separated by $n-1$ zeros of ψ_{n-1}. Now we consider the polynomial

$$\Phi := u_{2n} U_{2n} - v_{2n-1} T_{2n+1} = \sum_{\nu=0}^{n} a_\nu (T_{2\nu} U_{2n} - U_{2\nu-1} T_{2n+1}) .$$

From Abramowitz-Stegun (1970), formulas 22.7.25, 22.7.26, 22.12.4, we get

$$T_{2\nu} U_{2n} - U_{2\nu-1} T_{2n+1} = U_{2n-2\nu} , \quad \nu = 0, \ldots, n ,$$

and

$$\sum_{\mu=0}^{\nu} U_{2\mu}(x) = \begin{cases} 1\,, & \nu = 0 \\ \dfrac{1 - T_{2\nu+2}(x)}{2(1-x^2)} \geq 0 \quad \text{if} \quad |x| \leq 1\,, & \nu = 1, 2, \ldots \end{cases}$$

Finally, by partial summation we find

$$\begin{aligned} \Phi(x) &= \sum_{\nu=0}^{n} a_\nu U_{2n-2\nu}(x) = \sum_{\nu=0}^{n} a_{n-\nu} U_{2\nu}(x) \\ &= \sum_{\nu=0}^{n-1} (a_{n-\nu} - a_{n-\nu-1}) \sum_{\mu=0}^{\nu} U_{2\mu}(x) + a_0 \sum_{\mu=0}^{n} U_{2\mu}(x)\,. \end{aligned}$$

If $a_n > a_{n-1} \geq \cdots \geq a_0 \geq 0$, we find that $\Phi(x) > 0$ for $|x| \leq 1$. So we get from Lemma 3.1 the following

Theorem 3.2 *Let* $a_n > a_{n-1} \geq \cdots \geq a_0 \geq 0$. *Then the polynomial* $\varphi_n = \sum_{\nu=0}^{n} a_\nu T_\nu$ *has* n *simple zeros* x_k, $k = 1, \ldots, n$, *in* $(-1, 1)$, *which are separated by the* $n-1$ *zeros* y_k, $k = 1, \ldots, n-1$, *of* $\psi_{n-1} = \sum_{\nu=1}^{n} a_\nu U_{\nu-1}$.

Remark 3.3 With the aid of the Kakeya-Eneström theorem (cf. Reimer 1982, p. 101) we can get essentially Theorem 3.2. But this proof is based on arguments of complex function theory and therefore restricted to the case of Chebyshev polynomials which can be interpreted as the real resp. imaginary part of $\exp(int)$, $t = \arccos x$, $|x| \leq 1$. Our arguments are based on properties of orthogonal polynomials and can therefore be generalized to other cases e.g. the ultraspherical polynomials. We will come back to this problem elsewhere.

References

M. Abramowitz et al. (1965), Handbook of mathematical functions. Dover Publ., New York.

T.S. Chihara (1978), An introduction to orthogonal polynomials. Gordon and Breach, New York.

P. Henrici (1974), Applied and computational complex analysis, vol. 1. Wiley, New York.

A. Iserles and E.B. Saff (1989), Zeros of expansions in orthogonal polynomials. Math. Proc. Camb. Phil.-Soc. 105, 559–573.

F. Locher (1993), A stability test for real polynomials. To appear in Num. Math.

F. Peherstorfer (1990), Linear combinations of orthogonal polynomials generating positive quadrature formulas. Math. Comp. 55, 231–241.

M. Reimer (1982), Grundlagen der Numerischen Mathematik II. Akademische Verlagsgesellschaft, Wiesbaden.

Author's address:

Franz Locher
FernUniversität Hagen
Fachbereich Mathematik
Postfach 940
5800 Hagen, West-Germany

The Canonical Forms of a Lattice Rule

J. N. Lyness

Abstract. Much of the elementary theory of lattice rules may be presented as an elegant application of classical results. These include Kronecker group representation theorem and the Hermite and Smith normal forms of integer matrices. The theory of the canonical form is a case in point. In this paper, some of this theory is treated in a constructive rather than abstract manner. A step-by-step approach that parallels the group theory is described, leading to an algorithm to obtain a canonical form of a rule of prime power order. The number of possible distinct canonical forms is derived, and this is used to determine the number of integration lattices having specified invariants.

1. Notation Used to Describe and Classify Lattice Rules

An s-dimensional *lattice*, Λ, is a set of points having the property that, when $\mathbf{p}$ and $\mathbf{q}$ are members of Λ, so are $\mathbf{p}+\mathbf{q}$ and $\mathbf{p}-\mathbf{q}$. It may be defined by this property, together with a restriction that there are no points of accumulation. A very familiar lattice is the *unit* lattice Λ_0, which comprises all points $\mathbf{p} = (p_1, p_2, \ldots, p_s)$, all of whose components p_i are integers. An *integration* lattice is a lattice that contains the unit lattice Λ_0 as a sublattice. A *lattice rule* $Q(\Lambda)$ is a quadrature rule for $[0,1)^s$ that employs the points of $\Lambda \cap [0,1)^s$ as an abscissa set $A(Q)$ and assigns an equal weight to each. Some lattice rules are useful for integrating naturally periodic functions. Other lattice rules may be very inefficient. A representative selection of the literature on lattice rules may be found in Sloan (1992) and Niederreiter (1992).

The investigation of lattice rules is hampered by two features: first, the large numbers of different rules that are available, and second, a bewildering lack of uniqueness in the various convenient representations for investigating rules and for classifying them.

A classical approach to lattices is based on the generator matrix. It is readily shown that, given any s-dimensional lattice Λ, there exists a set of s generators $\mathbf{a}_1, \mathbf{a}_2, \ldots, \mathbf{a}_s$ such that all points

$$\mathbf{p} = \sum_{i=1}^{s} \lambda_i \mathbf{a}_i \qquad \lambda_i \text{ integer}$$

lie in the lattice, and all lattice points are of this form. The $s \times s$ matrix A whose j-th row is $\mathbf{a}_j$ $(j = 1, 2, \ldots, s)$ is referred to as a generator matrix of the lattice. When

Λ is an integration lattice, it is readily shown that all elements of A are rational and that $N = |\det A|^{-1}$ is the order of $A(Q)$, the abscissa set of $Q(\Lambda)$. An approach based particularly on the generator matrix $(A^{-1})^T$ of the reciprocal lattice $\Lambda^{\perp}$ has proved fruitful. However, inconvenient aspects of this approach include the absence of uniqueness of A and the difficulty of proceeding from the matrix A to a sum of function values.

The standard number theoretic rules of Korobov (1959) are also lattice rules. These are conventionally expressed in the form

$$Qf = \frac{1}{n}\sum_{j=1}^{n} \overline{f}\left(\frac{\mathbf{z}}{n}j\right). \tag{1.1}$$

Here $\mathbf{z} \in \Lambda_0$, and $\overline{f}(\mathbf{x})$ is a periodic continuation of $f(\mathbf{x})$ that coincides with $f(\mathbf{x})$ in $[0,1)^s$. As is conventional, we denote $\mathbf{x}$ modulo Λ_0 by $\{\mathbf{x}\}$. The components of $\{\mathbf{x}\}$ are the respective nonnegative fractional parts of the components of $\mathbf{x}$.

Only some lattice rules (those of rank 1) can be expressed in this form. On the other hand, all lattice rules may be expressed in a natural generalization of form (1.1) that we refer to as a t-cycle $D - Z$ rule form. This is

$$Qf = \frac{1}{d_1 d_2 \ldots d_t}\sum_{j_1=1}^{d_1}\sum_{j_2=1}^{d_2}\cdots\sum_{j_t=1}^{d_t} \overline{f}\left(\sum_{i=1}^{t}\frac{\mathbf{z}_i}{d_i}j_i\right), \tag{1.2}$$

where d_i are integers and $\mathbf{z}_i \in \Lambda_0$. Associated with this form are two integer matrices. $D = \mathrm{diag}\{d_1, d_2, \ldots, d_t\}$ is a $t \times t$ diagonal matrix, and Z is the $t \times s$ matrix whose j-th row is $\mathbf{z}_j$.

It is not difficult to show that this form represents a lattice rule. In fact, it is the rule of lowest order N that includes

$$\mathbf{c}_i = \{\mathbf{z}_i/d_i\} \qquad i = 1, 2, \ldots, t.$$

The lattice Λ is generated by these $\mathbf{c}_i$ together with the unit vectors $\mathbf{e}_k$, $k = 1, 2, \ldots, s$. Note that the m-panel product trapezoidal rule is a lattice rule. One of its s-cycle $D - Z$ forms has $D = mI$ and $Z = I$.

Form (1.2) is not unique. Moreover, it may be *repetitive*. That is, for some integer k, it may include only $d_1 d_2 \ldots d_t/k$ distinct abscissas, each repeated k times.

Definition 1.3 *The component* $\mathbf{z}/n$, *where* $\mathbf{z} = (\zeta_1, \zeta_2, \ldots, \zeta_s) \in \Lambda_0$ *is termed* proper *when* $\gcd(\zeta_1, \zeta_2, \ldots, \zeta_s, n) = 1$ *and* $\mathbf{c} = \mathbf{z}/n \in [0,1)^s$. *(Colloquially,* $\mathbf{z}/n$ *is in its lowest terms and the point is in the integration region.)*

If $\mathbf{z}/n$ in (1.1) is not proper, then the form is repetitive. This sort of trivial repetition is easy to recognize. However, possible repetition in the general t-cycle $D - Z$ form need not be at all obvious.

It is straightforward to show (see Sloan and Lyness 1989) that a necessary and sufficient condition for the rule form Qf in (1.2) to be repetitive is that there exist integers $j_1, j_2, \ldots, j_t$ with $j_i \in [0, d_i)$, not all zero, such that

$$\sum_{i=1}^{t} \frac{j_i \mathbf{z}_i}{d_i} \in \Lambda_0. \tag{1.4}$$

One classification of lattice rules based on the t-cycle form relies on the circumstance that the elements of the abscissa set $A(Q)$ form a group G under addition modulo Λ_0.

In Sloan and Lyness (1989), the Kronecker group representation theorem was applied to this group to show that a t-cycle nonrepetitive $D - Z$ form exists in which all elements d_i exceed 1, and $d_{i+1} \mid d_i$, and the $\mathbf{z}_i$ are linearly independent. Such a rule form was termed a *canonical form*. From the nomenclature of group theory, these particular values of d_i are termed *invariants*, and this particular value of t is termed the *rank* of the rule Q. The rank and invariants are unique to the lattice, but the choice of $\mathbf{z}_i$ in the canonical form is far from unique. However, no constructive approach for proceeding from a general t-cycle form to a canonical form was made available at that time.

This paper is concerned with finding a canonical form of a rule defined by a general t-cycle $D - Z$ form. We shall, in fact, show only how to find *nonrepetitive* forms in which $d_{i+1} \mid d_i$. We may then rely on Corollary 4.6 of Sloan and Lyness (1989) which assures us that a form that appears to be canonical is indeed canonical if it is nonrepetitive. (See Theorem 3.1 below.)

2. Decomposition and Reassembly

The accompanying table illustrates the underlying group theory.

G	F_1	F_2		F_s
S_1	E_{11}	E_{12}	$\ldots$	E_{1s}
S_2	E_{21}	E_{22}	$\ldots$	E_{2s}
$\vdots$				
S_q	E_{q1}	E_{q2}	$\ldots$	E_{qs}

Each entry is an Abelian group. Any group in the initial row (column) is the direct sum of the other groups in that row (column), and the groups in all but the initial column are cyclic groups. The table illustrates the decomposition of an *Abelian* group G of order $N = p_1^{\beta_1} p_2^{\beta_2} \ldots p_q^{\beta_q}$ into the direct sum of *cyclic* groups F_k. Here

$p_1, p_2, \ldots, p_q$ are distinct primes, and β_i are positive integers. The first stage is the decomposition of G into the direct sum

$$G = S_1 \oplus S_2 \oplus \ldots \oplus S_q \tag{2.1}$$

of its Sylow p groups. S_j, which is of order $p_j^{\beta_j}$, contains all elements of G whose order is any integer power of p_j. Any *noncyclic* Abelian group of prime power order may be decomposed into a direct sum of *cyclic* groups. Thus,

$$S_j = E_{j,1} \oplus E_{j,2} \oplus \ldots \oplus E_{j,s}, \tag{2.2}$$

where $E_{j,k}$ is a *cyclic* group of order $p_j^{\beta_{j,k}}$. These have been arranged in order so that $\beta_{j,k} \geq \beta_{j,k+1}$. Clearly, $\sum_{k=1}^{s} \beta_{j,k} = \beta_j$. There may be fewer nontrivial groups than indicated here. The theory is not compromised if some trivial groups $E_{j,k}$, which contain only the identity element, are included but ignored. For these, $\beta_{j,k} = 0$.

Finally, we apply the result that the direct sum of *cyclic* groups whose orders are mutually prime is also a cyclic group. Thus,

$$F_k = E_{1,k} \oplus E_{2,k} \oplus \ldots \oplus E_{q,k} \tag{2.3}$$

is a *cyclic* group of order $n_k = \prod_{j=1}^{q} p_j^{\beta_{jk}}$. Note that, since $\beta_{j,k} \geq \beta_{j,k+1}$, it follows that $n_{k+1} \mid n_k$.

What is illustrated here is the Kronecker decomposition of an Abelian group G into the direct sum of s cyclic groups F_k; $k = 1, 2, \ldots, s$. The nontrivial values of n_k are termed the invariants of G, and the number of these is termed the rank of G.

The above remarks comprise at most a schematic for a possible derivation of a famous theorem in group theory. For our purposes, as we shall see, we do not *need* group theory. We may use the schematic to derive a canonical form.

It appears that in our application, the decomposition (2.1) into Sylow p groups is very simple, as is the recomposition (2.3) of the cyclic groups $E_{j,k}$ into F_k. However, the middle stage (2.2) is nontrivial. The following examples illustrate the first and third operations. These are justified in the following theorem.

Theorem 2.4 *Let* $n = pq$, $(p, q) = 1$, *and* $\mathbf{z}, \mathbf{z}_1, \mathbf{z}_2 \in \Lambda_0$. *Then*

$$\sum_{j=1}^{n} \bar{f}\left(\frac{j\mathbf{z}}{n} + \epsilon\right) = \sum_{j_1=1}^{p} \sum_{j_2=1}^{q} \bar{f}\left(\frac{j_1\mathbf{z}}{p} + \frac{j_2\mathbf{z}}{q} + \epsilon\right)$$

and

$$\sum_{j_1=1}^{p} \sum_{j_2=1}^{q} \bar{f}\left(\frac{j_1\mathbf{z}_1}{p} + \frac{j_2\mathbf{z}_2}{q} + \epsilon\right) = \sum_{j=1}^{n} \bar{f}\left(\frac{j\mathbf{z}_3}{pq} + \epsilon\right),$$

where $\mathbf{z}_3 = q\mathbf{z}_1 + p\mathbf{z}_2$. *Moreover, if* $\mathbf{z}_1/p$ *and* $\mathbf{z}_2/q$ *are* proper *(see Definition (1.3) above), so is* $\{\mathbf{z}_3/pq\}$.

The proof of either part is trivial. Note that there are no minor restrictions. For example, $\mathbf{z}_2/q$ need not be in its lowest terms. The ϵ is notationally helpful in establishing extensions of these results.

As an example, we put the following rule into canonical form.

Example 2.5

$$Qf = \frac{1}{12.15} \sum_{j_1=1}^{12} \sum_{j_2=1}^{15} \overline{f}\left(\frac{(5,8,3)}{12} j_1 + \frac{(8,1,0)}{15} j_2 \right). \tag{2.5}$$

This may be reexpressed as

$$\frac{1}{4.3.3.5} \sum_{j_1=1}^{4} \sum_{j_2=1}^{3} \sum_{j_3=1}^{3} \sum_{j_4=1}^{5} \overline{f}(\mathbf{x}),$$

where

$$\mathbf{x} = \frac{(5,8,3)}{4} j_1 + \frac{(5,8,3)}{3} j_2 + \frac{(8,1,0)}{3} j_3 + \frac{(8,1,0)}{5} j_4.$$

This may be replaced by

$$\mathbf{x}' = \frac{(1,0,3)}{4} j_1 + \frac{(2,2,0)}{3} j_2 + \frac{(2,1,0)}{3} j_3 + \frac{(3,1,0)}{5} j_4.$$

The sum in (2.5), which contains 180 elements of G, has now been reexpressed as the direct sum of three Sylow p groups, with $p = 2$, 3, and 5, respectively. Those with $p = 2$ and 5 are already cyclic groups. The Sylow 3 group is the direct sum of two cyclic groups.

The intermediate stage comprises, in general, removing repetition from each of the Sylow p groups. In this case, it is clear by inspection that the groups of order 3 are distinct. That is, the sums over j_2 and j_3 include nine distinct terms and not three. Applying the second part of Theorem 2.4 twice, we find successively

$$\begin{aligned} Qf &= \frac{1}{180} \sum_{j_{12}=1}^{12} \sum_{j_3=1}^{3} \sum_{j_4=1}^{5} \overline{f}(\mathbf{x}'') \\ &= \frac{1}{180} \sum_{j_{124}=1}^{60} \sum_{j_3=1}^{3} \overline{f}(\mathbf{x}''') \\ &= \frac{1}{180} \sum_{k_1=1}^{60} \sum_{k_2=1}^{3} \overline{f}\left(\frac{(31,52,45)}{60} k_1 + \frac{(2,1,0)}{3} k_2 \right). \end{aligned} \tag{2.6}$$

Here

$$\mathbf{x}'' = \frac{(11,8,9)}{12} j_{12} + \frac{(2,1,0)}{3} j_3 + \frac{(3,1,0)}{5} j_4,$$

and

$$\mathbf{x}''' = \frac{(91,52,45)}{60} j_{124} + \frac{(2,1,0)}{3} j_3.$$

This rule form (2.6) is in canonical form.

The reader may have noticed that the overall result can be significantly altered by what appears at first sight to be a minor change in the problem. For example, if in Example 2.5 we replace the second vector (8,1,0) by (7,1,0), we find that the vector (2,1,0) in $\mathbf{x}'$ must be replaced by (1,1,0). The middle stage becomes

$$\sum_{j_2=1}^{3}\sum_{j_3=1}^{3}\bar{f}\left(\frac{(2,2,0)}{3}j_2+\frac{(1,1,0)}{3}j_3+\epsilon\right)=3\sum_{j_{23}=1}^{3}\bar{f}\left(\frac{(1,1,0)}{3}j_{23}+\epsilon\right).$$

The new form Qf is repetitive, and Q has one invariant instead of two.

A satisfactory procedure to obtain the canonical form cannot rely on the chance recognition of this sort of circumstance. Subsequent sections are devoted to providing an algorithm for handling this problem.

3. Rules of Prime Power Order

As a preliminary, we remove some trivial complications. We are interested only in sequential proper forms. Qf is *sequential* if $d_1 \geq d_2 \geq \ldots \geq d_t > 1$. The i-th component is *proper* if $\mathbf{z}_i/d_i$ is in its lowest terms and $\mathbf{c}_i = \mathbf{z}_i/d_i \in [0,1)^s$. The form is sequential proper when it is sequential and each element is proper. It is a trivial task to reexpress Qf in a t' cycle $D-Z$, sequential proper form.

A geometric view is that the rule may be defined by t points $\mathbf{c}_1, \mathbf{c}_2, \ldots, \mathbf{c}_t$ of orders $d_1, d_2, \ldots, d_t$, respectively. A sequential proper form may be immediately constructed by reordering the points so that $d_1 \geq d_2 \ldots \geq d_t > 1$, and setting $\mathbf{z}_i = d_i\{\mathbf{c}_i\}$.

It still may be repetitive (trivially if $\mathbf{c}_i = \mathbf{c}_{i+1}$ and $d_i = d_{i+1}$).

Theorem 4.5 and its corollary 4.6 in Sloan and Lyness 1989 assure us that a non repetitive sequential proper rule form in which $d_{i+1} \mid d_i$ is a canonical form. This is difficult to exploit as it is difficult in general to recognize a non repetitive form. However, the theory becomes much easier as soon as we restrict ourselves to prime power forms. In this section we deal with lattices and rule forms of prime power order; these are t cycle $D-Z$ rule forms in which

$$d_j = p^{\alpha_j} \qquad j = 1, 2, \ldots, t, \tag{3.1}$$

where p is a prime. For these, the divisibility condition is naturally satisfied and it follows:

Theorem 3.1 *A prime power sequential proper form is canonical if and only if it is non repetitive.*

In other words, a prevalent situation in which a rule form is both non repetitive and not canonical does not arise in the prime power order context.

In this section and in sections 4 and 5, we deal with lattices and rule forms of prime power order only. The problem of finding a canonical form reduces to that of recognizing and removing repetition.

Theorem 3.2 *Let Qf and $Q'f$ be prime power t-cycle $D - Z$ forms having the same prime p, and the same Z matrix. Let $D = diag(d_1, d_2, \ldots, d_t)$ and $D' = diag(p, p, \ldots, p) = pI$. Then Qf and $Q'f$ are either both repetitive or both non-repetitive.*

Proof. The condition for Qf to be repetitive is that there exist integers $j_i \in [0, d_i)$ $i = 1, 2, \ldots, t$ not all zero such that

$$\sum_{i=1}^{t} j_i \mathbf{z}_i / d_i \in \Lambda_0. \tag{3.3}$$

Let us suppose that Qf is repetitive and the above j_i exist. Let I' be $\{i \mid j_i \neq 0\}$. Let $j_i/d_i = \gamma_i/m_i$ in their lowest terms. Then $\gamma_i \neq 0$, $i \in I'$, and

$$\sum_{i \in I'} \gamma_i \mathbf{z}_i / m_i \in \Lambda_0. \tag{3.4}$$

Let $p^\lambda = \max_{i \in I'} m_i$. Let I'' be $\{i \mid m_i = p^\lambda\}$. Then multiplying (3.4) by $p^{\lambda-1}$ and subsuming all integers to the right, we obtain

$$\sum_{i \in I''} \gamma_i \mathbf{z}_i / p \in \Lambda_0. \tag{3.5}$$

This statement implies that $Q'f$ is repetitive.

The converse is marginally simpler to prove. If we set

$$\begin{array}{ll} j_i = \gamma_i d_i / p & i \in I'' \\ j_i = 0 & i \notin I'', \end{array}$$

we recover (3.3) from (3.5). □

Recasting this theorem geometrically is revealing.

Corollary 3.6 *Any nonrepetitive prime power t-cycle rule form includes precisely $p^t - 1$ distinct points of order p.*

Theorem 3.2 shows that the critical quantities in determining whether a prime power form is repetitive is the Z matrix. The integers d_i play a secondary role. This is confirmed in the following theorem.

Theorem 3.7 *A t-cycle $D - Z$ form of a prime power rule is repetitive if and only if the rank of Z modulo p is less than t.*

Proof: In view of Theorem 3.2, we need only establish this for a t-cycle $D' - Z$ form where

$$D' = \mathrm{diag}\{p, p, \ldots, p\}.$$

If this is repetitive, there exist $j_1, j_2, \ldots, j_t$, not all zero such that

$$\left\{\sum_{i=1}^{t} j_i \mathbf{z}_i / p\right\} = \mathbf{0}.$$

This implies that $\sum_{i=1}^{t} j_i \mathbf{z}_i = p\mathbf{u}$ for some $\mathbf{u} \in \Lambda_0$; this is the condition that the rank of Z modulo p is less than t. □

The simple results of this section may be readily incorporated into an algorithm that transforms a general p-power t-cycle $D - Z$ rule form into a sequential, proper, and non-repetitive form. In the final form, Z either is unimodular or contains a $t \times t$ unimodular submatrix.

This algorithm is a variant of a standard triangularization technique.

At the start of the j-th stage, we have

$$d_1 \geq d_2 \geq \ldots \geq d_j \geq \ldots \geq d_t > 1$$

all being powers of p, and $\mathbf{z}_i/d_i \; i = 1 \ldots j-1$ are all proper. Associated with $\mathbf{z}_1, \mathbf{z}_2, \ldots, \mathbf{z}_{j-1}$ we have distinct integers $i_1, i_2, \ldots, i_{j-1}$, which we term *column indices*. Moreover,

$$\left.\begin{array}{lll} Z_{r,i_c} = 1 & \text{when} & r = c \\ Z_{r,i_c} = 0 & \text{when} & r > c \end{array}\right\} \begin{array}{l} c = 1, 2, \ldots, j-1 \\ r = 1, 2, \ldots, t \end{array} .$$

The j-th stage comprises the following:

1. Let $\mathbf{z}_j = (\zeta_1, \zeta_2, \ldots, \zeta_s)$. Identify a component, say ζ_ℓ, for which $(\zeta_\ell, p) = 1$. Set $i_j = \ell$. Replace $\mathbf{z}_j$ by $k\mathbf{z}_j$, where $k = \zeta_\ell^{-1} (\bmod\ d_j)$ and then replace $\mathbf{z}_j$ by $\mathbf{z}'_j = (\zeta'_1, \zeta'_2, \ldots, \zeta'_s)$ with $\zeta'_i \in [0, d_j)$. (This leaves $\zeta'_\ell = 1$.)
2. For $i > j$, replace $\mathbf{z}_i$ by $\mathbf{z}_i - Z_{i,\ell}\mathbf{z}_j$. (This leaves $Z_{i,\ell} = 0$; $i > j$.)
3. Carry out trivial adjustments on Z and D necessary to leave a t'-cycle $D - Z$ sequential proper form.

Comments:

1. We have by hypothesis that $\mathbf{z}_j/d_j$ is proper. Thus, there is some component of $\mathbf{z}_j$, say the ℓ-th, that is not a multiple of p, and ℓ may be chosen for i_j in 1.

3. These adjustments include the following:

   a. Put any improper $\mathbf{z}_i/d_i$ into proper form $\mathbf{z}'_i/d'_i$.

b. Remove any rows $\mathbf{z}_i$ for which either $\mathbf{z}_i = \mathbf{0}$ or $d_i = 1$; naturally this step reduces the value of t.

c. If necessary, reorder the rows of Z so that the ordering of d_i is sequential.

The final Z matrix is an integer matrix. If its columns were permuted in accordance with the column indices, it would be upper triangular with unit diagonal. Thus, it contains a $t \times t$ unimodular submatrix (corresponding to retaining only the columns numbered $i_1, i_2, \ldots, i_t$).

Because of this unimodularity, it follows that when $t = s$, the matrix $A = D^{-1}Z$ is a generator matrix of Λ. In general, $t < s$, and $\tilde{A} = D^{-1}Z$ comprises the first t rows of a generator matrix. The remaining $s - t$ rows of the generator matrix may be chosen as unit vectors $\mathbf{e}_\lambda$, the λ being the elements of $[1, s]$ not assigned to be column indices.

Finally, note that Z is *not* unique. In general, the column indices can be chosen in many ways, and rows having the same d element may be interchanged.

4. Canonical Form Redundancy of Prime Power Forms

In the preceding section, we remarked that the Z matrix is not unique. In this section, we quantify this lack of uniqueness. We take an r-cycle prime power rule form Qf which is in canonical form; we see how many distinct reassignments of Z exist. Initially, we shall have to assume that the result depends on p, D, and Z. However, it will appear that it simply depends on p and D. Let

$$Qf = \frac{1}{n_1 n_2 \ldots n_r} \sum_{j_1=1}^{n_1} \sum_{j_2=1}^{n_2} \cdots \sum_{j_r=1}^{n_r} \bar{f}\left(\sum_{k=1}^{r} \frac{j_k \mathbf{z}_k}{n_k} \right) \tag{4.1}$$

be a canonical form of Q. Here as before,

$$n_1 \geq n_2 \geq \ldots \geq n_r > 1,$$

and each n_i is a power of p. Also, $\mathbf{z}_k \in [0, n_k)^s$, $k = 1, 2, \ldots, r$. Note that, since this is a canonical form, we know that $\mathbf{z}_k / n_k$ is in its lowest terms. Each member of the set of generators

$$\mathbf{c}_k = \mathbf{z}_k / n_k \tag{4.2}$$

is itself an element of $[0, 1)^s$.

Definition 4.3 $\bar{\bar{\mu}}_s(n_1, n_2, \ldots, n_s; N; Z)$ *is the number of distinct ways of assigning the Z-matrix so that the rule is the same. (Here $\mathbf{z}_k \in [0, n_k)^s$.)*

Geometrically, this is the number of different point sets $\mathbf{c}_1, \mathbf{c}_2, \ldots, \mathbf{c}_r$ that can be put into (4.1) leaving the same rule. These points are $\in [0, 1)^s$.

It will appear that $\bar{\bar{\mu}}_s$ is independent of Z, and we shall later drop the Z in the definition.

The construction of a formula for $\bar{\bar{\mu}}_s$ uses relatively straightforward concepts that nevertheless must be applied with care. We shall suppose that we assign the vectors $\mathbf{z}_j$ (or $\mathbf{c}_j = \mathbf{z}_j/n_j$) in turn, starting with $j = 1$ and ending with $j = r$. Clearly, every $\mathbf{c}_j$ has to be a member of the abscissa set. Thus, when we start the j-th stage, we may limit the choice for $\mathbf{c}'_j$ as follows:

$$\mathbf{c}'_j = \sum_{i=1}^{r} \lambda_i^j \mathbf{c}_i \qquad \lambda_i^j \in [0, n_i) \tag{4.4}$$

or

$$\mathbf{z}'_j = \sum_{i=1}^{r} \lambda_i^j \frac{n_j}{n_i} \mathbf{z}_i. \tag{4.5}$$

It is convenient to drop the { } symbol here. The restriction $\lambda_i^j \in [0, n_i)$ in (4.4) simply prevents duplication, since $\lambda_i^j \mathbf{c}_i \equiv (\lambda_i^j + n_i)\mathbf{c}_i$. When assigning $\mathbf{c}'_j$, it is clear that for those values of i for which $n_i < n_j$, we may choose any available λ_i^j, and hence there are n_i distinct choices for each such λ_i^j. However, for those values of i for which $n_i > n_j$, the choice has to be restricted so that $\mathbf{z}'_j$ in (4.5) turns out to be in Λ_0. That is, $\lambda_i^j(n_j/n_i)$ must be an integer. Thus, for these values of i, there are only n_j distinct choices for each λ_i^j. (Geometrically, this recognizes that $\mathbf{c}_j$ is of order n_j and therefore only multiples of $\mathbf{c}_i$ of this order or less may be included.) There remain values of i for which $n_i = n_j$.

To fix ideas, suppose $n_{j+1} > n_j > n_{j-1}$. Then there is only one value λ_j^j remaining to assign. Since $\mathbf{z}_j$ itself must be attainable, we can allow λ_j^j to take one of $n_j(1 - p^{-1})$ distinct values only.

In this case, then, the number of ways of assigning $\mathbf{z}'_j$ is the product over i of the number of ways of assigning λ_j^i. This gives

$$\prod_{i=1}^{j-1} n_j \times n_j \left(1 - \frac{1}{p}\right) \times \prod_{i=j+1}^{r} n_i.$$

Thus, when $n_{j-1} > n_j > n_{j+1}$, whatever the individual values of $\mathbf{z}_i$, $i = 1, 2, \ldots, r$, the number of ways of reassigning $\mathbf{z}_j$, without changing any other $\mathbf{z}_i$ is

$$n_j^j \prod_{i=j+1}^{r} n_i \left(1 - \frac{1}{p^{\lambda(j)}}\right) \quad \text{with } \lambda(j) = 1.$$

In general, we cannot expect all n_j to be distinct. Let us suppose

$$n_j > n_{j+1} = n_{j+2} = \ldots = n_{j+w} > n_{j+w+1} \tag{4.6}$$

and isolate the part of the calculation involving the assignment of $\mathbf{c}'_{j+1} \ldots \mathbf{c}'_{j+w}$ to the extent that this is affected by $\mathbf{c}_{j+1} \ldots \mathbf{c}_{j+w}$. For convenience, we suppress the subscript j and denote by n the common value n_{j+1}. We need to consider the number of possibilities for

$$\begin{aligned} \mathbf{c}'_1 &= \sum_{i=1}^{w} \lambda_i^1 \mathbf{c}_i \\ \mathbf{c}'_2 &= \sum_{i=1}^{w} \lambda_i^2 \mathbf{c}_i \\ &\dots \\ \mathbf{c}'_w &= \sum_{i=1}^{w} \lambda_i^w \mathbf{c}_i. \end{aligned} \tag{4.7}$$

These are assigned in the following order. First, λ_i^1, $i = 1, 2, \dots, w$ are assigned. We need $\mathbf{c}'_1$ to be of order n. To ensure this, λ_i^1 may be chosen in any way so long as at least one term $\lambda_i^1 \mathbf{c}_i$ is of order n. This is equivalent to choosing a point of order n in a w-dimensional space. The number of points is $n^w(1 - p^{-w})$. Later in the calculation, we have to assign $\mathbf{c}'_2$. However, $\mathbf{c}_1$ has already been assigned and is of order n. To ensure that $n\mathbf{c}_2$ is independent of $n\mathbf{c}_1$, one must ensure that at least one of $\lambda_i^2 \mathbf{c}_i$, $i = 2, 3, \dots, w$ is of order n. λ_1^2 is not restricted. The number of points available is $n^w(1 - p^{-w+1})$.

Continuing in this way, we find the number of ways of assigning $\mathbf{c}'_k$, $k = 1, 2, \dots, w$ to be

$$n^w(1 - p^{-w-1+k}).$$

We note again that the result does not depend in detail on the rule form. It depends only on p and on $n_1, n_2, \dots, n_r$. When several n_j are equal, it depends particularly on the pattern. It is notationally convenient to define an integer index.

Definition 4.8 *With respect to*

$$\begin{aligned} &n_1 \geq n_2 \geq n_3 \geq \dots \geq n_r, \quad \textit{the index} \\ &\lambda(j) = k - j, \end{aligned} \tag{4.8}$$

where k is the smallest integer for which $n_k < n_j$.

Theorem 4.9 *Given a rule form Qf, the number of nontrivial ways of reassigning $\mathbf{c}_j$ so that the rule remains unaltered is*

$$\prod_{i=1}^{r} n_{\max(i,j)} \left(1 - \frac{1}{p^{\lambda(j)}}\right). \tag{4.9}$$

This number, large as it is, refers only to the number of ways of reassigning $\mathbf{c}_j$. Since it is independent of the individual choice for the other generators $\mathbf{c}_i$ $(i \neq j)$, we obtain the total number of nontrivial assignments as the product of r corresponding terms. Thus, we have the following corollary.

Corollary 4.10

$$\bar{\bar{\mu}}_s(n_1, n_2, \ldots, n_s; N) = \prod_{j=1}^{r} n_j^j \prod_{i=j+1}^{r} n_i(1 - p^{-\lambda(j)}) \tag{4.10}$$
$$= N \prod_{t=1}^{r} n_t^{2t-2}(1 - p^{-\lambda(t)}).$$

5. The Number of Distinct Prime Power Rules Having Specified Invariants

In the preceding section we derived a formula for $\bar{\bar{\mu}}_s(n_1, n_2, \ldots, n_s; N)$. This is the number of distinct assignments for $\mathbf{c}_1, \mathbf{c}_2, \ldots, \mathbf{c}_r$ that give rise to the same rule Q. The formula is the same for any rule having these invariants. We may exploit this fact to derive a formula for the number of distinct rules $\nu_s(n_1, n_2, \ldots, n_s; N)$ having these invariants.

Definition 5.1 $\bar{\mu}_s(n_1, n_2, \ldots, n_s; N)$ *is the number of distinct ways of assigning the Z matrix so that the rule has these invariants. (Here $\mathbf{z}_k \in [0, n_k)^s$.)*

Geometrically, this is the number of different point sets $\mathbf{c}_1, \mathbf{c}_2, \ldots, \mathbf{c}_r$ that can be substituted into (4.1) to give a nonrepetitive rule form.

Comparing Definition 5.1 with Definition 4.3 shows that $\bar{\mu}_s$ is larger than $\bar{\bar{\mu}}_s$. Since the redundancy is the same for each distinct rule, it follows that

$$\nu_s(n_1, n_2, \ldots, n_s; N) = \frac{\bar{\mu}_s(n_1, n_2, \ldots, n_s; N)}{\bar{\bar{\mu}}_s(n_1, n_2, \ldots, n_s; N)}. \tag{5.2}$$

The determination of $\bar{\mu}_s$ is an exercise of the same type as the determination of $\bar{\bar{\mu}}_s$ but much easier in detail. (The formula does not bring in the pattern of n_i explicitly.)

The rule may be defined by an assignment of r points $\mathbf{c}_1, \mathbf{c}_2, \ldots, \mathbf{c}_r$. ($\mathbf{c}_i = \mathbf{z}_i/n_i$ is a point of order n_i.) $\bar{\mu}_s$ is the number of distinct point assignments. This is calculated as $\prod_{j=1}^{r} P_j$, where P_j is the number of ways of choosing $\mathbf{c}_j$ once $\mathbf{c}_1$, $\mathbf{c}_2$, and $\mathbf{c}_{j-1}$ have already been assigned.

The number of points of order n_j is $n_j^s(1 - p^{-s})$. (This is because when $n = p^\alpha$, the grid containing n^s points contains all points of orders $1, p, \ldots, p^\alpha$. The number of order precisely p^α is $(p^\alpha)^s - (p^{\alpha-1})^s$.) However, if $\mathbf{c}_1, \mathbf{c}_2, \ldots, \mathbf{c}_{j-1}$ have already been assigned, in view of Theorem 3.2, $\mathbf{c}_j$ must be chosen so that the order p points it introduces have not been included previously. There are in total $p^s - 1$ order p points. The number of these in the span of $\mathbf{c}_1, \mathbf{c}_2, \ldots, \mathbf{c}_{j-1}$ is $p^{j-1} - 1$. Thus, a proportion $(p^{j-1} - 1)/(p^s - 1)$ of the order p points is not available; alternatively stated, a proportion $(p^s - p^{j-1})/(p^s - 1) = (1 - p^{-s+j-1})/(1 - p^{-s})$ is available. It follows relatively painlessly that of the $n_j^s(1 - p^{-s})$ points of order n_j, only that proportion give rise to available order p points. Thus, $\mathbf{c}_j$ may be chosen in $n_j^s(1 - p^{j-1-s})$ ways. Hence,

$$\begin{aligned}\bar{\mu}_s &= \prod_{j=1}^{r} P_j = \prod_{j=1}^{r} n_j^s(1 - p^{j-1-s}) \\ &= N^s \prod_{j=1}^{r}(1 - p^{j-1-s}). \qquad (5.3)\end{aligned}$$

The final stage gives the following theorem.

Theorem 5.4

$$\nu_s(n_1, n_2, \ldots, n_s; N) = \prod_{j=1}^{\lfloor s/2 \rfloor} \left(\frac{n_j}{n_{s+1-j}}\right)^{s-2j+1} \prod_{j=1}^{r} \frac{(1 - 1/p^{s+1-j})}{(1 - 1/p^{\lambda(j)})}. \qquad (5.4)$$

Here $n_1 \geq n_2 \geq \ldots \geq n_r > 1$, and we have set $n_j = 1$ for all $j \in [r+1, s]$. The index $\lambda(j)$ is defined in (4.8).

With a single exception, $\nu_s > 1$. The exception is the s-dimensional m copy rule. Here all the invariants are equal, and the set $\lambda(j)$ comprises the first s positive integers.

The first three cases are

$$\nu_s(n_1, 1^{s-1}; N) = N^{s-1}(1 - 1/p^s)/(1 - 1/p) \qquad (5.5)$$

$$\nu_s(n_1, n_2, 1^{s-2}; N) = (N^{s-1}/n_2^2)(1 - 1/p^s)(1 - 1/p^{s-1})/(1 - 1/p^{j_1})(1 - 1/p), \qquad (5.6)$$

where

$$\begin{aligned} j_1 &= 1 \quad \text{when} \quad n_1 > n_2 > 1 \\ &= 2 \quad \text{when} \quad n_1 = n_2 > 1, \end{aligned}$$

$$\nu_s(n_1, n_2, n_3, 1^{s-3}; N) = \frac{N^{s-1}}{n_2^2 n_3^4} \frac{(1 - 1/p^s)(1 - 1/p^{s-1})(1 - 1/p^{s-2})}{(1 - 1/p^{k_1})(1 - 1/p^{k_2})(1 - 1/p)}, \qquad (5.7)$$

where

$$\begin{aligned} &k_1 = k_2 = 1 && \text{when} && n_1 > n_2 > n_3 > 1 \\ &k_1 = 3;\ k_2 = 2 && \text{when} && n_1 = n_2 = n_3 > 1 \\ &k_1 = 2;\ k_2 = 1 && \text{otherwise.} && \end{aligned}$$

These are valid only when $n_1 \geq n_2 \geq \ldots \geq n_r > 1$ and each n_i is a power of the same prime p.

These results have been obtained independently by Joe and Hunt (1992). Their general approach follows much the same lines as the one here. However, their derivation of $\bar{\mu}_s$ is embedded in group theory.

These results resemble a similar result (Lyness and Sørevik 1989) for $\nu_s(N)$, the number of distinct lattice rules of order N. When $N = p^\beta$, this is

$$\nu_s(N) = N^{s-1} \prod_{i=1}^{s-1} \left[\left(1 - \frac{1}{p^{i+\beta}}\right) \Big/ \left(1 - \frac{1}{p^i}\right)\right]. \qquad (5.8)$$

6. A General Formula for $\nu_s(\mathbf{n}; N)$

A simple application of the group decomposition described in Section 2 allows us to base a general formula for $\nu_s(\mathbf{n}; N)$ on the one valid only for prime power rules in Theorem 5.4. The decomposition (2.1) of G into the direct sum of Sylow p groups has the property that when G has invariants $\mathbf{n}(G)$, the invariants $\mathbf{n}(S_p)$ of each Sylow p group is known. The uniqueness of this decomposition leads to the result that the number of distinct abscissa sets for which G has invariants $\mathbf{n}(G)$ is the product of the several corresponding numbers for each component S_p.

Theorem 6.1 *Let the prime decomposition of N be*

$$N = p_1^{\beta_1} p_2^{\beta_2} \ldots p_q^{\beta_q}.$$

Let the invariants n_k of a rule Q of order N be given by

$$n_k = p_1^{\beta_{1k}} p_2^{\beta_{2k}} \ldots p_q^{\beta_{qk}} \qquad k = 1, 2, \ldots, s.$$

Then the number of distinct rules Q having these invariants is

$$\nu_s(n_1, n_2, \ldots, n_s, N) = \prod_{j=1}^{q} \nu_s(p_j^{\beta_{j1}}, p_j^{\beta_{j2}}, \ldots, p_j^{\beta_{js}}; \ p_j^{\beta_j}),$$

where an explicit expression for each of the factors on the right appears in (5.4) above.

In the statement of the above theorem, we have followed various conventions from earlier sections. In particular,

$$\beta_{j,i} \geq \beta_{j,i+1}.$$

This theorem is a natural extension of a result in Lyness and Sørevik (1989) to the effect that the number of distinct lattice rules of order N is given by

$$\nu_s(N) = \prod_{j=1}^{q} \nu_s(p_j^{\beta_j}).$$

7. Concluding Remarks

The general thrust of this article is to provide a straightforward and concrete approach to some of the basic structure of lattice rules. This approach has led through several areas which are new only in a marginal sense.

For example, spelling out the group theory and applying it as in Section 2 is new. Something like it was submitted in an earlier version of Sloan and Lyness (1989) but was excised by the referee. Again, the simple procedure in Section 3 for the reduction of a prime power rule to canonical form has not appeared before. What will soon be available is a more general process in which the Smith Normal Form of a generator matrix is used to obtain a general canonical form; see Lyness and Keast (1991) and Langtry (1991). However, in that process the simple underlying geometry is obscured.

The results of Sections 4 to 6 about the number of lattice rules with given invariants were available to the author in 1989 and promised in 1991. These were subsequently derived independently by Joe and Hunt (1992). In broad outline, their derivation parallels the one given here. However, in their work, $\bar{\bar{\mu}}_s$ appears as the result of a nontrivial argument based on group theory, whereas here it is derived in a direct way as a straightforward redundancy factor in a matrix representation.

This author hopes that this somewhat pedestrian exposition of these ideas will help enlighten an elegant branch of numerical quadrature.

Acknowledgment

This work was supported in part by Office of Scientific Computing, U.S. Department of Energy, under Contract W-31-109-Eng-38.

References

Joe S. and Hunt D. C. (1992), "The Number of Lattice Rules Having Given Invariants," Bull. Australian Math. Soc. 46, pp. 479–495. See also Applied Mathematics Preprint AM91/44, UNSW.

Korobov N. M. (1959), "The Approximate Computation of Multiple Integrals" (Russian), Dokl. Akad. Nauk. SSSR 124, pp. 1207–1210.

Lyness J. N. and Keast P. (1991), "Application of the Smith Normal Form to the Structure of Lattice Rules," Preprint MCS-P269-0891, Mathematics and Computer Science Division, Argonne National Laboratory, Argonne, Ill.

Lyness J. N. and Sørevik T. (1989), "The Number of Lattice Rules," BIT 29, pp. 527–534.

Langtry T. N. (1991), "The Determination of Canonical Forms for Lattice Quadrature Rules," private communication.

Niederreiter H. (1992), "Random Number Generation and Quasi-Monte Carlo Methods ,"CBMS-NSF 63, SIAM, Philadelphia.

Sloan I. H. (1992), "Numerical Integration in High Dimensions—The Lattice Rule Approach," in Numerical Integration, T. O. Espelid and A. Genz (eds.), 55–69, Kluwer Academic Publishers, The Netherlands.

Sloan I. H. and Lyness J. N. (1989), "The Representation of Lattice Quadrature Rules as Multiple Sums," Math. Comput. 52, pp. 81–94.

Dr. J. N. Lyness
Mathematics and Computer Science Division
Argonne National Laboratory
9700 South Cass Avenue
Argonne, IL 60439
lyness@mcs.anl.gov

ERROR ESTIMATES OF PRODUCT QUADRATURE FORMULAE*

GIUSEPPE MASTROIANNI AND PÉTER VÉRTESI**

Abstract. The authors give new error estimates of product quadrature formulae in weighted-L_1 norm.

1. Introduction.

Let $v^{\alpha,\beta}(x) = (1-x)^\alpha(1+x)^\beta$, $\alpha, \beta > -1$ be a Jacobi weight and $\{p_m(x)\}_m$ the corresponding sequence of orthonormal polynomials with positive leading coefficient. We denote by $x_k = x_{m,k}(v^{\alpha,\beta})$, $k = 1, 2, ..., m$, the zeros of $p_m(x)$.

Let also $u(x)$ be a generalized Jacobi weight ($u \in GJ$), i.e., u can be written as follows,

$$u(x) = v^{\gamma,\delta}(x) \prod_{k=1}^{r} |x - t_k|^{\gamma_k}, \tag{1.1}$$

where $v^{\gamma,\delta}$ is a Jacobi weight, $\gamma_k > -1$, and $-1 < t_1 < ... < t_r < 1$.

We consider now the quadrature formula

$$\int_{-1}^{1} f(x)u(x)dx = \sum_{k=1}^{m} w_{m,k} f(x_k) + e_m(f, u), \tag{1.2}$$

where:

i) f is a locally absolutely continuous function ($f \in AC_{LOC}$), i.e., f is absolutely continuous in every interval $[c, d] \subset (-1, 1)$;

---

*1990 *Mathematics Subject Classification (1985 Revision).* 41A55.

Key words and phrases. Numerical Quadrature, Orthogonal Polynomials, Generalized Jacobi weights.

**This material is based upon work supported by the Ministero dell'Universitá e della Ricerca Scientifica e Tecnologica (first author) and by the Hungarian National Foundation for Scientific Research Grant N.1910 (second author).

ii) $w_{m,k} = \int_{-1}^{1} l_{m,k}(x)u(x)dx$ and $l_{m,k}$ is the k-th fundamental Lagrange polynomial associated with the zeros x_k, $k = 1, ..., m$ of $p_m(x)$;

iii) $e_m(f, u)$, defined by the equation (1.2), is the error of the quadrature formula.

The estimates of the error of (1.2) are relevant both in theory and applications.
When the coefficients $w_{m,k}$, $k = 1, ..., m$ are all positive, the estimates of $e_m(f, u)$, under suitable conditions on u, are reported in several papers (see for instance [1],[4],[8],[13],[14] and other contributions in these proceedings). Moreover when the quadrature rule is positive and the weight u is defined by (1.1) an estimate of $e_m(f, u)$ comes from a very recent result in [11]. In fact, using the following notations

$$\varphi(x) = \sqrt{1-x^2}, \quad u_m(x) = v^{\gamma,\delta}(x) \prod_{k=1}^{r} (|x - t_k| + m^{-1})^{\gamma_k}$$

$$I_m = [-1, 1] \backslash [x_1, x_{2m}], \quad \|g\|_{L_1([a,b])} = \int_a^b |g(x)|dx, \quad -1 \le a < b \le 1$$

and $\|g\|_1 = \int_{-1}^{1} |g(x)|dx$, the estimate

$$|e_{2m}(f, u)| \le C \left[\frac{\|f^{(r)}\varphi^r u_m\|_{L_1([x_1,x_{2m}])}}{m(m-1)...(m-r+1)} + \|f^{(r)}\varphi^{2r} u_m\|_{L_1(I_m)} \right] \tag{1.3}$$

holds for every $f^{(r-1)} \in AC_{LOC}$ and for some positive constant C not depending on m and f. On the contrary, if the coefficients $w_{m,k}$, $k = 1, ..., m$ are not all positive, there is no estimate similar to (1.3) available in literature. In the following we prove that, under suitable assumptions on u and $v^{\alpha,\beta}$, inequality (1.3) also holds when $w_{m,k}$ are not all positive. As an application of this result, we derive a bound on the coefficients $w_{m,k}$.

2. Main Results.

Let C denote a positive constant which can take different values in different formulae. $\mathcal{P}_m$ is the set of algebraic polynomials of degree less than or equal to m. C^0 and L_1 are the classes of continuous and integrable functions in $[-1, 1]$ respectively. (We use all the notations of the previous Section 1).

Theorem 2.1: *Let $f^{(r-1)} \in AC_{LOC}$ and $r \geq 1$. Assume $u = v^{\gamma,\delta}$. If $\varphi^{2r} f^{(r)} u \in L_1$ and $u, v^{\alpha,\beta}$ satisfy*

$$(2.1) \qquad \begin{array}{lcl} -1 < \gamma - \frac{\alpha}{2} - \frac{1}{4} < 0 & & -1/2 \leq \alpha, \beta < 1/2 \\ & \text{and} & \\ -1 < \delta - \frac{\beta}{2} - \frac{1}{4} < 0 & & -1 < \gamma, \delta < 0 \end{array}$$

then there exists $m_0 \in \Re^+$, such that for $m > m_0$ and $r = 1, \dots, m$,

$$(2.2) \qquad |e_{2m}(f,u)| \leq C \left[\frac{\|f^{(r)} \varphi^r u\|_{L_1([x_1, x_{2m}])}}{m(m-1)...(m-r+1)} + \|f^{(r)} \varphi^{2r} u\|_{L_1(I_m)} \right],$$

where C does not depend on m and f.

Remark 1. By replacing conditions (2.1) with $uv^{-\frac{\alpha}{2}-\frac{1}{4}, -\frac{\beta}{2}-\frac{1}{4}} \in L_1$, $\alpha, \beta, \gamma, \delta, \gamma_k > -1$,from a result in [12], we have, $\forall f \in C^0$, $|e_m(f,u)| \leq C E_{m-1}(f)_\infty$, where C is independent of m and f, $E_{m-1}(f)_\infty$ is the best uniform approximation error by algebraic polynomials.

Remark 2. In the special case $u\varphi^r f^{(r)} \in L_1$ and leaving any other condition unchanged, we have

$$(2.3) \qquad |e_{2m}(f,u)| \leq \frac{C \|f^{(r)} \varphi^r u\|_1}{m(m-1)...(m-r+1)}.$$

From the last inequality, it follows that

$$(2.4) \qquad |e_m(f,u)| \leq \frac{C}{m} E_{m-2}(f')_{\varphi u, 1},$$

where $E_{m-2}(f')_{\varphi u,1} = \mathrm{Inf}_{P \in \mathcal{P}_{m-2}} \|(f' - P) u \varphi\|_1$ and C is independent of m and f. Estimates of $E_{m-2}(f')_{\varphi u,1}$ for several classes of functions may be found in [5].

Remark 3. In Theorem 2.1 we suppose $u = v^{\gamma,\delta}$.

In the hypothesis that $u(x) = v^{\gamma,\delta}(x) \prod_{k=1}^r |t_k - x|^{\gamma_k}$ (see (1.1)), it is possible to prove a weaker estimate of (2.2).

In fact, setting

$$u_m^*(x) = v^{\gamma,\delta}(x) \prod_{k=1}^r (|t_k - x| + m^{-1})^{\gamma_k} \log \frac{e}{|t_k - x| + m^{-1}},$$

if conditions (2.1) are satisfied and $-1 < \gamma_k < 0$, then

$$|e_{2m}(f,u)| \leq C\left[\frac{\|f^{(r)}\varphi^r u_m^*\|_{L_1([x_1,x_{2m}])}}{m(m-1)...(m-r+1)} + \|f^{(r)}\varphi^{2r}u_m^*\|_{L_1(I_m)}\right].$$

A consequence of Theorem 2.1 is the following corollary.

Corollary 2.2: *If* u *and* $v^{\alpha,\beta}$ *satisfy* (2.1), *then the bound*

$$|w_{2m,k}| \leq C\frac{\sqrt{1-x_k^2}}{m}u(x_k) \tag{2.5}$$

holds for some C *independent of* k *and* m.

Remark 4. If we denote by $\lambda_{m,k}(u)$ the k-th coefficient of the Gaussian quadrature formula with weight u, it is well known [6] (see also [7]) that there exist two positive constants C_1 and C_2, independent of k, m, $m > m_0$, such that

$$C_1 \leq \frac{m\lambda_{m,k}(u)}{\sqrt{1-x_k^2}u(x_k)} \leq C_2.$$

From (2.5) and the last inequality, it then follows $|w_{m,k}| \leq C\lambda_{m,k}(w)$. Hence the coefficients as well as the error of such a quadrature formula behave like the coefficients and the error of the Gaussian rule. We do not know whether this holds for any product convergent formula or not, but this is an open problem.

3.Proofs.

We recall some useful preliminary results, that can be found in [12]. Let us denote by γ_m the leading coefficient of $p_m(x)$ and by $\overline{p}_m(x)$ the m-th Jacobi polynomial with respect to the weight function $v^{\alpha+1,\beta+1}$. The k-th fundamental Lagrange polynomial $l_{m,k}$, based on the zeros of $p_m(x)$, may be written as follows,

$$l_{m,k}(x) = \frac{\gamma_{m-1}}{\gamma_m}\lambda_k p_{m-1}(x_k)\frac{p_m(x)}{x-x_k},$$

where $\lambda_k = \lambda_m(v^{\alpha,\beta}, x_k)$ and

$$\lambda_m(v^{\alpha,\beta}, x) = \left(\sum_{j=0}^{m-1} p_j^2(x)\right)^{-1}$$

is the n-th Christoffel function corresponding to the weight function $v^{\alpha,\beta}$.

Since $p_{m-1}(x_k) = a_m(1-x_k^2)\overline{p}_{m-1}(x_k)$, and $a_m > 0$, $\lim_{m\to\infty} a_m = 1$, one gets

$$l_{m,k}(x) = \frac{\gamma_{m-1}}{\gamma_m} a_m \lambda_k (1-x_k^2)\overline{p}_{m-1}(x_k)\frac{p_m(x)}{x-x_k}, \qquad \frac{\gamma_{m-1}}{\gamma_m} a_m \le c.$$

We shall need in the sequel the following bounds

$$m\lambda_k \le C\sqrt{1-x_k^2}\, v^{\alpha,\beta}(x_k), \qquad k = 1, ..., m \tag{3.2}$$

$$|p_m(x)| \le C(\sqrt{1-x}+m^{-1})^{-\alpha-\frac{1}{2}}(\sqrt{1+x}+m^{-1})^{-\beta-\frac{1}{2}}, \quad |x| \le 1 \tag{3.3}$$

with C independent of m, x_k and x.

The following lemma is basic for the proof of Theorem 2.1, and can be used in the study of convergence of other interpolation processes.

Lemma 3.1. *If $u = v^{\gamma,\delta}$ and $v^{\alpha,\beta}$ satisfy the conditions*

$$\begin{aligned} &-1 < \gamma - \frac{\alpha}{2} - \frac{1}{4} < 0, && -1/2 \le \alpha, \beta < \frac{1}{2} \\ &&& \textit{and} \\ &-1 < \delta - \frac{\beta}{2} - \frac{1}{4} < 0 && -1 < \gamma, \delta < 0, \end{aligned} \tag{3.4}$$

then, for $t \in [-1+m^{-2}, 1-m^{-2}]$,

$$\left|\int_{-1}^{1} p_m(x)\frac{u(x)}{x-t}dx\right| \le C\frac{u(t)}{\sqrt{v^{\alpha,\beta}(t)\varphi(t)}}, \tag{3.5}$$

where the integral is to be understood in the sense of Cauchy principal value, $\varphi(x) = \sqrt{1-x^2}$ and C is independent of m and t.

Proof. If t varies in any closed subset of $(-1,1)$, then Lemma 3.1 follows from [2, th 2.3 p.195]. Then, it is sufficient to prove (3.5) when $t \in [-1+m^{-2}, -3/4]$, $\quad m > 2$. The case $[3/4, 1-m^{-2}]$ is similar. We have:

$$\int_{-1}^{1} p_m(x)\frac{v^{\gamma,\delta}(x)}{x-t}dx = \left\{\int_{-1}^{2t+1} + \int_{2t+1}^{0} + \int_{0}^{1}\right\}\frac{p_m(x)}{x-t}v^{\gamma,\delta}(x)dx =: I_1 + I_2 + I_3.$$

If $x \in [2t+1, 0]$, it risults:

$$x - t \ge \frac{1+x}{2}, \quad 1-x \ge 1 \quad \text{and} \quad \frac{|p_m(x)|}{x-t}v^{\gamma,\delta}(x) \le C(1+x)^{\delta-\frac{\beta}{2}-\frac{5}{4}}.$$

Then

$$|I_2| \le C \int_{2t+1}^{0} (1+x)^{\delta-\frac{\beta}{2}-\frac{5}{4}} dx \le C(1+t)^{\delta-\frac{\beta}{2}-\frac{1}{4}},$$

from this, since $\delta - \beta/2 - 1/4 < 0$, it holds

$$|I_2| \le C \frac{v^{\gamma,\delta}(t)}{\sqrt{v^{\alpha,\beta}(t)\varphi(t)}}. \tag{3.6}$$

If $t \in [-1+m^{-2}, -3/4]$ and $x \ge 0$, then $x - t > 3/4$ and

$$|I_3| \le \frac{4}{3} \int_0^1 |p_m(x)| v^{\gamma,\delta}(x) dx \le C \le C \frac{v^{\gamma,\delta}(t)}{\sqrt{v^{\alpha,\beta}(t)\varphi(t)}}, \tag{3.7}$$

since $-1 < \gamma - \alpha/2 - 1/4 < 0$ and $-1 < \delta - \beta/2 - 1/4 < 0$.
Moreover

$$\begin{aligned} I_1 &= \int_{-1}^{2t+1} p_m(x) \frac{v^{\gamma,\delta}(x) - v^{\gamma,\delta}(t)}{x-t} dx + \frac{v^{\gamma,\delta}(t)}{v^{\alpha,\beta}(t)} \left[\int_{-1}^{1} p_m(x) \frac{v^{\alpha,\beta}(x)}{x-t} dx + \right. \\ &\quad \left. + \int_{2t+1}^{1} p_m(x) \frac{v^{\alpha,\beta}(x)}{x-t} dx + \int_{-1}^{2t+1} p_m(x) \frac{v^{\alpha,\beta}(t) - v^{\alpha,\beta}(x)}{x-t} dx \right] =: \\ &=: A_1 + \frac{v^{\gamma,\delta}(t)}{v^{\alpha,\beta}(t)} [A_2 + A_3 + A_4]. \end{aligned} \tag{3.8}$$

Since $t \in [-1+m^{-2}, -3/4]$, $x \in [-1, 2t+1]$ and $-1/2 \le \alpha < 1/2$, we obtain:

$$\left| \frac{v^{\gamma,\delta}(x) - v^{\gamma,\delta}(t)}{x-t} \right| \le C(1-t)^{\gamma} \left[(1+x)^{\delta} + \left| \frac{(1+x)^{\delta} - (1+t)^{\delta}}{x-t} \right| \right],$$

and $|p_m(x)| \le C(1-t)^{-\alpha/2-1/4}(1+x)^{-\beta/2-1/4}$.
Then

$$\begin{aligned} |A_1| &\le C(1-t)^{\gamma-\frac{\alpha}{2}-\frac{1}{4}} \int_{-1}^{2t+1} \left((1+x)^{\delta-\frac{\beta}{2}-\frac{1}{4}} + (1+x)^{-\frac{\beta}{2}-\frac{1}{4}} \left| \frac{(1+x)^{\delta} - (1+t)^{\delta}}{x-t} \right| \right) dx = \\ &= C \frac{v^{\gamma,\delta}(t)}{\sqrt{v^{\alpha,\beta}(t)\varphi(t)}} \int_0^2 \left(y^{\delta-\frac{\beta}{2}-\frac{1}{4}} + y^{-\frac{\beta}{2}-\frac{1}{4}} \left| \frac{y^{\delta}-1}{y-1} \right| \right) dy. \end{aligned} \tag{3.9}$$

The estimate of $|A_2|$ is known [3, th. 2.1, p.20] and

$$|A_2| \le C v^{\frac{\alpha}{2}-\frac{1}{4}, \frac{\beta}{2}-\frac{1}{4}}(t), \qquad |t| \le 1 - m^{-2}. \tag{3.10}$$

Moreover

$$(3.11)\quad |A_3| \leq \left| \left(\int_{2t+1}^{0} + \int_{0}^{1} \right) p_m(x) \frac{v^{\alpha,\beta}(x)}{x-t} dx \right| \leq$$
$$\leq C \int_{2t+1}^{0} (1+x)^{\beta/2-5/4} dx + C \int_{0}^{1} |p_m(x)| v^{\alpha,\beta}(x) dx \leq C v^{\frac{\alpha}{2}-\frac{1}{4},\frac{\beta}{2}-\frac{1}{4}}(t),$$

since $\alpha/2 - 1/4 \leq 0$ and $\beta/2 - 1/4 \leq 0$.

Finally, for $x \in [-1, 2t+1]$ and $t \in [-1+m^{-2}, -3/4]$, we obtain

$$\left| \frac{v^{\alpha,\beta}(x) - v^{\alpha,\beta}(t)}{x-t} \right| \leq C(1-t)^{\alpha} \left((1+x)^{\beta} + \left| \frac{(1+x)^{\beta} - (1+t)^{\beta}}{x-t} \right| \right)$$

and $|p_m(x)| \leq C(1-t)^{-\alpha/-1/4}(1+x)^{-\beta/2-1/4}$. Therefore

$$|A_4| \leq C(1-t)^{\frac{\alpha}{2}-\frac{1}{4}} \int_{-1}^{2t+1} \left((1+x)^{\frac{\beta}{2}-\frac{1}{4}} + (1+x)^{-\frac{\beta}{2}-\frac{1}{4}} \left| \frac{(1+x)^{\beta} - (1+t)^{\beta}}{x-t} \right| \right) dx \leq$$
$$\leq C v^{\frac{\alpha}{2}-\frac{1}{4},\frac{\beta}{2}-\frac{1}{4}}(t) \int_{0}^{2} \left(y^{\beta/2-1/4} + y^{-\frac{\beta}{2}-\frac{1}{4}} \left| \frac{y^{\beta}-1}{y-1} \right| \right) dy.$$

Then, remembering (3.8)-(3.11), we obtain

$$I_1 \leq C \frac{v^{\gamma,\delta}(t)}{\sqrt{v^{\alpha,\beta}(t)\varphi(t)}}.$$

The Lemma follows from (3.6) and (3.7). □

We now need two results [see 12,p.675]. If $P \in \mathcal{P}_{lm}$ with l a fixed positive integer , then

$$(3.12)\qquad \sum_{k=1}^{m} \lambda_k |P(x_k)| \leq C \int_{-1}^{1} |P(x)| v^{\alpha,\beta}(x) dx,$$

and C does not depend on m.

Let $u \in GJ$ and $P \in \mathcal{P}_m$; then

$$(3.13)\qquad \|Pu\|_{L_1} \leq C \|Pu\|_{L_1(B_m)},$$

where $B_m = [x_1, x_m] \setminus \cup_{k=1}^{r} (t_k - m^{-1}, t_k + m^{-1})$ and C is independent on m.

Lemma 3.2: *If $u = v^{\gamma,\delta}$ and $v^{\alpha,\beta}$ satisfy* (2.1)*, then, for any polynomial $P \in \mathcal{P}_{lm}$,*

$$(3.14)\qquad \sum_{k=1}^{m} |w_{m,k}| |P(x_k)| \leq C \int_{-1}^{1} |P(t)| u(t) dt,$$

with C independent of m.

Proof. Let $S_m(v^{\alpha,\beta}, uv^{-\alpha,-\beta}, x)$ be the m-th partial sum of the Fourier series of the function $uv^{-\alpha,-\beta}$ in the system $\{p_m(x)\}_m$. Since $w_{m,k} = \lambda_k S_m(v^{\alpha,\beta}, uv^{-\alpha,-\beta}, x_k)$, then setting $g_m(x) = \text{sgn} S_m(v^{\alpha,\beta}, uv^{-\alpha,-\beta}, x)$, we have

$$|w_{m,k}| = w_{m,k} g_m(x_k), \quad \textit{and} \quad |g_m(x)| \le 1.$$

Recalling (3.1) and the definition of $w_{m,k}$ in Sect. 1, one gets

$$\sum_{k=1}^{m} |w_{m,k}||P(x_k)| = \sum_{k=1}^{m} w_{m,k} g_m(x_k)|P(x_k)| =$$

$$= \frac{\gamma_{m-1}}{\gamma_m} a_m \sum_{k=1}^{m} \lambda_k (1-x_k^2) \overline{p}_{m-1}(x_k) g_m(x_k)|P(x_k)| \int_{-1}^{1} p_m(x) \frac{u(x)}{x-x_k} dx$$

Setting $Q(t) = \int_{-1}^{1} \frac{p_m(x)-p_m(t)}{x-t} u(x)dx \in \mathcal{P}_{m-1}$ and using (3.12),

$$\sum_{k=1}^{m} |w_{m,k}||P(x_k)| \le C \sum_{k=1}^{m} \lambda_k (1-x_k^2)|\overline{p}_{m-1}(x_k)||Q(x_k)||P(x_k)| \le$$

$$\le C \int_{-1}^{1} v^{\alpha+1,\beta+1}(t)|\overline{p}_{m-1}(t)||Q(t)||P(t)|dt.$$

From the definition of $\overline{p}_{m-1}$, by virtue of (3.3) and (3.13), it follows:

$$\sum_{k=1}^{m} |w_{m,k}||P(x_k)| \le C \int_{B_m} |P(t)| \sqrt{v^{\alpha,\beta}(t)\varphi(t)} |Q(t)|dt. \tag{3.15}$$

Furthermore,

$$|Q(t)| \le \left| \int_{-1}^{1} p_m(x) \frac{u(x)}{x-t} dx \right| + \left| p_m(t) \int_{-1}^{1} \frac{u(x)}{x-t} dx \right|,$$

where the integrals are to be understood in the sense of Cauchy principal value. Now, since $t \in B_m$ by virtue of Lemma 3.1,

$$\left| \int_{-1}^{1} p_m(x) \frac{u(x)}{x-t} dx \right| \le C \frac{u(t)}{\sqrt{v^{\alpha,\beta}(t)\varphi(t)}}$$

and, from (3.2) and [3,Lemma 3.1,p.24], we obtain

$$\left| p_m(t) \int_{-1}^{1} \frac{u(x)}{x-t} dx \right| \le C \frac{u(t)}{\sqrt{v^{\alpha,\beta}(t)\varphi(t)}}$$

The lemma then follows from (3.15) together with the last two inequalities. □

Proof of Theorem 2.1. Let

$$\Gamma_t(x) = (x-t)_0^+ = \begin{cases} 1 & \text{if } x > t \\ 0 & \text{if } x \le t \end{cases} \qquad x, t \in [-1,1]$$

and let u be the weight function defined by (1.1); two polynomials $Q^{\pm} \in \mathcal{P}_{2m-2}$ exist such that $Q^- \le \Gamma_t \le Q^+$ and $\int_{-1}^1 (Q^+(x) - Q^-(x))u(x)dx = \lambda_m(u,t)$, where $\lambda_m(u,t)$ is the m-th Christoffel function corresponding to u(see [9,p.106]). Assume $f \in AC_{LOC}$ $(r=1)$. By virtue of a well-known result by Peano,

$$|e_{2m}(f,u)| \le \int_{-1}^1 |e_{2m}(\Gamma_t,u)||f'(t)|dt, \tag{3.16}$$

where $e_{2m}(\Gamma_t,u)$ is the so-called Peano kernel.

Furthermore, for $|t| \le 1$,

$$|e_{2m}(\Gamma_t,u)| = |e_{2m}(\Gamma_t - Q^-,u)| =$$

$$= \left| \int_{-1}^1 (\Gamma_t(x) - Q^-(x))u(x)dx - \sum_{k=1}^{2m-1} w_{2m,k}(\Gamma_t(x_k) - Q^-(x_k)) \right| = |A - B|.$$

Now,

$$0 < A \le \int_{-1}^1 (Q^+(x) - Q^-(t))u(x)dx = \lambda_m(u,t)$$

$$|B| \le \sum_{k=1}^{2m-1} |w_{2m,k}|(Q^+(x_k) - Q^-(x_k)) \le$$

$$\le C \int_{-1}^1 (Q^+(x) - Q^-(x))u(x)dx = C\lambda_m(u,t),$$

and the last inequality follows from Lemma 3.2.

Therefore,

$$|e_{2m}(\Gamma_t,u)| \le C\lambda_m(u,t), \qquad |t| \le 1.$$

If $t \in [x_{2m},1]$, then $e_{2m}(\Gamma_t,u) = \int_t^1 u(x)dx \le C(1-t^2)u(t)$. The same result follows if $t \in [-1,x_1]$. As a consequence,

$$|e_{2m}(\Gamma_t,u)| \le C \begin{cases} \lambda_m(u,t) & \text{if } x_1 \le t \le x_{2m} \\ (1-t^2)u(t) & \text{if } t \in [-1,x_1] \cup [x_{2m},1] = I_m. \end{cases}$$

Since [6], $m\lambda_m(u,t) \le C\sqrt{1-t^2}u(t), \quad x_1 \le t \le x_{2m},\ m > 2$, C independent of m , from inequality (3.16), it follows that

$$(3.18) \qquad |e_{2m}(f,u)| \le C\left[\frac{\|f'\varphi u\|_{L_1[x_1,x_{2m}]}}{m} + \|f'\varphi^2 u\|_{L_1(I_m)}\right].$$

Then Theorem 3.1 holds if $f \in AC_{LOC}$ $(r = 1)$.

If, in particular, $\varphi f' \in L_1$, from (3.18) we have

$$(3.19) \qquad |e_{2m}(f,u)| \le \frac{C}{m}\|f'\varphi u\|_1.$$

Let now $f' \in AC_{LOC}$. In virtue of Peano's theorem, we obtain

$$(3.20) \qquad |e_{2m}(f,u)| \le \int_{-1}^{1} |e_{2m}(\Gamma_{t,1},u)||f"(t)|dt,$$

where

$$\Gamma_{t,p} = \frac{(x-t)_+^p}{p!}, \quad p \ge 1, \quad \Gamma_{t,0} = \Gamma_t.$$

It easily follows that

$$(3.21) \qquad |e_{2m}(\Gamma_{t,1},u)| \le C(1-t^2)^2 u(t), \quad t \in [-1,x_1] \cup [x_{2m},1] = I_m$$

We now have to estimate $e_{2m}(\Gamma_{t,1},u)$, when $t \in [x_1,x_{2m}]$. Using again Freud's Lemma, two polynomials $Q_1^{\pm} \in \mathcal{P}_{2m-4}$ turn out to exist such that $Q_1^- \le \Gamma_t \le Q_1^+$ and

$$\int_{-1}^{1} (Q_1^+(x) - Q_1^-(x))u(x)\varphi(x)dx = \lambda_{m-1}(u\varphi,t).$$

Set $q_1^{\pm}(x) = \int_{-1}^{x} Q_1^{\pm}(y)dy$. We get

$$|e_{2m}(\Gamma_{t,1},u)| = |e_{2m}(\Gamma_{t,1} - q_1^-,u)|.$$

Since $\Gamma'_{t,1}\varphi = \Gamma_t\varphi \in L_1$, we may use relation (3.19) and

$$|e_{2m}(\Gamma_{t,1},u)| \le \frac{C}{m}\|(\Gamma_{t,1} - q_1^-)'\varphi u\|_1 =$$

$$\frac{C}{m}\|(\Gamma_t - Q_1^-)\varphi u\|_1 \le \frac{C}{m}\|(Q_1^+ - Q_1^-)\varphi u\|_1 = \frac{c}{m}\lambda_{m-1}(\varphi u,t).$$

Since $\varphi u \in GJ$, $\quad (m-1)(\lambda_{m-1}(\varphi u, t)) \leq C\varphi^2(t)u(t), \quad t \in [x_1, x_{2m}]$, we obtain

$$|e_{2m}(\Gamma_{t,1}, u)| \leq \frac{C}{m(m-1)}\varphi^2(t)u(t).$$

Recalling now (3.20) and (3.21), we have

$$|e_{2m}(f,u)| \leq C\left[\frac{\|f"\varphi^2 u\|_{L_1([x_1,x_{2m}])}}{m(m-1)} + \|f"\varphi^4 u\|_{L_1(I_m)}\right].$$

Finally the theorem, which certainly holds for $r = 2$, follows by repeating the above reasonings. □

Proof of Corollary 2.2.

Since

$$w_{2m,k} = \lim_{t \to x_k^-} e_{2m}(\Gamma_t, u) - \lim_{t \to x_k^+} e_{2m}(\Gamma_t, u),$$

and

$$|e_{2m}(\Gamma_t, u)| \leq c\lambda_m(u,t), \quad |t| \leq 1,$$

Corollary 2.2 follows.□

Acknowledgements. The authors wish to thank the referee for many helpful suggestions.

References.

[1] H. Braß: *Quadraturverfahren*, Vandenhoeck and Ruprecht, Göttingen (1977).

[2] G. Criscuolo and G. Mastroianni: A Bound for generalized Jacobi Functions of second kind, *Calcolo,* **XXIV, II,** (1987), 193-198.

[3] G. Criscuolo and G. Mastroianni and P. Nevai: Associated Generalized Jacobi Functions and Polynomials, *J. of Math. Anal. and Appl.* **158**, 1, (1991), 15-34.

[4] R.A. De Vore and R. Scott: Error Bounds for Gaussian Quadrature and Weighted-L_1 Polynomial Approximation, *SIAM J. Numer. Anal.*, **21**, 2, (1984), 400-412.

[5] Z. Ditzian, V. Totik: Moduli of smoothness S.C.M., *Springer-Verlag*, Berlin (1987).

[6] T. Erdélyi and P.Nevai: Generalized Jacobi weights, Christoffel functions and zeros of orthogonal polynomials, *J.Approx.Theory* , (to appear).

[7] T. Erdélyi: Remez-type inequalities and their applications (preprint).

[8] K.J. Förster, K. Petras: On a Problem Proposed by H. Braß Concerning the Remainder Term in Quadrature for Convex Functions, *Numer. Math.* **57**, (1990), 763-777.

[9] G. Freud: *Orthogonal Polynomials*, AKADÉMIAI KIADÓ, Budapest (1971).

[10] D. Lubinski and P. Nevai: Markov-Bernstein Inequalities Revised, *Approx. Theory and its Appl.*, **314**, (1987), 98-118.

[11] G. Mastroianni: Christoffel functions and error of positive quadrature formulae (manuscript).

[12] P. Nevai: Mean Convergence of Lagrange Interpolation III, *Trans. Amer. Math. Soc.*, **282**, (1984), 669-698.

[13] K. Petras: Asymptotic Behaviour of Peano Kernel of Fixed Order, Numerical Integration III (H.Braß, G.Hämmerlin eds.), ISNM 85, Birkhäuser Verlag, Basel (1988), 186-198.

[14] K. Petras: Error Bounds for Gaussian and Related Quadrature and Application to R-Convex Functions, *SIAM J. Numer. Anal.*, **29**, 2, (1992), 578-585.

Giuseppe Mastroianni
Dipartimento di Matematica,
Universitá degli Studi della Basilicata
Via N.Sauro 85,
85100 Potenza - Italy
Péter Vértesi
Mathematical Institute
of the Hungarian Academy of Sciences
H-1364 Budapest P.O.B. 127 - Hungary

International Series of Numerical Mathematics, Vol. 112, © 1993 Birkhäuser Verlag Basel

Quasi-Monte Carlo Methods with Modified Vertex Weights

Harald Niederreiter and Ian H. Sloan

Abstract. Quasi-Monte Carlo methods for integration over the s-dimensional unit cube in which only one vertex is a quadrature point are here modified by distributing the corresponding weight between all the vertices. One particular rule of this kind, namely the one that integrates all multilinear functions exactly, is shown to be optimal, in the sense that among all vertex-modified rules that integrate constants exactly it yields the least value of an L_2 version of the discrepancy. The sum of the squares of the vertex weights in that rule, the "vertex variance", is shown to be related to the quality of the underlying quasi-Monte Carlo method. Numerical experiments confirm in a dramatic way the important role played by the vertex variance, even for the unmodified quasi-Monte Carlo method.

1. Introduction

In this paper we consider modifications of quasi-Monte Carlo rules for approximating

$$If = \int_{[0,1]^s} f(\mathbf{x})d\mathbf{x}, \tag{1.1}$$

with $s \geq 2$. The quasi-Monte Carlo rules we consider are of the form

$$Q_N f = \frac{1}{N}\sum_{j=0}^{N-1} f(\mathbf{x}_j), \tag{1.2}$$

where $\mathbf{x}_0 = \mathbf{0}$, and it is assumed that all other quadrature points $\mathbf{x}_1, \ldots, \mathbf{x}_{N-1}$ lie in the half-open unit cube $[0,1)^s$. For an introduction to quasi-Monte Carlo methods in general, see Niederreiter (1992).

One application we have in mind is to lattice rules. (For recent reviews of lattice rules, see Niederreiter 1992 and Sloan 1992.) In the usual presentations of lattice rules the integrand f is assumed to have a continuous extension to $\mathbf{R}^s$ which is 1-periodic with respect to each component of $\mathbf{x}$. No such assumption is made here, so one result of this paper is to widen the scope of lattice rules. In the applications to lattice rules we shall assume that the quadrature points $\mathbf{x}_1, \ldots, \mathbf{x}_{N-1}$ all lie in the interior of the unit cube, since otherwise there are theoretical reasons (see Price and Sloan 1992, Niederreiter and Sloan 1993) for modifying the weights at the boundary points too.

In this paper we modify the quasi-Monte Carlo method (1.2) by allowing all the vertices of the cube $[0,1]^s$ to be used as quadrature points. Thus we consider rules of the modified form

$$M_N f = \sum_{i_1,\ldots,i_s=0}^{1} w_{i_1\ldots i_s} f(i_1,\ldots,i_s) + \frac{1}{N}\sum_{j=1}^{N-1} f(\mathbf{x}_j), \tag{1.3}$$

where the vertex weights $w_{i_1\ldots i_s}$ are for the moment free, except for the condition that the sum of all vertex weights be $1/N$. One special case is $Q_N f$. Another, to which we shall pay some attention in this paper, is the symmetric choice

$$T_N f = \frac{1}{2^s N}\sum_{i_1,\ldots,i_s=0}^{1} f(i_1,\ldots,i_s) + \frac{1}{N}\sum_{j=1}^{N-1} f(\mathbf{x}_j). \tag{1.4}$$

(The symbol T may suggest the trapezoidal rule, which is included as a special case in (1.4) if $s = 1$.) In an earlier paper (Niederreiter and Sloan 1993) we showed that the rule T_N performs very well for certain 2-dimensional lattice rules based on Fibonacci numbers.

Another special case of (1.3) of interest to us in this paper is the rule

$$\overline{M}_N f = \sum_{i_1,\ldots,i_s=0}^{1} \overline{w}_{i_1\ldots i_s} f(i_1,\ldots,i_s) + \frac{1}{N}\sum_{j=1}^{N-1} f(\mathbf{x}_j), \tag{1.5}$$

in which the vertex weights $\overline{w}_{i_1\ldots i_s}$ are prescribed by requiring that

$$\overline{M}_N f = If \quad \text{if } f(\mathbf{x}) = x_1^{k_1}\ldots x_s^{k_s}, k_r = 0 \text{ or } 1, 1 \le r \le s. \tag{1.6}$$

In other words, the requirement is that $\overline{M}_N$ must integrate every multilinear polynomial exactly, i.e. every polynomial that is of degree at most 1 in each of the s variables. In Section 2 we shall show that the vertex weights in (1.5) can easily be calculated by an explicit formula.

Section 3 contains the first theoretical results of the paper. There we show first that all rules of the general form (1.3) satisfy an error bound of the form

$$|M_N f - If| \le t(M_N)V(f),$$

where $V(f)$, which depends only on f, is an L_2 measure of the variation of f, and $t(M_N)$ is a simple generalization of the L_2 discrepancy which is independent of f. Then we show, in Theorem 2, that among all rules of the form (1.3), the unique rule that minimises $t(M_N)$ is the rule $\overline{M}_N$ defined in (1.5) and (1.6). There is therefore justification in referring to $\overline{M}_N$ as the *optimal* vertex-modified rule. For $s = 2$ Theorems 1 and 2 are special cases of results obtained by Ritter (1970) by the use of multidimensional Peano-Sard theory, see Sard (1963). The proofs here use multidimensional integration by parts.

A significant quantity for any quadrature rule is the "variance", defined as the sum of the squares of the weights. (For an introduction to variance in the context of 1-dimensional rules of prescribed polynomial degree, see Förster 1988.) Here we restrict attention to the vertex weights (since all other weights are equal and unchanged), and so we define the *vertex variance* to be

$$v(M_N) = \sum_{i_1,\ldots,i_s=0}^{1} w_{i_1\ldots i_s}^2. \tag{1.7}$$

For example, the original rule Q_N has vertex variance $1/N^2$, while the symmetric rule T_N has the least possible vertex variance, namely $2^{-s}/N^2$.

For the particular case of the optimal rule $\overline{M}_N$ we write the vertex variance as

$$\overline{v}_N = v(\overline{M}_N) = \sum_{i_1,\ldots,i_s=0}^{1} \overline{w}_{i_1\ldots i_s}^2. \tag{1.8}$$

Clearly, the rule $\overline{M}_N$ will become unattractive if the vertex variance is large. (Later we shall see examples in which this does happen.) With this motivation, we prove in Section 4 that rules Q_N exist which are not only "good" in the usual sense of quasi-Monte Carlo methods, but which also yield satisfactorily "small" values of $\overline{v}_N$.

In Section 5 we show that the vertex variance $\overline{v}_N$ is related to the quality of the symmetric rule T_N: large values of $\overline{v}_N$ correspond to poor performance of T_N. Thus, the problems associated with large $\overline{v}_N$ cannot be overcome merely by changing to the rule T_N with least variance among all vertex-modified rules. Rather, we would recommend that rules with small $\overline{v}_N$ should be sought.

Finally, in Section 6 the constructions of this paper are applied to a package of 8-dimensional integrals. The results are startling. In brief, lattice rules do exist which are of good quality in a conventional sense for periodic integrands and which also have moderate values of $\overline{v}_N$, and these perform well, in conformity with the theory of this paper. But there also exist lattice rules which are excellent in the conventional sense but which have very large values of $\overline{v}_N$, and whose general performance is remarkably poor. The importance of controlling the vertex variance $\overline{v}_N$, whichever of the rules Q_N, T_N or $\overline{M}_N$ may be used, is therefore strikingly confirmed.

2. Weights for the rule $\overline{M}_N$

By definition, the rule $\overline{M}_N$ integrates all multilinear functions exactly. A convenient basis for the space of multilinear functions in s variables is provided by the 2^s functions defined by

$$l_{i_1\ldots i_s}(x_1,\ldots,x_s) = \prod_{r=1}^{s}(1-i_r-(-1)^{i_r}x_r),\ i_r \in \{0,1\}, 1 \le r \le s. \tag{2.1}$$

For example, for $s = 2$ the 4 basis functions are

$$l_{00}(x,y) = (1-x)(1-y),\ l_{10}(x,y) = x(1-y),\ l_{01}(x,y) = (1-x)y,\ l_{11}(x,y) = xy.$$

The particular virtue of this basis is that each member has the value 1 at one vertex and vanishes at all other vertices: we have

$$l_{i_1\dots i_s}(j_1,\dots,j_s) = \delta_{i_1j_1}\delta_{i_2j_2}\dots\delta_{i_sj_s}. \tag{2.2}$$

The rule $\overline{M}_N$ is exact for each member of this basis, so we have for $i_1,\dots,i_s = 0$ or 1,

$$\sum_{j_1,\dots,j_s=0}^{1} \overline{w}_{j_1\dots j_s} l_{i_1\dots i_s}(j_1,\dots,j_s) + \frac{1}{N}\sum_{j=1}^{N-1} l_{i_1\dots i_s}(\mathbf{x}_j) = \frac{1}{2^s},$$

or, using (2.2),

$$\overline{w}_{i_1\dots i_s} = \frac{1}{2^s} - \frac{1}{N}\sum_{j=1}^{N-1} l_{i_1\dots i_s}(\mathbf{x}_j). \tag{2.3}$$

This argument, in addition to giving us an easily computable expression for each vertex weight, also establishes in a direct way that the rule $\overline{M}_N$ is uniquely determined by the requirement that it integrate each multilinear function exactly.

An interesting interpretation of the vertex weights in $\overline{M}_N$ is available: from (2.2), (2.3) and the definitions of Q_N and T_N we have

$$\overline{w}_{i_1\dots i_s} - \frac{1}{N}\delta_{i_10}\dots\delta_{i_s0} = (I - Q_N)l_{i_1\dots i_s}, \tag{2.4}$$

$$\overline{w}_{i_1\dots i_s} - \frac{1}{2^sN} = (I - T_N)l_{i_1\dots i_s}. \tag{2.5}$$

That is, the departure of the weight $\overline{w}_{i_1\dots i_s}$ from the corresponding weight in the rule Q_N or T_N is just the quadrature error for the rule Q_N or T_N, respectively, applied to the multilinear function $l_{i_1\dots i_s}$.

3. Error bounds and the optimality property of $\overline{M}_N$

We first obtain an error bound for the general vertex-modified rule M_N by appropriately modifying a standard error bound for quasi-Monte Carlo methods. The bound makes use of the "local discrepancy"

$$g(x_1,\dots,x_s) = \frac{1}{N}A([0,x_1)\times\dots\times[0,x_s)) - x_1\dots x_s, \tag{3.1}$$

where $A([0,x_1)\times\dots\times[0,x_s))$ is the number of points of the rule Q_N that lie in the interval $[0,x_1)\times\dots\times[0,x_s)$. For a nonempty subset J of $\{1,2,\dots,s\}$ let $(\mathbf{x}_J,\mathbf{1})$ denote the vector $(v_1,\dots,v_s)$ with $v_j = x_j$ if $j\in J$ and $v_j = 1$ otherwise. If $J = \{j_1,\dots,j_r\}$, then we abbreviate the partial derivative $\partial^r f/\partial x_{j_1}\dots\partial x_{j_r} = f_{x_{j_1}\dots x_{j_r}}$ by f_J.

Theorem 1. *Assume that the partial derivative $f_{x_1\dots x_s}$ is continuous on $[0,1]^s$. Then*

$$|M_Nf - If| \le t(M_N)V(f), \tag{3.2}$$

where

$$V(f) = \left(\sum_J \int_0^1 \cdots \int_0^1 f_J(\mathbf{x}_J, \mathbf{1})^2 d\mathbf{x}_J\right)^{1/2}, \tag{3.3}$$

$$t(M_N) = \left(\sum_J \int_0^1 \cdots \int_0^1 (g(\mathbf{x}_J, \mathbf{1}) - c_J)^2 d\mathbf{x}_J\right)^{1/2}, \tag{3.4}$$

and both sums are over all nonempty subsets J *of* $\{1, 2, \ldots, s\}$. *The constants* c_J *are defined by*

$$c_J = \frac{1}{N} - \sum_{i_1=0}^{u_1(J)} \cdots \sum_{i_s=0}^{u_s(J)} w_{i_1 \ldots i_s} \tag{3.5}$$

with $u_k(J) = 0$ *if* $k \in J$ *and* $u_k(J) = 1$ *otherwise.*

Proof. We give the proof for the case $s = 2$. The proof for general s may be obtained in a similar way with (3.6) below replaced by the general formula in Zaremba (1968). By integration by parts, Zaremba (1968) shows that

$$Q_N f - If = -\int_0^1 g(x,1)df(x,1) - \int_0^1 g(1,y)df(1,y) + \int_0^1 \int_0^1 g(x,y)df(x,y). \tag{3.6}$$

Now for $s = 2$ the constants given by (3.5) are, with the simpler notation c_1 for $c_{\{1\}}$, etc., related to the weights by

$$c_1 = \frac{1}{N} - w_{00} - w_{01},\ c_2 = \frac{1}{N} - w_{00} - w_{10},\ c_{12} = \frac{1}{N} - w_{00},$$

and on solving for the weights,

$$w_{00} = \frac{1}{N} - c_{12},\ w_{10} = c_{12} - c_2,\ w_{01} = c_{12} - c_1,$$

$$w_{11} = \frac{1}{N} - w_{00} - w_{10} - w_{01} = -c_{12} + c_1 + c_2.$$

Thus

$$\begin{aligned} M_N f - Q_N f &= (w_{00} - \frac{1}{N})f(0,0) + w_{10}f(1,0) + w_{01}f(0,1) + w_{11}f(1,1) \\ &= c_1(f(1,1) - f(0,1)) + c_2(f(1,1) - f(1,0)) - c_{12}(f(1,1) - f(1,0) - f(0,1) + f(0,0)) \\ &= c_1 \int_0^1 df(x,1) + c_2 \int_0^1 df(1,y) - c_{12} \int_0^1 \int_0^1 df(x,y), \end{aligned}$$

and so, on combining with (3.6),

$$\begin{aligned} M_N f - If = & - \int_0^1 (g(x,1) - c_1)df(x,1) - \int_0^1 (g(1,y) - c_2)df(1,y) \\ & + \int_0^1 \int_0^1 (g(x,y) - c_{12})df(x,y), \end{aligned}$$

or, since the partial derivative f_{xy} is continuous,

$$\begin{aligned} M_N f - If = & -\int_0^1 (g(x,1)-c_1)f_x(x,1)dx - \int_0^1 (g(1,y)-c_2)f_y(1,y)dy \\ & + \int_0^1\int_0^1 (g(x,y)-c_{12})f_{xy}(x,y)dxdy. \end{aligned} \tag{3.7}$$

By applying the Cauchy-Schwarz inequality first to each term and then to the sum of 3 terms, we obtain

$$\begin{aligned} |M_N f - If| \le & \Big(\int_0^1 (g(x,1)-c_1)^2 dx + \int_0^1 (g(1,y)-c_2)^2 dy \\ & + \int_0^1\int_0^1 (g(x,y)-c_{12})^2 dxdy\Big)^{1/2} \cdot \\ & \cdot \Big(\int_0^1 f_x(x,1)^2 dx + \int_0^1 f_y(1,y)^2 dy + \int_0^1\int_0^1 f_{xy}(x,y)^2 dxdy\Big)^{1/2}, \end{aligned}$$

which is the inequality in the theorem for the special case $s = 2$. □

With the particular choice $M_N = Q_N$ every number c_J vanishes, and the bound (3.2) reduces to one given by Zaremba (1968). In that case, $t(Q_N)$ is the version of the L_2 discrepancy of the point set $\mathbf{x}_0, \ldots, \mathbf{x}_{N-1}$ which incorporates the projections on lower-dimensional faces of $[0,1]^s$.

Theorem 2. *Let Q_N be a given quasi-Monte Carlo rule in s dimensions. Among all the vertex-modified rules M_N, the unique rule that minimises $t(M_N)$, defined by* (3.4), *is the rule $\overline{M}_N$.*

Proof. In the typical term of (3.4) the unique choice of c_J that minimises

$$\int_0^1 \cdots \int_0^1 (g(\mathbf{x}_J, \mathbf{1}) - c_J)^2 d\mathbf{x}_J$$

is

$$\overline{c}_J = \int_0^1 \cdots \int_0^1 g(\mathbf{x}_J, \mathbf{1}) d\mathbf{x}_J. \tag{3.8}$$

We now show that with this choice the rule M_N integrates all multilinear functions exactly. If f is such a function, then each partial derivative in (3.7) or its higher-dimensional generalisation is a constant, and so may be taken outside of the corresponding integral. With the choice (3.8) each integral in (3.7) or its generalisation then vanishes, so $M_N f = If$. Since $\overline{M}_N$ is the only vertex-modified rule that integrates all multilinear functions exactly, we see that $\overline{M}_N$ is the unique rule that minimises $t(M_N)$. □

4. Existence of rules with small vertex variance $\overline{v}_N$

We recall that for a rule Q_N given by (1.2) the corresponding optimal vertex-modified rule $\overline{M}_N$ has vertex variance

$$\overline{v}_N = v(\overline{M}_N) = \sum_{i_1,\ldots,i_s=0}^{1} \overline{w}_{i_1\ldots i_s}^2.$$

The rule $\overline{M}_N$ is specified by the requirement of integrating all multilinear polynomials exactly, a condition that need not necessarily imply a small vertex variance $\overline{v}_N$. However, we shall prove in this section that there exist acceptable choices for the rule Q_N such that the vertex variance $\overline{v}_N$ is small.

We employ a standard quantity from the general theory of quasi-Monte Carlo methods, namely the *star discrepancy* of the nodes $\mathbf{x}_0, \ldots, \mathbf{x}_{N-1}$ of Q_N which is defined by

$$D_N^*(\mathbf{x}_0, \ldots, \mathbf{x}_{N-1}) = \sup_{\mathbf{x} \in [0,1]^s} |g(\mathbf{x})|,$$

where $g(\mathbf{x}) = g(x_1, \ldots, x_s)$ is as in (3.1).

Theorem 3. *We have*

$$\overline{v}_N \leq K_s D_N^*(\mathbf{x}_0, \ldots, \mathbf{x}_{N-1})^2 \tag{4.1}$$

with a constant K_s depending only on the dimension s.

Proof. It follows from (2.4) that

$$\overline{w}_{i_1 \ldots i_s} = (I - Q_N) l_{i_1 \ldots i_s} + \frac{1}{N} \delta_{i_1 0} \ldots \delta_{i_s 0}.$$

Now we apply Theorem 1. If $i_1, \ldots, i_s$ are not all 0, then this yields

$$|\overline{w}_{i_1 \ldots i_s}| \leq (2^s - 1) D_N^*(\mathbf{x}_0, \ldots, \mathbf{x}_{N-1}),$$

and if $i_1 = \ldots = i_s = 0$, then we get

$$|\overline{w}_{0 \ldots 0}| \leq (2s - 1)^{1/2} D_N^*(\mathbf{x}_0, \ldots, \mathbf{x}_{N-1}) + \frac{1}{N}.$$

Since $\mathbf{x}_0 = \mathbf{0}$, we have

$$D_N^*(\mathbf{x}_0, \ldots, \mathbf{x}_{N-1}) \geq \overline{\lim}_{n \to \infty} |g(\frac{1}{n}, \ldots, \frac{1}{n})| \geq \frac{1}{N}.$$

Thus we obtain, for $s \geq 2$,

$$|\overline{w}_{i_1 \ldots i_s}| \leq (2^s - 1) D_N^*(\mathbf{x}_0, \ldots, \mathbf{x}_{N-1})$$

for all $i_1, \ldots, i_s$, which implies (4.1). □

It is well known (see Chapter 3 in Niederreiter 1992) that for any dimension s and for any prescribed $N \geq 2$ there exist points $\mathbf{x}_0 = \mathbf{0}, \mathbf{x}_1, \ldots, \mathbf{x}_{N-1} \in [0,1)^s$ such that

$$D_N^*(\mathbf{x}_0, \ldots, \mathbf{x}_{N-1}) = O(N^{-1} (\log N)^{s-1}), \tag{4.2}$$

where here and in the following the implied constants in the Landau symbols depend only on s. It follows then from Theorem 3 that for any s and N there exists a rule Q_N with

$$\overline{v}_N = O(N^{-2} (\log N)^{2s-2}).$$

We remark that in view of the Koksma-Hlawka inequality (see, for example, Niederreiter 1992, Theorem 2.11) a rule Q_N satisfying (4.2) yields small integration errors for all integrands of bounded variation in the sense of Hardy and Krause. It should be noted that even among lattice rules we can find small values of $\overline{v}_N$; according to a result on p. 115 of Niederreiter (1992) there exists, for any s and N, a lattice rule Q_N such that

$$D_N^*(\mathbf{x}_0, \ldots, \mathbf{x}_{N-1}) = O(N^{-1}(\log N)^s),$$

and so

$$\overline{v}_N = O(N^{-2}(\log N)^{2s}).$$

5. The quality of the rule T_N is related to $\overline{v}_N$

We see in this section that the departure of $\overline{v}_N$ from its minimum value $1/(2^s N^2)$ is a measure of the failure of the symmetric rule T_N. First, from (1.8) and (2.5) we have

$$\begin{aligned} \overline{v}_N - \frac{1}{2^s N^2} &= \sum_{i_1,\ldots,i_s=0}^{1} \overline{w}^2_{i_1\ldots i_s} - \frac{1}{2^s N^2} \\ &= \sum_{i_1,\ldots,i_s=0}^{1} \left(\overline{w}_{i_1\ldots i_s} - \frac{1}{2^s N}\right)^2 \\ &= \sum_{i_1,\ldots,i_s=0}^{1} \left((I - T_N) l_{i_1\ldots i_s}\right)^2, \end{aligned} \tag{5.1}$$

in which the right-hand side is just the sum of the squares of the quadrature errors for T_N applied to the typical integrands provided by our basis of multilinear functions.

A deeper relation is given by Theorem 4 below, which connects $\overline{v}_N - 1/(2^s N^2)$ to the L_2 discrepancy bound for the rule T_N. First we prove the following technical lemma. The notation $\sim$ here and in the following indicates that there exist positive constants C_1 and C_2 depending only on s such that the ratio of the left to the right sides is bounded below by C_1 and above by C_2.

Lemma 1. *If $a_{i_1\ldots i_s}$, for $i_1, \ldots, i_s = 0$ or 1, are real numbers, not all 0, whose sum is 0, then with the notation in Theorem 1 we have*

$$\sum_J \left(\sum_{i_1=0}^{u_1(J)} \cdots \sum_{i_s=0}^{u_s(J)} a_{i_1\ldots i_s} \right)^2 \sim \sum_{i_1,\ldots,i_s=0}^{1} a^2_{i_1\ldots i_s}.$$

Proof. We give the proof for the case $s = 2$; the result for higher dimensions follows in a similar way. For $s = 2$ the result to be proved is

$$(a_{00} + a_{01})^2 + (a_{00} + a_{10})^2 + a_{00}^2 \sim a_{00}^2 + a_{01}^2 + a_{10}^2 + a_{11}^2,$$

or, since $a_{00} + a_{01} + a_{10} + a_{11} = 0$,

$$(a_{00} + a_{01})^2 + (a_{00} + a_{10})^2 + a_{00}^2 \sim a_{00}^2 + a_{01}^2 + a_{10}^2 + (a_{00} + a_{01} + a_{10})^2.$$

The ratio of the left side to the right is a rational function which has no zeros or poles in $\mathbf{R}^3$ or at infinity, thus there exist positive constants C_1 and C_2 bounding it below and above. □

Theorem 4. *There exist positive constants C_1 and C_2 depending only on s such that*

$$C_1\left(\overline{v}_N - \frac{1}{2^s N^2}\right) \le t(T_N)^2 - t(\overline{M}_N)^2 \le C_2\left(\overline{v}_N - \frac{1}{2^s N^2}\right).$$

Proof. It follows from (3.4) and (3.8) that for any vertex-modified rule M_N we have

$$t(M_N)^2 - t(\overline{M}_N)^2 = \sum_J (\overline{c}_J - c_J)^2,$$

or with the aid of (3.5) and Lemma 1,

$$t(M_N)^2 - t(\overline{M}_N)^2 = \sum_J \left(\sum_{i_1=0}^{u_1(J)} \cdots \sum_{i_s=0}^{u_s(J)} (\overline{w}_{i_1\ldots i_s} - w_{i_1\ldots i_s})\right)^2 \sim \sum_{i_1,\ldots,i_s=0}^{1} (\overline{w}_{i_1\ldots i_s} - w_{i_1\ldots i_s})^2.$$

Choosing now $M_N = T_N$, so that $w_{i_1\ldots i_s} = 1/(2^s N)$, and using (5.1), the right-hand side becomes $\overline{v}_N - 1/(2^s N^2)$. □

6. Numerical experiments with lattice rules

Except in 2 dimensions, there are no known effective lattice-rule constructions, so one must resort to a search over some class of rules. In the context of periodic integrands it is usual to minimise a quantity such as $P_\alpha = Q_N f_\alpha - 1$ for some $\alpha > 1$, where

$$f_\alpha(\mathbf{x}) = \prod_{r=1}^{s}\left(1 + 2\sum_{h=1}^{\infty} \frac{\cos 2\pi h x_r}{h^\alpha}\right). \tag{6.1}$$

The motivation is that (see Sloan and Kachoyan 1987, Section 7) P_α is the maximum quadrature error in the class of functions f whose Fourier coefficients satisfy

$$|\hat{f}(\mathbf{h})| \le (\overline{h}_1 \ldots \overline{h}_s)^{-\alpha} \quad \text{for all } \mathbf{h} = (h_1, \ldots, h_s) \in \mathbf{Z}^s,$$

with $\overline{h} = \max(1, |h|)$. If α is an even integer, then the quantity in parentheses in (6.1) is expressible in terms of Bernoulli polynomials; for example if $\alpha = 2$, then

$$1 + 2\sum_{h=1}^{\infty} \frac{\cos 2\pi h x}{h^2} = 1 + 2\pi^2(x^2 - x + \frac{1}{6}),\ 0 \le x \le 1.$$

In the present work we seek a rule for which P_2 has a small value (since among the functions we want to be able to handle are those which have a continuous but not smooth periodic extension), and for which also the vertex variance $\overline{v}_N$ has a small value. The preceding theory gives us three motivations for worrying about the vertex variance $\overline{v}_N$. First, the rule $\overline{M}_N$ lacks robustness if $\overline{v}_N$ is large. Second, by Theorem 4 the L_2 discrepancy of the rule T_N becomes large if $\overline{v}_N$ is large. And third, the individual vertex weights in $\overline{M}_N$ are closely related to the quadrature errors of the rules Q_N and T_N for the very typical nonperiodic integrands $l_{i_1 \ldots i_s}$ by (2.4) and (2.5).

As a numerical example we consider 8-dimensional quadrature for a fixed prime number N of points and search for good rules over the class of all lattice rules of the form

$$Q_N f = \frac{1}{N} \sum_{j=0}^{N-1} f\left(\left\{\frac{j}{N}(1, z, z^2, \ldots, z^7)\right\}\right),$$

with $1 \leq z \leq \lfloor N/2 \rfloor$. In Table 1 we show some "good" rules obtained in this search, together with values of $P_2, \overline{v}_N$ and other statistics related to the vertex weights. The first rule has the best value of P_2 out of all those investigated in the search, but has a strikingly bad value of $\overline{v}_N$. It also has some large negative weights, since $\sum |\overline{w}| = 1201/N$, compared to $\sum \overline{w} = 1/N$. The second rule has a slightly larger value of P_2 but a much better value of $\overline{v}_N$. The third rule has the smallest value of $\overline{v}_N$ out of all those examined in the search, but a somewhat larger value of P_2. The fourth rule has a very respectable value of $\overline{v}_N$ but a remarkably large value of P_2 (lying in the worst 5% of P_2 values). Possibly Rule 2 is an acceptable compromise. (In forming that view, we looked also at the maximum absolute value of the vertex weights $\overline{w}_{i_1 \ldots i_s}$: given that the interior points have an associated weight of $1/N$, we take the view that a maximum absolute value of $2/N$ might be acceptable, while one of $20/N$ is not.)

In Table 2 these four rules are applied to the Genz package of test integrals (Genz 1984 and 1987). In that package there are six classes of integrands, with the principal characteristic of each class identified by the descriptor in the table. Each integrand has two kinds of parameters, the first determining the degree of difficulty, the second the location of peaks etc. We followed Genz's recommendations for the "difficulty" parameters, except that for the "corner peak" family we have used the "easier" value of $h = 100$ instead of Genz's recommended value of $h = 600$ because with the original value all rules found the family too difficult to give useful information. To obtain the results in the table, 20 integrands from each family were sampled. The last row of the table gives the average of the median number of significant digits over all six families.

Several conclusions might be drawn from Table 2. The most striking is that whichever of the rules Q_N, T_N or $\overline{M}_N$ may be used, it seems that it really does pay to worry about the size of the vertex variance $\overline{v}_N$: for Rule 1 with a good value of P_2 but a poor value of $\overline{v}_N$ performs consistently poorly on each of the six families of integrands.

The second conclusion is that it is probably worthwhile to pay attention also to the size of P_2, in that Rule 4 performs worse on the whole than Rule 2: since the vertex variance of Rule 4 is smaller than that of Rule 2, it is reasonable to attribute this deterioration to the size of P_2. Yet it is remarkable that the effect is so mild, given that P_2 is 3 times larger.

Third, it may be fair to conclude that the performance of the version $\overline{M}_N$ is generally better than that of T_N, particularly in the case of Rule 1, for which $\overline{v}_N$ is large (so this is consistent with Theorem 4), except that $\overline{M}_N$ fails badly with Rule 1 for the case of the "corner peak". The latter failure is easily explained in terms of the lack of robustness of $\overline{M}_N$ when $\overline{v}_N$ is large: very large errors can then result when the integrand has a large value at a vertex, as with the "corner peak" functions.

Fourth, it might reasonably be observed that the superiority of the rule $\overline{M}_N$ over the rule T_N is marginal when the vertex variance is moderate, as in Rules 2 and 3, and that its use in other cases can be dangerous. So our tentative recommendation would be to seek a rule for which the vertex variance $\overline{v}_N$ is small, but then in practical applications to use only the much simpler symmetric rule T_N.

Table 1. "Good" 8-dimensional rules with $N = 10,007$.

Rule #	z	P_2	$N^2\overline{v}_N$	$N \max\lvert\overline{w}\rvert$	$N\sum\lvert\overline{w}\rvert$
1	3335	9.149	8715.05	21.00	1200.84
2	236	9.253	59.00	1.74	96.68
3	1855	10.848	22.69	0.83	64.26
4	1206	28.038	30.33	0.99	71.71

Table 2. Tests of the rules of Table 1 on Genz package.

	Median number of significant digits											
	Rule 1			Rule 2			Rule 3			Rule 4		
	Q_N	T_N	$\overline{M}_N$	Q_N	T_N	$\overline{M}_N$	Q_N	T_N	$\overline{M}_N$	Q_N	T_N	$\overline{M}_N$
Oscillatory	1.51	1.51	2.19	2.88	2.96	4.09	3.49	3.76	4.03	2.94	3.00	3.87
Product Peak	2.26	2.26	2.49	3.82	3.90	3.70	3.58	3.63	3.62	3.24	3.24	3.29
Corner Peak	1.04	0.99	0.68	1.94	2.53	2.51	2.00	2.89	2.60	1.92	2.66	2.51
Gaussian	1.71	1.71	1.77	2.94	2.97	3.00	2.62	2.62	2.61	2.39	2.39	2.39
Continuous	3.30	3.30	3.67	4.61	4.60	4.97	4.79	4.97	4.78	4.45	4.50	4.53
Discontinuous	1.44	1.42	1.44	2.79	2.83	2.81	2.95	2.87	3.10	3.23	3.19	3.16
Average	1.88	1.87	2.04	3.16	3.30	3.51	3.24	3.46	3.46	3.03	3.16	3.29

7. Conclusion

Perhaps the most surprising aspect of the 8-dimensional calculations in the preceding section is the extraordinary variability of the vertex variance $\overline{v}_N$: two rules which are almost identical from a standard way of assessing lattice rules for periodic functions turn out to have values of the vertex variance differing by more than two orders of magnitude – and as it happens, the rule that is the best of its family from the traditional point of view has the worst variance.

If we remember that the vertex variance $\overline{v}_N$ is just a sum of squares of vertex weights, and that the individual vertex weights are essentially just the errors in the symmetric variant of the rule applied to a very typical integrand (namely a function which is linear in each variable and has the value 1 at one vertex and 0 at the others), then it is perhaps not surprising that a rule that fails to integrate these well is not good for more general integrands, and vice versa. Surprising or not, this is certainly what is observed in Table 2.

Finally, the numerical experiments suggest that the vertex variance $\overline{v}_N$ is a key indicator of performance, even taken on its own. Of the 5,000 odd 8-dimensional rules considered in the preceding section, if we had retained just the one with the least vertex variance, then we would have obtained, as in Rule 3 of Table 2, one or two orders of magnitude better accuracy than if we had used the best rule (Rule 1) according to a traditional criterion. Whether a similar result will hold generally, only further experience will tell.

Acknowledgements

The second author acknowledges the support of the Australian Research Council and the Austrian Academy of Sciences.

References

Förster K.-J. (1988), On the minimal variance of quadrature formulas. In *Numerical Integration III*, eds. H. Braß and G. Hämmerlin. Internat. Series of Numer. Math., Vol. **85**, Birkhäuser Verlag, Basel, pp. 37–48.

Genz A. (1984), Testing multidimensional integration routines. In *Tools, Methods and Languages for Scientific and Engineering Computation*, eds. B. Ford, J.C. Rault and F. Thomasset. North-Holland, Amsterdam, pp. 81–94.

Genz A. (1987), A package for testing multiple integration subroutines. In *Numerical Integration: Recent Developments, Software and Applications*, eds. P. Keast and G. Fairweather. D. Reidel Publishing, Dordrecht, pp. 337–340.

Niederreiter H. (1992), *Random Number Generation and Quasi-Monte Carlo Methods.* SIAM, Philadelphia.

Niederreiter H. and Sloan I.H. (1993), Integration of non-periodic functions of two variables by Fibonacci lattice rules. *J. Comp. Appl. Math.*, to appear.

Price J.F. and Sloan I.H. (1992), Pointwise convergence of multiple Fourier series: sufficient conditions and an application to numerical integration. *J. Math. Anal. Appl.* **169**, 140–156.

Ritter K. (1970), Two-dimensional spline functions and best approximations of linear functionals. *J. Approximation Th.* **3**, 352–368.

Sard A. (1963), *Linear Approximation.* American Math. Society, Providence.

Sloan I.H. (1992), Numerical integration in high dimensions – the lattice rule approach. In *Numerical Integration: Recent Developments, Software and Applications*, eds. T.O. Espelid and A. Genz. Kluwer Academic Publishers, Dordrecht, pp. 55–69.

Sloan I.H. and Kachoyan P.J. (1987), Lattice methods for multiple integration: theory, error analysis and examples. *SIAM J. Numer. Anal.* **24**, 116–128.

Zaremba S.K. (1968), Some applications of multidimensional integration by parts. *Ann. Polon. Math.* **21**, 85–96.

Authors' addresses:

Prof.Dr. H. Niederreiter
Institute for Information Processing
Austrian Academy of Sciences
Sonnenfelsgasse 19
A–1010 Vienna
Austria

Prof. I.H. Sloan
School of Mathematics
University of New South Wales
Sydney, NSW 2033
Australia

Gaussian Quadrature Formulae for Splines *

Geno Nikolov

Abstract

Explicit quadrature formulae of Gauss, Radau and Lobatto type are found for spaces of polynomial splines of degree 1 (arbitrary knots) and 2 (the case of equidistant knots). Error analysis for the Sobolev spaces W_∞^r, $r = 2, 3$ is given. It is shown that the Gauss type quadratures for splines with equidistant knots are asymptotically optimal.

1. Introduction

The classical way for construction of quadrature formulae is based on the concept of algebraic degree of precision(ADP). The quadrature formula

$$\int_a^b f(x)dx \approx \sum_{i=1}^{s} a_i f(x_i)$$

is said to have algebraic degree of precision m if m is the largest non-negative integer such that if $f \in \pi_m$ the left hand side and the right hand side of the above formula are equal (as usual, π_m is the set of all algebraic polynomials of degree not exceeding m). A judicious choice of the nodes x_i provides quadrature formulae with high ADP and minimal number of point evaluations - such as the well known quadratures of Gauss, Radau, Lobatto. A good reason for the ADP approach is the Weierstrass theorem about density of the algebraic polynomials in space of continuous functions. However, there are a lot of suggestions in the literature (see e.g. de Boor (1978)) that polynomial splines are somewhat better tool for approximation than polynomials. In other words, sometimes it is preferable to use low degree piecewise polynomial approximants with more knots than simple polynomials of higher degree. This is a motivation for investigation of quadrature formulae of Gauss, Radau and Lobatto type, associated with eligible linear spaces of polynomial splines.

The fundamental question of existence and uniqueness of such quadratures is closely connected with the problem of existence and uniqueness of monosplines with preassigned maximal number of zeros. It was solved by Karlin and Micchelli (1972). The same problem is considered in more general setting for weak Tchebysheff systems by Bojanov et al.(1992).

The purpose of this note is to present explicit Gauss, Radau and Lobatto quadrature formulae for spaces of polynomial splines of degree 1 and 2 with equidistant

*Research supported by the Ministry of Education and Science under grant no. MM-15

knots. It is shown that these formulae are asymptotically optimal in the Sobolev spaces W^r_∞, $r = 2, 3$.

2. Preliminaries

We recall in this section some well known properties of spline functions. A spline function of degree $r-1$ with knots $x_1 < x_2 < \cdots < x_s$ is called every function $f \in C^{r-2}(\mathbf{R})$, such that its restriction on (x_i, x_{i+1}), $i = 0, \ldots, s$ is a polynomial of degree $r-1$ (here $x_0 := -\infty$, $x_{s+1} := \infty$). Every such a function can be represented in the form

$$f(t) = a_0 + a_1 t + \cdots + a_{r-1} t^{r-1} + \sum_{i=1}^{s} c_i (t - x_i)_+^{r-1},$$

where $(a_i)_0^{r-1}$ and $(c_i)_1^s$ are real or complex numbers and

$$(t - x_i)_+^{r-1} = \begin{cases} (t - x_i)^{r-1} & \text{if } t \geq x_i, \\ 0 & \text{otherwise.} \end{cases}$$

The set of spline functions of degree $r-1$ and knots $(x_i)_1^s$ forms a linear space X of dimension $r+s$ with a basis

$$\{1, t, \ldots, t^{r-1}, (t - x_1)_+^{r-1}, \ldots, (t - x_s)_+^{r-1}\}.$$

Another, more convenient basis is that one formed by B-splines.

A B-spline of degree $r-1$ and knots $x_0, \ldots, x_r$ is called the function

$$B_0(t) = (\bullet - t)_+^{r-1}[x_0, x_1, \ldots, x_r],$$

where $g[x_0, \ldots, x_r]$ is the usual r-th divided difference of the function g at the nodes $x_0, \ldots, x_r$. To obtain a basis of B-splines for the space X we add arbitrary knots $x_{-(r-1)} < x_{-(r-2)} < \cdots < x_{-1} < x_0 < x_1$ and $x_s < x_{s+1} < \cdots < x_{s+r}$. Denote

$$B_i(t) := (\bullet - t)_+^{r-1}[x_{i-r}, x_{i-r+1}, \ldots, x_i], \ i = 1, \ldots, s+r.$$

The following are basic properties of B-splines, which will be used in the sequel.

i) The functions $\{B_i(t)\}_{i=1}^{s+r}$ form a basis of X on the interval (x_0, x_{s+1}) ;

ii) $B_i(t) > 0$ if $t \in (x_{i-r}, x_i)$ and $B_i(t) = 0$ otherwise ;

iii) $\int_{-\infty}^{\infty} B_i(t)\, dt = 1/r$.

The definition of spline functions can be extended to allow multiple knots, i.e., $x_1 \leq x_2 \leq \ldots \leq x_s$. Let $t_1 < t_2 < \cdots < t_l$ and $\mathbf{t} := ((t_1, \nu_1), \ldots, (t_l, \nu_l)) \equiv (x_1, \ldots, x_s)$, where ν_i is the number of x's equal to t_i, $(\nu_i < r)$. In this case a spline function of degree $r-1$ with knots $\mathbf{t}$ is called every function

$$f \in C^{r-1-\nu_i}(t_{i-1}, t_{i+1}), \ i = 1, \ldots, l \ (t_0 := -\infty, \ t_{l+1} := \infty),$$

which restriction on (t_i, t_{i+1}) belongs to π_{r-1}, $i = 0, \dots, l$. The definition of and properties $(i) - (iii)$ of the B-splines remain unchanged .

The following theorem is a special case of Theorem 7.1 due to Karlin and Micchelli (1972):

Theorem A 1. *There exists a unique quadrature formula of the type*

$$\int_0^1 f(x)dx \approx Af(0) + Bf(1) + \sum_{k=1}^{n} a_k f(\tau_k) =: Q(f),$$

with $0 < \tau_1 < \dots < \tau_n < 1$; *i.e.,* $A, B, \{a_k\}$ *and* $\{\tau_k\}$ *exist such that* $\int_0^1 f(x)dx = Q(f)$ *for any* f *from the family of functions*

$$\{1, x, \dots, x^{r-1}, (x - x_1)_+^{r-1}, \dots, (x - x_l)_+^{r-1}\}, \quad (0 < x_1 < \dots < x_l < 1),$$

where l=a+b+$2n$-r, *with* a *(and* b*) being equal to 1 in the case* $A \neq 0$ $(B \neq 0)$, *and* a (b) *equal to 0 in the opposite case.*

Corresponding to this quadrature formula is the unique monospline of degree r *involving* n *knots in (0,1), determined by the conditions*

$$K^{(j)}(0) = 0 \ for \ j \in \{0, 1, \dots, r - 1 - a\},$$

$$K^{(j)}(1) = 0 \ for \ j \in \{0, 1, \dots, r - 1 - b\},$$

$$K(x_i) = 0 \ for \ i = 1, 2, \dots, l.$$

This monospline has the form

$$K(t) = \frac{t^r}{r!} - A\frac{t^{r-1}}{(r-1)!} - \sum_{k=1}^{n} a_k \frac{(t - \tau_k)_+^{r-1}}{(r-1)!}$$

and the coefficient B *is determined by* $B = K^{(r-1)}(1)$.

Moreover, for the remainder of this quadrature formula we have

$$\int_0^1 f(x)dx - Q(f) = (-1)^r \int_0^1 K(x) f^{(r)}(x)dx$$

The theorem remains true if the monospline $K(t)$ has multiple zeros; for example, if x_i is a zero of multiplicity λ $(1 \leq \lambda \leq r - 1)$ then the quadrature formila is exact for $(x - x_i)_+^{r-1}, (x - x_i)_+^{r-2}, \dots, (x - x_i)_+^{r-\lambda}$.

Before concluding this section, set

$$W_\infty^r =: \{f \in C^{r-1}[0,1], f^{(r-1)} \text{ is abs. cont. in } (0,1), \quad \operatorname*{supvrai}_{t \in (0,1)} |f^{(r)}(t)| \leq M\}.$$

Then evidently for the remainder of the above quadrature formula we have

$$\sup_{f \in W_\infty^r} \left| \int_0^1 f(x)dx - Q(f) \right| = M \int_0^1 |K(x)| dx. \tag{2.1}$$

3. Gauss type quadrature formulae for polygonal functions

Consider the following sets of knots

$$0 = x_0 < x_1 < \ldots < x_{2n-1} = 1,$$

$$0 = x_0' < x_1' < \ldots < x_{2n}' = 1,$$

$$0 = x_0'' < x_1'' < \ldots < x_{2n+1}'' = 1.$$

Let X, X' and X'' be the spaces of continuous piecewise linear functions, defined by

$$X =: \{f \in C[0,1],\ f_{|(x_i,x_{i+1})} \in \pi_1,\ i = 0,\ldots,2n-2\},$$

$$X' =: \{f \in C[0,1],\ f_{|(x_i',x_{i+1}')} \in \pi_1,\ i = 0,\ldots,2n-1\},$$

$$X'' =: \{f \in C[0,1],\ f_{|(x_i'',x_{i+1}'')} \in \pi_1,\ i = 0,\ldots,2n\}.$$

Clearly X, X' and X'' are of dimensions $2n, 2n+1$ and $2n+2$, respectively. Our first theorem presents Gauss, Radau and Lobatto type quadratures, associated with these spaces.

Theorem 1. *a) The Gauss quadrature formula, associated with space X is*

$$I(f) := \int_0^1 f(x)\,dx \approx \sum_{k=1}^{n} a_k f(\tau_k) =: Q_G^n(f),$$

where

$$a_1 = \frac{x_1 + x_2 - 2x_0}{2}, \quad \tau_1 = \frac{x_1 x_2 - x_0^2}{x_1 + x_2 - 2x_0}, \tag{3.1}$$

$$\begin{aligned} a_k &= \frac{x_{2k-1} + x_{2k} - x_{2k-3} - x_{2k-2}}{2}, \\ \tau_k &= \frac{x_{2k-1}x_{2k} - x_{2k-3}x_{2k-2}}{x_{2k-1} + x_{2k} - x_{2k-3} - x_{2k-2}}, \qquad k = 2,3,\ldots,n-1, \end{aligned} \tag{3.2}$$

$$a_n = \frac{2x_{2n-1} - x_{2n-2} - x_{2n-3}}{2}, \quad \tau_n = \frac{x_{2n-1}^2 - x_{2n-3}x_{2n-2}}{2x_{2n-1} - x_{2n-2} - x_{2n-3}}; \tag{3.3}$$

b) The Radau quadrature formula, associated with space X' is

$$I(f) := \int_0^1 f(x)\,dx \approx \sum_{k=1}^{n} a_k' f(\tau_k') + a f(1) =: Q_R^n(f),$$

where

$$a_1' = \frac{x_1' + x_2' - 2x_0'}{2}, \quad \tau_1' = \frac{x_1' x_2' - x_0'^2}{x_1' + x_2' - 2x_0'}, \tag{3.4}$$

$$(3.5)\qquad a_k' = \frac{x_{2k-1}' + x_{2k}' - x_{2k-3}' - x_{2k-2}'}{2},$$
$$\tau_k' = \frac{x_{2k-1}'x_{2k}' - x_{2k-3}'x_{2k-2}'}{x_{2k-1}' + x_{2k}' - x_{2k-3}' - x_{2k-2}'}, \qquad k = 2,3,\ldots,n,$$

$$(3.6)\qquad a = \frac{x_{2n}' - x_{2n-1}'}{2};$$

c) The Lobatto quadrature formula, associated with space X'' is

$$I(f) := \int_0^1 f(x)\,dx \approx \sum_{k=1}^{n} a_k'' f(\tau_k'') + bf(0) + cf(1) =: Q_L^n(f),$$

where

$$(3.7)\qquad a_k'' = \frac{x_{2k}'' + x_{2k+1}'' - x_{2k-1}'' - x_{2k-2}''}{2}$$
$$\tau_k'' = \frac{x_{2k}''x_{2k+1}'' - x_{2k-2}''x_{2k-1}''}{x_{2k}'' + x_{2k+1}'' - x_{2k-1}'' - x_{2k-2}''}, \qquad k = 1,\ldots,n,$$

$$(3.8)\qquad b = \frac{x_1'' - x_0''}{2}, \quad c = \frac{x_{2n+1}'' - x_{2n}''}{2}.$$

Proof. We prove only part *a)*, the proof of the remaining parts is identical. Add two additional knots $x_0 < 0$ and $x_{2n} > 1$ in order to obtain a basis of X $(B_i(t))_{i=1}^{2n}$, where $B_i = (\bullet - t)_+[x_{i-2}, x_{i-1}, x_i]$. According to property *ii)* each of the intervals (x_{i-1}, x_{i+1}), $i = 2,\ldots,2n-3$ as well as (x_0, x_1) and (x_{2n-2}, x_{2n-1}) must contain at least one node. On the other hand, our quadrature formula has exactly n nodes, therefore $\tau_k \in (x_{2k-2}, x_{2k-1})$, $k = 1,\ldots,n$. Clearly,

$$(3.9)\qquad B_i(t) = \begin{cases} \dfrac{t - x_{i-2}}{(x_i - x_{i-2})(x_{i-1} - x_{i-2})} & \text{if } x_{i-2} \le t < x_{i-1}, \\[2ex] \dfrac{x_i - t}{(x_i - x_{i-2})(x_i - x_{i-1})} & \text{if } x_{i-1} \le t \le x_i, \\[2ex] 0 & \text{, otherwise.} \end{cases}$$

The requirement for the quadrature formula to be exact for B_i implies

$$(3.10)\qquad \int_0^1 B_{2k-1}(t)dt = a_k B_{2k-1}(\tau_k) = 1/2, \; k = 2,\ldots,n$$
$$\int_0^1 B_{2k}(t)dt = a_k B_{2k}(\tau_k) = 1/2, \; k = 1,\ldots,n-1,$$

(here we use property *iii)*), and

$$(3.11)\qquad \int_0^1 B_1(t)dt = a_1 B_1(\tau_1) = \frac{x_1 - x_0}{2(x_1 - x_{-1})},$$

$$(3.12)\qquad \int_0^1 B_{2n}(t)dt = a_n B_{2n}(\tau_n) = \frac{x_{2n-1} - x_{2n-2}}{2(x_{2n} - x_{2n-2})}.$$

By straightforward calculations we get (3.1)-(3.8) on the basis of (3.9)-(3.12). Notice that the nodes $(\tau_k)_2^{n-1}$ are exactly the abscissae of the intersection points of graphs of B_{2k-1} and B_{2k}. The proof of Theorem 1 is completed.

We present below two special cases of Theorem 1. The first corollary deals with the equidistant case, i.e., when $x_k = k/(2n-1)$, $x'_k = k/(2n)$ and $x''_k = k/(2n+1)$. Set

$$R_G^n := \sup_{f\in W_\infty^2} |I(f) - Q_G^n(f)|,\ R_R^n := \sup_{f\in W_\infty^2} |I(f) - Q_R^n(f)|,\ R_L^n := \sup_{f\in W_\infty^2} |I(f) - Q_L^n(f)|.$$

Corollary 1. *The Gauss, Radau and Lobatto quadrature formulae for splines of* 1^{-st} *degree with equidistant knots are*

$$(3.13)\quad \int_0^1 f(x)\,dx \approx \frac{2}{2n-1}\left[\frac{3}{4}\left(f\left(\frac{2}{6n-3}\right) + f\left(\frac{6n-5}{6n-3}\right)\right) + \sum_{k=2}^{n-1} f\left(\frac{4k-3}{4n-2}\right)\right],$$

$$(3.14)\qquad \int_0^1 f(x)\,dx \approx \frac{3}{4n} f\left(\frac{1}{3n}\right) + \frac{1}{n}\sum_{k=2}^{n} f\left(\frac{4k-3}{4n}\right) + \frac{1}{4n} f(1),$$

and

$$(3.15)\qquad \int_0^1 f(x)\,dx \approx \frac{2}{2n+1}\left[\frac{1}{4}(f(0) + f(1)) + \sum_{k=1}^{n} f\left(\frac{4k-1}{2(2n+1)}\right)\right],$$

respectively. The errors of these quadratures are

$$R_G^n = \frac{n-1}{4(2n-1)^3}M,\quad R_R^n = \frac{3n-1}{96n^3}M \ \ \textit{and}\ \ R_L^n = \frac{3n+1}{12(2n+1)^3}M.$$

Proof. Formulae (3.13)-(3.15) follow from Theorem 1. According to (2.1), the error bounds are equal to $M\int_0^1 |K(x)|dx$, where $K(t)$ is the corresponding monospline. For instance, for the Gauss quadrature formula (3.13) this monospline is

$$K(t) = \begin{cases} t^2/2 & \text{if}\quad 0 \le t \le \tau_1, \\ (t - \frac{2k-1}{2n-1})(t - \frac{2k}{2n-1})/2 & \text{if}\quad \tau_k \le t \le \tau_{k+1},\ k = 1, \dots, n-1, \\ (t-1)^2/2 & \text{if}\quad \tau_n \le t \le 1. \end{cases}$$

Using Theorem A1, one can easily find the explicit expressions for the remaining two monosplines. Then we apply (2.1) to get desired error bounds. The proof of Corollary 1 is completed.

Remark 1. The optimal quadrature formula in W_∞^2 with n nodes is known (see e.g. Nikolski(1979)). The error bound is

$$R_{opt}^n = \frac{M}{32(n-1+\sqrt{3}/2)^2} = \frac{M}{32}(n^{-2} + O(n^{-3})).$$

Corollary 1 shows that the rules (3.13)-(3.15) have the same major error term, and in this sense are almost optimal. Additional advantages are: rational nodes and weihgts, positive coefficients, almost Tchebysheff type (and hence small variance). Formulae (3.1)-(3.3) show that Gauss quadrature is of Tchebysheff type if and only if the knots (x_i) satisfy $x_{2k-1}+x_{2k}=k/n$, $k=1,\ldots,n-1$. In the next corollary we present a special case of such a choice.

Corollary 2. *The Gauss quadrature formula for space X with knots $x_0=0$, $x_{2n-1}=1$ and*

$$x_{2k-1}=\frac{4k-1}{4n},\quad x_{2k}=\frac{4k+1}{4n},\quad k=1,\ldots,n-1$$

is

$$\int_0^1 f(x)\,dx \approx \frac{1}{n}\left[f\left(\frac{15}{32n}\right)+f\left(1-\frac{15}{32n}\right)+\sum_{k=2}^{n-1} f\left(\frac{2k-1}{2n}\right)\right] =: Q_{GT}^n(f). \tag{3.16}$$

For the error of this rule we have

$$R_{GT}^n=\sup_{f\in W_\infty^2}|I(f)-Q_{GT}^n(f)|=\left(\frac{1}{32}n^{-2}+\frac{29}{3072}n^{-3}\right)M. \tag{3.17}$$

Remark 2. Clearly (3.16) is exactly the midpoint rule slightly repaired. At the expence of a small variation of the end nodes we get asymptotically optimal for W_∞^2 rule. In Brass(1977) similar improvement of the midpoint rule is presented due to Watermann, but the alteration there is in the weights, therefore it is not of Tchebysheff type.

4. Gauss type quadratures for quadratic splines

4.1. Radau quadrature for C-piecewise parabolas. Let $x_k=k/n$ for $k=-1,0,\ldots,n+1$, and

$$Y=\{f: f\in C[0,1], f_{|(x_i,x_{i+1})}\in\pi_2,\ i=1,\ldots,n-1\}.$$

Since this space has dimension $2n+1$, we will obtain only the Radau quadrature formula. The B-splines that form a basis of Y are

$$\begin{aligned} B_{2k-1}(t)&=(\bullet-t)_+^2[x_{k-2},x_{k-1},x_{k-1},x_k],\ k=1,\ldots,n+1,\\ B_{2k}(t)&=(\bullet-t)_+^2[x_{k-1},x_{k-1},x_k,x_k],\ k=1,\ldots,n.\end{aligned} \tag{4.1}$$

Clearly,

$$B_{2k-1}(t)=\begin{cases} n^3(t-x_{k-2})^2/2 & \text{if } x_{k-2}\le t\le x_{k-1},\\ n^3(t-x_k)^2/2 & \text{if } x_{k-1}\le t\le x_k,\\ 0, \text{ otherwise.} & \end{cases} \tag{4.2}$$

$$(4.3)\qquad B_{2k}(t)=\begin{cases}2n^3(t-x_{k-1})(x_k-t) & \text{if } x_{k-1}\le t\le x_k,\\ 0,\ \text{otherwise}.\end{cases}$$

If the desired quadrature formula is of the form

$$\int_0^1 f(t)\,dt\approx\sum_{k=0}^{n-1}a_kf(\tau_k)+b_nf(1),$$

then as in the proof of Theorem 1 we conclude that $\tau_k\in(x_k,x_{k+1})$, $k=0,\dots,n-1$. For convenience's sake we substitute

$$\theta_k=:n\tau_k-k,\ k=0,1,\dots,n-1.$$

With this notation from (4.2)-(4.3) we have

$$(4.4)\qquad \begin{aligned}B_{2k+1}(\tau_{k-1})&=n\theta_{k-1}^2/2,\\ B_{2k+1}(\tau_k)&=n(1-\theta_k)^2/2,\\ B_{2k+2}(\tau_k)&=2n\theta_k(1-\theta_k).\end{aligned}$$

Then, from property *iii)* of B-splines and exactness condition we get

$$\int_0^1 B_0(t)\,dt=a_0B_0(\tau_0)=1/6,\quad \int_0^1 B_1(t)\,dt=a_0B_1(\tau_0)=1/3,$$

which combined with (4.4) yields

$$\theta_0=\frac{1}{3},\quad a_0=\frac{3}{4n}.$$

Further, if θ_k is found, the requirement

$$(4.5)\qquad \int_0^1 B_{2k+2}(t)\,dt=a_kB_{2k+2}(\tau_k)$$

and the last equation of (4.4) imply

$$(4.6)\qquad a_k=\frac{1}{6n\theta_k(1-\theta_k)},\quad k=0,1,\dots,n-1.$$

Again by exactness condition we have

$$(4.7)\qquad \int_0^1 B_{2k+1}(t)\,dt=a_{k-1}B_{2k+1}(\tau_{k-1})+a_kB_{2k+1}(\tau_k)=1/3.$$

Now from relations (4.4)-(4.7) we get

$$B_{2k+2}(\tau_k)/B_{2k+1}(\tau_k)=4\theta_k/(1-\theta_k)=4(1-\theta_{k-1})/(4-5\theta_{k-1}),$$

hence

(4.8) $$\theta_k = \frac{1-\theta_{k-1}}{5-6\theta_{k-1}}, \quad k = 1, 2, \ldots, n-1.$$

Finally, from $\int_0^1 B_{2n+1}(t)\,dt = a_{n-1}B_{2n+1}(\tau_{n-1}) + b_n B_{2n+1}(1) = 1/6$, we get

$$b_n = \frac{2-3\theta_{n-1}}{6n(1-\theta_{n-1})}.$$

The sequence $\{\theta_k\}$, derived by $\theta_0 = 1/3$ and the recurrent relation (4.8) has an explicit representation

$$\theta_k = \frac{S_k}{S_k + S_{k+1}},$$

where $S_k = (2-\sqrt{3})^k + (2+\sqrt{3})^k$. Moreover, the sequence $\{\theta_k\}$ is monotone decreasing and it is easy to see that

$$\theta_k - \frac{3-\sqrt{3}}{6} = \frac{\sqrt{3}-1}{S_k + S_{k+1}}, \quad \text{and} \quad |a_k - \frac{1}{n}| = \frac{2}{n(S_{2k+1}+4)}.$$

The associated monospline $K(t)$ in this case does not change the sign in (0,1) since it has only double zeros $x_k = k/n$, $k = 1, \ldots, n-1$. Therefore, applying Theorem A1, we get

$$\int_0^1 |K(t)|\,dt = -\int_0^1 K(t)\,dt = \frac{1}{3}\sum_{k=0}^{n-1} a_k\tau_k^3 + \frac{1}{3}b_n - \frac{1}{12}.$$

Then, using convergence of the nodes $\{\tau_k\}$ and the weights $\{a_k\}$, it is not difficult to obtain on the basis of the last equality

$$\int_0^1 |K(t)|\,dt = \frac{\sqrt{3}}{216}n^{-3} + O(n^{-4}).$$

Thus we established the following theorem

Theorem 2. *The Radau quadrature formula for space*

$$Y = \{f : f \in C[0,1],\ f_{|(\frac{i}{n},\frac{i+1}{n})} \in \pi_2,\ i = 1, \ldots, n-1\}$$

is

$$\int_0^1 f(x)\,dx \approx \sum_{k=0}^{n-1} a_{k,n} f\left(\frac{k+\theta_k}{n}\right) + b_n f(1) =: Q^n_{R,2}(f),$$

where $\theta_0 = 1/3$ *and* $\theta_k = (1-\theta_{k-1})/(5-6\theta_{k-1})$, $k = 1, \ldots, n-1$,

$$a_{k,n} = \frac{1}{6n\theta_k(1-\theta_k)}, \quad k = 0, \ldots, n-1,$$

and

$$b_n = \frac{2-3\theta_{n-1}}{6n(1-\theta_{n-1})}.$$

For the error of this quadrature we have

$$\sup_{f\in W^3_\infty} \left| \int_0^1 f(x)\,dx - Q^n_{R,2}(f)\right| = \left(\frac{\sqrt{3}}{216}n^{-3} + O(n^{-4})\right) M.$$

4.2. Gauss quadrature formula for C^1 piecewise parabolas. Let $m = 2n-2$ $(n \ge 2)$ and $x_k = k/m$, $k = -2, -1, \dots, 2n$, then it follows from *i)* that B-splines

$$B_k(t) = (\bullet - t)^2_+[x_{k-3}, x_{k-2}, x_{k-1}, x_k],\ k = 1, \dots, 2n$$

form a basis for space

$$Y_1 = \left\{f \in C^1[0,1],\, f_{|(x_{i-1},x_i)} \in \pi_2,\, i = 1, \dots, 2n-2\right\}.$$

We construct in this section the Gauss quadrature for Y_1 (clearly $\dim Y_1 = 2n$)

$$\int_0^1 f(x)\,dx \approx \sum_{k=1}^n a_{k,n} f(\tau_{k,n}) \tag{4.9}$$

(henceforth we drop the indices n in the weights and the nodes).
Since Y_1 is invariant with respect to transformation $x \to 1-x$, it follows from uniqueness of Gauss quadrature that it must be of symmetric type.
For $k = 1, \dots, [n/2]$ we seek the nodes $\tau_k \in (x_{2k-2}, x_{2k-1})$. Set for simplicity

$$\tau_k = \frac{2k-1-\delta_k}{m},\quad k = 1, \dots, [n/2],$$

then our assumption becomes $\delta_k \in (0,1)$. Moreover, it is easy to see on the basis of definition of B-splines that

$$\begin{aligned} &B_{2k-1}(\tau_k) = m\delta_k^2/6,\ k = 1, \dots, [(n-1)/2],\\ &B_{2k}(\tau_k) = m(1 + 2\delta_k - 2\delta_k^2)/6,\ k = 1, \dots, [(n-1)/2],\\ &B_{2k-1}(\tau_{k-1}) = m(1-\delta_{k-1})^2/6,\ k = 2, \dots, [(n+1)/2].\end{aligned} \tag{4.10}$$

In addition,

$$\int_0^1 B_1(t)\,dt = 1/18,\ \int_0^1 B_2(t)\,dt = 5/18 \text{ and } \int_0^1 B_k(t)\,dt = 1/3 \text{ for } k = 3, \dots, 2n-2.$$

As in the proof of Theorem 2, the requirement (4.9) to be exact for B_1 and B_2 implies

$$B_2(\tau_1)/B_1(\tau_1) = (1 + 2\delta_1 - 2\delta_1^2)/\delta_1^2 = 5,$$

hence

$$\delta_1 = (1 + 2\sqrt{2})/7,\ a_1 = (2\sqrt{2}-1)^2/(3m).$$

Further, the requirement (4.9) to be exact for B_k, $k = 3, \dots, [(n-1)/2]$ becomes

$$a_{k-1}B_{2k-1}(\tau_{k-1}) + a_k B_{2k-1}(\tau_k) = 1/3,\quad a_k B_{2k}(\tau_k) = 1/3.$$

For k=2 we get

$$\delta_2^2/(1+2\delta_2-2\delta_2^2) = 1-2(\sqrt{2}-1)^2/3 =: \alpha_2,$$

and for $k=3,\dots,[(n-1)/2]$

$$\delta_k^2/(1+2\delta_k-2\delta_k^2) = (4\delta_{k-1}-3\delta_{k-1}^2)/(1+2\delta_{k-1}-2\delta_{k-1}^2) =: \alpha_k,$$

therefore we obtain successively for $k=2,\dots,[(n-1)/2]$

$$\delta_k = (\alpha_k + \sqrt{\alpha_k(3\alpha_k+1)})/(2\alpha_k+1),$$

(4.11)

$$a_k = 2/[m(1+2\delta_k-2\delta_k^2)].$$

We use some symmetry arguments to find remaining weights and nodes. Consider separately the cases of even and odd n:

Case A: $n=2l$. We choose $\tau_{l+1}=1-\tau_l$, $a_{l+1}=a_l$, and determine a_l and δ_l from the conditions

$$a_{l-1}B_{2l-1}(\tau_{l-1})+a_lB_{2l-1}(\tau_l)=1/3,\ a_lB_{2l}(\tau_l)+a_{l+1}B_{2l}(\tau_{l+1})=1/3.$$

Taking into account $B_{2l}(\tau_{l+1}) = B_{2l+1}(\tau_l)$, we get from the above equations, (4.10) and (4.11)

$$\frac{\delta_l^2}{2-\delta_l^2} = \frac{4\delta_{l-1}-3\delta_{l-1}^2}{1+2\delta_{l-1}-2\delta_{l-1}^2} =: \alpha_l,\ a_l = \frac{2}{m(2-\delta_l^2)},$$

hence

$$\delta_l = \sqrt{\frac{2\alpha_l}{1+\alpha_l}},\ a_l = a_{l+1} = \frac{1+\alpha_l}{m}.$$

By the symmetry, we have

(4.12) $$a_k = a_{n+1-k},\ \tau_k = 1-\tau_{n+1-k},\ k=l+2,\dots,n.$$

Case B: $n=2l+1$. In this case a_l and δ_l are determined by (4.11), while $\tau_{l+1}=1/2$ and a_{l+1} have to satisfy

$$a_lB_{2l+1}(\tau_l)+a_{l+1}B_{2l+1}(\tau_{l+1})=1/3,$$

which yields

$$a_{l+1} = \frac{2}{m}\alpha_{l+1},\ \alpha_{l+1} = \frac{4\delta_l-3\delta_l^2}{1+2\delta_l-2\delta_l^2}.$$

The remaining weights and nodes are determined again by (4.12). Thus we proved

Theorem 3. *Let $m=2n-2$ $(n\ge 2)$, $x_k=k/m$, $k=0,\dots,m$, and*

$$Y_1 = \left\{f\in C^1[0,1],\ f_{|(x_{i-1},x_i)}\in\pi_2,\ i=1,\dots,m\right\}.$$

The Gauss quadrature formula for space Y_1 *is*
a) for odd n $(n{=}2l{+}1)$

$$\int_0^1 f(x)\,dx \approx \sum_{k=1}^{l} a_k[f(\tau_k)+f(1-\tau_k)] + a_{l+1}f(1/2) =: Q_{G,2}^{2l+1}(f),$$

where

$$\tau_k = \frac{2k-1-\delta_k}{m},\quad k=1,\dots,l \tag{4.13}$$

and $(\delta_k)_1^l$ and $(a_k)_1^l$ *can be determined successively by following relations:*

$$\delta_1 = \frac{1+2\sqrt{2}}{7},\quad a_1 = \frac{(2\sqrt{2}-1)^2}{3m},\quad \alpha_2 = 1-\frac{2(\sqrt{2}-1)^2}{3}, \tag{4.14}$$

$$\alpha_k = \frac{4\delta_{k-1}-3\delta_{k-1}^2}{1+2\delta_{k-1}-2\delta_{k-1}^2},\quad k=3,\dots,[(n+1)/2], \tag{4.15}$$

$$\delta_k = \frac{\alpha_k+\sqrt{\alpha_k(3\alpha_k+1)}}{2\alpha_k+1},\quad k=2,\dots,[(n-1)/2], \tag{4.16}$$

$$a_k = \frac{2}{m(1+2\delta_k-2\delta_k^2)},\quad k=2,\dots,[(n-1)/2], \tag{4.17}$$

$$a_{l+1} = \frac{2\alpha_{l+1}}{m};$$

b) for even n $(n{=}2l)$

$$\int_0^1 f(x)\,dx \approx \sum_{k=1}^{l} a_k[f(\tau_k)+f(1-\tau_k)] =: Q_{G,2}^{2l}(f),$$

where $(\tau_k)_1^l$ *are defined by* (4.13); δ_1, a_1 and α_2 *are determined by* (4.14); $(\alpha_k)_3^l$, $(\delta_k)_2^{l-1}$ and $(a_k)_2^{l-1}$ *are expressed by* (4.15), (4.16) and (4.17), *respectively; finally,*

$$\delta_l = \sqrt{\frac{2\alpha_l}{1+\alpha_l}},\quad a_l = \frac{1+\alpha_l}{m}.$$

Our next theorem indicates that the above quadrature formula seems to be quite good in W_∞^3.

Theorem 4. *The error of the Gauss quadrature associated with space* Y_1 *is*

$$R_{G,2}^n := \sup_{f\in W_\infty^3} |I(f)-Q_{G,2}^n(f)| = (\frac{1}{192}n^{-3}+O(n^{-4}))M, \tag{4.18}$$

i.e., this quadrature is asymptotically optimal in W_∞^3.

The constant $\frac{1}{192}$ in (4.18) is best possible for the more narrow class $\tilde{W}^3_\infty$ of 1-periodic functions and therefore for W^3_∞ too (for best quadrature formulae in spaces $\tilde{W}^r_\infty$ see Motorni(1973)). The basic ingredients of the proof of Theorem 4 are the following two lemmas

Lemma 1. *The sequence $(\delta_k)_{k=1}^{[n/2]}$ is monotone increasing and (if it is continued in the same recurrent way) has a limit 1. Moreover, for $k=1,\ldots,[n/2]$ we have*

$$0 < \tau_k - x_{2k-2} < \frac{1}{m} 2^{-2^{k-1}}, \tag{4.19}$$

$$0 < \frac{2}{m} - a_k < \frac{4}{m} 2^{-2^{k-1}}. \tag{4.20}$$

Proof. Since $1-\alpha_k = (1-\delta_{k-1})^2/(1+2\delta_k - 2\delta_k^2)$, we have

$$0 < 1 - \alpha_k < (1-\delta_{k-1})^2.$$

On the other hand, $1-\delta_k = g(1) - g(\alpha_k) = g'(\xi)(1-\alpha_k)$, where

$$g(x) = \frac{x + \sqrt{x(3x+1)}}{2x+1}.$$

Since $g'(x) \in (0,1)$ for $x \in (0.5, 1)$, we get

$$0 < 1 - \delta_k < (1-\delta_{k-1})^2 < \ldots < (1-\delta_1)^{2^{k-1}} < 2^{-2^{k-1}}, \tag{4.21}$$

which proves monotonicity of sequence (δ_k). In addition, for $k = 1, 2, \ldots, [n/2]$ we have

$$0 < \tau_k - x_{2k-2} = \frac{1-\delta_k}{m} < \frac{1}{m} 2^{-2^{k-1}},$$

hence (4.19) is proved. Similarly we can get (4.20). Thus we observe here second order convergence for the weights and the nodes of this quadrature, in contrast with those of the Radau quadrature in the previous section, where geometric convergence was established. Lemma 1 is proved.

Lemma 2. *The nodes and weights of the Gauss quadrature in Theorem 3 satisfy the following relations*

$$a) \quad \sum_{k=1}^{l} a_k = \begin{cases} \frac{1}{2} - \frac{1}{m} + o(n^{-4}) & \text{for } n = 2l+1 - \text{odd}, \\ \frac{1}{2} & \text{for } n = 2l - \text{even}; \end{cases} \tag{4.22}$$

$$b) \quad \sum_{k=1}^{l} a_k \tau_k = \begin{cases} \frac{1}{8} - \frac{1}{2m} + \frac{1}{3m^2} + o(n^{-4}) & \text{for } n = 2l+1 - \text{odd}, \\ \frac{1}{8} - \frac{1}{6m^2} + o(n^{-4}) & \text{for } n = 2l - \text{even}. \end{cases} \tag{4.23}$$

Proof. (4.22) follows from the fact that our quadrature is exact for $f(x) = 1$ (for odd n we also use the inequalities $0 < \frac{2}{m} - a_{l+1} = \frac{2}{m}(1 - \alpha_{l+1}) < \frac{2^{1-2^l}}{m}$. To prove (4.23) we introduce function

$$\varphi(x) = \begin{cases} -x + \frac{1}{2} + (-1)^n \frac{1}{2m} & \text{for } 0 \le x \le x_{2l-1}, \\ \frac{m}{2}(x - x_{2l})^2 & \text{for } x_{2l-1} \le x \le x_{2l}, \\ 0, & \text{otherwise.} \end{cases} \tag{4.24}$$

Since $\int_0^1 \varphi(t)\,dt$ coincides with the right hand side of the Gauss quadrature, after comparison, taking into account (4.22) and Lemma 1, we get easily (4.23). Lemma 2 is proved.

We sketch proof of Theorem 4. By (2.1) and symmetry arguments we have

$$R^n_{G,2} = 2M \int_0^{\frac{1}{2}} |K(t)|\,dt = 2M \int_0^{x_{n-1}} |K(t)|\,dt.$$

Then, using the fact that $K(t)$ changes the sign at points $(x_k)_1^{m-1}$ only, and the representation

$$K(t) = \frac{t^3}{6} - \frac{1}{2}\sum_{k=1}^{j} a_k(\tau_k - t)^2 \text{ for } t \in (\tau_j, \tau_{j+1}),\ j = 0, \ldots, l,$$

we obtain easily

$$\begin{aligned} 2\int_0^{x_{n-1}} |K(t)|\,dt &= \frac{1}{192} - \frac{1}{6}\sum_{j=1}^{s}(x_{2j}^4 - x_{2j-1}^4) \\ &-\frac{2}{3}\sum_{k=1}^{s} a_k \sum_{j=k}^{s}[(\tau_k - x_{2j})^3 - (\tau_k - x_{2j-1})^3] + \frac{1}{3}\sum_{k=1}^{l} a_k(\tau_k - \frac{1}{2})^3, \end{aligned} \tag{4.25}$$

where $s = l$ for $n = 2l + 1$ and $s = l - 1$ for $n = 2l$. To estimate the right hand side of (4.25) we apply Lemma 2 and, based on Lemma 1, some identities of the type

$$\sum_{k=1}^{s} \psi_k(\delta_k) = \sum_{k=1}^{s} \psi_k(1) + O(1) \text{ for } s = 1, 2, \ldots \tag{4.26}$$

true for any sequence of differentiable functions $(\psi_k)_1^\infty$ satisfying $|\psi_k'(t)| \le c k^3$ for all $t \in (0.5, 1)$ with a constant c independent on k. We omit the details.

The results obtained here suggest validity of the following conjecture

Conjecture. *For every integers $r \ge 2$ and $n \ge r/2$ let $I(f) \approx Q^n_{G,r-1}(f)$ be the n-point Gauss quadrature formula for space of polynomial splines of degree $r - 1$ with knots $x_k = k/(2n + 1 - r)$, $k = 1, \ldots, 2n - r$. Let*

$$R^n_{G,r-1} := \sup_{f \in W^r_\infty} |I(f) - Q^n_{G,r-1}(f)|, \quad R^n_{opt,r} := \inf_{Q^n} \sup_{f \in W^r_\infty} |I(f) - Q^n(f)|,$$

where Q^n *is the class of all n-point quadratures. Then*

$$\lim_{n\to\infty} \frac{R^n_{G,r-1}}{R^n_{opt,r}} = 1.$$

Notice that we already proved this conjecture for r=2 and r=3 (see Corollary 1 and Theorem 4). For W^1_∞ the midpoint rule is optimal, but it evidently is the Gauss quadrature for space $\{f : f_{|(\frac{k}{n},\frac{k+1}{n})} \in \pi_1,\ k = 0,\ldots,n-1\}$.
It is possible using similar arguments as above to construct also the Lobatto quadrature for smooth splines of degree 2 with equidistant knots as well as a numerical algorithm for computation of the Gauss quadrature for cubic splines with equidistant knots.

References

[1] Bojanov B., Grozev G. and Žhensykbaev A. (1992), Generalized Gaussian Quadrature Formulae for Weak Tchebyshev Systems, in "Optimal Recovery", B. Bojanov and H. Wozniakowski eds., 1992 Nova Science Publishers,Inc., pp. 115-140.

[2] de Boor C. (1978), "A Practical Guide to Splines", Springer Verlag

[3] Brass H.(1977), "Quadraturverfahren", Göttingen: Vandenhoeck & Ruprecht.

[4] Karlin S. and Micchelli C. (1972), The Fundamental Theorem of Algebra for Monosplines Satisfying Boundary Conditions, Isr. J. Math. v.11, pp.405-451..

[5] Motorni V.P.(1973), The best quadrature formula of the form $\sum_{k=1}^n p_k f(x_k)$ for some classes of periodic differentiable functions, Docl.Acad.Nauk USSR 211, 5, pp.1060-1062 (in Russian).

[6] Nikolski S.M. (1979), "Quadrature Formulae" 3-rd ed., Nauka, Moscow (in Russian)

Author's address:

Geno Nikolov
Department of Mathematics
University of Sofia
Boulevard James Boucher 5
1126 Sofia, Bulgaria

Quadrature Formulas For Convex Classes of Functions

Erich Novak

Abstract. In this paper we study problems of optimal recovery in the case of a nonsymmetric convex class of functions. We concentrate on the reconstruction of linear functionals, i.e., the problem of optimal numerical integration. We begin with a class of convex functions. We prove that adaption cannot help in the worst case, but considerably helps in the case of Monte Carlo methods.
We give examples for linear problems on a convex set where (deterministic) adaptive methods are much better than nonadaptive ones.

1. Introduction

No numerical method, say, no quadrature formula, can be good for every function f on some domain. We need some a priori information about f of the form $f \in F$. Usually one assumes that f is an element of a certain Banach space X and so might have certain smoothness properties.
Then it is our task, for example, to find good quadrature formulas such that

$$\left| \int f(t)\, dt - \sum_{i=1}^{n} a_i f(t_i) \right| \leq c_n \cdot \|f\|_X$$

holds with as small c_n as possible. This means that we make a worst case analysis on the symmetric and convex set

$$F = \{f \in X \mid \|f\|_X \leq 1\}.$$

This approach is the usual one in numerical analysis, at least if the solution operator is linear. Also most of the known results on optimal recovery and closely related problems on n-widths usually are studied under the assumption that the set F of problem

elements is convex and symmetric. In many cases, however, we have a different type of a priori information. We give some examples.
Sometimes we know that f is positive because, for example, f is a certain density function. In this case we should consider sets of the type

$$F = \{f \in X \mid \|f\|_X \leq 1,\ f \geq 0\}.$$

Observe that such a set is still convex, but not symmetric. In other cases we might know in advance that f is a monotone or convex function. This also leads us to study convex classes of functions that are nonsymmetric. The geometric extra-information, given by positivity, monotonicity, or convexity, is very important in some cases. It often helps to find an effective numerical method, even if the problem is ill-posed without this information. Examples of this kind can be found in Gorenflo, Vessella (1991).
Therefore it is usually not a good idea to just ignore the additional information about f. However, it may seem that it is still enough to study symmetric and convex sets – at least modulo some minor details. Let us again consider the case where we want to approximate an integral or linear functional by some quadrature formula. By taking $F - F$, defined by

$$F - F = \{f_1 - f_2 \mid f_1, f_2 \in F\},$$

we clearly get a symmetric set and for each convex set F we get the error estimate

$$\inf_{S_n} \Delta_{\max}^{F}(S_n) \leq \inf_{S_n} \Delta_{\max}^{F-F}(S_n) \leq 2 \inf_{S_n} \Delta_{\max}^{F}(S_n).$$

Here the maximal error (over F or over $F - F$) is defined in the usual way and the infimum runs through all methods of the form

$$(1) \qquad S_n(f) = a_0 + \sum_{i=1}^{n} a_i f(t_i)$$

with nonadaptively chosen knots $t_1, \ldots, t_n$. It is known that methods of the form (1) always are optimal in the class of all nonadaptive methods, see Sukharev (1986). In the symmetric case the constant a_0 is not needed and we even know that such linear methods are optimal in the class of all adaptive methods that use n function values, see Bakhvalov (1971).
So we know that optimal error bounds for F and for $F - F$ differ at most by a factor 2 in the case of nonadaptive methods and adaption does not help for $F - F$. Therefore we can get much better error bounds on F only if we allow adaptive methods. We will see later that for some convex F adaptive methods actually are much better than nonadaptive ones.

2. Historical background

Consider the set of monotone functions, given by

$$F_m = \{f : [0,1] \to [0,1] \mid f \text{ monotone}, f(0) = 0, f(1) = 1\}.$$

The problem of optimal numerical integration for the space F_m was studied by Kiefer (1957) and Novak (1992). The problem is to find a formula based solely on function evaluation at n points of $[0,1]$, which approximates $S(f) = \int_0^1 f(t)\,dt$ with small error over the class F_m. The knots t_i may be chosen adaptively, i.e., sequentially. Kiefer proved that the best method (with respect to the maximal error on F_m) is given by the trapezoidal rule

$$S_n^T(f) = \frac{1}{2n+2} + \frac{1}{n+1} \sum_{i=1}^{n} f\left(\frac{i}{n+1}\right).$$

Hence we have an affine and nonadaptive algorithm which is optimal in the worst case. The maximal error of this method is $1/(2n+2)$. Observe that adaption does not help in the worst case. This is known for arbitrary linear $S : F \to \mathbf{R}$ in the case that F is convex and symmetric.
In Novak (1992) it is proved that for the class F_m adaptive Monte Carlo methods are much better than nonadaptive ones. Similar results hold for the average error with respect to the Ulam measure: while nonadaptive methods give the optimal rate $n^{-\log 6/\log 4}$ one can construct adaptive methods with a rate $n^{-3/2}$. This means that adaption helps in different stochastic settings while adaption does not help in the worst case. These results are quite different from the known worst case results for Hölder- or Sobolev spaces or the average case results for Gaussian measures. For recent results and references concerning the average case see the paper of Ritter, Wasilkowski, Woźniakowski (1993) in this volume.
Also in some other papers linear problems (such as integration or optimal reconstruction in L_∞-norm) have been studied for certain nonsymmetric convex classes of functions. We mention the papers Braß (1982), Förster, Petras (1990), Gal, Micchelli (1980), Glinkin (1983, 1984), Heindl (1982), Papageorgiou (1992), Petras (1993), Schnorr (1992), Sonnevend (1983), Sukharev (1986) and Zwick (1988).
In this paper we are mainly interested in optimal quadrature formulas. Therefore we do not define certain Gelfand-type widths for nonsymmetric classes that are closely related to more general problems of optimal recovery and approximation theory, see Novak (1993). We study the following problem in the worst case setting for linear problems. Can adaption help much on a convex class of functions?
We could not find the answer to this question in the literature. Much is known about linear problems

$$S : X \to Y,$$

when studied on a symmetric and convex set $F \subset X$ in the worst case, see Traub, Wasilkowski and Woźniakowski (1988). Adaption for nonconvex sets is studied, for example, by Huerta (1986) and Wasilkowski and Gao (1992).
A slight superiority of adaptive methods can be proven in some cases even if F is symmetric, see Kon and Novak (1989, 1990). It is well known, however, that adaption cannot help much in that case. Although adaptive methods are widely used, most theoretical results show that adaption does not help under various conditions, see, for example, Bakhvalov (1971), Gal and Micchelli (1980), Micchelli, Rivlin (1977, 1985), Traub and Woźniakowski (1980), Traub, Wasilkowski and Woźniakowski (1988), and Wasilkowski and Woźniakowski (1984).
It is known that randomized methods are much better than nonrandomized (or pure) methods for numerical integration and many classes of functions. For standard classes (unit balls of Sobolev or Hölder spaces), however, adaptive Monte Carlo methods are only slightly better than nonadaptive ones. See Novak (1988a) for a survey of the results of Bakhvalov (1959) and Novak (1988b). Adaptive Monte Carlo methods for the class F_m are much better than nonadaptive Monte Carlo methods while nonadaptive Monte Carlo methods are only slightly better than (adaptive or nonadaptive) deterministic methods, see Novak (1992).
Our starting point is the optimal integration of convex functions. We prove that adaption does not help in the worst case, but considerably helps for Monte Carlo methods. Then we study more general problems and give examples for linear problems on a convex set, where adaptive methods are much better than nonadaptive ones. We conclude this section with two remarks:
– We recently learnt that N. P. Korneichuk (1993) independently found another example of a convex set, where adaption helps for a linear problem in the worst case.
– In this paper we only study linear problems, i.e., the solution operator $S : X \to Y$ is linear. Our remark that sometimes the study of problem sets F is interesting for convex but nonsymmetric F also applies for nonlinear S, however.
In the case of ordinary differential equations, for example, it is important to distinguish between $y' = -f(x,y)$ and $y' = f(x,y)$. Both equations might have a very different behavior, the stability characteristics of the problem are often not captured in an adequate way by the norm of f. This was recently emphasized in the paper by Iserles and Söderlind (1992).

3. Optimal integration of convex functions

A survey on the optimal integration of convex functions is given in Petras (1993). Here we only explain some recent results which were found together with Andreas Schnorr, see the thesis Schnorr (1992) for detailed proofs. Let

$$F_{uv} = \{f : [0,1] \to \mathbf{R} \mid f \text{ convex}, \ f'(0) \geq u, \ f'(1) \leq v\},$$

where $u \leq v$. We begin with a result of Zwick (1988) who found the optimal nonadaptive method of the form

$$S_n(f) = \phi(f(t_1), \ldots, f(t_n)),$$

i.e., the knots may be chosen arbitrarily in advance but may not depend on f.

Proposition 1 (Zwick 1988)
Let $n \in \mathbf{N}$ and $t_i = (2i-1)/(2n)$. Then the formula

$$S_n(f) = \frac{1}{16n^2} \cdot (v-u) + \frac{1}{n} \sum_{i=1}^{n} f(t_i)$$

is optimal in the class of nonadaptive formulas. The maximal error of S_n on the class F_{uv} is given by

$$\Delta_{\max}(S_n) = \frac{v-u}{16n^2}.$$

□

Glinkin (1983) studied some special adaptive methods for this problem, i.e., methods of the form

$$S_n(f) = \phi(f(t_1), \ldots, f(t_n)),$$

where the chosen knot t_k may depend on $f(t_1), \ldots, f(t_{k-1})$. The following result shows, however, that adaption does not help for this problem.

Proposition 2
Let $n \in \mathbf{N}$ and $t_i = (2i-1)/(2n)$. Then the formula

$$S_n(f) = \frac{1}{16n^2} \cdot (v-u) + \frac{1}{n} \sum_{i=1}^{n} f(t_i)$$

is even optimal in the class of all adaptive methods.

□

Sketch of the proof: Assume that for some function $f^* \in F_{uv}$ we use n arbitrary knots t_i. Without loss of generality we assume that $0 \leq t_1 < t_2 < \ldots < t_n \leq 1$. Let $f^*(t_i) = y_i$. How can we compute the endpoints of the interval

$$J := \left\{ \int_0^1 f(t)\,dt \mid f \in F_{uv},\ f(t_i) = y_i \text{ for all } i \right\}?$$

The answer to this question seems to be needed for Proposition 2. A slightly different problem was studied by Heindl (1982). These problems, however, seem to be difficult and therefore we need a trick to prove Proposition 2. We consider the subspace $F^*_{uv} := C^1 \cap F_{uv}$ and the extended information

$$N^*(f) = (f(t_1), f'(t_1), \ldots, f(t_n), f'(t_n)).$$

We can ask the same question: How can we compute the endpoints of the interval

$$J^* := \left\{ \int_0^1 f(t)\, dt \mid f \in F^*_{uv},\ N^*(f) = N^*(f^*) \right\}?$$

Now the answer is much simpler because there is a smallest and a biggest convex function determined by $N^*(f)$. The length of J^* clearly depends on f^* and the knots t_i, but the following is true.

Lemma: Define $f^* \in F^*_{uv}$ by

$$f^*(t) = (v-u)\frac{t^2}{2} + ut.$$

Then the length l of the interval J^* is given by

$$l = (v-u)\left(\frac{t_1^3 + (1-t_n)^3}{2} + \sum_{i=1}^{n-1} \frac{(t_{i+1}-t_i)^3}{8} \right)$$

and

$$l \geq \frac{v-u}{8n^2}.$$

□

The proof of this lemma is elementary, and Proposition 2 follows.

□

Remark

Actually our proof shows that Zwick's method is optimal in the class of all adaptive methods using extended information of the form

$$N^*(f) = (f(t_1), f'(t_1), \ldots, f(t_n), f'(t_n)).$$

The main difficulty of this proof was to find the 'worst' function f^* which is used in the lemma. In the case of a convex and symmetric set F we always can take $f^* = 0$ as the worst function, see Bakhvalov (1971).

□

In the rest of this section we deal with Monte Carlo methods. We do not repeat known results about stochastic methods for other function classes, see Novak (1988a, 1992) and Traub, Wasilkowski, Woźniakowski (1988). We present results for methods of the form

$$S_n^\omega(f) = \phi^\omega(f(t_1^\omega), \dots, f(t_n^\omega)),$$

i.e., the knots t_i^ω are random variables (which can be chosen nonadaptively or adaptively) and also the ϕ^ω is randomly chosen. Then, as usual, the error of S_n^ω is defined in a 'worst case stochastic sense',

$$\Delta_{\max}(S_n^\omega) = \max_{f \in F} E(|\int_0^1 f(t)\,dt - S_n^\omega(f)|).$$

The proofs of the following results are similar to the proofs given in Novak (1992) for the class of monotone functions, therefore we skip some details.

Proposition 3
Let S_n^ω be any nonadaptive stochastic method for the integration problem on the class F_{uv}. Then the lower bound

$$\Delta_{\max}(S_n^\omega) \ge \frac{1}{128n^2}(v-u)$$

holds. Thus the (stochastic) error of nonadaptive stochastic methods can be only slightly smaller than the worst case error of deterministic methods.

□

Sketch of the proof: We divide the interval $I = [0,1]$ into $2n$ disjoint intervals I_j of the same length $1/2n$. For a nonadaptive method S_n^ω there is at least one little interval I_{j^*} such that no knot t_i^ω is in I_{j^*} with probability at least 1/2. It is easy to construct two piecewise-linear functions $f_1, f_2 \in F_{uv}$ with $f_1(t) = f_2(t)$ for $t \notin I_{j^*}$ and

$$\int_0^1 f_1(t)\,dt - \int_0^1 f_2(t)\,dt = \frac{v-u}{32n^2}.$$

Then we have $S_n^\omega(f_1) = S_n^\omega(f_2)$ with probability at least 1/2 and we get the desired result.

□

The following result, however, shows that adaptive Monte Carlo methods are much better (in a stochastic sense) than nonadaptive Monte Carlo methods or adaptive determinisitic methods. This is quite different than known results for unit balls of Hölder or Sobolev spaces, but it is similar to the results of Novak (1992) for the class of monotone functions.

Proposition 4
Let S_n^ω be any adaptive stochastic method for the integration problem on the class F_{uv}. Then the lower bound

$$\Delta_{\max}(S_n^\omega) \geq \frac{\sqrt{2}}{128}(v-u)n^{-5/2}$$

holds. This lower bound gives the optimal order of convergence because we can construct methods with

$$\Delta_{\max}(S_n^\omega) = O((v-u)\cdot n^{-5/2}).$$

□

Sketch of the proof: The lower bound can be proved by using Bakhvalov's technique. For $n \in \mathbf{N}$ we divide $[0,1]$ into $2n$ intervals I_j of length $1/2n$. Let $f^*(t) = (v-u)t^2/2 + ut$.
On I_j we can define two functions $\phi_{j,1}$ and $\phi_{j,2}$ with the following properties:
- $\phi_{j,1}$ is affine, $\phi_{j,2}$ is continuous and piecewise affine with two pieces;
- both functions coincide with f^* at the endpoints of the interval I_j, moreover, $\phi'_{j,2}$ coincides with $(f^*)'$ at the endpoints, too.

The functions $\phi_{j,i}$ are uniquely determined by these properties and

$$\int_{I_j} (\phi_{j,1}(t) - \phi_{j,2}(t))\, dt = (v-u)/(64n^3).$$

Each of the 2^{2n} functions on $[0,1]$ that coincides with one of the $\phi_{j,k}$ on I_j is convex and therefore in the class F_{uv}. Now consider the equidistribution μ on this set of 2^{2n} functions. One can estimate the expectation, with respect to μ, of the error of an arbitrary adaptive deterministic quadrature formula S_n to obtain

$$E_\mu(|\int_0^1 f(t)\, dt - S_n(f)|) \geq \frac{\sqrt{2}}{128}(v-u)n^{-5/2}.$$

By applying Fubini's theorem we obtain the desired lower bound.
We shortly describe an adaptive Monte Carlo method with the optimal order of convergence. Let $m \in \mathbf{N}$.
First step. Define $a_i = (2i-1)/(2m)$ for $i = 1, \ldots, m$ and compute $f(a_i)$ for each i. This step certainly is nonadaptive and deterministic. We also put $a_0 = 0$ and $a_{m+1} = 1$.
Second step. In each interval $[a_i, a_{i+1}]$ we determine the area F_i between the pointwise largest and smallest function (in the given interval) that fits the data at the points a_i. It can be proved that $\sum_i F_i \leq 3(v-u)/(8m^2)$. In the case

$$\frac{3(k-1)(v-u)}{8m^3} < F_i \leq \frac{3k(v-u)}{8m^3}$$

we divide the interval $[a_i, a_{i+1}]$ in k smaller intervals of the same size. In this way we obtain $r(f) \leq 2m + 1$ of the smaller intervals I_k. This step is adaptive but still deterministic. We do not compute any function values in this step.
Third step. Now we take t_k randomly according to the normalized Lebesgue measure in I_k; the random variables t_k are independent. We also define $t'_k \in I_k$ by the condition that $(t_k + t'_k)/2$ is the midpoint of I_k. We finally take the adaptive Monte Carlo method

$$S_n^\omega(f) = \sum_{k=1}^{r(f)} \frac{\lambda(I_k)}{2}(f(t_k) + f(t'_k)).$$

Observe that we use at most $n = m + 4m + 2$ function values. It can be proven that this method has the optimal order of convergence.

4. Optimal recovery: can adaption help?

In the rest of this paper we only study deterministic methods. We want to know whether adaption can help for linear problems on a convex set of functions. Let X and Y be Banach spaces and let $F \subset X$ be a convex set. We assume that $S : X \to Y$ is linear and continuous. We study the problem of optimal recovery of $S(f)$ for $f \in F \subset X$, if only (adaptive or nonadaptive) information of the form

$$N(f) = (L_1(f), L_2(f), \dots, L_n(f))$$

is available. Each method is of the form $S_n = \phi \circ N$ with some $\phi : \mathbf{R}^n \to Y$ and we want to minimize the maximal error

$$\Delta_{\max}(S_n) = \sup_{f \in F} \|S(f) - S_n(f)\|.$$

We assume that the L_i are arbitrary linear continuous functionals $L_i : X \to \mathbf{R}$. In the adaptive case the choice of L_i may depend on $L_1(f), \dots, L_{i-1}(f)$. See, for example, Traub, Wasilkowski, Woźniakowski (1988) for the exact definitions and known results. We want to compare the numbers

$$\inf \Delta_{\max}(S_n) \qquad \text{and} \qquad \inf \Delta_{\max}(S_n^{\mathrm{ad}}),$$

where S_n runs through all nonadaptive methods and S_n^{ad} runs through all adaptive methods. The following example shows that adaptive methods may be much better than nonadaptive methods.

Example 1. Let $X = Y = l_\infty$ and

$$F = \{x \in X \mid x_i \geq 0, \ \sum_{i=1}^{\infty} x_i \leq 1, \ x_k \geq x_{2k}, \ x_k \geq x_{2k+1}\}.$$

Assume that we want to reconstruct $x \in F$ in l_∞-norm using (adaptively or non-adaptively) linear functionals as information. That is $S = Id$. Let e^i be the sequence defined by $e^i_k = \delta_{ik}$. For $m \in \mathbf{N}$ we obtain

$$e^i/m \in F - F, \qquad i = 1, \dots, 2^{m-1}.$$

Now we can use a deep result (see Tikhomirov 1990, p. 192) on the Gelfand-diameters of the octahedron $O_N = \{x \in \mathbf{R}^N \mid \sum |x_i| \le 1\}$ in l_∞-norm, namely

$$d^N(O_{2N}) \asymp 1/\sqrt{N}.$$

Therefore for $m \in \mathbf{N}$ we conclude

$$d^{2^{m-2}}(F - F) \ge \frac{1}{m} \cdot d^{2^{m-2}}(O_{2^{m-1}}) \ge \frac{c'}{2^{m/2}m}.$$

From this we easily derive the lower bound

$$\inf_{S_n} \Delta_{\max}(S_n) \ge \frac{c}{\sqrt{n}\log n}$$

for the error of nonadaptive methods. Now we describe an adaptive method which is much better. For simplicity we assume here that n is odd. By δ_i we mean the functional $\delta_i(x) = x_i$. First we describe the functionals L_i which are of the form $L_i = \delta_{l_i}$. Take $L_1 = \delta_1$. Suppose that $L_i(x) = x_{l_i}$ are already computed for $1 \le i \le 2k - 1$. Define

$$J = \{j \in \{l_1, \dots, l_{2k-1}\} \mid 2j \notin \{l_1, \dots, l_{2k-1}\}\}$$

and

$$j^* = \min\{j \in J \mid x_j = \max_{j \in J} x_j\}.$$

Take $L_{2k} = \delta_{2j^*}$ and $L_{2k+1} = \delta_{2j^*+1}$. We consider the adaptive information

$$N_n(x) = (L_1(x), \dots, L_n(x)).$$

From the definition of N_n follows that

$$N_n(x) = (x_{l_1}, \dots, x_{l_n})$$

with pairwise different l_i. The number n is odd, say $n = 2m + 1$. Then we already have picked up the largest $(m + 1)$ coordinates of x. Since $x \in F$, we conclude that

$$x_k \le 1/(m + 2)$$

for all k which do not coincide with one of the l_i. Hence we can reconstruct x from the information N_n up to an error of $1/(2m + 4) = 1/(n + 3)$, i.e., we have found a method with

$$\Delta_{\max}(S_n^{\mathrm{ad}}) \le 1/(n + 3).$$

□

5. On adaptive quadrature formulas

In most practical cases, X is a Banach space of functions and only certain linear functionals are available as information. Here we only study the case that all functionals

$$\delta_x : f \in X \mapsto f(x) \in \mathbf{R}$$

are continuous and form the set of available functionals, i.e, each L_i is of the form $L_i = \delta_{x_i}$. Hence we study methods of the form

$$S_n^{(\mathrm{ad})}(f) = \phi(f(x_1), \dots, f(t_n)).$$

We consider the case $S : X \to \mathbf{R}$, i.e., we are looking for optimal quadrature formulas. It is rather easy to find an example where adaptive quadrature formulas are slightly better than nonadaptive ones. In the following example the set F is 3-dimensional and $n = 2$, so it is the simplest possible example.

Example 2. We put

$$F = \{f : [0,1] \to \mathbf{R} \mid f_{|[i/3,(i+1)/3[} = c_i, \quad i = 0,1,2; \quad c_0 \in [-1,1],$$
$$c_1 \in [-(1+c_0)/2, (1+c_0)/2], \quad c_2 \in [-(1-c_0)/2, (1-c_0)/2]\}$$

and

$$S(f) = \int_0^1 f(t)\,dt.$$

Then F is a convex set and $S(f) = (c_0 + c_1 + c_2)/3$ for $f \in F$. It is easy to prove that each nonadaptive formula $S_2(f) = \phi(f(t_1), f(t_2))$ has the worst case error of at least $1/3$ while there is an adaptive formula with error $1/6$ on F.

□

One can find more dramatic examples. For simplicity we again choose a finite dimensional example. Given this, however, one also could construct more difficult infinitely dimensional examples.

Example 3. Let

$$F = \{x \in \mathbf{R}^{2^N-1} \mid x_i \ge 0, \sum_{i=1}^{2^N-1} x_i \le 1, \ x_k \ge x_{2k}, \ x_k \ge x_{2k+1}\}.$$

Then $F \subset X$ is convex. We consider the linear functional

$$S(x) = \sum_{i=1}^{2^N-1} x_i.$$

We only study the simplest case $n = 2^N - 2$. Assume that we want to reconstruct $S(x) \in \mathbf{R}$ using (adaptively or nonadaptively) linear functionals of the form $L_i(x) = x_{l_i}$. For nonadaptive methods we obtain the error bound

$$\inf_{S_n} \Delta_{\max}(S_n) \geq \frac{1}{2N} \approx \frac{1}{\log n}.$$

Now we describe an adaptive method which is much better. First we evaluate

$$x_1, x_2, \ldots, x_{2^{N-1}-1}.$$

At least one of these numbers, say x_l, satisfies

$$x_l \leq \frac{1}{2^{N-1} - 1}.$$

Then we compute all remaining coordinates of x with the exception x_{2l}. Now it is easy to define a method S_n^{ad} which uses this information and satisfies

$$\Delta_{\max}(S_n^{\mathrm{ad}}) \leq \frac{1}{2^N - 2} = \frac{1}{n}.$$

□

Acknowledgment

I am grateful for remarks of Knut Petras, Klaus Ritter, and Henryk Woźniakowski concerning this paper. The author is supported by a Heisenberg scholarship of the DFG.

References

N. S. Bakhvalov (1959): On approximate computation of integrals. Vestnik MGU, Ser. Math. Mech. Astron. Phys. Chem. **4**, 3–18. [Russian]

N. S. Bakhvalov (1971): On the optimality of linear methods for operator approximation in convex classes of functions. USSR Comput. Math. Math. Phys. **11**, 244–249. [Russian original: Zh. vychisl. Mat. mat. Fiz. **11**, 244–249]

H. Braß (1982): Zur Quadraturtheorie konvexer Funktionen. In: G. Hämmerlin (ed.), Numerical integration, pp. 34–47, ISNM **57**.

K.-J. Förster, K. Petras (1990): On a problem proposed by H. Braß concerning the remainder term in quadrature for convex functions. Numer. Math. **57**, 763–777.

S. Gal, C. A. Micchelli (1980): Optimal sequential and non-sequential procedures for evaluating a functional. Appl. Anal. **10**, 105–120.

I. A. Glinkin (1983): Optimal algorithms for integrating convex functions. Math. USSR Comput. Maths. Math. Phys. **23**(2), 6–12. [Russian original: Zh. vychisl. Mat. mat. Fiz. **23**(2), 267–277]

I. A. Glinkin (1984): Best quadrature formula in the class of convex functions. Math. Notes **35**, 368–374. [Russian original: Mat. Zametki **35**, 697–707]

R. Gorenflo, S. Vessella (1991): Abel Integral Equations. Lecture Notes in Mathematics **1461**, Springer-Verlag.

G. Heindl (1982): Optimal quadrature of convex functions. In "Numerical Integration", ISNM **57**, 138–147.

I. P. Huerta (1986): Adaption helps for some nonconvex classes. J. Complexity **2**, 333-352.

A. Iserles, G. Söderlind (1992): Global bounds on numerical error for ordinary differential equations. Numerical Analysis Reports, University of Cambridge, England, to appear in J. Complexity

J. Kiefer (1957): Optimum sequential search and approximation methods under minimum regularity assumptions. J. Soc. Indust. Appl. Math. **5**, 105–136.

M. A. Kon, E. Novak (1989): On the adaptive and continuous information problems. J. Complexity **5**, 345–362.

M. A. Kon, E. Novak (1990): The adaption problem for approximating linear operators. Bull. Amer. Math. Soc. **23**, 159–165.

N. P. Korneichuk (1993): Optimization of active algorithms for recovery of functions from the class Lip-α. To appear in J. Complexity

C. A. Micchelli, T. J. Rivlin (1977): A survey of optimal recovery. In: Optimal estimation in approximation theory, C. A. Micchelli and T. J. Rivlin (eds.), Plenum Press, New York.

C. A. Micchelli, T. J. Rivlin (1985): Lectures in optimal recovery. Lecture Notes in Math. **1129**, Springer.

E. Novak (1988a): Deterministic and stochastic error bounds in numerical analysis. Lecture Notes in Mathematics **1349**, Springer.

E. Novak (1988b): Stochastic properties of quadrature formulas. Numer. Math. **53**, 609–620.

E. Novak (1992): Quadrature formulas for monotone functions. Proc. Amer. Math. Soc. **115**, 59–68.

E. Novak (1993): Optimal recovery and n-widths for convex classes of functions. In preparation.

A. Papageorgiou (1992): Integration of monotone functions of several variables. To appear in J. Complexity

K. Petras (1993): Quadrature theory of convex functions. These proceedings.

K. Ritter, G. W. Wasilkowski, H. Woźniakowski (1993): On multivariate integration for stochastic processes. These proceedings.

A. Schnorr (1992): Optimale Verfahren zur Integration konvexer Funktionen. Diplomarbeit, Mathematisches Institut Erlangen.

G. Sonnevend (1983): An optimal sequential algorithm for the uniform approximation of convex functions on $[0,1]^2$. Appl. Math. Optim. **10**, 127–142.

A. G. Sukharev (1986): On the existence of optimal affine methods for approximating linear functionals. J. Complexity **2**, 317–322.

V. M. Tikhomirov (1990): Approximation theory. In: Encyclopaedia of Mathematical Sciences, Vol. 14, Springer-Verlag. [The Russian original was published by Publisher VINITI, Moscow 1987, R. V. Gamkrelidze, ed.]

J. F. Traub, G. W. Wasilkowski, H. Woźniakowski (1988): Information-based complexity. Academic Press.

J. F. Traub, H. Woźniakowski (1980): A general theory of optimal algorithms. Academic Press.

G. W. Wasilkowski, F. Gao (1992): On the power of adaptive information for functions with singularities. Math. of Comp. **58**, 285–304.

G. W. Wasilkowski, H. Woźniakowski (1984): Can adaption help on the average?, Numer. Math. **44**, 169–190.

D. Zwick (1988): Optimal quadrature for convex functions and generalizations. In: Numerical Integration III, H. Braß, G. Hämmerlin (eds.), ISNM **85**, 310–315.

Erich Novak
Mathematisches Institut
Universität Erlangen
Bismarckstr. 1 1/2
W-8520 Erlangen
Germany
novak@mi.uni-erlangen.de

International Series of Numerical Mathematics, Vol. 112, © 1993 Birkhäuser Verlag Basel

On positive quadrature formulas

Franz Peherstorfer

Abstract. In this paper we show how to get in a simple and unified way most of the known characterizations of positive interpolatory quadrature formulas and obtain also some new results.

1. Introduction

Let σ be a positive measure on $[-1,+1]$ normed by $\int_{-1}^{+1} d\sigma = 1$ and such that the support of $d\sigma$ contains an infinite set of points. We call an interpolatory quadrature formula (abbreviated q.f.) of the form

$$(1.1) \qquad \int_{-1}^{+1} f(x)d\sigma(x) = \sum_{j=1}^{n} d_j f(x_j) + R_n(f),$$

where $-1<x_1<x_2<\ldots<x_n<1$ and $R_n(f) = 0$ for $f \in \mathbb{P}_{2n-1-m}$ ($\mathbb{P}_n$ denotes as usual the set of polynomials of degree at most n), $0\le m\le n$, a $(2n-1-m,n,d\sigma)$ q.f.. If in addition all quadrature weights d_j, $j = 1,\ldots,n$ are positive then it is called a positive $(2n-1-m,n,d\sigma)$ q.f.. Furthermore we say that a polynomial $t_n \in \mathbb{P}_n$ generates a $(2n-1-m,n,d\sigma)$ q.f. if t_n has n simple zeros $x_1<x_2<\ldots<x_n$ in $(-1,1)$ and if the interpolatory q.f. based on the nodes x_j, $j = 1,\ldots,n$, is a $(2n-1-m,n,d\sigma)$ q.f.. For $q_n \in \mathbb{P}_n \setminus \mathbb{P}_{n-1}$ we denote by $q_{n-1}^{[1]}$ the associated polynomial of q_n with respect to $d\sigma$, i.e.

$$(1.2) \qquad q_{n-1}^{[1]}(y) = \int_{-1}^{+1} \frac{q_n(y) - q_n(x)}{y - x} \, d\sigma(x).$$

Note that $q_{n-1}^{[1]} \in \mathbb{P}_{n-1}$. With the help of (1.2) we obtain for the weights d_j of the interpolatory q.f. (1.1) the known formula

$$(1.3) \qquad d_j = \int_{-1}^{+1} \frac{t_n(x)}{(x-x_j)t_n'(x_j)} d\sigma(x) = \frac{t_{n-1}^{[1]}(x_j)}{t_n'(x_j)} \quad \text{for } j = 1,\dots,n,$$

where

$$(1.4) \qquad t_n(x) = \prod_{j=1}^{n} (x-x_j).$$

Hence the positivity of the d_j's is equivalent to the interlacing property of the zeros of t_n and $t_{n-1}^{[1]}$. Taking into consideration the well known fact (see e.g. Brass 1977) that a polynomial t_n with n simple zeros generates a $(2n-1-m,n,d\sigma)$ q.f. if and only if it is orthogonal to $\mathbb{P}_{n-1-m}$ with respect to $d\sigma$ we get the following first characterization of positive q.f..

Lemma 1.1. *Let* $n,m \in \mathbb{N}_0$. *Then* t_n *generates a positive* $(2n-1-m,n,d\sigma)$ *q.f. if and only if* t_n *is orthogonal to* $\mathbb{P}_{n-1-m}$ *with respect to* $d\sigma$, t_n *has* n *simple zeros in* $(-1,1)$ *and the zeros of* t_n *and* $t_{n-1}^{[1]}$ *interlace.*

Next let us denote by p_n the monic polynomial of degree n which is orthogonal to $\mathbb{P}_{n-1}$ with respect to $d\sigma$, i.e.

$$(1.5) \qquad \int_{-1}^{+1} x^j p_n(x) d\sigma(x) = 0 \quad \text{for } j = 0,\dots,n-1,$$

and let us recall that (p_n) satisfies a recurrence relation of the form

$$(1.6) \qquad p_n(x) = (x-\alpha_n)p_{n-1}(x) - \lambda_n p_{n-2}(x) \quad n = 1,2,\dots,$$

where $p_{-1} := 0$, $p_0 := 1$ and the λ_n's have the property to be positive. Returning to the problem of characterizing positive q.f. we see, observing that a polynomial t_n orthogonal to $\mathbb{P}_{n-1-m}$ with respect to $d\sigma$ can be represented in the form, $\mu_1,\dots,\mu_m \in \mathbb{R}$.

$$(1.7)\qquad t_n(x) = \sum_{j=o}^{m} \mu_j p_{n-j}(x),$$

that this problem can be considered as the problem of finding necessary and sufficient conditions on the numbers μ_j from (1.7) such that $\sum_{j=o}^{m} \mu_j p_{n-j}$ and $\sum_{j=o}^{m} \mu_j p_{n-j-1}^{[1]}$ have n resp. n-1 simple interlaced zeros in (-1,1). This explains in a certain way that the problem under consideration is a nontrivial one, since even the problem of finding conditions such that a linear combination of orthogonal polynomials of the form (1.7) has all its zeros simple and located in (-1,1) is not solved so far.

The goal of this paper is to show how to get in a simple and unified way most of the known characterizations of positive quadrature formulas. One of the reasons that this might be of interest consists in the fact that several results on this topic were obtained by proofs being not very transparent. Let us note that in this paper we are mainly interested in general necessary and sufficient conditions guaranteeing positivity of the q.f., in particular of q.f. with high degree of exactness, and that we do not discuss positivity of q.f. based on nodes which are the zeros of special orthogonal polynomials like the Jacobi polynomials. For questions of this type see Askey (1986) and the references given there, Locher (1973) for a simple approach when positivity fails and Sottas (1988). This paper is organized as follows. In Section 2 we put together all facts on polynomials orthogonal with respect to a positive definite sequence which are needed in the sequel. Then in Section 3 we show how to get with the help of the results of Section 2 different characterizations of positive quadrature formulas. Finally in Section 4 we discuss several consequences of the given characterizations, among others we obtain new simple conditions implying the positivity of the q.f. if t_n is given in the form (1.7) and get new facts on the location of the nodes of a positive q.f..

2. Polynomials orthogonal with respect to a positive definite sequence

Let $N \in \mathbb{N} \cup \{\infty\}$ and let $\{c_j\}_o^N$ be a sequence of real numbers. To such a sequence

$\{c_j\}_o^N$ we associate the following linear functional $\alpha_c : \mathbb{P}_N \to \mathbb{R}$ defined by

$$\alpha_c\left(\sum_{j=o}^{N} a_j x^j\right) = \sum_{j=o}^{N} a_j c_j .$$

where $\mathbb{P}_\infty$ denotes the set of all real polynomials. The sequence $\{c_k\}_o^m$, $0 \leq m \leq N$, is called positive definite on $[-1,1]$ if for arbitrary $q \in \mathbb{P}_m$, $q \not\equiv 0$,

$$q \geq 0 \text{ on } [-1,1] \text{ implies } \alpha_c(q) > 0.$$

Henceforth in this section we assume that $\{c_j\}_o^N$ *is positive definite on* $[-1,1]$. We say that the polynomial $P_n(x) = x^n + \ldots$, $0 \leq n \leq [\frac{N+1}{2}]$, is orthogonal with respect to $\{c_j\}_o^m$, $2n-1 \leq m \leq N$, resp. with respect to the linear functional α_c if

$$\alpha_c(x^k P_n(x)) = 0 \quad \text{for } k = 0,\ldots,n-1,$$

i.e.

$$\sum_{j=o}^{n} a_j c_{k+j} = 0 \quad \text{for } k = 0,\ldots,n-1. \tag{2.1}$$

Now let us state some known facts on polynomials orthogonal with respect to $\{c_j\}_o^N$ resp. α_c which will be needed in the following and can be found in Akhiezer and Krein (1962), Akhiezer (1965) and Brezinski (1980). It is well known that P_n can be represented in the form, $1 \leq n \leq [\frac{N+1}{2}]$,

$$P_n(x) = \frac{1}{\Delta_{n-1}} \cdot \begin{vmatrix} c_o & c_1 & \cdots & c_n \\ \vdots & \vdots & & \vdots \\ c_{n-1} & c_n & \cdots & c_{2n-1} \\ 1 & x & \cdots & x^n \end{vmatrix} \tag{2.2}$$

where

$$\Delta_n := \begin{vmatrix} c_o & c_1 & \cdots & c_n \\ \vdots & \vdots & & \vdots \\ c_n & c_{n-1} & \cdots & c_{2n} \end{vmatrix} \tag{2.3}$$

and that the P_n's satisfy the recurrence relation

$$(2.4)\qquad P_n(x) = (x-\gamma_n)P_{n-1}(x) - \beta_n P_{n-2}(x) \quad n = 1,2,\ldots,[\tfrac{N+1}{2}]$$

where $P_{-1}(x) := 0$, $P_o(x) := 1$, and

$$(2.5)\qquad \beta_n = \frac{\Delta_{n-1}\Delta_{n-3}}{\Delta_{n-2}^2} > 0, \qquad \gamma_n = \frac{\Delta'_{n-1}}{\Delta_{n-1}} - \frac{\Delta'_{n-2}}{\Delta_{n-2}},$$

and

$$(2.6)\qquad \Delta'_{n-1} := \begin{vmatrix} c_o & c_1 & \cdots & c_{n-2} & c_n \\ c_1 & c_2 & \cdots & c_{n-1} & c_{n+1} \\ \vdots & \vdots & & \vdots & \vdots \\ c_{n-2} & c_{n-1} & \cdots & c_{2n-4} & c_{2n-2} \\ c_{n-1} & c_n & \cdots & c_{2n-3} & c_{2n-1} \end{vmatrix}$$

from which it follows (compare also Natanson 1965, § 2) that all zeros of P_n, $1 \le n \le [\frac{N+1}{2}]$, are real, simple and located in $(-1,1)$ and that the zeros of P_n and P_{n-1} interlace.

An important role will also play the associated polynomials of P_n of order k denoted by $P_n^{(k)}$, $n,k, \in \mathbb{N}$, $n+k \le [\frac{N+1}{2}]$, and defined by the shifted recurrence relation

$$(2.7)\qquad P_n^{(k)}(x) = (x-\gamma_{n+k})P_{n-1}^{(k)}(x) - \beta_{n+k}P_{n-2}^{(k)}(x), \quad n = 1,\ldots,[\tfrac{N+1}{2}]-k,$$

where $p_{-1}^{(k)}(x) := 0$ and $P_o^{(k)}(x) := 1$. The associated polynomials of different order are related to each other by the known formula, $0 \le k \le [\frac{N+1}{2}]$, $1 \le n \le [\frac{N+1}{2}]-k$,

$$(2.8)\qquad P_n^{(k)} = (x-\gamma_{k+1})P_{n-1}^{(k+1)} - \beta_{k+2}P_{n-2}^{(k+2)}.$$

Furthermore we obtain from (2.7) and (2.8) by induction arguments that for $1 \le j \le n-1$

$$(2.9)\qquad P_n^{(k)} = P_j^{(n+k-j)}P_{n-j}^{(k)} - \beta_{n+k+1-j}P_{j-1}^{(n+k+1-j)}P_{n-j-1}^{(k)}$$

and that this representation is unique for $2j \leq n$; i.e. if $2j \leq n$ and

$$P_n^{(k)} = u_j P_{n-j}^{(k)} - v_{j-1} P_{n-j-1}^{(k)},$$

where $u_j \in \mathbb{P}_j$, $v_{j-1} \in \mathbb{P}_{j-1}$, then

$$u_j = P_j^{(n+k-j)} \quad \text{and} \quad v_{j-1} = \beta_{n+k+1-j} P_{j-1}^{(n+k+1-j)},$$

which follows by considering (2.9) at the zeros of $P_{n-j}^{(k)}$. The associated polynomial of order one, also called simply associated polynomial, has a representation of the form

$$\text{(2.10)} \qquad \text{const. } P_{n-1}^{(1)}(y) = \alpha_c\left(\frac{P_n(y)-P_n(x)}{y-c}\right),$$

since the polynomial at the right hand side satisfies the same recurrence relation as $P_{n-1}^{(1)}$. With the help of (2.10) the following characterization of the polynomial orthogonal with respect to the sequence $\{c_j\}_o^N$ can be shown (see Peherstorfer 1990, Lemma 3, and also Natanson 1965)

Lemma 2.1. *Let* $0 \leq n \leq [\frac{N+1}{2}]$ *and let* $P_n(x) = x^n + \ldots$ *and* $Q \in \mathbb{P}_{n-1}$. *Then* P_n *is orthogonal with respect to the sequence* $\{c_j\}_o^N$ *and* $Q = \text{const } P_{n-1}^{(1)}$ *if and only if*

$$\text{(2.11)} \qquad \frac{Q(z)}{P_n(z)} = \sum_{j=o}^{2n-1} c_j z^{-(j+1)} + O\left(\frac{1}{z^{2n+1}}\right) \quad \text{for } |z| \to \infty.$$

Note that it follows immediately from (2.11) that

$$\text{(2.12)} \qquad P_{n-1}P_{n-1}^{(1)} - P_n P_{n-2}^{(1)} = \prod_{j=2}^{n} \beta_j > 0$$

which implies by the interlacing property of the zeros of P_n and P_{n-1} that the zeros of P_n and $P_{n-1}^{(1)}$ are also interlaced. Furthermore let us point out that in view of (2.7) all the zeros of $P_n^{(k)}$ are real, simple and interlace with those one of $P_{n-1}^{(k)}$ and, because of (2.8) using induction arguments with respect to k, also with those one of $P_{n-1}^{(k+1)}$.

Finally let us mention the simple fact that all results apply to the polynomials p_n orthogonal with respect to $d\sigma$ since obviously $c_j = \int_{-1}^{+1} x^j d\sigma(x)$, $j = 0,1,2,\ldots$ is a positive definite sequence. In view of (1.2) and (2.10) we also have

$$p_{n-1}^{(1)} = p_{n-1}^{[1]} \quad \text{for } n \in \mathbb{N}. \tag{2.13}$$

In the following we will also need the fact that the associated polynomials $(p_n^{(k)})$ of order k, $k \in \mathbb{N}_0$, are by Favard's Theorem (see Natanson 1965) orthogonal with respect to a positive measure, denoted by $\sigma^{(k)}$, the support of which is, since all zeros of $p_n^{(k)}$ lie in (-1,+1), contained in [-1,+1], i.e.,

$$\int_{-1}^{+1} x^j p_n^{(k)}(x) d\sigma^{(k)}(x) = 0 \quad \text{for } j = 0,\ldots,n-1. \tag{2.14}$$

3. The Characterization Theorem

In order to get our main characterization we need the following Lemma, which is essentially known.

Lemma 3.1. *Let* $\ell \in \mathbb{N}$ *and let* $g_\ell(x) = x^\ell + \ldots, g_{\ell-1}(x) = x^{\ell-1} + \ldots$ *be two aribtrary polynomials whose zeros are simple, interlaced and located in* (-1,+1). *Then the polynomials* $g_\ell, g_{\ell-1}$ *can be generated uniquely by a recurrence relation of the form*

$$g_j(x) = (x-\gamma_j) g_{j-1}(x) - \beta_j g_{j-2}(x) \quad j = 1,\ldots,\ell. \tag{3.1}$$

$g_{-1}(x) := 0$ *and* $g_0(x) := 1$, *with*

$$\beta_j > 0 \text{ and } g_j(\pm 1) = (\pm 1)^j \quad \text{for } j = 1,\ldots,\ell \tag{3.2}$$

where the g_j*'s,* $j = 1,\ldots,\ell$, *are monic polynomials of degree j having all zeros simple and located in* (-1,1) *and the zeros of* g_j *and* g_{j-1} *interlace.*

Proof. Wendroff (1961) has proved that g_ℓ and $g_{\ell-1}$ can be generated by a recurrence relation of the form (3.1) with $\beta_j > 0$, $j = 2,\ldots,\ell$. Since the zeros of g_ℓ and $g_{\ell-1}$ interlace and are located in $(-1,+1)$ it follows by induction arguments, using (3.1) and the fact that $\beta_j>0$, that the zeros of g_j and g_{j-1}, $j = 1,\ldots,\ell$, have also these properties.□.

We divide the characterization of positive q.f. in two theorems. The first one will be proved by elementary methods using heavily the results of Section 2 while for the second theorem a few basic facts on polynomials orthogonal on the unit circle are used.

Characterization Theorem 3.2. *Let* $n,m \in \mathbb{N}_0, n\geq m$, *and let* t_n *be a monic polynomial of degree n. The following statements are equivalent:*

(a) t_n *generates a positive* $(2n-1-m,n,d\sigma)$ *q.f.*

(b) t_n *is orthogonal with respect to a sequence* $\{c_j\}_0^{2n-1}$ *which is positive definite on* $[-1,1]$ *and satisfies the conditions*

$$c_j = \int_{-1}^{+1} x^j d\sigma(x) \quad \text{for} \quad j = 0,\ldots,2n-1-m \tag{3.3}$$

(c) t_n *can be generated by a recurrence relation of the form*

$$t_j(x) = (x-\tilde{\alpha}_j)t_{j-1}(x) - \tilde{\lambda}_j t_{j-2}(x) \quad j = 1,\ldots,n, \tag{3.4}$$

$t_{-1}(x) := 0$, $t_o(x) := 1$, *with* $\tilde{\alpha}_j \in \mathbb{R}$ *and* $\tilde{\lambda}_j>0$ *for* $j = 1,\ldots,n$, *and*

$$\tilde{\alpha}_j = \alpha_j \text{ for } j = 1,\ldots,n-[\tfrac{m+1}{2}] \text{ and } \tilde{\lambda}_j = \lambda_j \text{ for } j = 1,\ldots,n-[\tfrac{m}{2}] \tag{3.5}$$

and

$$\operatorname{sgn} t_j(\pm 1) = (\pm 1)^j \quad \text{for } j = 1,\ldots,n. \tag{3.6}$$

(d) Put $\ell := [(m+1)/2]$. *There are polynomials* $g_\ell(x) = x^\ell+\ldots, g_{\ell-1}(x) = x^{\ell-1}+\ldots$, *whose zeros are simple, interlaced and located in* $(-1,1)$ *such that*

$$t_n(x) = g_\ell(x)p_{n-\ell}(x) - \tilde{\lambda}_{n+1-\ell}g_{\ell-1}(x)p_{n-\ell-1}(x) \tag{3.7}$$

with $\tilde{\lambda}_{n+1-\ell} > 0$ *and* $\tilde{\lambda}_{n+1-\ell} = \lambda_{n+1-\ell}$ *if* $m = 2\ell-1$ *and* $t_n(\pm 1) = (\pm 1)^n$.

Proof. (a) => (b). Since t_n generates a positive $(2n-1-m,n,d\sigma)$ q.f. we have by Lemma 1.1 that the zeros of t_n and $t_{n-1}^{[1]}$ strictly interlace. Hence partial fraction expansion together with (1.3) gives

$$\frac{t_{n-1}^{[1]}(z)}{t_n(z)} = \sum_{j=1}^{n} \frac{d_j}{z-x_j} = \sum_{k=o}^{\infty} \frac{c_k}{z^{k+1}} \qquad \text{for } z \in \mathbb{C}\setminus[-1,+1]. \tag{3.8}$$

where

$$c_k := \sum_{j=1}^{n} d_j x_j^{\,k} \quad \text{for } k = 0,1,2,\ldots. \tag{3.9}$$

Obviously, since $d_j>0$ for $j = 1,\ldots,n$ and $-1<x_1<\ldots<x_n<1$, the sequence $\{c_k\}_o^\infty$ is positive definite on $[-1,1]$ and, by (3.8) and Lemma 2.1, t_n is orthogonal with respect to $\{c_k\}_o^{2n-1}$. Taking into consideration the fact that the q.f. has degree of exactness $2n-1-m$ relation (3.3) follows immediately from (3.9).

(b) => (c). Since $\{c_k\}_o^{2n-1}$ is positive definite on $[-1,1]$ it follows from Section 2 that the polynomials t_j, $j = 0,\ldots,n$, orthogonal with respect to $\{c_k\}_o^{2n-1}$ satisfy a recurrence relation of the form (3.4) with $\tilde{\lambda}_j>0$, $j = 2,\ldots,n$, and have all their zeros simple and located in $(-1,1)$. Hence (3.4) and (3.6) are fulfilled and (3.5) remains only to be shown. Since by (2.5) $\tilde{\alpha}_j$ resp. $\tilde{\lambda}_j$ depend on $c_o,\ldots,c_{2j-1}$ resp. $c_o,\ldots,c_{2j-2}$ only, relation (3.5) follows immediately because of (3.3).

(c) => (d). The implication follows from relation (2.9) and what has been said there by setting

$$g_\ell = t_\ell^{(n-\ell)} \quad \text{and} \quad g_{\ell-1} = t_{\ell-1}^{(n+1-\ell)}$$

and recalling the properties of the zeros of the associated polynomials mentioned in Section 2.

(d) => (a) In view of Lemma 3.1 g_ℓ and $g_{\ell-1}$ can be generated uniquely by a recurrence relation of the form

$$g_j = (x-\tilde{\alpha}_{n+1-j})g_{j-1} - \tilde{\lambda}_{n+2-j}g_{j-2} \quad j = 1,\ldots,\ell, \tag{3.10}$$

where $g_{-1}(x) := 0$, $g_o(x) := 1$ and where

$$\tilde{\lambda}_{n+2-j} > 0 \quad \text{and} \quad \operatorname{sgn} g_j(\pm 1) = (\pm 1)^j \quad \text{for } j = 1,\ldots,\ell$$

Now let us consider the polynomials $\tilde{t}_j$, $j = 0,\ldots,n$, generated by

$$(3.11) \qquad \tilde{t}_j = (x - \tilde{\alpha}_j)\tilde{t}_{j-1} - \tilde{\lambda}_j\tilde{t}_{j-2} \quad j = 1,\ldots,n,$$

$\tilde{t}_{-1}(x) := 0, \tilde{t}_o(x) := 1$, where

$$(3.12) \qquad \tilde{\alpha}_j = \alpha_j \quad \text{and} \quad \tilde{\lambda}_j = \lambda_j \quad \text{for } j = 1,\ldots,n-\ell,$$

$\tilde{\lambda}_{n+1-\ell}$ is given by (3.7) and the remaining $\tilde{\alpha}_j$'s and $\tilde{\lambda}_j$'s are given by (3.10). Then it follows, note because of (3.10) we have

$$(3.13) \qquad g_j = \tilde{t}_j^{(n-j)} \quad j = 1,\ldots,\ell,$$

that

$$\tilde{t}_n = g_j\tilde{t}_{n-j} - \tilde{\lambda}_{n+1-j}g_{j-1}\tilde{t}_{n-1-j} \quad j = 1,\ldots,\ell$$

and moreover, taking into account (3.12)

$$\tilde{t}_n = g_\ell p_{n-\ell} - \tilde{\lambda}_{n+1-\ell}g_{\ell-1}p_{n-\ell-1} = t_n.$$

For the associated polynomial $\tilde{t}_{n-1}^{(1)}$ of $\tilde{t}_n$ defined by

$$\tilde{t}_j^{(1)} = (x-\tilde{\alpha}_{j+1})t_{j-1}^{(1)} - \tilde{\lambda}_{j+1}\tilde{t}_{j-2}^{(1)} \quad j = 1,\ldots,n-1,$$

$\tilde{t}_{-1}^{(1)}(x) := 0$, $\tilde{t}_o^{(1)}(x) := 1$, we get in an analogous way using (2.9), (3.12) and (3.13) that

$$\tilde{t}_{n-1}^{(1)} = g_\ell p_{n-\ell-1}^{(1)} - g_{\ell-1}p_{n-\ell-2}^{(1)}$$

On the other hand if we take the associated polynomial of t_n given in (3.7) with respect to $d\sigma$ then it follows with the help of the orthogonality property of $p_{n-\ell}$ and $p_{n-\ell-1}$ and the fact that $n \geqq 2\ell$

$$t_{n-1}^{[1]} = g_\ell p_{n-\ell-1}^{(1)} - g_{\ell-1}p_{n-\ell-2}^{(1)} = \tilde{t}_{n-1}^{(1)}.$$

Since by the results of Section 2 $\tilde{t}_n = t_n$ and $\tilde{t}_{n-1}^{(1)} = t_{n-1}^{[1]}$ have n resp. n-1 simple, interlaced zeros in (-1,1) the implication follows from

Lemma 1.1. □

Next let us give other characterizations. We denote by $\Phi_n(z) = z^n+\dots$, $n \in \mathbb{N}_0$, the polynomial orthogonal on $[0,2\pi]$ with respect to the positive measure

$$\psi(\varphi) = \begin{cases} -\sigma(\cos\varphi) & \text{for } \varphi \in [0,\pi], \\ \sigma(\cos\varphi) & \text{for } \varphi \in (\pi,2\pi], \end{cases}$$

i.e.

$$\int_0^{2\pi} e^{-ik\varphi}\Phi_n(e^{i\varphi})d\psi(\varphi) = 0 \quad \text{for } k = 0,\dots,n-1.$$

Note if σ is absolutely continuous on $[-1,+1]$ and $\sigma'(x) = w(x)$ then ψ is absolutely continuous with $\psi'(\varphi) = w(\cos\varphi)|\sin\varphi|$ for $\varphi \in [0,2\pi]$. It is well known, see e.g. Szegö (1967) that the Φ_n's satisfy a recurrence relation of the type

$$\Phi_n(z) = z\Phi_{n-1}(z) - a_{n-1}\Phi^*_{n-1}(z) \quad \text{for } n \in \mathbb{N},$$

where $a_n \in (-1,1)$ for $n \in \mathbb{N}_0$ and where $\Phi^*_n(z) = z^n\Phi_n(z^{-1})$ denotes the reciprocal polynomial of Φ_n. The following relations hold between polynomials p_n resp. $p^{(1-x^2)}_{n-1}$ orthogonal on $[-1,+1]$ with respect to $d\sigma$ resp. $(1-x^2)d\sigma$:

$$2^{n-1}p_n(x) = \mathrm{Re}\{z^{-n+1}\Phi_{2n-1}(z)\} = \frac{\mathrm{Re}\{z^{-n}\Phi_{2n}(z)\}}{(1-a_{2n-1})}, \tag{3.16}$$

$$2^{n-1}p^{(1-x^2)}_{n-1}(x) = \frac{\mathrm{Im}\{z^{-n+1}\Phi_{2n-1}(z)\}}{\sin\varphi} = \frac{\mathrm{Im}\{z^{-n}\Phi_{2n}(z)\}}{(1+a_{2n-1})\sin\varphi}, \tag{3.17}$$

where $x = \frac{1}{2}(z+z^{-1})$, $z = e^{i\varphi}$, $\varphi \in [0,\pi]$. The parameters are given by

$$a_{2n-1} = 1-(u_n+v_n) \quad \text{and} \quad a_{2n} = (v_n-u_n)/(v_n+u_n),$$

where

$$u_n = p_{n+1}(1)/p_n(1) \quad \text{and} \quad v_n = -p_{n+1}(-1)/p_n(-1).$$

Characterization Theorem 3.2. *Let* $n,m \in \mathbb{N}_0$, $n \geq m$, *and let* t_n *be a monic polynomial of degree* n. *The following statements are equivalent:*

(a) t_n *generates a positive* $(2n-1-m,n,d\sigma)$ *q.f.*

(e) *Put* $\ell := [(m+1)/2]$. *There are polynomials* r_ℓ, $s_{\ell-1}$ *of degree* ℓ *resp.*

$\ell-1$, *whose zeros are simple, interlaced and located in* $(-1,1)$ *such that*

(3.19) $$t_n(x) = r_\ell(x)p_{n-\ell}(x) - s_{\ell-1}(x)p_{n-\ell-1}(x)$$

with $t_n(\pm1) = (\pm1)^n$. *If* $m = 2\ell-1$ *then* r_ℓ *resp.* $s_{\ell-1}$ *has leading coefficient* $(1-a_{2(n-\ell)-1})/2$ *resp.* $(1+a_{2(n-\ell)-1})/2$.

(f) *There is a real polynomial* $q_m(z) = z^m+\ldots$ *having all its zeros in the open unit disk* $\{z \in \mathbb{C} : |z|<1\}$ *such that*

(3.20) $$2^{n-1}t_n(x) = \mathrm{Re}\{z^{-n+1}q_m(z)\Phi_{2n-1-m}(z)\},$$

$x = \frac{1}{2}(z+z^{-1})$, $z = e^{i\varphi}$, $\varphi \in [0,\pi]$.

Proof. We show that Theorem 3.2 d) <=> (e) <=> (f).
(d) <=> (e). Put

$$e^{\pm}_{n-\ell} = p_{n-\ell-1}(\pm1)/p_{n-\ell}(\pm1),$$

(3.21) $$2s_{\ell-1}(x) = (e^{+}_{n-\ell} - e^{-}_{n-\ell})\,\tilde{\lambda}_{n+1-\ell}g_{\ell-1}(x),$$

and

(3.22) $$2r_\ell(x) = 2g_\ell(x) - [(e^{+}_{n-\ell} - e^{-}_{n-\ell})x + (e^{+}_{n-\ell} + e^{-}_{n-\ell})]\tilde{\lambda}_{n+1-\ell}g_{\ell-1}(x).$$

Representing $(x^2-1)p^{(1-x^2)}_{n-\ell-1}$ in the form (see Szegö (1967) Theorem 2.5)

$$(x^2-1)p^{(1-x^2)}_{n-\ell-1}(x) = p_{n+1-\ell}(x) - \mu_{1,n-\ell}p_{n-\ell}(x) - \mu_{2,n-\ell}p_{n-\ell-1}(x)$$

$$= (x-\alpha_{n+1-\ell}-\mu_{1,n-\ell})p_{n-\ell}(x)-(\mu_{2,n-\ell}+\lambda_{n+1-\ell})p_{n-\ell-1}(x)$$

we get by straightforward calculation that

$$\mu_{2,n-\ell} + \lambda_{n+1-\ell} = \frac{2}{e^{+}_{n-\ell}-e^{-}_{n-\ell}} \text{ and } \mu_{1,n-\ell} + \alpha_{n+1-\ell} = -\frac{e^{+}_{n-\ell}+e^{-}_{n-\ell}}{e^{+}_{n-\ell}-e^{-}_{n-\ell}}$$

which gives again by straightforward calculation that

$$r_\ell p_{n-\ell} + s_{\ell-1}\cdot(x^2-1)p^{(1-x^2)}_{n-\ell-1} = g_\ell p_{n-\ell} - \tilde{\lambda}_{n+1-\ell}g_{\ell-1}p_{n-\ell-1}.$$

Concerning the statement on the leading coefficients of $s_{\ell-1}$ and r_ℓ

resp. on $\tilde{\lambda}_{n+1-\ell}$ if $m = 2\ell-1$ we observe that from the recurrence relation of $p_{n+1-\ell}$ at the points $\pm$ 1 it follows by eliminating $\alpha_{n+1-\ell}$ that

$$\lambda_{n+1-\ell}(e^{+}_{n-\ell} - e^{-}_{n-\ell}) = 2 - u_{n-\ell} - v_{n-\ell} = 1+a_{2(n-\ell)-1}$$

which gives in view of (3.21) and (3.22) the assertion.

(e) <=> (f). Let us set

$$q_m(z) = (2z)^{\ell}\{r_\ell(\tfrac{1}{2}(z+z^{-1})) + \frac{(z-z^{-1})}{2}\, s_{\ell-1}(\tfrac{1}{2}(z+z^{-1}))\} \quad \text{if } m = 2\ell$$

and

$$q_m(z) = 2(2z)^{\ell-1}\{\frac{r_\ell(\frac{1}{2}(z+z^{-1}))}{(1-a_{2(n-\ell)-1})} + \frac{(z-z^{-1})}{2}\,\frac{s_{\ell-1}(\frac{1}{2}(z+z^{-1}))}{(1+a_{2(n-\ell)-1})}\} \quad \text{if } m = 2\ell-1$$

i.e. for $m = 2\ell$

$$r_\ell(x) = 2^{-\ell}\mathrm{Re}\{e^{-i\ell\varphi}q_m(e^{i\varphi})\} \quad \text{and} \quad s_{\ell-1}(x) = 2^{-\ell}\,\frac{\mathrm{Im}\{e^{-i\ell\varphi}q_m(e^{-i\varphi})\}}{\sin\varphi}$$

and for $m = 2\ell-1$

$$r_\ell(x) = (1-a_{2(n-\ell)-1})2^{-\ell}\mathrm{Re}\{e^{-i(\ell-1)\varphi}q_m(e^{i\varphi})\}$$

and

$$s_{\ell-1}(x) = (1+a_{2(n-\ell)-1})2^{-\ell}\,\frac{\mathrm{Im}\{e^{-i(\ell-1)\varphi}q_m(e^{i\varphi})\}}{\sin\varphi}.$$

Then it follows with the help of (3.16) and (3.17) that (3.20) resp. (3.19) holds. Thus the properties of the zeros of q_m resp. $s_{\ell-1}$ and r_ℓ remain to be shown. Considering the net change in arg $e^{-i\ell\varphi}q_m(e^{i\varphi})$ as φ varies from 0 to 2π resp. the net change in arc $\tan(s_{\ell-1}(x)/r_\ell(x))$ as x varies from +1 to -1 denoted by $\Delta_0^{2\pi}\ldots$ resp. $\Delta_{+1}^{-1}\ldots$ we have by symmetry with respect to π

$$\Delta_0^{2\pi}\arg e^{-i\ell\varphi}q_m(e^{i\varphi}) = \Delta_0^{2\pi}\text{arc tan}\,\frac{\mathrm{Im}\{e^{-i\ell\varphi}q_m(e^{i\varphi})\}}{\mathrm{Re}\{e^{-i\ell\varphi}q_m(e^{i\varphi})\}}$$

$$= 2\,\Delta_{+1}^{-1}\,\text{arc tan}\,\frac{s_{\ell-1}(x)}{r_\ell(x)}$$

which gives, recalling the fact that $\Delta_0^{2\pi}\arg q_m(e^{i\varphi})$ is equal to the number of zeros of q_m in $|z|<1$ if q_m has no zero on $|z| = 1$, the assertion.□

Theorem 3.2 b) is implicitly contained in the proof of Theorem 1 resp. Corollary 1 in Peherstorfer (1981) and Theorem 3.2c) has been given in Peherstorfer (1984) Theorem 2.1. An alternative approach to Theorem 3.2b) and c), roughly speaking the onedimensional restriction of a characterization of interpolatory cubature formulas, has been given by H.J. Schmid (1989). A slightly weaker version of Theorem 3.2d) is due to Sottas and Wanner (1982) who described positive q.f. having all their nodes real and simple but not necessarily located in (-1,1). Compare also Theorem 2 in the author's paper (1981) which is Theorem 3.2f) above, which obviously implies immediately, up to the important interlacing property of the zeros of g_ℓ and $g_{\ell-1}$, that relation (3.7) hold.

4. Some consequences of the Characterization Theorem.

From Theorem 3.2 c) we obtain immediately the following result of Michelli and Rivlin (1973) which was a substantial extension of a result of Fejer (1933). By the way let us also mention that Förster (1991) in studying the convergence behaviour of $(2n-3,n,d\sigma)$q.f. got another approach to characterize positive $(2n-3,n,d\sigma)$ q.f..

Corollary 4.1. *Let* $n \in \mathbb{N}$, $n\geq 2$, *and let* t_n *be a monic polynomial of degree* n. *Then* t_n *generates a positive* $(2n-3,n,d\sigma)$ *q.f. if and only if* $t_n(x) = (x-\tilde{\alpha})p_{n-1}(x)-\tilde{\beta}p_{n-2}(x)$ *with* $\tilde{\beta}>0$ *and* $\operatorname{sgn} t_n(\pm 1) = (\pm 1)^n$.

Next let us observe that a linear combination of orthogonal polynomials $\sum_{j=0}^{m} \mu_j p_{n-j}$, $n,m \in \mathbb{N}$, $\mu_1,\ldots,\mu_m \in \mathbb{R}$, can be represented by (2.9) in the following way

$$(4.1)\qquad \sum_{j=0}^{m} \mu_j p_{n-j} = \Big(\sum_{j=0}^{m} \mu_j p_{m-j}^{(n-m)}\Big)p_{n-m} - \lambda_{n+1-m}\Big(\sum_{j=0}^{m} \mu_j p_{m-j-1}^{(n-m+1)}\Big)p_{n-m-1},$$

where the representation is unique if $n\geq 2m$. With the help of this relation we get in view of Theorem 3.2 d) the following conditions for the positivity of a q.f..

Corollary 4.2. *Let* $n,m \in \mathbb{N}$, $n\geq 2m$ *and let* $\mu_o,\dots,\mu_m \in \mathbb{R}, \mu_o \neq 0$. a) $t_n := \sum_{j=0}^{m} \mu_j p_{n-j}$ *generates a positive* $(2n-1-m,n,d\sigma)$ *q.f. if and only if all zeros of* $\sum_{j=o}^{m} \mu_j p_{m-j}^{(n-m)}$ *and* $\sum_{j=o}^{m-1} \mu_j p_{m-j-1}^{(n-m+1)}$ *are simple, interlaced and located in* $(-1,1)$ *and* $t_n(\pm 1) = (\pm 1)^n$.

b) $t_n := \sum_{j=o}^{m} \mu_j p_{n-j}$ *generates a positive* $(2n-1-m,n,d\sigma)$ *q.f. if and only if* $\sum_{j=o}^{m} \mu_j p_{m-j}^{(n-m)}$ *generates a positive* $(m-1,m,d\sigma^{(n-m)})$ *q.f. on* $[-1,+1]$ *where* $d\sigma^{(n-m)}$ *is defined in* (2.14), *and* $t_n(\pm 1) = (\pm 1)^n$.

Proof. a) follows from Theorem 3.2 d) in conjunction with the well known fact that an interpolatory q.f. is uniquely determined by its nodes.
b) follows immediately from a) and Lemma 1.1.□

Let us note that the problem of checking zero properties of two polynomials of degree n resp. n-1 has thus been reduced by Corollary 4.2. to polynomials of degree m and m-1. Corollary 4.2 becomes particulary simple if $d\sigma(x) = (1-x^2)^{\pm 1/2}$, compare Peherstorfer (1984), Corollary 2, and (1990), Corollary 2, because in this case $d\sigma^{(n)}=(1-x^2)^{1/2}dx$ for $n \in \mathbb{N}$. Positive q.f., in particular with preassigned nodes, based on zeros of linear combinations of orthogonal polynomials have also been studied by Shohat (1937) and Thron (1984).
Concerning the location of the nodes of a positive q.f. we have the following extension of a result of Bernstein (1936) (see Braß 1977, Satz 61).

Corollary 4.3. *Suppose that* $t_n \in \mathbb{P}_n, n \in \mathbb{N}$ *has* n *simple zeros in* $(-1,1)$ *and that* t_n *is of the form* $t_n = \tilde{g}_\ell p_{n-\ell} - \tilde{g}_{\ell-1} p_{n-\ell-1}$, $n\geq 2\ell$, *where* $\tilde{g}_\ell$, $\tilde{g}_{\ell-1}$ *are polynomials of degree* ℓ *and* $\ell-1$ *respectively. Then* t_n *generates a positive* $(2n-1-2\ell,n,d\sigma)$ *q.f. if and only if the zeros of* t_n *and* $\tilde{g}_{\ell-1}p_{n-\ell}$ *and of* t_n *and* $\tilde{g}_\ell p_{n-\ell-1}$ *interlace.*

Proof. Since $n \geqq 2\ell$ we get from the representation of t_n, using the orthogonality property of $p_{n-\ell}$ and $p_{n-\ell-1}$ and relation (2.13) that

$$t_{n-1}^{[1]} = \tilde{g}_\ell p_{n-\ell-1}^{(1)} - \tilde{g}_{\ell-1} p_{n-\ell-2}^{(1)}$$

and thus we have at the zeros y of t_n using (2.12) that

$$t_{n-1}^{[1]}(y) = \text{const.}\tilde{g}_{\ell-1}(y)/p_{n-\ell}(y) \quad \text{and} \quad t_{n-1}^{[1]}(y) = \text{const.}\tilde{g}_\ell(y)/p_{n-\ell-1}(y)$$

where const. is positive, which gives by Lemma 1.1 the assertion. □

Finally let us mention that from Theorem 3.2f) further results on the distribution of the nodes of positive q.f. as well as on the asymptotic positivity of q.f. based on linear combinations of orthogonal polynomials can be obtained. Furthermore from Corollary 4.2 a new characterization of Gauss-Kronrod quadrature can be derived. Due to lack of space all these facts will be given elsewhere.

References

Akhiezer N.I.(1965), The Classical Moment Problem, Hafner Publ.Co., New York.

Akhiezer N.I. and Krein M.G.(1962), Some Questions in the Theory of Moments, Translations of Mathematical Monographs, Vol.50, Amer.Math.Soc. Providence,R.I..

Askey R.(1986), Positive quadrature methods and positive polynomial sums, Approximation Theory, V(C.K.Chui, L.L.Schumaker, and J.D.Ward, eds.), Academic Press, New York, 1-30.

Braß H.(1977), Quadraturverfahren, Vandenhoeck u.Ruprecht, Göttingen-Zürich.

Brézinski C.(1980), Padé-Type Approximation and General Orthogonal Polynomials, ISNM 50, Birkhäuser, Basel.

Fejer L.(1933), Mechanische Quadraturen mit positiven Cotesschen Zahlen, Math.Z. 37, 287-309.

Förster K.-J.(1991), On quadrature formulae near Gaussian quadrature, Numerical Integration (T.O.Espelid, A.Genz, eds.), NATO ASI Series C, Vol.357, Kluwer, Dordrecht, 45-54.

Locher F.(1973), Norm bounds of quadrature processes, SIAM J. Numer.Anal. 10, 553-558.

Micchelli C.A. and Rivlin T.J.(1973), Numerical integration rules near Gaussian quadrature, Israel J.Math.16, 287-299.
Natanson I.P.(1965), Constructive Function Theory, Vol.II, F.Ungar Publ. Co., New York.
Peherstorfer F.(1981), Characterization of positive quadrature formulas, SIAM J.Math.Anal. 12, 935-942.
Peherstorfer F.(1984), Characterization of quadrature formulas. II. SIAM J.Math.Anal.15, 1021-1030.
Peherstorfer F.(1990a), Linear combinations of orthogonal polynomials generating positive quadrature formulas, Math.Comp.55, 231-241.
Peherstorfer F.(1990b), On Bernstein-Szegö orthogonal polynomials on several intervals, SIAM J.Math.Anal.21, 461-482.
Schmid H.J.(1989), A note on positive quadrature rules, Rocky Mountain J. Math.19(1989), 395-404.
Shohat J.(1937), On mechanical quadratures, in particular, with positive coefficients, Trans.Amer.Math.Soc.42, 461-496.
Sottas G.(1988), Positivity domain of ultraspherical type quadrature formulas with Jacobi abscissas: numerical investigations. Numerical integration III (Braß H. and Hämmerlin G., eds.), ISNM 85, Birkhäuser, Basel.
Szegö G.(1967), Orthogonal Polynomials, 3th ed., Amer Math.Soc. Colloquium Publications.
Thron W.J.(1984), Quadrature formulas and moment problems.
Wendroff B.(1961), On orthogonal polynomials, Proc.Amer.Math.Soc.12, 654-655.

Franz Peherstorfer
Institut für Mathematik
J. Kepler Universität Linz
A-4040 Linz
Austria

International Series of Numerical Mathematics, Vol. 112,

Quadrature Theory of Convex Functions

A Survey and Additions

Knut Petras

Abstract. This article presents an overview as well as some new results on the quadrature theory of convex functions. The application of the classical quadrature rules such as trapezoidal and midpoint rules or positive rules of interpolatory type, including Gaussian quadrature, are discussed. Comparisons with quadrature formulae being optimal for subclasses of convex functions are given. Several further approaches are mentioned in the last two sections.

1. Introduction

The positivity, monotony or convexity of a function seem to be natural observations. Such properties may follow directly from the underlying problem, while bounds for derivatives of an unknown function are in general not available. It can be shown that the convexity in general yields smaller error bounds in numerical integration than both the other mentioned properties. We will investigate quadrature errors for convex functions f on $[-1,1]$, which means in particular that f is bounded.

Brass (1982a) has constructed a quadrature rule $(Q_n)_{n\in\mathbb{N}}$ for which an error of the order $O(n^{-2})$ can always be guaranteed for convex functions, while no smaller order is available for all convex functions. All further quadrature rules have to compete with such a method.

In Sect. 3, we consider the quadrature formulae, which aroused the first and the most interest in literature on quadrature of convex functions – the trapezoidal and the midpoint rule. They are found to be appropriate methods mainly if the first derivative at the boundary points ± 1 exists, since their errors are not of the order $O(n^{-2})$ otherwise. Sect. 4 presents methods for the computation of error constants in certain subclasses of convex functions as well as the respective optimal formulae. This shows that the error $O(n^{-2})$ is indeed available for all convex functions. For several reasons it is the authors opinion that these optimal rules should not be used in practice. Positive quadrature formulae of high algebraic degree seem to be an appropriate alternative (cf. Sect. 5). Exemplarily, in Sect. 6, the errors of Gaussian formulae are investigated more detailed. In Sect. 7, two methods of estimating the error a-posteriori are mentioned – a simple linear method and the central algorithm. Various generalizations of numerical integration of convex functions by linear quadrature formulae may be found in Sect. 8.

2. Notation

We denote by Q_n a (linear) quadrature formula, i.e.,

$$Q_n[f] = \sum_{\nu=1}^{n} a_\nu f(x_\nu), \tag{2.1}$$

where the a_ν are called weights and the x_ν are called nodes of Q_n. Such quadrature formulae shall approximate the integral

$$I[f] = \int_{-1}^{1} f(x)\,dx \tag{2.2}$$

(each finite interval may easily be transformed onto $[-1, 1]$). The remainder functional is given by $R_n = I - Q_n$. We define the algebraic degree $m = \deg(Q_n)$ by

$$R_n[p] \begin{cases} = 0 & \text{if } p \in \mathbb{P}_m \\ \neq 0 & \text{if } p \in \mathbb{P}_{m+1} \setminus \mathbb{P}_m, \end{cases} \tag{2.3}$$

where $\mathbb{P}_s$ is the space of all polynomials of degree less than or equal to s. Q_n is called positive, if all its weights are nonnegative.

The popular quadrature formulae, which we will consider in the following, are trapezoidal formulae Q_n^{Tr}, midpoint formulae Q_n^{Mi}, Clenshaw-Curtis formulae Q_n^{CC}, Gaussian formulae Q_n^{G}, Lobatto formulae Q_n^{Lo} and Radau formulae $Q_n^{\mathrm{Ra},1}$ resp. $Q_n^{\mathrm{Ra},-1}$ involving the node 1 resp. -1.

Let K be a class of functions and let Q_n be a quadrature formula. Then, we denote by $\varrho(Q_n, \mathrm{K})$ the corresponding error constant, i.e.,

$$\varrho(Q_n, \mathrm{K}) = \sup_{f \in \mathrm{K}} |R_n[f]|. \tag{2.4}$$

A quadrature formula, $Q_n^{\mathrm{opt}} = Q_n^{\mathrm{opt(K)}}$, is called optimal, if it has the smallest error constant for the class K among all quadrature formulae with n nodes. An obvious measure for the quality of a given quadrature formula Q_n is

$$\mathrm{qual}(Q_n, \mathrm{K}) = \frac{\varrho(Q_n, \mathrm{K})}{\varrho(Q_n^{\mathrm{opt}}, \mathrm{K})} \tag{2.5}$$

(cf. Brass, 1988). If this quantity is equal to 1, then Q_n is optimal. If it increases, the quadrature formula becomes worse.

Let L be a linear functional, which vanishes on $\mathbb{P}_{s-1}$ and is continuous on $A^{s-1}[-1,1]$, the space of all functions with a $s-2^{\mathrm{nd}}$ absolutely continuous derivative on $[-1,1]$. Then, for all $f \in A^s[-1,1]$,

$$(2.6) \qquad L[f] = \int_{-1}^{1} f^{(s)}(x) K_s(L,x)\,dx, \qquad \text{where } K_s(L,x) = L\left[\frac{(\cdot - x)_+^{s-1}}{(s-1)!}\right]$$

(cf. Sard 1948). $K_s(L,\cdot)$ is the so-called Peano kernel. If $\deg(Q_n) \geq s-1$, then, the first s Peano kernels of R_n exist. The divided difference, $\mathrm{dvd}(x_0,\ldots,x_s)$, is also a linear functional, which has s Peano kernels. L is called positive (negative) definite of order s, if $f \in C^s[-1,1]$ and $f^{(s)}(x) > 0$ for all $x \in [-1,1]$ imply that $L[f] > 0$ (< 0). Positive definiteness is equivalent with $K_s(L,\cdot)$ being nonnegative throughout $[-1,1]$ and positive for an argument in $[-1,1]$.

The definitions of the quadrature formulae mentioned above, as well as properties of Peano kernels may be found, e.g., in Brass (1977) or Davis and Rabinowitz (1984).

A function f is called convex on $[-1,1]$ if and only if, for all $a, b \in [-1,1]$, the chord with endpoints $(a, f(a))$ and $(b, f(b))$ lies nowhere underneath the graph of f. This implies that there is at least one line of support at each $x \in (-1,1)$, and that each divided difference $\mathrm{dvd}(a,b,c)[f]$, where $a,b,c \in [-1,1]$, is nonnegative.

3. Trapezoidal and midpoint formulae

For convex integrands, trapezoidal and midpoint formulae have been the first quadrature formulae under consideration in the literature. The reason is that R_n^{Tr} is negative and R_n^{Mi} is positive for all convex functions. This follows easily from the above characterizations of convex functions. Using only this definiteness and $Q_{2n+1}^{\mathrm{Tr}} = (Q_{n+1}^{\mathrm{Tr}} + Q_n^{\mathrm{Mi}})/2$, one may easily obtain the following results for convex functions f (for (3.1), (3.3) and (3.4), cf. Brass, 1977, pp.60 ff. and pp.176 ff.); the comparison,

$$(3.1) \qquad 0 \leq R_n^{\mathrm{Mi}}[f] \leq -R_{n+1}^{\mathrm{Tr}}[f],$$

the stopping inequality,

$$(3.2) \qquad |R_{2n+1}^{\mathrm{Tr}}[f]| \leq Q_{n+1}^{\mathrm{Tr}}[f] - Q_{2n+1}^{\mathrm{Tr}}[f], \qquad \text{(McGawn, 1972)}$$

and the error estimates

$$(3.3) \qquad R_n^{\mathrm{Mi}}[f] \leq \frac{1}{2n}\left(f(-1) - f(-1+\frac{1}{n}) - f(1-\frac{1}{n}) + f(1)\right),$$

$$(3.4) \qquad |R_{n+1}^{\mathrm{Tr}}[f]| \leq \frac{1}{n}\left(f(-1) - f(-1+\frac{1}{2n}) - f(1-\frac{1}{2n}) + f(1)\right). \qquad \text{(Wolfe 1959)}$$

Further estimates of R_n^{Tr} and R_n^{Mi} for convex integrands are listed in Brass' book (1977, pp.60 ff. and pp.176 ff.).

Deeper results are the following inverse theorems, which show that (3.3) and (3.4) yield sharp estimates.

(Gaier 1970) For each nonlinear convex function f, there exists a positive constant c such that, for all $n\,(\geq 2)$,

$$|R_n^{\mathrm{Tr}}[f]| + |R_{n+1}^{\mathrm{Tr}}[f]| \geq c \cdot n^{-2}. \tag{3.5}$$

(Brass 1978) Let f be convex, then,

$$f \in \mathrm{Lip}\,\beta \;\Leftrightarrow\; R_n^{\mathrm{Tr}}[f] = O(n^{-1-\beta}). \tag{3.6}$$

(Brass 1979) Let $\sigma_n = o(n^{-1})$, then there is a convex function f satisfying $|R_n^{\mathrm{Tr}}[f]| \geq \sigma_n$.

Particularly, these results mean that $O(n^{-2})$ is only attained, if $f'(-1)$ and $f'(1)$ exist and that we never obtain a better order than $O(n^{-2})$ in nontrivial cases. Similar results as (3.5) and (3.6) also hold for midpoint formulae (for (3.6), cf. Brass, 1978, and for (3.5), we use that $K_2(R_n^{\mathrm{Mi}} + R_{n+1}^{\mathrm{Mi}}, x) \geq d \cdot n^{-2}$ holds with a fixed positive constant d, if x remains in a fixed closed subinterval of $(-1,1)$). Thus, we state:

If we may choose the nodes arbitrarily, trapezoidal and midpoint formulae should only be used for a convex function f, if it is known that the first derivative of f at -1 and 1 exists.

We will see that the definiteness is no proper reason to apply midpoint or trapezoidal formulae, since simple upper and lower bounds for the integral may also be obtained with quadrature formulae, which are not definite of order 2, such as the Gaussian formulae. We simply have to add an appropriate multiple of a divided difference (see eq. (4.5)). Nevertheless, there is an interesting property of the trapezoidal formula in comparison with the central algorithm (cf. Sect. 7).

4. Optimal formulae and error constants for classes of convex functions

For the whole class $\mathrm{K}^{\mathrm{conv}}$ of convex functions on $[-1,1]$, no finite error constant exists, since each positive constant multiple of a convex function is again convex. Thus, we have to define subclasses of convex functions in order to obtain optimal formulae. Glinkin (1984) calculated the optimal formulae for the classes

$$\mathrm{K}_{u,v}^{\mathrm{Gl}} = \mathrm{K}^{\mathrm{conv}} \cap \{f \mid \; f'_+(-1) \geq u \text{ and } f'_-(1) \leq v \;\}. \tag{4.1}$$

They are complicated and we only note here, that the respective error constants are slightly greater than $(v-u)/(4n^2)$. Glinkin's classes are examples that an affine formula may yield a smaller error constant than the optimal linear formula. Zwick (1988) calculated the simple optimal affine formulae and showed that their error constants are equal to $(v-u)/(4n^2)$. However, not each convex function has a finite derivative at 1 and -1 such that error estimates for Glinkin's classes do not apply to the whole class $\mathrm{K}^{\mathrm{conv}}$. It is also easy, to prove that the optimal linear and affine formulae do not have an error of the order $O(n^{-2})$ for all convex functions. Of wider applicability is the class defined by Brass (1982a),

$$\mathrm{K}^{\mathrm{Br}} = \mathrm{K}^{\mathrm{conv}} \cap \{f \mid \|f\|_\infty \le 1\}. \tag{4.2}$$

The optimal error constants are slightly smaller than $4/n^2$, such that the optimal formulae $Q_n^{\mathrm{opt(Br)}}$ always satisfy

$$|R_n^{\mathrm{opt(Br)}}[f]| \le \frac{4}{n^2}\|f\|_\infty, \tag{4.3}$$

if f is convex, i.e., the error is always of the order $O(n^{-2})$. Brass also showed that there is no null sequence $(\varepsilon_n)_{n\in\mathbb{N}}$ such that a corresponding quadrature rule $(Q_n)_{n\in\mathbb{N}}$ satisfies $R_n[f] = O(n^{-2}\varepsilon_n)$ for all convex functions f. The situation changes for adaptive stochastic rules (cf. Novak, 1993).

The third class of convex functions, which may be found in the literature (cf. Förster and Petras 1990), is

$$\mathrm{K}^{\mathrm{FP}} = \mathrm{K}^{\mathrm{conv}} \cap \{f \mid f(1) - 2f(0) + f(-1) \le 1\}, \tag{4.4}$$

yielding error estimates of the form

$$|R_n[f]| \le 2\varrho(Q_n, \mathrm{K}^{\mathrm{FP}})\,\mathrm{dvd}(-1,0,1)[f]. \tag{4.5}$$

The latter class is similar to Brass' class, but it is easier to calculate the corresponding error constants as well as to compute a divided difference instead of the supremum norm. (The explicit calculation of the supremum norm is not necessary for Brass' optimal formula, see Brass, 1982a, Theorem 2).

For all three function classes, we may compute the error constant using Peano kernels or the following generalization,

$$D^-(R_n, x) = R_n[(x-\cdot)_+]/(1+x) \quad \text{and} \quad D^+(R_n, x) = R_n[(\cdot - x)_+]/(1-x), \tag{4.6}$$

as well as

$$\left\{\begin{matrix} M(R_n) \\ m(R_n) \end{matrix}\right\} = \left\{\begin{matrix} \sup \\ \inf \end{matrix}\right\} \left\{D^-(R_n, y) + D^+(R_n, x) \mid -1 < y \le x < 1\right\}. \tag{4.7}$$

Let Q_n be a linear quadrature formula and let $p_\nu(x) = x^\nu$.

i) (Zwick, 1988) If $\deg(Q_n) \geq 0$, *then,*

$$\varrho(Q_n, K^{\mathrm{Gl}}_{u,v}) = (v-u)\|D^+(R_n,\cdot) + \frac{uR_n[p_1]}{v-u}\|_\infty. \tag{4.8}$$

ii)

$$\varrho(Q_n, K^{\mathrm{Br}}) = \max\{|R_n[p_0]|, 2M(R_n) - R_n[p_0], R_n[p_0] - 2m(R_n)\}. \tag{4.9}$$

iii) (Förster and Petras, 1990) If $\deg(Q_n) \geq 1$, *then,*

$$\varrho(Q_n, K^{\mathrm{FP}}) = \sup_{x\in(-1,1)} \frac{|K_2(R_n,x)|}{1-|x|}. \tag{4.10}$$

The assumptions on the degree in i) and iii) are necessary for the existence of finite error constants.

Proof of *ii)* We readily see that each continuous function in K^{Br} may be approximated arbitrarily precise by a spline s of the form

$$s(t) = b \cdot \sum_{\nu=1}^{k} b_\nu s^+(y_\nu, t) + c \cdot \sum_{\nu=1}^{l} c_\nu s^-(z_\nu, t) + d, \tag{4.11}$$

where $s^+(x,t) = (t-x)_+/(1-x)$ and $s^-(x,t) = (x-t)_+/(1+x)$, $-1 < z_\nu \leq x^* \leq y_\nu < 1$, with $d \in [-1,1]$ and with positive coefficients b, b_ν, c, c_ν satisfying $\sum_{\nu=1}^k b_\nu = \sum_{\nu=1}^l c_\nu = 1$, and $b, c \leq 1-d$. Let now be $\Delta^+(x^*) = \sup_{x\geq x^*} D^+(x)$, $\Delta^-(x^*) = \sup_{x\leq x^*} D^-(x)$ and let $\delta^+(x^*)$ and $\delta^-(x^*)$ be the respective infima, then, we obtain

$$R_n[s] \begin{cases} \leq b\cdot\Delta^+(x^*) + c\cdot\Delta^-(x^*) + dR_n[1] \\ \geq b\cdot\delta^+(x^*) + c\cdot\delta^-(x^*) + dR_n[1] \end{cases} \tag{4.12}$$

$$\begin{cases} \leq \max\{b,c\}\cdot M(R_n) + dR_n[1] \\ \geq \min\{b,c\}\cdot m(R_n) + dR_n[1] \end{cases} \tag{4.13}$$

$$\begin{cases} \leq (1-d)M(R_n) + dR_n[1] \\ \geq (1-d)m(R_n) + dR_n[1]. \end{cases} \tag{4.14[6)}$$

The terms in (4.14) are linear functions with argument d and therefore attain their extreme values at the boundary, i.e., for $d = 1$ or $d = -1$. The modifications for discontinuous functions in K^{Br} may be done as in Förster and Petras (1990, proof of Theorem 1). The right hand side of (4.9) is attained either for $f(x) \equiv 1$ or for the piecewise linear interpolant f of points $(-1,a)$, $(x^{(1)},-1)$, $(x^{(2)},-1)$ and $(1,b)$, where $|a| = |b| = 1$. (If $x^{(1)} = -1$ and $a = 1$, the function value at -1 may be equal to 1, while $\lim_{x \to -1} f(x) = -1$; a similar modification is done for $x^{(2)} = 1$.) □

The optimal formula in K^{FP} had not been determined yet. We state:

Let $h_n = (n-1+\sqrt{2})^{-2}$. Then, the unique optimal quadrature formula $Q_n^{\mathrm{opt(FP)}}$ for the class K^{FP} is symmetric (i.e., $x_\nu = -x_{n+1-\nu}$ and $a_\nu = a_{n+1-\nu}$). For $n \geq 2$, the nodes and weights are given by

(4.15) $$a_{(n+1)/2} = 4h_n(\sqrt{2} + n - \frac{3}{2}) \text{ for odd } n, \qquad a_1 = h_n(\sqrt{2}+1)^2,$$

and for all further nodes and weights with $\nu \leq n/2$, we have

(4.16) $$x_\nu = -1 + h_n(\sqrt{2} + 2(\nu - 1))^2, \qquad a_\nu = 4h_n(\sqrt{2} + 2(\nu - 1)).$$

The corresponding error constant is

(4.17) $$\varrho(Q_n^{\mathrm{opt(FP)}}, \mathrm{K}^{\mathrm{FP}}) = h_n.$$

Furthermore, $Q_1^{\mathrm{opt(FP)}}[f] = 2f(0)$ and $\varrho(Q_1^{\mathrm{opt(FP)}}, \mathrm{K}^{\mathrm{FP}}) = 1 =: h_1$.

PROOF: Let

(4.18) $$Q_n[f] = \sum_{\nu=1}^{n} b_\nu f(y_\nu).$$

Between two of the nodes y_ν, the second Peano kernel of R_n is a polynomial of degree 2 with main coefficient 1/2. This observation and some calculation show that, if $|K_2(R_n, x)| \leq h_n(1 + x)$ for all $x \in [-1, 1]$, we have

(4.19) $$y_1 \leq -1 + 2h_n \qquad \text{and} \qquad 1 + y_{\nu+1} \leq (\sqrt{1+y_\nu} + 2\sqrt{h_n})^2.$$

If furthermore, for some $\nu < n/2$, $y_\nu \neq x_\nu$ or $b_\nu \neq a_\nu$ holds, strict inequality in (4.19) follows. Otherwise, we always have equality.

Thus, if $Q_n \neq Q_n^{\mathrm{opt(FP)}}$ and $\varrho(Q_n, K^{\mathrm{FP}}) \leq \varrho(Q_n^{\mathrm{opt(FP)}}, K^{\mathrm{FP}})$, then we obtain $y_{\lfloor (n+1)/2 \rfloor} \leq x_{\lfloor (n+1)/2 \rfloor}$ and, analogously, $y_{\lfloor (n+2)/2 \rfloor} \geq x_{\lfloor (n+2)/2 \rfloor}$ with strict inequality in at least one of the latter relations. For odd n, we would obtain $y_{(n+1)/2} \neq y_{(n+1)/2}$ and for even n, the graph of the second Peano kernel of R_n would intersect the graph of $g(x) = -h_n(1 - |x|)$ in $(y_{n/2}, y_{(n+2)/2})$. Therefore, $Q_n^{\mathrm{opt(FP)}}$ is the unique optimal formula in the class K^{FP}. □

Eq. (4.10) has the following interpretation. Let h denote the symmetric 'hat function' on $[-1,1]$, $h(x)=(1-|x|)$. Then, the error constant is the least positive constant c, such that $c\cdot h(x)\geq K_2(R_n,x)$ and $-c\cdot h(x)\leq K_2(R_n,x)$ for all $x\in[-1,1]$. For obvious reasons, the second Peano kernel of the optimal formula defined in (4.15)-(4.17) touches the corresponding upper and lower hat in each subinterval between two of its nodes (cf. Fig.1).

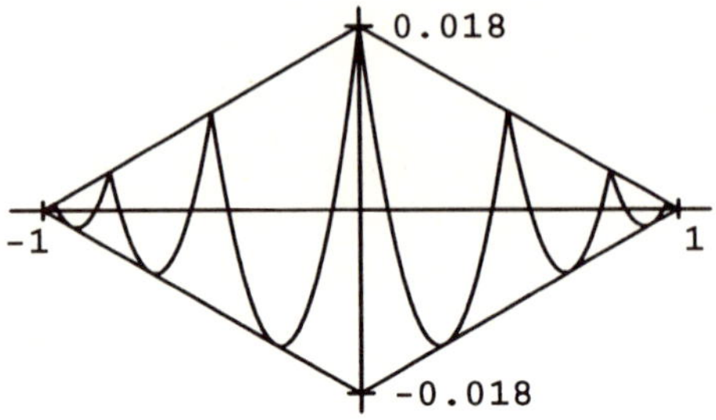

Fig.1: $K_2(R_7^{\mathrm{opt(FP)}},\cdot)$ and the corresponding hat functions

Another sight on the optimality in the class K^{FP} may be more useful and yields a correspondence between K^{Br} and K^{FP}. If we want to determine a number c such that $c\cdot f\in\mathrm{K}^{\mathrm{FP}}$, we have to compute the function values at -1, 0 and 1. Hence we may also use them in the quadrature formula without additional costs. In this situation it is not sensible to use a quadrature formula with an even number n of nodes, since it would have the same error constant as the optimal formula with $n-1$ nodes. For odd n, $Q_n^{\mathrm{opt(Br)}}$ is defined as follows,

$$Q_n^{\mathrm{opt(Br)}}[f]=c_n\left\{nf(0)+2\sum_{\nu=1}^{(n-1)/2}\nu\left[f(-1+\nu^2h)+f(1-\nu^2h)\right]\right\}, \tag{4.20}$$

where $c_n=4(n+1)^{-2}$. Analogously to the proof of the optimality of $Q_n^{\mathrm{opt(FP)}}$ in K^{FP}, we can prove the following.

Let n be odd and let

$$\widehat{Q}_{n+2}^{\mathrm{opt(FP)}}[f]=Q_n^{\mathrm{opt(Br)}}+\frac{c_n}{4}\{f(-1)+2f(0)+f(1)\}. \tag{4.21}$$

Then, the optimal quadrature formula in K^{FP} among all quadrature formulae with $n+2$ nodes including -1, 0 and 1 is $\widehat{Q}_{n+2}^{\mathrm{opt(FP)}}$ and its error constant equals $c_n/4$. The optimal affine formula Q_n^{aff} in K^{FP} having n nodes is

$$Q_n^{\mathrm{aff}}[f]=Q_n^{\mathrm{opt(Br)}}+c_nf(0)+\frac{c_n}{4}. \tag{4.22}$$

Its error constant also equals $c_n/4$.

Using his optimal formula $Q_n^{\mathrm{opt(Br)}}$ and the corresponding optimal error bound, Brass (1982a) pointed out that the integral of a convex function f is in the interval $[Q_n^{\mathrm{opt(Br)}}[f]+c_nf(0),Q_n^{\mathrm{opt(Br)}}+c_n(f(-1)+f(1))/2]$. The formula $\widehat{Q}_{n+2}^{\mathrm{opt(FP)}}$ yields exactly the midpoint of this interval.

Using the optimal affine formula is not very sensible. If we know the value of $f(-1)-2f(0)+f(1)$, we should better multiply the affine part with this quantity such that we obtain a better error estimate if $f(-1)-2f(0)+f(1)<1$. This is equivalent to the application of the linear formula $\widehat{Q}_{n+2}^{\text{opt(FP)}}$. Furthermore, the optimal linear formula defined in (4.15)-(4.17) gives a better estimate than Q_n^{aff}, whenever $f(-1)-2f(0)+f(1)<c_n/h_n$, where $c_n/h_n>1-(4-2\sqrt{2})\sqrt{c_n}$.

As mentioned above, Glinkin's and Zwick's optimal formulae have the disadvantage that not for all convex functions, the order of the error is $O(n^{-2})$. Furthermore, for $f\in C^3[-1,1]\cap \mathrm{K}^{\text{conv}}$, we have

$$R_n^{\text{opt(Br)}}[f]=\frac{2}{3n^2}\Big\{f(1)+4f(0)+f(-1)\Big\}+O(n^{-3}) \tag{4.20}$$

(see Brass 1982a) as well as

$$R_n^{\text{opt(FP)}}[f]=\frac{1}{3n^2}\Big\{f(1)-2f(0)+f(-1)\Big\}+O(n^{-3}). \tag{4.21}$$

However smooth the function is, we have generally $R_n[f]\asymp n^{-2}$ for both these latter optimal formulae. We will see in the next section that, in some sense, there are better quadrature formulae.

5. Positive quadrature formulae of high degree.

Since subclasses of convex functions may be chosen arbitrarily, optimal formulae also have somewhat the character of arbitrariness. Another objection is that they generally do not converge faster than n^{-2}. In contrast, positive quadrature formulae of high algebraic degree converge the faster the smoother the integrand is. Brass (1982a) has shown that the errors of the (positive) Clenshaw-Curtis formulae also tend to zero at least as n^{-2} for all convex functions. He posed the problem to extend this result to Gaussian quadrature (Brass, 1982b). A much more general result has been proved.

(Förster and Petras 1990) Let $\deg(R_n)\geq 2m-1$*, then, for every convex function* f *on* $[-1,1]$,

$$|R_n[f]|\leq \frac{6\pi}{2m+1}\left(\sin\frac{2\pi}{2m+1}\right)\{f(1)-2f(0)+f(-1)\}. \tag{5.1}$$

For positive quadrature formulae of interpolatory type, we can therefore guarantee the same order of the error as for the optimal formulae. Nevertheless, the guaranteed error constants are about $3\pi^2$ times as large as the optimal error constants. A method is known (Petras, 1992a) to compute pointwise the best possible bounds of the second Peano kernels of quadrature remainders, if positivity and $\deg(Q_n) \geq m$ are assumed. In Fig. 2, the functions C_m and c_m, defined by

$$\left\{ \begin{matrix} C_m(x) \\ c_m(x) \end{matrix} \right\} = \left\{ \begin{matrix} \sup \\ \inf \end{matrix} \right\} \{ K_2(R_n, x) \mid \deg(Q_n) \geq m,\ Q_n \text{ is positive} \} \tag{5.2}$$

are drawn for $m = 7$.

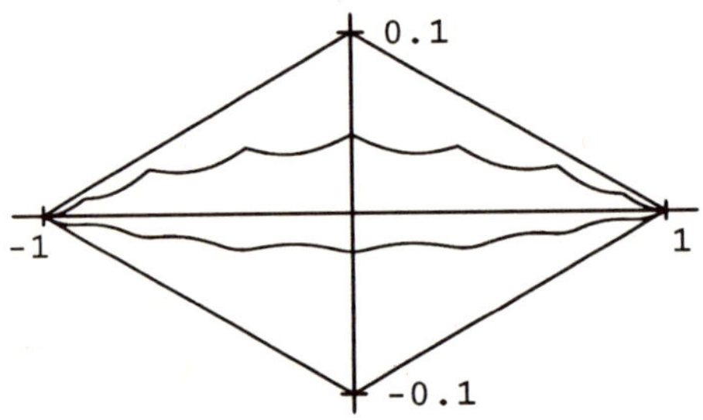

Fig.2: C_7, c_7 and the corresponding hat functions

Regarding more numerical examples, it turns out, that the lower and upper hat functions of minimal height which enclose all second Peano kernels for formulae of degree greater or equal to m is determined by the behaviour of the functions C_m and c_m at the boundary of the basic interval. This would imply, that the bound for the error constant would be equal to the greatest possible weight at -1 and 1. If this behaviour could be proved for all m, the validity of the following conjecture would have been shown.

Conjecture. *For all positive quadrature formulae Q_n, which satisfy $\deg(Q_n) \geq m$, we have*

$$\varrho(Q_n, \mathrm{K}^{\mathrm{FP}}) \leq \varrho(Q^{\mathrm{Lo}}_{(m+3)/2}, \mathrm{K}^{\mathrm{FP}}) = \frac{8}{(m+3)(m+1)} \tag{5.3}$$

if m is odd, and

$$\varrho(R_n, \mathrm{K}^{\mathrm{FP}}) \leq \varrho(R^{\mathrm{Ra},\pm 1}_{(m+2)/2}, \mathrm{K}^{\mathrm{FP}}) = \frac{8}{(m+2)^2} \tag{5.4}$$

if m is even. These bounds would obviously be best possible.

If this conjecture would hold, then, the quality (2.5) of each positive quadrature formula of interpolatory type in the classes K^{Br} and K^{FP} would be less than $8 + 16n^{-2}$.

6. Gaussian quadrature.

By eq. (4.10), the partitioning of the Gaussian nodes and the fact that $K_2(R_n^{\mathrm{G}},\cdot)$ has exactly two sign changes on each subinterval between consecutive nodes, we may readily obtain that $R_n^{\mathrm{G}}[f] = O(n^{-2})$ for all convex functions (cf. Fig.3). To obtain the asymptotically best error constants we have to use far more detailed investigations on the second Peano kernel. With known bounds (cf. Petras, 1993) we have the following result for the above-mentioned classes of convex functions,

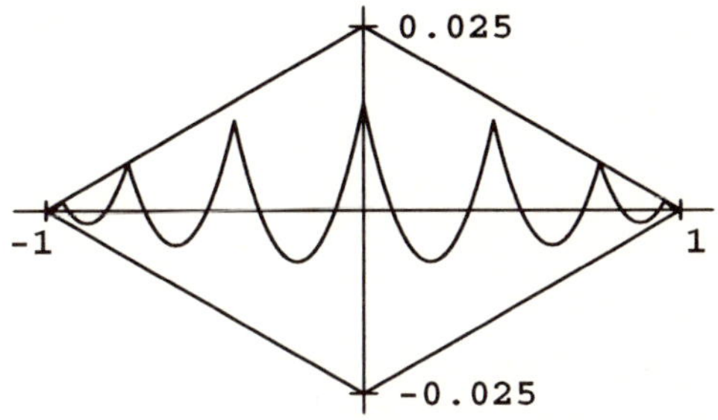

Fig.3: $K_2(R_7^{\mathrm{G}},\cdot)$ and the corresponding hat functions

$$(6.1)\quad n^2\varrho(Q_n^{\mathrm{G}},\mathrm{K}^{\mathrm{Gl}}) \le \frac{\pi^2}{12}(v-u) \quad \text{and} \quad N^2\varrho(Q_n^{\mathrm{G}},\mathrm{K}^{\mathrm{FP}}) = \frac{N^2}{4}\varrho(Q_n^{\mathrm{G}},\mathrm{K}^{\mathrm{Br}}) \le \frac{\pi^2}{6},$$

where $N = n + 1/2$. The constants on the right-hand sides are best possible. These inequalities also mean that

$$(6.2)\qquad \mathrm{qual}(Q_n^{\mathrm{G}},\mathrm{K}^{\mathrm{Gl}}) \le \frac{\pi^2}{3} = 3.2898\ldots,$$

$$(6.3)\qquad \mathrm{qual}(Q_n^{\mathrm{G}},\mathrm{K}^{\mathrm{FP}}) \le \frac{\pi^2}{6} = 1.6449\ldots \quad \text{and} \quad \mathrm{qual}(Q_n^{\mathrm{G}},\mathrm{K}^{\mathrm{Br}}) \le \frac{\pi^2}{6}\frac{n}{n-1}.$$

Thus, the Gaussian quadrature rule seems to be a reasonable method for the integration of convex functions. We therefore want to investigate the Gaussian error under additional assumptions on the integrand. It may sometimes appear to be a pathological case, if the convex integrand has corners, i.e., jump discontinuities in the first derivative. In the following, we discuss the distinction of convex functions having corners and those which have none.

Let f be convex.

i) If f' is discontinuous in $(-1,1)$, then,

$$(6.4)\qquad |R_n^{\mathrm{G}}[f]| + |R_n^{\mathrm{Lo}}[f]| \ne o(n^{-2}).$$

ii) If f' is continuous in $(-1,1)$, then, the sequence of the values $n^2|R_n^{\mathrm{G}}[f]|$ is Cesàro summable to zero, i.e.,

$$(6.5)\qquad \lim_{m\to\infty}\frac{1}{m}\sum_{n=1}^{m} n^2|R_n^{\mathrm{G}}[f]| = 0.$$

iii) There exists a convex function f with a continuous derivative in $(-1,1)$, which satisfies

$$R_n^{\mathrm{G}}[f] \neq o(n^{-2}). \tag{6.6}$$

iv) If f' is absolutely continuous on each closed subinterval of $(-1,1)$, then,

$$R_n^{\mathrm{G}}[f] = o(n^{-2}). \tag{6.7}$$

The proof will appear elsewhere, since such results are available for more general classes than convex functions.

7. Linear a-posteriori estimates and optimal recovery

We will mention two ways of estimating the quadrature error for convex functions a-posteriori. The first is a simple method to reduce the a-posteriori error bound and the second is the solution of the optimal recovery problem.

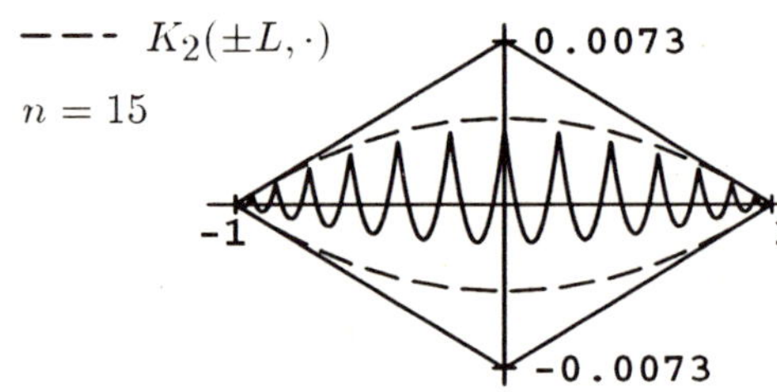

Fig.4: Gaussian Peano kernel with hat functions and Peano kernels of L and $-L$

The simple linear method will be introduced exemplarily for the Gaussian formulae. It uses the observation, that there is always a gap between the Peano kernel and the peak of the corresponding hat function. Therein, we fit a polygonal line with the nodes of Q_n^{G} as corners (cf. Fig.4). This polygonal line may be chosen as the second Peano kernel of the following linear functional L,

$$L[f] = \frac{\pi^2}{12n^2}\left(1+\frac{\ln n}{2n}\right)\sum_{\nu=1}^{n}(1-x_\nu^2)(x_{\nu+1}-x_{\nu-1})\,\mathrm{dvd}(x_{\nu-1},x_\nu,x_{\nu+1})[f], \tag{7.1}$$

where $x_0=-1$ and $x_{n+1}=1$. Since $K_2(-L,x)\le K_2(R_n^{\mathrm{G}},x)\le K_2(L,x)$ holds for all $x\in[-1,1]$ (which may be deduced with some computational effort from Petras, 1993), we have the definiteness of order 2 of $R_n^{\mathrm{G}}-L$ and $R_n^{\mathrm{G}}+L$, and therefore:

Suppose f is convex, then,

$$\begin{aligned}|R_n^{\mathrm{G}}[f]| \le \frac{\pi^2}{12n^2}\left(1+\frac{\ln n}{2n}\right)\cdot\Big\{&4\,\mathrm{dvd}(-1,0,1)[f]\\ &-\sum_{\nu=1}^{n}(1-|x_\nu|)^2(x_{\nu+1}-x_{\nu-1})\,\mathrm{dvd}(x_{\nu-1},x_\nu,x_{\nu+1})[f]\Big\}.\end{aligned} \tag{7.2}$$

If it could be proved that the factor $(1 + \ln n/(2n))n^{-2}$ could be replaced by $(n + 1/2)^{-2}$, we would have a proper improvement of the bound in (4.10) and (6.1). Applying the same method to the trapezoidal formulae, we may obtain the result of Wolfe (see eq. (3.4)).

Another method is the central algorithm (cf. Traub and Woźniakowski, 1980, p. 14). For given function values $f(x_1), \ldots, f(x_n)$, it is the aim to calculate the quantity $Q^c[f] = (\alpha + \beta)/2$, where

$$\text{(7.3)} \qquad \begin{Bmatrix} \alpha \\ \beta \end{Bmatrix} = \begin{Bmatrix} \min \\ \max \end{Bmatrix} \left\{ \int_{-1}^{1} g(x)\,dx \mid g \in K^{\text{conv}} \quad \text{and} \quad g(x_i) = f(x_i) \right\}.$$

A first approach to determine this value was given by Heindl (1982). A general theorem, which might not only be helpful for convex functions, as well as explicit formulae for equidistant nodes and small n are given in that article. For $n \leq 5$, these formulae are linear, for $n \leq 7$, they may be given in a closed form, and for greater n, distinctions of cases are necessary. Heindl also indicated an optimization algorithm to compute the quantity $Q^c[f]$. These ideas could be developed further by Niessner (1985), who stated a finite algorithm to compute $Q^c[f]$ explicitly.

The computation of $Q^c[f]$ is rather complicated. For practical purposes it may be more helpful to determine the upper bound β in (7.3) simply by using the generalized trapezoidal formula with the x_ν as nodes. Thus one merely obtains an additional factor 2 in the error estimate in comparison with the best possible method, the central algorithm.

8. Generalizations

8a) Weighted quadrature. The good behaviour of quadrature formulae of high algebraic degree only depends on the functional I and not on the exactness for polynomials. To show this, and to show that $O(n^{-2})$ is not always attainable for a fixed choice of a quadrature rule, we regard integrals with weight functions w,

$$\text{(8.1)} \qquad I[f] = \int_{-1}^{1} f(x)w(x)\,dx.$$

It has been shown (cf. Förster and Petras, 1990) that for bounded weight functions, the error constants for the respective Gaussian rule are of the same order, $O(n^{-2})$, as those of the optimal rule. In the case of unbounded ultraspherical weight functions, the situation changes. $O(n^{-2})$ is still the error order of the optimal rule, but not of the respective Gaussian rule. Furthermore, there are weight functions, such that the corresponding error constants of any rule cannot be of the order $O(n^{-2})$.

8b) r-convex functions. A function is called r-convex on $[-1, 1]$, if all its divided differences with $r + 1$ nodes in $[-1, 1]$ are nonnegative, i.e., a convex function is 2-convex.

Köhler (1990) gave a generalization of the inequality of Wolfe (cf. eq. (3.4)) and of the inverse theorem of Brass (cf. eq. (3.6)) to r-convex functions. These results show that r-convexity yields reasonable error bounds for compound quadrature formulae being definite of order r, if the $r-1^{\text{st}}$ derivative of the integrand exists and is bounded. They also show the limits of compound quadrature formulae for convex functions.

However, also quadrature formulae with high algebraic degree including the Gaussian formulae may not always compete with the respective optimal quadrature formulae, if $r > 2$ and no further restrictions are placed on the integrand. Then, for all quadrature formulae of interpolatory type mentioned in Sect. 2, we can only guarantee that $R_n[f] = O(n^{-2})$, while there are rules, which always yield $R_n[f] = O(n^{-r})$ (cf. Petras, 1992b).

8c) Adaptive and stochastic rules. For a special type of adaptive algorithms, the so-called one-step-optimal algorithms, Glinkin (1983) showed that no improvement upon linear rules is possible for Glinkin's class (4.1). Several further methods are investigated by Novak (1993) in these proceedings. Exemplarily, we only mention, that (at least) for Glinkin's class, it is possible to define a combination of adaptive and Monte Carlo methods such that the expectation of the error is of the order $O(n^{-5/2})$. For further reference, see Novak's article.

8d) Convex functions in more dimensions. Not much has been written on quadrature errors for convex functions in two or more dimensions. A first step has been done by Allasia and Giordano (1983) who generalized the trapezoidal and the midpoint rule for the interval to triangles. The generalization of the midpoint formula is the so-called centroid method (cf. Good and Gaskins, 1971). For this method, the region of integration is divided into subregions and the integral over each subregion is estimated by its volume times the function value at the centroid. On each triangular subregion, the generalized trapezoidal formula integrates the linear function, which interpolates the integrand at the three corners. Obviously, the centroid method yields a lower bound and the generalized trapezoidal formula yields an upper bound for the integral. Allasia and Giordano also showed some monotonicity properties of their rules.

Acknowledgement. I am grateful to H. Brass and E. Novak for helpful suggestions concerning this paper.

References

ALLASIA, G., GIORDANO, C. (1983), Approssimazioni bilaterali di Integrali doppi. *Calcolo* 20, pp. 73–83

BRASS, H. (1977), *Quadraturverfahren*, Göttingen: Vandenhoeck & Ruprecht

BRASS, H. (1978), Umkehrsätze beim Trapezverfahren. *Aequationes Math.* 18, pp. 338–344

BRASS, H. (1979), Der Wertebereich des Trapezverfahrens. In: Hämmerlin, G. (ed.), *Numerische Integration*, ISNM 45, Basel: Birkhäuser, pp. 98–108

BRASS, H. (1982a), Zur Quadraturtheorie konvexer Funktionen. In: Hämmerlin, G. (ed.), *Numerical Integration*, ISNM 57, Basel: Birkhäuser, pp. 34–47

BRASS, H. (1982b), Problem 2. In: Hämmerlin, G. (ed.), *Numerical Integration*, ISNM 57, Basel: Birkhäuser, pp. 268–269

BRASS, H. (1988), Universal quadrature rules in the space of periodic functions. In: Brass, H. and Hämmerlin, G. (eds.), *Numerical Integration III*, ISNM 85, Basel: Birkhäuser, pp. 16–24

DAVIS, P.J., RABINOWITZ, P. (1984), *Methods of Numerical Integration*, London: Academic Press

FÖRSTER, K.-J., PETRAS, K. (1990), On a problem proposed by H. Brass concerning the remainder term in quadrature for convex functions. *Numer. Math.* 57, pp. 763–777

GAIER, D. (1970), Saturation bei Spline-Approximation und Quadratur. *Numer. Math.* 16, pp. 129–140

GLINKIN, (1983), Optimal algorithms for integrating convex functions. *U.S.S.R. Comput. Math. and Math. Phys.* 23, pp. 6–12

GLINKIN, (1984), Best quadrature formula in the class of convex functions. *Math. Notes* 36, pp. 368–374

GOOD, I.J., GASKINS, R.A. (1971), The centroid method of numerical integration, *Numer. Math.* 16, pp. 343-359

HEINDL, G. (1982), Optimal quadrature of convex functions. In: Hämmerlin, G. (ed.), *Numerical Integration*, ISNM 57, Basel: Birkhäuser, pp. 138–147

KÖHLER, P. (1990), On the error of compound quadrature formulas for r-convex functions. *Aequationes Math.* 39, pp. 6–18

MCGAWN, M.M. (1972), *Exit criteria for the numerical solution of differential equations.* M.S. Thesis, University of Wyoming, Laramie

NIESSNER, H. (1985), *Zur optimalen Quadratur konvexer Funktionen.* Doctoral Thesis, Gesamthochschule Wuppertal

NOVAK, E. (1993), Quadrature formulas for convex classes of functions. These Proceedings

PETRAS, K. (1992a), One sided L1-Approximation and bounds for Peano kernels. In: Espelid, T.O., Genz, A. (eds.), *Numerical Integration*, NATO ASI Series C: Vol. 357, Dordrecht: Kluwer, pp. 165–174

PETRAS, K. (1992b), Error bounds for Gaussian and related quadrature and applications to r-convex functions. *SIAM J. Numer. Anal.* 29, pp. 578–585

PETRAS, K. (1993), Gaussian quadrature formulae – second Peano kernels, nodes weights and Bessel functions. To appear in *Calcolo*

SARD, A. (1948), Integral representations of remainders. *Duke Math. J.* 15, pp. 333–345

TRAUB AND WOŹNIAKOWSKI (1980), *A General Theory of Optimal Algorithms*, New York: Academic Press

WOLFE, J.M. (1959), An adjusted trapezoidal rule using function values within the range of integration. *Amer. Math. Monthly* 66, pp. 125–127

ZWICK, D. (1988), Optimal quadrature for convex functions and generalizations. In: Brass, H., Hämmerlin, G. (eds.), *Numerical Integration III*, ISNM 85, Basel: Birkhäuser, pp. 310–315

Knut Petras
Institut für Angewandte Mathematik
Technische Universität Braunschweig
Pockelsstr. 14
W-3300 Braunschweig, Germany
e-mail: i1040214@dbstu1.rz.tu-bs.de

On Multivariate Integration for Stochastic Processes

Klaus Ritter, Grzegorz W. Wasilkowski, and Henryk Woźniakowski

Abstract. We present bounds on the minimal average case errors of quadrature formulas that use n function values for multivariate integration. The error bounds are derived in terms of smoothness properties of the covariance function of the underlying stochastic process.

1. Introduction

We study multivariate integration for stochastic processes. We are particularly interested in the minimal average case error of quadrature formulas that use n function values. Recently, this problem was solved for the isotropic Wiener process (Brownian motion in Lévy's sense), see Wasilkowski (1991), for the Wiener (Brownian) sheet process, see Woźniakowski (1991), and for the folded Wiener sheet process, see Paskov (1991).

In this paper we consider general stochastic processes characterized by their covariance functions K. We study the convergence of the minimal errors in terms of smoothness of K. We prove that for smooth K the minimal errors go quickly to zero. It may, however, happen that for nonsmooth K the minimal error is zero even for $n = 1$. Thus, smoothness is a sufficient but not a necessary condition for fast convergence of the minimal errors.

We consider two types of smoothness of K. The first type is called the global smoothness and assumes that K is r times differentiable and its rth derivatives satisfy a Hölder-like condition with exponent α. Then, for d dimensional integration, the minimal errors go to zero at least as fast as $n^{-1/2-(r+\alpha/2)/d}$. This upper bound is sharp for some processes including the isotropic Wiener process for which $r = 0$ and $\alpha = 1$. Thus, if d is large relative to $r + \alpha/2$ then the rate of convergence is close to $n^{-1/2}$. This is the worst possible rate of convergence for stochastic processes with a covariance function which is integrable along the diagonal.

The second type of smoothness is called the mixed derivative smoothness and assumes that K is r_i times differentiable with respect to the ith variable and the partial derivatives satisfy a Hölder-like condition with exponent $2\alpha_i$. Then, for d dimensional integration, the minimal errors go to zero at least as fast as $n^{-\gamma-1/2}\log^{(\nu-1)(\gamma+1)} n$, where $\gamma = \min_i(r_i + \alpha_i)$ and ν is equal to the number of variables for which $r_i + \alpha_i = \gamma$.

Hence, in this case, the dimension d only mildly affects the rate of convergence. The rate of convergence, modulo a logarithmic power, is the same as for the folded Wiener sheet process, see Paskov (1991).

Our proof technique is based on known relations between the average case and worst case settings. We also use relations between integration and function approximation in the average case setting. (See Sections 2 and 3.1 for precise statements and references.) These relations are used in this paper to derive bounds on the minimal errors. They can also be used to derive good sample points of quadrature formulas. Although we do not address this issue in the paper, we add that grid points are good sample points for the global smoothness of K, see Theorem 1, and Novak (1988). For the mixed derivative smoothness of K, good sample points can be derived form hyperbolic cross points, see Theorem 2, Temlyakov (1987), and Paskov (1991).

In the concluding section we discuss extensions to nonlinear quadrature formulas which may use adaptive sample points. Complexity issues and the randomized setting are briefly considered. We also indicate how our results can be applied to the probabilistic setting and for different error criteria.

In this paper we present only upper bounds on the minimal errors. For some special covariance functions these bounds are sharp, sometimes modulo a logarithmic power. In general, getting sharp bounds on the minimal errors requires more specific assumptions on the regularity of K. An example of such assumptions is provided by tensor products of scalar covariance functions with each factor satisfying the Sacks-Ylvisaker regularity conditions, see Sacks and Ylvisaker (1966, 1968, 1970a, 1970b), as well as Wahba (1971), and Benhenni and Cambanis (1992). We will report these sharp bounds in a future paper.

2. Basic Definitions

Let F be a class of integrable functions $f : D \to \mathbb{R}$ defined on the d dimensional unit cube $D = [0,1]^d$. Let

$$\mathcal{I}(f) = \int_D f(t)\,dt, \quad \forall f \in F,$$

be the multivariate integration operator. We approximate $\mathcal{I}(f)$ by computing function values at some points from D. In this paper we mainly deal with quadrature (cubature) formulas, i.e., linear algorithms of the form

$$Q(f) = \sum_{i=1}^{n} a_i\, f(x_i)$$

for some real weights a_i and some points x_i from D. We consider the error between $\mathcal{I}(f)$ and $Q(f)$ in the average case setting. (The extensions to nonlinear algorithms Q and other settings are discussed in Section 4.) In the average case setting, it is assumed that f is a measurable stochastic process with mean 0 and a covariance function

$$K(x,t) = \mathrm{E}(f(x)f(t)),$$

where E stands for the expectation. We assume that

$$\int_D K(t,t)\,dt \;<\; +\infty, \tag{1}$$

which is equivalent to $E(\|f\|_2^2) < \infty$, and we can therefore assume that $F \subseteq L_2(D)$.

The average case error of Q is defined as

$$e^{\text{avg}}(Q) \;=\; \mathrm{E}\left((\mathcal{I}(f) - Q(f))^2 \right)^{1/2},$$

i.e., it is the standard deviation of the random variable $\mathcal{I}(f) - Q(f)$. Observe that (1) implies that $e^{\text{avg}}(Q)$ is well defined and finite.

Instead of starting with a process f one may start with a function $K : D \times D \to \mathbb{R}$. Any such K is the covariance function of some stochastic process f iff K is nonnegative definite, i.e., the matrix $(K(t_i,t_j))_{i,j=1}^m \geq 0$ for any m and points t_i from D. Moreover, if $t \mapsto K(t,t)$ is continuous almost everywhere on D then a measurable process with covariance function K exists, see Gihman and Skorohod (1974) for details.

It is well known that the average case error of Q coincides with the worst case error of Q over the unit ball of the Hilbert space H_K whose reproducing kernel is the covariance function K. This property has been used in many papers, see e.g., Kimeldorf and Wahba (1970a, 1970b), Larkin (1972), Micchelli and Wahba (1981), Parzen (1962), Paskov (1991), Sacks and Ylvisaker (1966, 1968, 1970a, 1970b), Traub et. al. (1988, Section 2.2 of Chapter 7), Wahba (1971, 1990), Wasilkowski (1991), Woźniakowski (1992), and Ylvisaker (1975).

More precisely, the Hilbert space H_K is the completion of the set of all finite linear real combinations of $K(\cdot,t_1),\ldots,K(\cdot,t_k)$. The completion is with respect to the norm $\|\cdot\|_K$ generated by the inner product $\langle\cdot,\cdot\rangle_K$ defined by $\langle K(\cdot,t_1), K(\cdot,t_2)\rangle_K = K(t_1,t_2)$. In the Hilbert space H_K we have the reproducing property

$$f(t) \;=\; \langle f, K(\cdot,t)\rangle_K, \quad \forall t \in D, \forall f \in H_K.$$

Then the average case and worst case errors of Q are related as follows

$$e^{\text{avg}}(Q) \;=\; e^{\text{wor}}(Q;H_K) := \sup_{\|f\|_K \leq 1} |\mathcal{I}(f) - Q(f)| \;=\; |\mathcal{I}(g) - Q(g)|^{1/2}, \tag{2}$$

where the function g is given as

$$g(x) \;=\; \mathcal{I}\,(K(x,\cdot)) \;-\; Q\,(K(x,\cdot)) \;=\; \int_D K(x,t)\,dt \;-\; \sum_{i=1}^n a_i\,K(x,x_i). \tag{3}$$

In fact, (2) can be easily checked by direct computation of $e^{\text{avg}}(Q)$ and $e^{\text{wor}}(Q;H_k)$.

Suppose we can compute function values at arbitrary points. It is then natural to choose the points $x_1, x_2, \ldots, x_n$ and the weights $a_1, a_2, \ldots, a_n$ such that the average case error (or, equivalently, the worst case error over the unit ball of H_K) is minimized. Obviously, this minimal error depends on the covariance function K. Therefore we denote

$$e^{\text{avg}}(n,K) \;=\; \inf_{\{a_i,x_i\}_{i=1}^n} \mathrm{E}\left(\left(\mathcal{I}(f) \;-\; \sum_{i=1}^n a_i\, f(x_i) \right)^2 \right)^{1/2}$$

as the *nth minimal average case error.*

We illustrate our discussion by two examples.

Example 1: Isotropic Wiener process (Brownian motion in Lévy's sense). In this case we have

$$K(x,t) = (\|x\| + \|t\| - \|x-t\|)/2,$$

where $\|\cdot\|$ is the Euclidean norm of $\mathbb{R}^d$. The Hilbert space H_K has been characterized by Molchan (1967) for odd d and Ciesielski (1975) for arbitrary d. In particular, the restriction $f|_D$ of any function $f \in C_0^\infty(\mathbb{R}^d)$ with $f(0) = 0$ belongs to H_K. For any such function we have

$$\|f|_D\|_K^2 \le a_d \int_{\mathbb{R}^d} \left|(-\Delta)^{(d+1)/4} f(t)\right|^2 dt,$$

where Δ is the Laplace operator and a_d is a constant. (For $d+1$ not divisible by 4, $\Delta^{(d+1)/4}$ is understood in the generalized form.) Furthermore, $\|f|_D\|_K^2$ equals the minimum of the above r.h.s. over all extensions to functions in $C_0^\infty(\mathbb{R}^d)$.

The nth minimal average case error was found in Wasilkowski (1991) by analyzing the worst case in the Hilbert space H_K. We have

$$e^{\text{avg}}(n,K) = \Theta\left(n^{-1/2-1/(2d)}\right).$$

Observe that for large d, the exponent of n is close to $-1/2$. As we shall see in Section 4.5, $-1/2$ is the maximal exponent which can occur for multivariate integration in the average case setting with the covariance function satisfying (1). □

Example 2: Wiener (Brownian) sheet process. In this case we have

$$K(x,t) = \prod_{i=1}^{d} \min\{x_i, t_i\}$$

for $x = [x_1, \dots, x_d]$ and $t = [t_1, \dots, t_d]$. Let $f^{(\vec{1})} = \partial^d f/(\partial t_1 \cdots \partial t_d)$ with $\vec{1} = [1, 1, \dots, 1]$. Then the Hilbert space H_K is given by

$$H_K = \{f \in C(D) : f^{(\vec{1})} \in L_2(D),\ f(x) = 0 \text{ if } x_i = 0 \text{ for some } 1 \le i \le d\}$$

with the inner product $\langle f, h\rangle_K = \int_D f^{(\vec{1})}(t)\, h^{(\vec{1})}(t)\, dt$, see e.g., Paskov (1991). We have

$$e^{\text{avg}}(Q) = \left(\int_D \left(t_1 t_2 \cdots t_d - \sum_{i=1}^{n} a_i\, \kappa_{[0,t)}(\vec{1} - x_i)\right)^2 dt\right)^{1/2},$$

where $\kappa_{[0,t)}$ is the characteristic (indicator) function of the interval $[0,t) = [0,t_1) \times [0,t_2) \times [0,t_d)$. For $a_i = 1/n$, the right hand side is the L_2 discrepancy of the points $\vec{1} - x_1, \vec{1} - x_2, \dots, \vec{1} - x_n$, see Woźniakowski (1991).

Then one can use the worst case results for multivariate integration in the space H_K, see Bykovskij (1985), Frolov (1976), and Temlyakov (1990), or one can use the L_2 discrepancy results due to Roth (1954, 1980) and Frolov (1980) to conclude that

$$e^{\mathrm{avg}}(n, K) = \Theta\left(n^{-1}\,(\log n)^{(d-1)/2}\right).$$

Observe that this quantity depends only mildly on the dimension d. The average case error of minimal order can be achieved by using the weights $a_i = 1/n$ and the points x_i which are derived from Hammersley points. □

3. Smooth Covariance Functions

In this section we study the nth minimal average case error $e^{\mathrm{avg}}(n, K)$ in terms of smoothness of the covariance function K. One can hope that $e^{\mathrm{avg}}(n, K)$ goes quickly to zero for smooth K. As we shall see this is indeed the case. It may, however, happen that even for nonsmooth K, the nth minimal average case error goes quickly to zero or even is zero. Thus, smoothness of K is a sufficient but, in general, not necessary condition for fast convergence of $e^{\mathrm{avg}}(n, K)$.

We first show that for some nonsmooth K we can even have $e^{\mathrm{avg}}(1, K) = 0$, see also Section 3.4. Indeed, assume that $K(x,t) = k(s)\,k(t)$ for some (nonsmooth) nonzero function k from $L_2(D)$. Then the Hilbert space $H_K = \mathrm{span}\{k\}$ is a one dimensional space. Let x_1 be a point for which $k(x_1)$ is nonzero. Define $Q(f) = a_1\, f(x_1)$ with $a_1 = \int_D k(t)\,dt / k(x_1)$. It is easy to check that now the function g in (3) is zero and therefore $e^{\mathrm{avg}}(Q) = 0$. Thus, this integration problem can be solved even exactly by using only one function value although the covariance function K need not be smooth.

We now show that for smooth K the nth minimal average case error goes quickly to zero. In Section 3.1, we recall a relation between the integration and function approximation problems. This relation will enable us to derive upper bounds on the nth minimal average case errors for two specific kinds of smoothness, see Sections 3.2 and 3.3.

3.1 Function Approximation Problem. As for the integration problem, consider the approximation of f by linear algorithms

$$Q(f) = \sum_{i=1}^{n} g_i(\cdot) f(x_i)$$

with $g_i \in L_2(D)$. The average case error of Q for the function approximation problem is then defined by

$$e^{\mathrm{avg}}(Q; \mathrm{App}) = \mathrm{E}\left(\|f - Q(f)\|_2^2\right)^{1/2}.$$

The average case error of Q is bounded by the worst case error of Q in the unit ball of the Hilbert space H_K, see Woźniakowski (1992). That is, for every g_i and every x_i we have

$$e^{\mathrm{avg}}(Q; \mathrm{App}) \le e^{\mathrm{wor}}(Q; H_k, \mathrm{App}) := \sup_{\|f\|_K \le 1} \|f - Q(f)\|_\infty. \tag{4}$$

Observe that in the average case, we measure the error by L_2 norm, whereas, in the corresponding worst case by L_∞ norm.

The *nth minimal average case error* of App is defined by

$$e^{\text{avg}}(n, K; \text{App}) = \inf_{\{g_i, x_i\}_{i=1}^n} \mathrm{E}\left(\|f - \sum_{i=1}^n g_i(\cdot) f(x_i)\|_2^2 \right)^{1/2}.$$

To distinguish between the errors for the integration and function approximation problems, we sometimes write $e^{\text{avg}}(n, K; \text{Int})$ instead of $e^{\text{avg}}(n, K)$.

The nth minimal average case errors of Int and App are related, see Wasilkowski (1991). That is, for every K and every n we have

$$\sqrt{n}\, e^{\text{avg}}(2n, K; \text{Int}) \leq e^{\text{avg}}(n, K; \text{App}). \tag{5}$$

Characterizing the space H_K for a given kernel K need not be an easy task. If, however, the space H_K can be continuously embedded into another space W for which the corresponding results concerning the worst case L_∞ function approximation are available then (4) and (5) can be used to derive upper bounds on the average case errors. This is exactly the approach we take in the next two subsections.

3.2 Global Smoothness. In this subsection we assume that K is several times differentiable and its partial derivatives satisfy a Hölder-like condition.

More precisely, let $\vec{v} = [v_1, \ldots, v_d]$ be a vector with nonnegative integer components and let $|\vec{v}| = \sum_{i=1}^d v_i$. For a given nonnegative integer r and a number $\alpha \in [0,1]$ assume that K satisfies the following conditions:

$$K \in C^{r,r}(D \times D), \tag{6}$$

and for any $\vec{v}$ with $|\vec{v}| = r$,

$$\sup_{x,h:\, x, x+h \in D} \frac{K^{(\vec{v},\vec{v})}(x+h, x+h) - 2K^{(\vec{v},\vec{v})}(x+h, x) + K^{(\vec{v},\vec{v})}(x,x)}{\|h\|^\alpha} =: c_{\vec{v}} < +\infty. \tag{7}$$

Let $C^{r,\beta}(D)$ denote the Banach space of functions defined on D that have continuous derivatives of order up to r, and any rth derivative satisfies the Hölder condition with exponent β.

Lemma 1 *If (6) and (7) hold then $H_K \subseteq C^{r,\alpha/2}$. More precisely, for any $g \in H_K$*

$$\|g^{(\vec{v})}\|_\infty \leq \|g\|_K \max_{|\vec{w}| \leq r} \sup_{x \in D} \sqrt{K^{(\vec{w},\vec{w})}(x,x)}, \qquad \forall\, |\vec{v}| \leq r, \tag{8}$$

and, for $|\vec{v}| = r$, $g^{(\vec{v})}$ satisfies the Hölder condition with exponent $\alpha/2$ and constant bounded by $\|g\|_K \sqrt{c_{\vec{v}}}$.

Proof: It is well known that (6) implies $H_K \subset C^r(D)$, and that we have

$$g^{(\vec{v})}(x) = \left\langle g, K^{(\vec{0},\vec{v})}(\cdot, x) \right\rangle_K, \qquad \forall\, |\vec{v}| \leq r,\ x \in D,$$

for $g \in H_K$, see, e.g., Sacks and Ylvisaker (1970b, p. 123) for the case $d = 1$. Thus $|g^{(\vec{v})}(x)| \le \|g\|_K \|K^{(\vec{0},\vec{v})}(\cdot,x)\|_K = \|g\|_K \left(K^{(\vec{v},\vec{v})}(x,x)\right)^{1/2}$. Furthermore,

$$\begin{aligned}
&\left(g^{(\vec{v})}(x+h) - g^{(\vec{v})}(x)\right)^2 \\
&\le \|g\|_K^2 \left(K^{(\vec{v},\vec{v})}(x+h,x+h) - 2\,K^{(\vec{v},\vec{v})}(x+h,x) + K^{(\vec{v},\vec{v})}(x,x)\right) \\
&\le \|g\|_K^2 \, c_{\vec{v}} \, \|h\|^\alpha.
\end{aligned}$$

□

We are ready to prove the following theorem.

Theorem 1 *If K satisfies (6) and (7) then*

$$\begin{aligned}
e^{\mathrm{avg}}(n,K;\mathrm{Int}) &= O\left(n^{-1/2-(r+\alpha/2)/d}\right), \\
e^{\mathrm{avg}}(n,K;\mathrm{App}) &= O\left(n^{-(r+\alpha/2)/d}\right).
\end{aligned}$$

Proof: Due to Lemma 1, the unit ball in H_K is a subset of a ball in the space $C^{r,\alpha/2}(D)$. Hence, due to (4) we have $e^{\mathrm{avg}}(n,K;\mathrm{App}) = O\left(e^{\mathrm{wor}}(n,C^{r,\alpha/2};\mathrm{App})\right)$, where

$$e^{\mathrm{wor}}(n,C^{r,\alpha/2};\mathrm{App}) = \inf_{\{g_i,x_i\}_{i=1}^n} \sup \left\{ \|f - \sum_{i=1}^n g_i(\cdot) f(x_i)\|_\infty : \|f\|_{C^{r,\alpha/2}} \le 1 \right\}.$$

It is known, see Novak (1988, p. 34), that $e^{\mathrm{wor}}(n,C^{r,\alpha/2};\mathrm{App}) = O\left(n^{-(r+\alpha/2)/d}\right)$. This and (5) complete the proof. □

We remark that for the isotropic Wiener process, see Example 1, as well as for the Wiener sheet process, see Example 2, the assumptions (6) and (7) hold with $r = 0$ and $\alpha = 1$. However, the upper bounds of Theorem 1 are sharp only in the case of the isotropic Wiener process, see Wasilkowski (1991). In Section 3.3 we study different smoothness assumptions which lead to sharp bounds for the Wiener sheet process.

Suppose we want to guarantee that the average case error is at most ε. This can be done by computing $n(\varepsilon)$ function values where $n(\varepsilon) = O(\varepsilon^{2/(1+(2r+\alpha)/d)})$ for Int, and $n(\varepsilon) = O(\varepsilon^{d/(r+\alpha/2)})$ for App. Note that for fixed r and d tending to infinity, the upper bound on $n(\varepsilon)$ for the integration problem has exponent close to 2. As already mentioned, this is the worst exponent which can occur in the average case setting for covariance functions satisfying (1), see Section 4.5. For the function approximation problem, the upper bound on $n(\varepsilon)$ depends exponentially on d and for some K (like isotropic Wiener process) this bound is sharp. This shows that the function approximation problem in many variables can be computationally very hard even in the average case setting.

3.3 Mixed Derivative Smoothness. In this subsection we assume that K is several times differentiable with respect to each variable and its partial derivatives satisfy a Hölder-like condition. As we shall see, this type of smoothness yields small bounds on the average case error for integration as well as function approximation even for large d.

Let $\vec{r} = [r_1, r_2, \ldots, r_d]$ be a vector with nonnegative integer components. Let $\vec{\alpha} = [\alpha_1, \alpha_2, \ldots, \alpha_d]$ be a vector with $\alpha_i \in (0,1]$, and let $\vec{m} = [m_1, m_2, \ldots, m_d] \in \{0,1\}^d$. By $\Delta_u^{\vec{m}}$ we mean the composition of one-dimensional forward differences of order m_i with step u_i for the i-th variable. Define the set

$$\Omega_{\vec{m}} = \{(y,h) \in D \times \mathbb{R}^d : 0 \le y_i + m_i h_i \le 1\}.$$

The smoothness conditions of the covariance function K are given by the following two requirements:

$$K \in C^{\vec{r},\vec{r}}(D \times D), \tag{9}$$

and

$$\sup_{(y,h)\in\Omega_{\vec{m}}} \frac{\left|\Delta_{h,h}^{\vec{m},\vec{m}} K^{(\vec{k},\vec{k})}(y,y)\right|}{\prod_{i=1}^d |h_i|^{2\alpha_i m_i}} =: c_{\vec{m}} < +\infty, \tag{10}$$

for all $\vec{m} \in \{0,1\}^d$ and every $\vec{k}$ with integer components $k_i = r_i$ if $m_i = 1$ and $1 \le k_i \le r_i$ if $m_i = 0$. In (10), we use the convention $0/0 = 0$.

Example 3: Tensor Products. We illustrate (9) and (10) for a specific class of covariance functions K. Suppose that $K(x,t) = \prod_{i=1}^d K_i(x_i,t_i)$ with one dimensional covariance functions K_i which are r_i times continuously differentiable, $K_i \in C^{r_i,r_i}([0,1] \times [0,1])$. This means that K is a tensor product $K = \bigotimes_{i=1}^d K_i$ of smooth covariance kernels K_i. Obviously such K satisfies (9). Assume additionally that there exist positive constants c_i such that the difference $\Delta_{h,h}^{1,1}$ of $K^{(r_i,r_i)}$ is uniformly bounded by $c_i|h|^{2\alpha_i}$, i.e.,

$$\left|K_i^{(r_i,r_i)}(y+h,y+h) - 2K_i^{(r_i,r_i)}(y+h,y) + K_i^{(r_i,r_i)}(y,y)\right| \le c_i\,|h|^{2\alpha_i}$$

for $y, y+h \in [0,1]$. Clearly, this implies (10).

Note that the Wiener sheet process, see Example 2, is of this special form with $K_i(x_i,t_i) = \min\{x_i,t_i\}$. In this case, $\vec{r} = \vec{0}$ and $\vec{\alpha} = [1/2, \ldots, 1/2]$.

In terms of the stochastic process f, the assumptions of this example mean that the derivatives $f^{(\vec{k})}$ exist (in the mean square sense) for any $\vec{k} \le \vec{r}$ and the variances of $f^{(r_i\vec{e}_i)}(y + h\vec{e}_i) - f^{(r_i\vec{e}_i)}(y)$ are of the order $O(|h|^{2\alpha_i})$ uniformly in y. Here $\vec{e}_i$ is the ith unit vector in $\mathbb{R}^d$. □

Let $W^{\vec{r},\vec{\alpha}}(D)$ be the Banach space of functions g defined on D with $g^{(\vec{k})}$ continuous if $\vec{k} \le \vec{r}$ and $|\Delta_h^{\vec{1}} g^{(\vec{r})}(x)| = O(\prod_{i=1}^d |h_i|^{\alpha_i})$ uniformly in x. Using the same proof techniques as in Lemma 1 it is easy to show the following.

Lemma 2 *If (9) and (10) hold then $H_K \subseteq W^{\vec{r},\vec{\alpha}}$. More precisely, for any $g \in H_K$*

$$\|g^{(\vec{v})}\|_\infty \le \|g\|_K \max_{|\vec{w}|\le r} \sup_{x\in D} \sqrt{K^{(\vec{w},\vec{w})}(x,x)}, \qquad \forall \vec{v} \le \vec{r},$$

and

$$\sup_{(y,h)\in\Omega_{\vec{1}}} \frac{\left|\Delta_h^{\vec{1}} g^{(\vec{r})}(y)\right|}{\prod_{i=1}^d |h_i|^{\alpha_i}} \le \|g\|_K \sqrt{c_{\vec{1}}}.$$

We are ready to show how fast the nth minimal average case errors go to zero for K that is smooth in the sense of (9) and (10).

Theorem 2 *If K satisfies (9) and (10) then*

$$\begin{aligned} e^{\text{avg}}(n, K; \text{Int}) &= O\left(n^{-\gamma-1/2} \log^{(\nu-1)(\gamma+1)} n\right), \\ e^{\text{avg}}(n, K; \text{App}) &= O\left(n^{-\gamma} \log^{(\nu-1)(\gamma+1)} n\right), \end{aligned}$$

where γ denotes the minimal regularity with respect to all variables and ν equals the number of variables with minimal regularity,

$$\gamma = \min_{1\le i\le d} (r_i + \alpha_i) \qquad \text{and} \qquad \nu = \#\{i : r_i + \alpha_i = \gamma\}.$$

Proof: Assume first that K satisfies the following boundary condition:

$$K^{(\vec{0},\vec{k})}(x, y) = 0 \quad \forall x \in D,\, y \in \partial D,\, \vec{k} \le \vec{r}. \tag{11}$$

Then any $f \in H_K$ vanishes together with its derivatives on the boundary of D. We can then extend any function from H_K to a periodic function from the space $W^{\vec{r},\vec{\alpha}}(\mathbb{R}^d)$ by setting

$$f(x_1, \ldots, x_{i-1}, x_i + 1, x_{i+1}, \ldots, x_d) = -f(x).$$

Due to Lemma 2, the extended functions satisfy $|\Delta_h^{\vec{1}} f^{(\vec{r})}(x)| / \prod_{i=1}^d |h_i|^{\alpha_i} \le c\|f|_D\|_K$ and $|f^{(\vec{v})}(x)| \le c\|f|_D\|_K$ for any $x, h \in \mathbb{R}^d$ and any $\vec{v} \le \vec{r}$ with a suitable constant c independent of f. Furthermore, they have period 2 in each variable and their integrals over $[0, 2]$ vanish for all variables. The worst case setting for this class of functions has been studied by Temlyakov (1987). He proved that there exist points $z_i = [z_{i,1}, \ldots z_{i,d}] \in [0, 2]^d$ and functions $v_i \in C([0, 2]^d)$ such that the error of $Q(f) = \sum_{i=1}^n f(z_i) v_i$ for the function approximation problem satisfies

$$\sup_{y\in[0,2]^d} \left|f(y) - Q(f)(y)\right| = O\left(n^{-\gamma} \log^{(\nu-1)(\gamma+1)} n\right) \sup_{y,h\in\mathbb{R}^d} \frac{\left|\Delta_h^{\vec{1}} f^{(\vec{r})}(y)\right|}{\prod_{i=1}^d |h_i|^{\alpha_i}}$$

To apply the above estimate to our class of functions defined over $D = [0, 1]^d$, we need to switch to points x_i from D and functions f_i defined on D. Define $x_i = [x_{i,1}, \ldots, x_{i,d}] \in D$ and $f_i \in C(D)$ by

$$x_{i,j} = \begin{cases} z_{i,j}, & \text{if } z_{i,j} \in [0, 1], \\ z_{i,j} - 1, & \text{if } z_{i,j} \in (1, 2], \end{cases}$$

and

$$f_i(y) = (-1)^{\sum_{j=1}^d (z_{i,j} - x_{i,j})} \cdot v_i(y).$$

Let $\overline{Q}(f) = \sum_{i=1}^n f(x_i)f_i$. Then

$$\sup_{x\in D} |f(x) - \overline{Q}(f)(x)| = \sup_{y\in[0,2]^d} |f(y) - Q(f)(y)| = O\left(n^{-\gamma} \log^{(\nu-1)(\gamma+1)} n\right)$$

for any f from the unit ball in H_K. This estimate and (5) complete the proof for K satisfying the boundary condition (11).

Assume now that K does not satisfy (11). In this case we apply a well known periodization technique, see e.g, Bykovskij (1985). For $y = [y_1, y_2, \ldots, y_d]$, let

$$\psi_i(y_i) = \frac{\int_0^{y_i} (u(1-u))^{2(r_i+1)}\, du}{\int_0^1 (u(1-u))^{2(r_i+1)}\, du} \qquad \text{and} \qquad \Psi(y) = [\psi_1(y_1), \ldots, \psi_d(y_d)].$$

Note that $\psi_i(0) = 0$, $\psi_i(1) = 1$, ψ_i is increasing, and is arbitrarily many times differentiable. Note also that $(\psi_i')^{1/2}(u)$ is proportional to $(u(1-u))^{r_i+1}$, and therefore $(\Psi')^{1/2}$ is also arbitrarily many times differentiable. Define

$$f_\Psi(y) = \Psi'(y)^{1/2} \cdot f(\Psi(y)) \qquad \text{with} \qquad \Psi'(y) = \prod_{i=1}^d \psi_i'(y_i).$$

For any $Q(f) = \sum_{i=1}^n f(x_i)g_i$ and any $f \in F$, the change of variables $x = \Psi(y)$ yields $\|f - Q(f)\|_2 = \|f_\Psi - Q_\Psi(f_\Psi)\|_2$, where

$$Q_\Psi(h)(y) = \sum_{i=1}^n h(\Psi^{-1}(x_i))\overline{g}_i(y), \quad \text{with} \quad \overline{g}_i(y) = \left(\Psi'(y)/\Psi'(\Psi^{-1}(x_i))\right)^{1/2} g_i(\Psi(y)).$$

The stochastic process f_Ψ has zero mean and the covariance function K_Ψ given by

$$K_\Psi(x,t) = \left(\Psi'(x) \cdot \Psi'(t)\right)^{1/2} \cdot K(\Psi(x), \Psi(t)).$$

From $\|f - Q(f)\|_2 = \|f_\Psi - Q_\Psi(f_\Psi)\|_2$ we conclude that

$$e^{\text{avg}}(n, K; \text{App}) = e^{\text{avg}}(n, K_\Psi; \text{App}). \tag{12}$$

The regularity of Ψ and of $\Phi := (\Psi')^{1/2}$, together with (9), imply that the kernel K_Ψ also satisfies (9). Moreover, K_Ψ satisfies the boundary condition (11), since $\Psi^{(\vec{k})}$ vanishes at the boundary of D if $\vec{k} \le \vec{r}$. The condition (10) and regularity of Ψ also imply that $\Delta_{h,h}^{\vec{1},\vec{1}} K_\Psi^{(\vec{r},\vec{r})}(y,y)/\prod_{i=1}^d |h_i|^{\alpha_i}$ is uniformly bounded on $\Omega_{\vec{1}}$. (The proof of this claim is rather tedious yet straightforward and therefore we omit it.) Thus, we can apply the first part of the proof for the covariance function K_Ψ and the upper bounds presented in the theorem hold for K_Ψ. Due to (12), the same upper bounds hold for K. This completes the proof. □

As we know, the Wiener sheet process, see Example 2, satisfies (9) and (10) with $\vec{r} = 0$ and $\vec{\alpha} = [1/2, \ldots, 1/2]$. The upper bounds of Theorem 2 coincide, modulo a power of $\log n$, with those reported in Woźniakowski (1991, 1992). For folded Wiener sheet processes the bounds of Theorem 2 are also sharp, modulo a power of $\log n$, see Paskov (1991) for the integration problem and Woźniakowski (1992) for the function approximation problem.

Suppose that $r_i + \alpha_i$ are uniformly bounded from below, $r_i + \alpha_i \geq a > 0$. Then $\gamma \geq a$ and the upper bounds of Theorem 2 depend weakly on d. Indeed, the only dependence on d is through $\nu \leq d$ which determines the power of $\log n$. Thus, if we want to guarantee the average case error at most ε we have to compute at most $n(\varepsilon)$ function values, where $n(\varepsilon) = O\left(\varepsilon^{2/(1+2\gamma)}(\log 1/\varepsilon)^{(d-1)(\gamma+1)/(\gamma+1/2)}\right)$ for Int, and $n(\varepsilon) = O\left(\varepsilon^{1/\gamma}(\log 1/\varepsilon)^{(d-1)(\gamma+1)/\gamma}\right)$ for App. Hence, the power of $1/\varepsilon$ does not depend on d and the dependence on d is only through $\log 1/\varepsilon$. This is in a sharp contrast with Theorem 1 where a different type of smoothness of K is assumed.

3.4 Nonsmooth K. We stress that the conclusions of Theorems 1 and 2 may hold for covariance operators K that need not be smooth. Indeed, as in the proof of Theorem 2, for a (nonsmooth) K we change variables $x = \Psi(y)$ for some Ψ. Then we get a new covariance function K_Ψ satisfying (12). If K_Ψ is smooth, i.e., it satisfies the regularity assumptions (6-7) or (9-10) then Theorem 1 or Theorem 2 holds for the (nonsmooth) covariance function K.

To illustrate this, consider the process f defined over $D = \mathbb{R}_+^d$ with

$$K(x,y) = \prod_{i=1}^{d} \left(\alpha_i\ (x_i y_i)^{(\alpha_i-1)/2} \exp\left(-(x_i^{\alpha_i} + y_i^{\alpha_i})/2\right) \min\left\{e^{-x_i^{\alpha_i}}, e^{-y_i^{\alpha_i}}\right\}\right)$$

for $\alpha_i \in (0,1]$. Thus, D is unbounded and K is singular on ∂D for $\alpha_i < 1$. However, for $x_i = \psi_i(t_i) = (\ln 1/t)^{1/\alpha_i}$ we get $K_\Psi(t,z) = \prod_{i=1}^d \min\{t_i, z_i\}$ with $t, z \in D_\Psi = [0,1]^d$. In the average case case setting, due to (12), the integration and function approximation problems for such nonsmooth K with unbounded domain D are equivalent to the integration and function approximation problems, respectively, with the Wiener sheet process over the finite domain $[0,1]^d$.

4. Extensions

In this section we briefly indicate how the results of the previous sections can be generalized in a number of directions.

4.1 Nonlinear Algorithms. So far we have assumed that the multivariate integral $\mathcal{I}(f)$ is approximated by a *linear* algorithm $Q(f) = \sum_{i=1}^n a_i f(x_i)$. It is natural to ask if we can decrease the error by using *nonlinear* algorithms. That is, if we approximate $\mathcal{I}(f)$ by $\phi(f(x_1), f(x_2), \ldots, f(x_n))$ for some nonlinear mapping $\phi : \mathbb{R}^n \to \mathbb{R}$.

This problem has been studied in many papers in different settings. For instance, in the worst case setting the error of ϕ in a class $F_0 \subset F$ is defined by

$$e^{\mathrm{wor}}(\phi) = \sup_{f \in F_0} |\mathcal{I}(f) - \phi(f(x_1), f(x_2), \ldots, f(x_n))|.$$

Smolyak (1965), this result can also be found in Traub et. al. (1988, p. 76), proved that if F_0 is convex and balanced ($f \in F_0$ implies that $-f \in F_0$) then nonlinear algorithms are not better than linear algorithms. That is,

$$\inf_{\phi} e^{\text{wor}}(\phi) = \inf_{Q-\text{linear}} e^{\text{wor}}(Q).$$

The result of Smolyak has been generalized for a linear operator $\mathcal{I}$ and information $f(x_1), f(x_2), \ldots, f(x_n)$ replaced by $L_1(f) + \eta_1, L_2(f) + \eta_2, \ldots, L_n(f) + \eta_n$ with a linear functional L_i which is computed with deterministic or stochastic noise η_i. In many cases, nonlinear algorithms are not better than linear algorithms, see Packel (1988) and Traub et. al. (1988, Chapter 4) for a survey. There are, however, some counterexamples as well. In fact, it may even happen that the worst case error of any linear algorithm is infinity whereas there exist nonlinear algorithms with finite worst case error and this error can even be arbitrarily small, see Werschulz and Woźniakowski (1986).

We now turn to the average case setting. For nonlinear algorithms, the basic relation (2) between the average case and worst case settings does *not* hold. In fact, it may even happen that there exists a *nonlinear* algorithm with zero *average* error whereas the *worst* case error over the unit ball of H_K (which is, of course, convex and balanced) of this and all other nonlinear algorithms is positive and is at least as large as the worst case error of some linear algorithm.

To guarantee that nonlinear algorithms are not better than linear algorithms we have to know more about the stochastic process. Namely, we have to know the probability measure μ which defines the process. Knowing μ, the average case error of a (nonlinear) algorithm is defined as the standard deviation of the random variable $\mathcal{I}(f) - \phi(f(x_1), f(x_2), \ldots, f(x_n))$,

$$e^{\text{avg}}(\phi; \mu) = \left(\int_F \left(\mathcal{I}(f) - \phi(f(x_1), f(x_2), \ldots, f(x_n))\right)^2 \mu(df) \right)^{1/2}.$$

The power of nonlinear algorithms strictly depends on the measure μ. As already mentioned for some measures μ it may happen that $e^{\text{avg}}(\phi; \mu) = 0$ for some nonlinear ϕ and $e^{\text{avg}}(Q; \mu) > 0$ for all linear Q.

For what measures μ are nonlinear algorithms no better than linear algorithms? This holds, in particular, if μ is *Gaussian* and F is a separable Banach space, see e.g., Traub et. al. (1988, Chapter 6). Basic information about Gaussian measures may be found in Kuo (1975), Gihman and Skorohod (1974), and Vakhania et. al. (1987). Gaussian measures are, in a sense, least informative. Namely, if we consider all probability measures μ on F with mean zero and the same covariance function $K_\mu = K$ then

$$\sup_{\mu:\ K_\mu = K} \inf_{\phi} e^{\text{avg}}(\phi; \mu) = \inf_{\phi} e^{\text{avg}}(\phi; \mu_G) = \inf_{Q-\text{linear}} e^{\text{avg}}(Q; \mu_G), \tag{13}$$

where μ_G is a Gaussian measure with zero mean and covariance function K, see Kadane and Wasilkowski (1985).

The average case setting has been studied for general operators $\mathcal{I}$ and information $L_1(f) + \eta_1, L_2(f) + \eta_2, \ldots, L_n(f) + \eta_n$ with a linear functional L_i and deterministic or stochastic noise η_i. Typically, nonlinear algorithms are not better than linear algorithms if a Gaussian measure is used. A survey may be found in Traub et. al. (1988).

4.2 Adaption. So far we have assumed that the information is of the form $f(x_1), f(x_2), \dots, f(x_n)$, where the points x_i are given simultaneously. Such information is called *nonadaptive* and has obvious application for parallel computation. On the other hand, the choice of the point x_i can depend on the already computed information $f(x_1), f(x_2), \dots, f(x_{i-1})$. Also the number n of function values can depend on the computed function values. Such information is called *adaptive*. Adaption requires sequential computation since we have to wait with the decision whether to compute more than $(i-1)$ function values and with the eventual choice of x_i until $f(x_1), f(x_2), \dots, f(x_{i-1})$ are computed.

Can adaption be better than nonadaption? That is, can we decrease the error of approximation of $\mathcal{I}(f)$ by using adaptive information?

Again, the answer depends on the setting. In the *worst* case setting, Bakhvalov (1971) proved that adaption does not help for approximating linear functionals (such as $\mathcal{I}$) as long as F_0 is convex and balanced. This has been generalized for linear operators by Gal and Micchelli (1980), and Traub and Woźniakowski (1980); for operators adaption can help, although, only by at most a factor of two. An example of a linear operator for which adaption helps was found by Kon and Novak (1989). We stress that adaption may be very powerful if F_0 is convex but not balanced, see Novak (1992b).

In the average case setting, the answer depends on the measure. For Gaussian measures and linear operators, adaption does not help through adaptive choice of the evaluation points but may help through adaptive choice of the number of function values. More precisely, adaption does not help much for adaptive choice of n if the squares of the nth minimal average case errors e_n^2 of linear algorithms are semi-convex, see Wasilkowski (1986). (A sequence $\{a_i\}$ is semi-convex if there exist two positive constants a and b such that $a\,b_i \le a_i \le b\,b_i$ and $b_i \le (b_{i-1} + b_{i+1})/2$, $\forall\, i \ge 2$.) Recently, Plaskota (1992) found a linear problem with some restrictions on the choice of the evaluation points such that the sequence e_n^2 is not semi-convex and the adaptive choice of n is very powerful. As in Section 4.1, Gaussian measures are also least informative for algorithms that use adaptive information.

For non-Gaussian measures adaption may be very powerful. For specific results concerning scalar integration with non-Gaussian measures, we refer to Novak (1992a), and Wasilkowski and Gao (1992).

4.3 Complexity. Complexity is the minimal cost needed to compute an approximation of the solution with error at most ε. The definition of cost and error depends on the setting. For instance, in the average case setting we take the expected cost and error of computing $Q(f)$. The cost of computing $Q(f)$ consists of the cost of n function values and the number of arithmetic operations and comparisons needed to compute $Q(f)$. For multivariate integration with a Gaussian measure, nonlinear algorithms do not help and adaption can help only by a multiplicative constant if $e^{\text{avg}}(n,K)^2$ is a semi-convex sequence. Observe that for all sharp bounds presented in this paper semi-convexity holds. Hence, the average case complexity is $\Theta(n)$, where n is the smallest integer for which $e^{\text{avg}}(n, K) \le \varepsilon$. For instance, for the multivariate integration problem, the average case complexity is $\Theta\left(\varepsilon^{-2/(1+1/d)}\right)$ or $\Theta\left(\varepsilon^{-1}(\log 1/\varepsilon)^{(d-1)/2}\right)$ depending on whether the isotropic Wiener or the Wiener sheet process is considered.

Gaussian measures are also least informative from the complexity point of view. That is, if the average case cost and errors are measured by the corresponding expectations for the worst case measure μ with fixed $K_\mu = K$ then again the Gaussian measure μ_G is the most difficult. Thus, the average case complexity for μ_G provides an upper bound on the average case complexity for any μ with $K_\mu = K$.

4.4 Probabilistic Setting. Multivariate integration can also be studied in the *probabilistic* setting in which we would like to guarantee the error not to exceed ε with probability at least $1 - \delta$. That is, we now look for an algorithm which can only fail for functions from a set of measure at most δ. Thus, the parameter δ measures the reliability of an algorithm.

For a Gaussian measure, the probabilistic and average case settings are closely related, see Traub et. al. (1988, Chapter 8). For linear functionals ($\mathcal{I}$ in particular), the probabilistic (ε, δ) complexity is essentially the same as the average case complexity for $\varepsilon/\psi^{-1}(1-\delta)$, where $\psi(x) = \sqrt{2/\pi}\int_0^x \exp(-t^2/2)\,dt$ and $\psi^{-1}(1-\delta) = \sqrt{2\log 1/\delta}(1 + o(1))$ as $\delta \to 0$. Observe a very weak dependence on δ. Hence, for multivariate integration for the isotropic Wiener process, the probabilistic complexity is

$$\Theta\left(\varepsilon^{-2/(1+1/d)}\,(\log 1/\delta)^{1/(1+1/d)}\right),$$

whereas for the Wiener sheet process the probabilistic complexity is

$$\Theta\left(\varepsilon^{-1}\,(\log 1/\delta)^{1/2}\,(\log 1/\varepsilon\, +\, \log\log 1/\delta)^{(d-1)/2}\right).$$

4.5 Randomized Setting. Multivariate integration can also be studied in the *randomized* setting in which, in particular, we can randomly choose points x_i and weights a_i of the quadrature formula $Q(f) = \sum_{i=1}^n a_i f(x_i)$. The error in the randomized setting is defined as the expected error of randomization for a worst case function f from a given class F_0. The error of the classic Monte Carlo algorithm is $O(n^{-1/2})$ if $\int_D f^2(t)\,dt$ is uniformly bounded for $f \in F_0$. This bound can be improved for smooth functions f. Bakhvalov (1959) found minimal error bounds for a number of Sobolev classes of functions. Recent results for different problems in the randomized setting may be found in Heinrich (1992), Mathé (1991), Novak (1988), and Wasilkowski (1989).

It is well known that the nth minimal average case error $e^{\text{avg}}(n, K)$ for an *arbitrary* K with $\int_D K(t,t)\,dt \le \beta$ is bounded by the error of the Monte Carlo algorithm for functions with $\int_D f^2(t)\,dt \le \beta$. Hence, $e^{\text{avg}}(n, K) = O(n^{-1/2})$. This means that the average case error of some linear algorithm for multivariate integration goes to zero at least as quickly as $n^{-1/2}$. As we know, for the isotropic Wiener process and large d, the behavior of $e^{\text{avg}}(n, K)$ is almost as bad as $\Theta(n^{-1/2})$. This shows that the isotropic Wiener process almost maximizes the difficulty of multivariate integration in the average case setting.

4.6 Different Error Criteria. So far we have defined the error between the multivariate integral $\mathcal{I}(f)$ and the quadrature formula $Q(f)$ by the *absolute* error criterion $|\mathcal{I}(f) - Q(f)|$. Different error criteria can also be used. For instance, the *normalized* error criterion $|\mathcal{I}(f) - Q(f)|/\|f\|$ or the *relative* error criterion $|\mathcal{I}(f) - Q(f)|/|\mathcal{I}(f)|$ can be used, see Traub et. al. (1988, Chapter 9) for these and other error criteria.

The normalized and absolute error criteria are closely related. More interesting is the relative error criterion for a Gaussian measure, see Jackowski and Woźniakowski (1987). It is surprising, but easy to show, that multivariate integration with the relative error criterion *cannot* be solved neither in the worst case setting nor in the average case setting (assuming that $\mathcal{I}$ cannot be expressed as a finite linear combination of function values.)

This follows from the fact that the worst case as well as the average case error of any algorithm that uses finitely many (adaptive or nonadaptive) function values is at least one. Hence, it is impossible to compute an approximation with error less than 1 at finite cost and therefore the worst case and average case complexity are infinite.

Only in the probabilistic setting we can solve multivariate integration with the relative error criterion for a Gaussian measure. The probabilistic (ε, δ) complexity is roughly the same as the average case complexity with the absolute error criterion with ε replaced by $\pi\,\varepsilon\,\delta/2$. Observe that now the parameter δ plays the same role as the parameter ε contrary to the probabilistic setting with the absolute error criterion. We also add that the probabilistic (ε, δ) complexity is achieved by *a nonlinear* algorithm which is simple to implement.

For example, for the isotropic Wiener process the probabilistic complexity of multivariate integration with the relative error criterion is $\Theta\left((\varepsilon\,\delta)^{-2/(1+1/d)}\right)$, whereas for the Wiener sheet process we have $\Theta\left((\varepsilon\,\delta)^{-1}\,(\log 1/\varepsilon + \log 1/\delta)^{(d-1)/2}\right)$.

Acknowledgments

We thank for valuable comments from E. Novak and J. F. Traub.

The first author was partially supported by the Deutsche Forschungsgemeinschaft (DFG). Part of this work was done while the first author was visiting the Department of Computer Science, University of Kentucky. The second author was partially supported by the the National Science Foundation under Grant CCR-91-14042, and the third by AFOSR-91-0347.

References

BAKHVALOV, N. S. (1959) "On approximate computation of integrals" (in Russian), *Vestnik MGU, Ser. Math. Mech. Astron. Phys. Chem.* **4**, 3–18.

BAKHVALOV, N. S. (1971) "On the optimality of linear methods for operator approximation in convex classes of functions," *USSR Comput. Math. Math. Phys.* **11**, 244–249.

BENHENNI, K., AND S. CAMBANIS (1992) "Sampling designs for estimating integrals of stochastic processes," *The Annals of Statistics* **20**, 161–194.

BYKOVSKIJ, V. A. (1985) "On exact order of optimal quadrature formulas for spaces of functions with bounded mixed derivatives" (in Russian), *Report of Dalnevostochnoi Center of Academy of Sciences, Vladivostok, USSR.*

CIESIELSKI, Z. (1975) "On Lévy's Brownian motion with several-dimensional time," *Lecture Notes in Math.* **472**, (A. Dold and B. Eckman, eds.,) pp. 29–56, Springer-Verlag, New York.

FROLOV, K. K. (1976) " Upper bounds on quadrature formulas errors for some classes of functions," *Dokl. Akad. Nauk SSSR* **231**, 818–821.

FROLOV, K. K. (1980) "Upper bounds on discrepancy in L_p, $2 \leq p < \infty$," *Dokl. Akad. Nauk SSSR* **252**, 805–807.

GAL, S., AND C. A. MICCHELLI (1980) "Optimal sequential and non-sequential procedures for evaluating a functional," *Appl. Anal.* **10**, 105–120.

GIHMAN, I. I., AND A. V. SKOROHOD (1974) *The Theory of Stochastic Processes I,* Springer-Verlag, Berlin.

HEINRICH, S. (1992) "Lower bounds for the complexity of Monte Carlo function approximation," *J. Complexity* **8**, 277–300.

JACKOWSKI, T., AND H. WOŹNIAKOWSKI (1987) "Complexity of approximation with relative error criterion in worst, average and probabilistic settings," *J. of Complexity* **3**, 114–134.

KADANE, J. B., AND G. W. WASILKOWSKI (1985) "Average case ϵ-complexity in computer science – a bayesian view," in *Bayesian Statistics 2, Proc. 2nd Valencia International Meeting, 1983* (J. M. Bernardo et. al., eds.,) pp. 361–374. Elsevier Science Publisher B. V. (North Holland).

KIMELDORF, G. S., AND G. WAHBA (1970a) "A correspondence between Bayesian estimation on stochastic processes and smoothing by splines," *Ann. Math. Stat.* **41**, 495–502.

KIMELDORF, G. S., AND G. WAHBA (1970b) "Spline functions and stochastic processes," *Sankhya Ser. A* **32**, 173–180.

KON, M. A, AND E. NOVAK (1989) "On the adaption problem" *BAMS (new series)* **23**, 159–165.

KUO, H. (1975) *Gaussian Measures in Banach Spaces,* Lecture Notes in Mathematics, **463**, Springer-Verlag, Berlin.

LARKIN, F. M. (1972) "Gaussian measure in Hilbert space and application in numerical analysis," *Rocky Mountain J. Math.* **2**, 379–421.

MATHÉ, P. (1991) "Random approximation of Sobolev embeddings," *J. Complexity* **7**, 261–281.

MICCHELLI, C. A., AND G. WAHBA (1981) "Design problems for optimal surface interpolation," *Approximation Theory and Applications* (Z. Ziegler, ed.,) pp. 329–347, Academic Press, New York.

MOLCHAN, G. M. (1967) "On some problems concerning Brownian motion in Lévy's sense," *Theory of Prob. Appl.* **12**, 682–690.

NOVAK, E. (1988) *Deterministic and Stochastic Error Bounds in Numerical Analysis,* Lecture Notes in Math., **1349**, Springer-Verlag, Berlin, 1988.

NOVAK, E. (1992a) "Quadrature formulas for monotone functions," *Proc. Amer. Math. Soc.* **115**, 59–68.

NOVAK, E. (1992b) "Quadrature, optimal recovery, and widths for convex classes of functions," Preprint.

PACKEL, E. W. (1988) "Do linear probms have linear optimal algorithms?" *SIAM Review* **30**, 388–403.

PARZEN, E. (1962) "An approach to time series analysis," *Ann. Math. Stat.* **32**, 951–989.

PASKOV, S. (1991) "Average Case Complexity of Multivariate Integration for Smooth Functions" *Dept. of Computer Science, Columbia University,* to appear in *J. Complexity.*

PLASKOTA, L. (1992) "A note on varying cardinality in the average case setting," to appear in *J. Complexity.*

ROTH, K. F. (1954) "On irregularities of distribution," *Mathematika* **1**, 73–79.

ROTH, K. F. (1980) "On irregularities of distribution, IV," *Acta Arith.* **37**, 67–75.

SACKS, J., AND D. YLVISAKER (1966) "Designs for regression with correlated errors," *Ann. Math. Stat.* **37**, 68–89.

SACKS, J., AND D. YLVISAKER (1968) "Designs for regression problems with correlated errors; many parameters," *Ann. Math. Stat.* **39**, 49–69.

SACKS, J., AND D. YLVISAKER (1970a) "Design for regression problems with correlated errors III," *Ann. Math. Stat.* **41**, 2057–2074.

SACKS, J., AND D. YLVISAKER (1970b) "Statistical design and integral approximation," *Proc. 12th Bienn. Semin. Can. Math. Congr.*, 115–136.

SKOROHOD, A. V. (1974) *Integration in Hilbert Space,* Springer-Verlag, New York.

SMOLYAK, S. A. (1965) "On optimal restoration of functions and functionals of them" (in Russian), Candidate Dissertation, Moscow State University.

TEMLYAKOV, V. N. (1987) "Approximate recovery of periodic functions of several variables," *Math. USSR Sbornik* **56**, 249–261.

TEMLYAKOV, V. N. (1990) "On a way of obtaining lower estimates for the errors of quadrature formulas errors", *Math. USSR Sbornik* **181**, 1403–1413, (in Russian), *Math. USSR Sbornik* **71**, 1992, 247-257, (in English).

TRAUB, J. F., G. W. WASILKOWSKI, AND H. WOŹNIAKOWSKI (1988) *Information-Based Complexity,* Academic Press, New York.

TRAUB, J. F., AND H. WOŹNIAKOWSKI (1980) *A General Theory of Optimal Algorithms,* Academic Press, New York.

VAKHANIA, N. N., V. I. TARIELADZE, AND S. A. CHOBANYAN (1987) *Probability Distributions on Banach Spaces,* Reidel, Dordrecht.

WAHBA, G. (1971) "On the regression design problem of Sacks and Ylvisaker," *Ann. Math. Stat.* **42**, 1035–1043.

WAHBA, G. (1990) *Spline Models for Observational Data,* SIAM-NSF Regional Conference Series in Appl. Math. **59**.

WASILKOWSKI, G. W. (1986) "Information of varying cardinality," *J. of Complexity* **2**, 204–228.

WASILKOWSKI, G. W. (1989) "Randomization for continuous problems," *J. of Complexity* **5**, 195–218.

WASILKOWSKI, G. W. (1991) "Integration and approximation of multivariate functions: average case complexity with isotropic Wiener measure," to appear in *Bull. AMS.*

WASILKOWSKI, G. W., AND F. GAO (1991) "On the power of adaptive information for functions with singularities," *Mathematics of Computation* **58**, 285–304.

WERSCHULZ, A. G., AND H. WOŹNIAKOWSKI (1986) "Are linear algorithms always good for linear problems?" *Aequationes Math.* **31**, 202–212.

WOŹNIAKOWSKI, H. (1991) "Average case complexity of multivariate integration," *Bull. AMS (new series)* **24**, 185–194.

WOŹNIAKOWSKI, H. (1992) "Average case complexity of linear multivariate problems, Part 1: Theory, Part 2: Applications," to appear in *J. Complexity* **8**.

YLVISAKER, D. (1975) "Designs on random fields," in *A survey of statistical design and linear models,* (J. Srivastava, ed.), 593–607, North-Holland.

Authors' Addresses:

K. RITTER, MATHEMATISCHES INSTITUT, UNIVERSITÄT ERLANGEN-NÜRNBERG, W-8520 ERLANGEN, GERMANY.

G. W. WASILKOWSKI, DEPARTMENT OF COMPUTER SCIENCE, UNIVERSITY OF KENTUCKY, LEXINGTON, KY 40506, USA.

H. WOŹNIAKOWSKI, DEPARTMENT OF COMPUTER SCIENCE, COLUMBIA UNIVERSITY, NY, NY 10027, USA; AND INSTITUTE OF APPLIED MATHEMATICS, UNIVERSITY OF WARSAW, POLAND.

International Series of Numerical Mathematics, Vol. 112, © 1993 Birkhäuser Verlag Basel

Rational Hermite Interpolation and Quadrature

Claus Schneider

Abstract. Rational Hermite interpolation is used in two different ways in order to derive and analyze quadrature rules. One approach yields quadratures of Gaussian–type whereas the other one generalizes Engels' dual quadratures exhibiting the close connection between rational Hermite interpolation and quadrature in general.

1. Rational Hermite Interpolation

In this section we shortly collect some facts about rational interpolation which will be used later. Let $n \in \mathbf{N}, t_{1n} < t_{2n} < \cdots < t_{nn}$ knots in $[a,b], \Omega_n(t) := \prod_{i=1}^{n}(t - t_{in}), \alpha_{in} \in \mathbf{R}, \beta_{in} \in \mathbf{R} \setminus \{0\}, i = 1(1)n$, and $f \in C^1[a,b]$. Then the function

$$r_n(f;t) = \frac{\sum_{i=1}^{n}\left(\{\frac{\alpha_{in}}{t-t_{in}} + \frac{\beta_{in}}{(t-t_{in})^2}\}f(t_{in}) + \frac{\beta_{in}}{t-t_{in}}f'(t_{in})\right)}{\sum_{i=1}^{n}\left(\frac{\alpha_{in}}{t-t_{in}} + \frac{\beta_{in}}{(t-t_{in})^2}\right)} \tag{1.1}$$

interpolates f and f' at the knots $t_{in}, i = 1(1)n$. (1.1) is called the *barycentric representation* of a rational Hermite interpolant. Henrici (1982) studied barycentric formulas for polynomial and trigonometric interpolation (cf. also Werner 1984), whereas Schneider and Werner (1986 / 1991) discussed rational and general rational Hermite interpolation, respectively. The following results are proved in these papers but they are given here once more, because they will be used in the sections to come. Let

$$q_n(t) \;:=\; \Omega_n^2(t)\sum_{i=1}^{n}\left(\frac{\alpha_{in}}{t-t_{in}} + \frac{\beta_{in}}{(t-t_{in})^2}\right) \tag{1.2}$$

be the "denominator polynomial" of r_n. Then $q_n \in \mathbf{P}_{2n-1}$ and r_n can obviously be written in the form

$$r_n(f;t) = \frac{\Omega_n^2(t)}{q_n(t)}\sum_{i=1}^{n}\left(\{\frac{\alpha_{in}}{t - t_{in}} + \frac{\beta_{in}}{(t - t_{in})^2}\}f(t_{in}) + \frac{\beta_{in}}{t - t_{in}}f'(t_{in})\right) \tag{1.3}$$

Furthermore, if $q_n(t) \neq 0$, then the remainder is given by

$$f(t) - r_n(f;t) = \frac{\Omega_n^2(t)}{q_n(t)}(fq_n)[t_{1n}, t_{1n}, \cdots, t_{nn}, t_{nn}, t], \tag{1.4}$$

where the square brackets in (1.4) denote divided differences. With $q_n \equiv 1$, the polynomial case is easily recovered.

2. Gaussian Quadratures

A quadrature rule is now simply derived by integrating r_n instead of f, i.e., we approximate $I(w;f) := \int_a^b w(t)f(t)dt$ ($w \in L^1[a,b], w > 0$ a.e.) by

$$Q_n(w;f) := \int_a^b w(t)r_n(f;t)dt. \tag{2.1}$$

Obviously, this approach would not be useful, if q_n changes sign in $[a,b]$. Therefore, we assume q_n to be positive in the interval of integration. Later on, we have to check whether this condition holds in the special cases to be considered. Then the quadrature rule Q_n is given by the following weights (cf. (1.3)) :

$$\begin{aligned} Q_n(w;f) &= \sum_{i=1}^{n}(A_{in}f(t_{in}) + B_{in}f'(t_{in})), \\ B_{in} &:= \beta_{in}\int_a^b \frac{w(t)\Omega_n^2(t)}{q_n(t)(t-t_{in})}dt, \\ A_{in} &:= \beta_{in}\int_a^b \frac{w(t)\Omega_n^2(t)}{q_n(t)(t-t_{in})^2}dt + \frac{\alpha_{in}}{\beta_{in}}B_{in}, i = 1(1)n. \end{aligned} \tag{2.2}$$

The remainder term directly follows from (1.4) :

$$\begin{aligned} E_n(w;f) &:= I(w;f) - Q_n(w;f) \\ &= \int_a^b w(t)\frac{\Omega_n^2(t)}{q_n(t)}(fq_n)[t_{1n}, t_{1n}, \cdots, t_{nn}, t_{nn}, t]dt \\ &= \frac{(fq_n)^{(2n)}(\xi)}{(2n)!}\int_a^b w(t)\frac{\Omega_n^2(t)}{q_n(t)}dt \end{aligned} \tag{2.3}$$

if $f \in C^{2n}[a,b]$ and $q_n(t) \neq 0$ for all $t \in [a,b]$. Clearly, for a quadrature it would be quite useful to have vanishing weights $B_{in}, i = 1(1)n$. In case of a *polynomial* r_n, this just yields Markoff's (1885) nice way to derive Gaussian quadrature. Here we have to search knots t_{in} such that Ω_n is orthogonal to P_{n-1} with respect to the more complicated weight w/q_n. But we do have the freedom to choose q_n. E.g., one could fix q_n in order to make the remainder small or nice or both. Or q_n could be chosen such that functions having a singularity close (with respect to $\mathbb{C}$) to $[a,b]$ are integrated sufficiently well (cf. W.Gautschi's contribution to these proceedings). Furthermore, there is another choice arising from rational interpolation. Namely, we could *fix the knots* t_{in} in advance. Then we would have to search for a denominator polynomial q_n such that the given Ω_n is orthogonal to P_{n-1} with respect to the weight w/q_n. Anyway, vanishing B_{in} do not help only avoiding the evaluation of $f'(t_{in})$ but they are also leading to positive $A_{in}, i = 1(1)n$, (cf. (2.2)) if q_n is positive because $\beta_{in} = q_n(t_{in})/\{\Omega_n'(t_{in})\}^2$. In order to illustrate the possibilities more concretely, we will give some examples :

(I)

Fix q_n such that $q_n > 0$ on $[a,b]$ *independently* of the knots, compute t_{in} such that $B_{in} = 0$, and compute $A_{in}, i = 1(1)n$:

a) $q_n \equiv 1$: Gaussian quadrature.

b) $q_n(t) := \prod_{i=1}^{m_1}(b + b_i - t)\prod_{i=1}^{m_2}(t + a_i - a)$,
m_1, m_2 fixed and $m_1 + m_2 < 2n$, $a_i, b_i > 0$. Such rules would be appropriate for functions f having poles like $1/q_n$ close to the endpoints a, b.

c) Complex conjugate poles could be treated similarly.

d) $\epsilon \in (0,1]$, $[a,b] = [-1,1]$, and $q_n(t) := \prod_{i=1}^{n}(1 - \epsilon t t_{in})^2$ is Wilf's (1961) quadrature fitting in our setting after minor modifications.

e) Finally, $q_n(t) := \frac{1}{n}\sum_{i=1}^{n}(t - t_{in})^2$, $n > 1$, leads to a Lobatto-like simple remainder, because
$(fq_n)[t_{1n}, t_{1n}, \cdots, t_{nn}, t_{nn}, t]$
$= \frac{1}{n}\sum_{i=1}^{n} f[t_{1n}, t_{1n}, \cdots, t_{i-1,n}, t_{i-1,n}, t_{i+1,n}, t_{i+1,n}, \cdots, t_{nn}, t_{nn}, t]$
$= f^{(2n-2)}(\xi)/(2n-2)!$.

Examples a)–c) may be treated within the theory of classical orthogonal polynomials whereas d), e) are leading to implicitly defined orthogonal polynomials (cf. Engels 1980).

(II)

Fix $t_{in} \in [a,b]$ and compute q_n such that $B_{in} = 0, i = 1(1)n$:
Example: Let $[a,b] = [-1,1], w \equiv 1, n = 4, t_{i4} \in \{-1, -1/3, 1/3, 1\}$. Here we have included the endpoints to the set of knots. Hence, $B_{14} = B_{44} = 0$ iff $\beta_{14} = \beta_{44} = 0$, i.e., iff there is no interpolation of f' at -1 and $+1$. Therefore we set $\beta_{14} = \beta_{44} = 0$. Then (cf. (1.2)) $q_4(t) = (t^2 - 1)s(t)$ with an $s \in \mathbf{P}_5$. Now we are looking for a polynomial s such that the integrals over $(t^2-1)(t^2-1/9)(t-\sigma)/s(t)$ vanish for $\sigma \in \{-1/3, 1/3\}$, i.e., $B_{24} = B_{34} = 0$. The solution of this problem with *minimal degree* is given by $s_2(t) := t^2 + c^2$ where c solves the transcendental equation
$7 + 9c^2 = \frac{1}{c}\arctan(\frac{1}{c})(1+c^2)(1+9c^2)$, i.e., $|c| \approx 0.395365$.
An elementary computation yields the partial fraction decomposition of
$q_4(t)/\Omega_4^2(t) = (t^2-1)s_2(t)/\Omega_4^2(t)$
and hence the coefficients
$\beta_{24} = \beta_{34} = -9(1+9c^2)/32$,
$\alpha_{44} = -\alpha_{14} = 81(1+c^2)/128$,
$\alpha_{24} = -\alpha_{34} = 27(5-27c^2)/128$.

Now the weights A_{24}, A_{34} can be computed according to (2.2). This formula is not useful for A_{14}, A_{44} because of the vanishing β_{14}, β_{44}. Therefore we have to compute these weights directly from (1.3) and (2.1). Anyway, the results are the well-known $A_{i4} \in \{1/4, 3/4\}$. In order to study the remainder $E_4(1; f)$ we observe that
$(fq_4)[-1,-1,\frac{-1}{3},\frac{-1}{3},\frac{1}{3},\frac{1}{3},1,1,t] =$
$(fs_2)[-1,\frac{-1}{3},\frac{-1}{3},\frac{1}{3},\frac{1}{3},1,t] =$
$\frac{1-3c^2}{4} f[\frac{-1}{3},\frac{-1}{3},\frac{1}{3},\frac{1}{3},t] + \frac{3}{8}(1+c^2)\{f[-1,\frac{-1}{3},\frac{-1}{3},\frac{1}{3},t] + f[\frac{-1}{3},\frac{1}{3},\frac{1}{3},1,t]\} =$
$f^{(4)}(\xi_t)/4!$
because all coefficients are positive. Here we used the identities
$s_2(t) = t^2 + c^2 = \frac{1-3c^2}{4}(t^2-1) + \frac{3}{8}(1+c^2)\{(t-\frac{1}{3})(t-1) + (t+\frac{1}{3})(t+1)\}$
and Leibniz' rule for divided differences. Hence,
$E_4(1; f) = \frac{-16(fs_2)^{(6)}(\eta)}{135\cdot 6!} = \frac{-3}{80} \cdot (\frac{2}{3})^5 f^{(4)}(\xi)$.
Thus, this Newton–Cotes formula may be generated by rational Hermite interpolation too. It is probably an open problem if any quadrature with positive weights can be constructed by this procedure.

This example directly leads to the following

Modifications

i) Include one or both endpoints — say $t_{\cdot n}$ — to the knots. Then the respective weight $B_{\cdot n}$ vanishes iff $\beta_{\cdot n} = 0$, i.e., there is no interpolation of f' at $t_{\cdot n}$. And the quadrature will be of Radau– or Lobatto–type.

ii) Let $\iota \in \{1, 2, \cdots, n\}$ and set $\beta_{\iota n} := 0$, $q_n(t) := t - t_{\iota n}$ or — equivalently — interpolate f at $t_{1n}, \cdots, t_{nn}$ and f' at the same knots but *not* at $t_{\iota n}$. Then Shohat (1937) proved that the roots of $p_n + \gamma p_{n-1}$ ($\gamma \neq 0$, p_n, p_{n-1} the orthogonal polynomials with respect to w on $[a, b]$) lead to vanishing $B_{jn}, j \neq \iota$, and at least $n-1$ roots are in (a, b). Clearly the remainder formula (2.3) holds for that case too yielding

$$E_n(w; f) = \int_a^b w(t) \frac{\Omega_n^2(t)}{t - t_{\iota n}} f[t_{1n}, t_{1n}, \cdots, t_{\iota-1,n}, t_{\iota-1,n}, t_{\iota,n}, t_{\iota+1,n}, t_{\iota+1,n}, \cdots, t_{nn}, t_{nn}, t]dt \qquad (2.4)$$

On the other hand, (2.4) is exactly the remainder of Engels' (1982) dual quadrature as will be shown in the next section. Hence, dual quadrature introduced via polynomial Hermite interpolation is just a special case of Gaussian quadrature derived from rational Hermite interpolation. But dual quadrature using rational functions will yield some new results.

3. Dual Quadratures

Engels (1982) proposed to interpolate F (a primitive of f) and f instead of f and f', to integrate the *derivative* of the interpolant, and to compute knots such that the weights of $F(t_{in})$ vanish. In our setting this means that $I(w;f)$ is approximated by

$$\begin{aligned} Q_n^\star(w;f) &:= \int_{-1}^{1} w(t) r_n'(F;t)dt \\ &=: \sum_{i=1}^{n} \{A_{in}^\star f(t_{in}) + B_{in}^\star F(t_{in})\}. \end{aligned} \tag{3.1}$$

Obviously, these formulas are only useful if $B_{in}^\star = 0$. Replacing f by F and f' by f immediately yields the following formulas (cf. (1.3)) :

$$\begin{aligned} A_{in}^\star &:= \beta_{in} \int_{-1}^{1} w(t) \left(\frac{\Omega_n^2(t)}{q_n(t)(t-t_{in})}\right)' dt, \\ B_{in}^\star &:= \beta_{in} \int_{-1}^{1} w(t) \left(\frac{\Omega_n^2(t)}{q_n(t)(t-t_{in})^2}\right)' dt + \frac{\alpha_{in}}{\beta_{in}} A_{in}^\star, i = 1(1)n, \end{aligned} \tag{3.2}$$

$$E_n^\star(w;f) = \int_{-1}^{1} w(t) \left(\frac{\Omega_n^2(t)}{q_n(t)}(Fq_n)[t_{1n}, t_{1n}, \cdots, t_{nn}, t_{nn}, t]\right)' dt. \tag{3.3}$$

A quadrature rule $Q_n^\star$ with vanishing weights $B_{in}^\star, i = 1(1)n$, is called a *dual quadrature.* In the sequel, we are going to consider only two special cases:

i) $q_n \equiv 1$ in order to proof the result about Engels' approach.

ii) $w \equiv 1$ in order to describe the intrinsic connection between the rational dual quadrature approach and quadrature in general. (This has been already performed by the author (1982) — but unfortunately far too complicated).

Theorem 1

Let $q_n \equiv 1, p_j \perp \mathsf{P}_{j-1}$ with respect to $w, j \in \{n, n-1\}$. Then

i) $B_{in}^\star = 0 \Leftrightarrow p_n(t_{in}) + \gamma p_{n-1}(t_{in}) = 0$
for some $\gamma \in \mathsf{R}$ and $i = 1(1)n$.
($\gamma = 0$: dual quadrature = Gaussian quadrature).

ii) For a dual quadrature the remainder $E_n^\star(w;f)$ given in (3.3) is equal to the remainder $E_n(w;f)$ in (2.4).

iii) Engels' dual quadratures are Shohat's quadratures.

Proof:

i) Merschen (1983).

ii) Let τ be an abbreviation for the string $t_{1n}, t_{1n}, \cdots, t_{nn}, t_{nn}$. Then
$E_n^\star(w; f) = \int_{-1}^{1} w(t)\Omega_n^2(t)F[\tau, t, t]dt + 2\int_{-1}^{1} w(t)\Omega_n(t)\Omega_n'(t)F[\tau, t]dt.$
Using $f[u_1, \cdots, u_m] = \sum_{i=1}^{m} F[u_1, \cdots, u_m, u_i]$ (which is easy to prove by induction) yields
$E_n^\star(w; f) = \int_{-1}^{1} w(t)\Omega_n^2(t)f[\tau, t]dt - 2\int_{-1}^{1} w(t)\Omega_n^2(t)\sum_{i=1}^{n}\{F[\tau, t, t_{in}] - \frac{F[\tau,t]}{t-t_{in}}\}dt.$
The sum in the second term is just $-\sum_{i=1}^{n} \frac{F[\tau,t_{in}]}{t-t_{in}}$.
Hence, there remains $2\sum_{i=1}^{n} F[\tau, t_{in}] \int_{-1}^{1} w(t)\frac{\Omega_n^2(t)}{t-t_{in}}dt$.
But the integrals do not depend on i because Shohat's and Engels' quadrature as well integrate $\frac{\Omega_n^2(t)}{(t-t_{in})(t-t_{jn})}, i \neq j$, exactly — the result being zero. Therefore,
$E_n^\star(w; f) = \int_{-1}^{1} w(t)\frac{\Omega_n^2(t)}{t-t_{\iota n}}\{f[\tau, t](t - t_{\iota n}) + f[\tau]\}dt$
for any $\iota \in \{1, \cdots, n\}$ which immediately yields the assertion and another form of the remainder.

iii) By i) both quadratures have the same knots. By ii) the functions $f_i(t) := (\frac{\Omega_n(t)}{t-t_{in}})^2$ are integrated exactly by both rules. Thus, the weights coincide too because $B_{in}^\star = 0$ for dual quadratures. □

Theorem 2

Let $w \equiv 1, \Omega_n^2(1)/q_n(1) = \Omega_n^2(-1)/q_n(-1)$, and $\lambda_n := \Omega_n^2(1)/q_n(1) \neq 0$. Then

$$B_{in}^\star = 0, i = 1(1)n \Leftrightarrow \alpha_{in} = \frac{2t_{in}\beta_{in}}{t_{in}^2 - 1}, i = 1(1)n. \tag{3.4}$$

Proof: If $w \equiv 1$ then the integrals given for the weights in (3.2) can be computed directly : $A_{in}^\star = 2\beta_{in}\lambda_n/(1 - t_{in}^2), B_{in}^\star = 4\beta_{in}\lambda_n t_{in}/(1 - t_{in}^2)^2 + \alpha_{in}A_{in}^\star/\beta_{in}, i = 1(1)n$. □

Remarks:

a) The previous result excludes the endpoints from the set of knots.

b) If the α_{in} are chosen according to (3.4) then
$q_n(t) = \Omega_n^2(t)\sum_{i=1}^{n} \frac{\beta_{in}/(1-t_{in}^2)}{(t-t_{in})^2}\{(t - t_{in})^2 + (1 - t^2)\}.$

Furthermore, it is straightforward to prove the following results.

Corollary 1

Let $B^{\star}_{in} = 0, i = 1(1)n$. Then

1) $\beta_{in}/(1-t_{in}^2) > 0,\ i = 1(1)n \Rightarrow q_n(t) > 0$ for all $t \in [-1,1]$.

2) $\lambda_n = \Omega_n^2(1)/q_n(1) = \Omega_n^2(-1)/q_n(-1) = 1/\sum\limits_{i=1}^{n} \frac{\beta_{in}}{1-t_{in}^2}$.

3) $A^{\star}_{in} = \lambda_n \frac{2\beta_{in}}{1-t_{in}^2},\ i = 1(1)n$ which is positive if the $\beta_{in}/(1-t_{in}^2)$ do not change sign.

4) $\sum\limits_{i=1}^{n} A^{\star}_{in} = 2$.

5) $E_n^{\star}(1;f) = 2\lambda_n (Fq_n)[-1,\tau,1]$.

6) $r_n(F;t) = \dfrac{\sum\limits_{i=1}^{n} A^{\star}_{in}\{(1+\frac{1-t^2}{(t-t_{in})^2})F(t_{in}) + \frac{1-t_{in}^2}{t-t_{in}} f(t_{in})\}}{2+(1-t^2)\sum\limits_{i=1}^{n} \frac{A^{\star}_{in}}{(t-t_{in})^2}}$.

Remainder formulas like 5) have been derived by the author (1983) by other means which could be generalized to the case of a non-constant w (1988). For a constant weight, Saff and Snader (1987) derived another proof via complex analysis. But here we want to emphasize the connection between quadrature and rational Hermite interpolation. Namely, if we consider an arbitrary quadrature rule Q_n integrating constants exactly with weights $A^{\star}_{in}$ and knots $t_{in} \in (-1,1)$, then r_n defined by 6) with just these weights and knots yields $r_n(F;1) - r_n(F;-1) = \sum\limits_{i=1}^{n} A^{\star}_{in} f(t_{in})$, i.e., Q_n. If the weights are positive then even $Q_n = \int\limits_{-1}^{1} r_n'(F;t)dt$ holds. Hence, quadrature and dual quadrature via rational Hermite interpolation are essentially the same (if $w \equiv 1$). Actually the same result holds if one endpoint is used as a knot, e.g. -1 :

Theorem 3

Let $w \equiv 1, t_{1n} = -1, t_{in} \notin \{-1,1\}, i = 2(1)n$, and $F(t) := \int\limits_{-1}^{t} f(\rho)d\rho$, i.e., $F(-1) = 0$. Then

$$\begin{array}{llll} B^{\star}_{in} = 0, i = 2(1)n & \Leftrightarrow & \alpha_{in} = \frac{\beta_{in}}{t_{in}-1}, i = 2(1)n \\ B^{\star}_{1n} = 0 & \Leftrightarrow & \alpha_{1n} = \frac{\beta_{1n}}{t_{1n}-1} + 2\frac{q_n(1)}{\Omega_n^2(1)}. \end{array} \tag{3.5}$$

Proof: Evaluation of the integrals given in (3.2) and the definition of β_{1n}. □

Corollary 2

Let the assumptions of Theorem 3 hold. Then

1) $A^\star_{in} = \frac{\Omega^2_n(1)}{q_n(1)} \cdot \frac{\beta_{in}}{1-t_{in}}, i = 1(1)n.$

2) $\frac{\Omega^2_n(1)}{q_n(1)} = 2/\sum_{i=1}^{n} \frac{\beta_{in}}{1-t_{in}} \Rightarrow \sum_{i=1}^{n} A^\star_{in} = 2$ and the weights are positive if the $\frac{\beta_{in}}{1-t_{in}}$ have constant sign.

3) Under the assumption in 2) and if $B^\star_{in} = 0, i = 1(1)n$, we have
$$r_n(F;t) = \frac{\sum_{i=1}^{n} A^\star_{in}\{\frac{1-t}{(t-t_{in})^2}F(t_{in}) + \frac{1-t_{in}}{t-t_{in}}f(t_{in})\}}{\frac{2}{1+t} + (1-t)\sum_{i=1}^{n}\frac{A^\star_{in}}{(t-t_{in})^2}}.$$

4) If $B^\star_{in} = 0, i = 1(1)n$, then $r_n(F;1) - r_n(F;-1) = \sum_{i=1}^{n} A^\star_{in} f(t_{in}) - \underbrace{F(-1)}_{=0}$.

5) If $B^\star_{in} = 0, i = 1(1)n$, then
$E^\star_n(1;f) = F(1) - r_n(F;1) = \frac{\Omega^2_n(1)}{q_n(1)}(Fq_n)[-1,-1,t_{2n},t_{2n},\cdots,t_{nn},t_{nn},1].$

If, finally, both endpoints are knots then two rational Hermite interpolants have to be mixed in order to construct a quadrature by the dual approach.

References

Engels H. (1980), Numerical Quadrature and Cubature. Academic Press, London.

Engels H. and Merschen A. (1982), Duale Quadraturen. In: Numerical Integration (G. Hämmerlin, ed.), ISNM 57, Birkhäuser, Basel, pp. 48–54.

Henrici P. (1982), Essentials of Numerical Analysis. J. Wiley, New York.

Markoff A. (1885), Sur la méthode de Gauss pour le calcul approché des intégrales. Math.Ann. 25, pp. 427–432.

Merschen B.A. (1983), Duale Quadraturformeln und ihre Eigenschaften. Doctoral thesis, Aachen.

Saff E.B. and Snader J.C. (1987), The Error for Quadrature Methods : A Complex Variables Approach. AMM 94, pp. 175–180.

Schneider C. (1982), Quadrature and Generalized Hermite Interpolation. In: Numerical Integration (G. Hämmerlin, ed.), ISNM 57, Birkhäuser, Basel, pp. 212–221.

Schneider C. (1983), Error Analysis for Numerical Integration – an Algorithmic Approach. Habilitationsschrift, Mainz.

Schneider C. (1988), Error Bounds for the Numerical Evaluation of Integrals with Weights. In: Numerical Integration III (H. Braß, G. Hämmerlin, eds.), ISNM 85, Birkhäuser, Basel, pp. 226–236.

Schneider C. and Werner W. (1986), Some New Aspects of Rational Interpolation. Math.Comp. 47, pp. 285–299.

Schneider C. and Werner W. (1991), Hermite Interpolation : The Barycentric Approach. Computing 46, pp. 35–51.

Shohat J. (1937), On Mechanical Quadratures, in particular, with Positive Coefficients. Trans.AMS 42, pp. 461–496.

Werner W. (1984), Polynomial Interpolation : Lagrange versus Newton. Math.Comp. 43, pp. 205–217.

Wilf H. S. (1961), Exactness Conditions in Numerical Quadrature. Numer.Math. 6, pp. 315–319.

Author's Address:

Claus Schneider
Fachbereich Mathematik
Johannes Gutenberg–Universität Mainz
Saarstraße 21
D–W 6500 Mainz, Germany
schneider@mat.mathematik.uni-mainz.de

International Series of Numerical Mathematics, Vol. 112, © 1993 Birkhäuser Verlag Basel

A NEW VARIABLE TRANSFORMATION FOR NUMERICAL INTEGRATION

Avram Sidi

Computer Science Department, Technion, Haifa

Abstract

Presently, variable transformations are used to enhance the performance of lattice rules for multidimensional integration. The transformations that are in the literature so far are of either polynomial or exponential nature. Following a short survey of some of the transformations that have been found to be effective, we propose a new transformation, denoted the $\sin^m$-transformation, that is neither polynomial nor exponential, but trigonometric, in nature. This transformation is also a representative of a general class of variable transformations that we denote $\mathcal{S}_m$. We analyze the effect of transformations in $\mathcal{S}_m$ within the framework of one-dimensional integration, and show that they have some very interesting and useful properties. Present results indicate that transformations in $\mathcal{S}_m$ can be more advantageous than known polynomial transformations, and have less underflow and overflow problems than exponential ones. Indeed, the various numerical tests performed with the $\sin^m$-transformation support this. We end the paper with numerical examples through which some of the theory is verified.

1 Introduction

Let $f(x)$ be a function defined on the interval $[0,1]$, and define

$$I[f] = \int_0^1 f(x)dx. \tag{1.1}$$

Let $Q_n[f]$ stand for either the trapezoidal rule $T_n[f]$ or the midpoint rule $M_n[f]$ for $I[f]$, which are given by

$$T_n[f] = \frac{1}{n}\left[\frac{1}{2}f(0) + \sum_{j=1}^{n-1} f\left(\frac{j}{n}\right) + \frac{1}{2}f(1)\right] \tag{1.2}$$

and

$$M_n[f] = \frac{1}{n}\sum_{j=1}^{n} f\left(\frac{j-1/2}{n}\right). \tag{1.3}$$

Then, provided $f(x)$ is differentiable on $[0,1]$ $2N+1$ times, we have the Euler-Maclaurin formula

$$I[f] - Q_n[f] = \sum_{\mu=1}^{N} c_\mu [f^{(2\mu-1)}(1) - f^{(2\mu-1)}(0)] n^{-2\mu} + R_N(n; f), \tag{1.4}$$

where

$$R_N(n; f) = O(n^{-2N-1}) \text{ as } n \to \infty. \tag{1.5}$$

The c_μ in (1.4) are constants independent of $f(x)$ and n, and are related to the Bernoulli numbers and polynomials. For details, see, e.g., Davis and Rabinowitz (1984).

In general, the error $I[f] - Q_n[f]$ is $O(n^{-2})$ as $n \to \infty$, by (1.4). When $f^{(2\mu-1)}(1) - f^{(2\mu-1)}(0) = 0$, $1 \le \mu \le k \le N-1$, it is obvious that $I[f] - Q_n[f]$ is $O(n^{-2k-2})$ as $n \to \infty$. This kind of an improvement can be achieved by a variable transformation of the form

$$I[f] = \int_0^1 f(\psi(t))\psi'(t)dt, \tag{1.6}$$

where $\psi(t)$ is an increasing function of t on $[0,1]$ satisfying

$$\psi(0) = 0 \text{ and } \psi(1) = 1, \tag{1.7}$$

and has a sufficient number of derivatives that vanish at $t = 0$ and $t = 1$. Thus applying the trapezoidal rule or the midpoint rule to the integral on the right hand side of (1.6) results in very accurate approximations. For the sake of clarity we shall denote the trapezoidal and midpoint rule approximations on the transformed integral in (1.6) by $\hat{T}_n[f]$ and $\hat{M}_n[f]$, respectively, and let $\hat{Q}_n[f]$ stand for either one of them. We thus have

$$\hat{T}_n[f] = \frac{1}{n} \sum_{j=1}^{n-1} f\left(\psi\left(\frac{j}{n}\right)\right) \psi'\left(\frac{j}{n}\right) \tag{1.8}$$

and

$$\hat{M}_n[f] = \frac{1}{n} \sum_{j=1}^{n} f\left(\psi\left(\frac{j-1/2}{n}\right)\right) \psi'\left(\frac{j-1/2}{n}\right). \tag{1.9}$$

Comparing (1.8) with (1.2), we notice that the $j = 0$ and $j = n$ terms in the former are missing. These two terms are automatically zero in (1.8) since $\psi'(0) = \psi'(1) = 0$.

The idea of employing variable transformations seems to have been suggested first by Korobov (1963) in connection with the numerical approximation of integrals on the unit hypercube by lattice rules. The transformation of Korobov (1963) is a polynomial one, and is given by

$$\psi(t) = (2m+1)\binom{2m}{m} \int_0^t [u(1-u)]^m du. \tag{1.10}$$

Later Sag and Szekeres (1964) proposed the tanh-transformation

$$\psi(t) = \frac{1}{2}\tanh\left(-\frac{c}{2}\left(\frac{1}{t} - \frac{1}{1-t}\right)\right) + \frac{1}{2}, \quad c > 0. \tag{1.11}$$

(Actually, $c = 1$ in Sag and Szekeres(1964).) For this transformation

$$\psi^{(i)}(0) = \psi^{(i)}(1) = 0, \text{ all } i = 1, 2, \dots . \tag{1.12}$$

In a historical paper Iri et al (1970) proposed the transformation

$$\psi(t) = \frac{\int_0^t \phi(u)du}{\int_0^1 \phi(u)du}, \tag{1.13}$$

where

$$\phi(t) = \exp\left(-\frac{c}{t(1-t)}\right) \text{ with } c > 0. \tag{1.14}$$

(Actually $c = 1$ in Iri et al (1970).) This has been known as the IMT-transformation in the literature. We note that (1.12) is satisfied by this transformation too.

We finally mention the double exponential transformation of Mori (1978), for which

$$\psi(t) = \frac{1}{2}\tanh\left(a \sinh\left(b\left(\frac{1}{1-t} - \frac{1}{t}\right)\right)\right) + \frac{1}{2}, \quad a, b > 0, \tag{1.15}$$

which also satisfies (1.12).

For a more complete list of references we refer the reader to the paper of Beckers and Haegemans (1991). In this paper the above mentioned transformations and a few others are compared with respect to their numerical efficiency when applied in conjunction with lattice rules (see, e.g., Sloan (1985) and Sloan and Kachoyan (1987)), to multiple integrals on the unit hypercube. One of the conclusions of this paper is that while polynomial transformations may be very well suited for well behaved integrands, IMT and tanh-transformations may be better for integrands with singularities on the boundary of the hypercube, although overflows and underflows may occur due to the fact that many of the abscissas may be extremely close to the surfaces of the hypercube because of the exponential nature of these transformations. (We also recall that a large number of the $\psi\left(\frac{j}{n}\right)$ and $\psi\left(\frac{j-1/2}{n}\right)$ in (1.8) and (1.9) are clustered in two very small regions, one to the right of 0 and the other to the left of 1, with quite a few of them being very close to 0 or 1. This may create overflow or underflow problems in some cases.)

In Section 2 of the present work we propose a new variable transformation $\psi(t)$ that is neither polynomial nor exponential, but trigonometric, in nature. This transformation is also a representative of a general class of variable transformations that we denote $\mathcal{S}_m$, and define properly again in Section 2. Any transformation $\psi(t)$ in $\mathcal{S}_m$, although not a polynomial itself, *behaves* polynomially at the end points $t = 0$ and $t = 1$, which ensures that the $\psi\left(\frac{j}{n}\right)$ and $\psi\left(\frac{j-1/2}{n}\right)$ in (1.8) and (1.9) cannot get too close to 0 or to 1, thus reducing the possibility of overflow and underflow. In Section 3 we analyze the effect of transformations in $\mathcal{S}_m$ in general, and of the $\sin^m$-transformation in particular, within the framework of one-dimensional integration of regular integrands. We do this by examining the Euler-Maclaurin expansions of the rules $\hat{Q}_n[f]$ in detail. It turns out that these expansions have a very surprising structure, and show that transformations in $\mathcal{S}_m$ are more effective than the analogous polynomial transformation of (1.10) in one-dimensional integration. (The sense of the analogy will be explained at the end of Section 2.) In Section 4 we provide a similar analysis for one-dimensional integrals of functions that have singularities at the end points of the integration interval. We have also observed numerically that the rules $\hat{Q}_n[f]$ with the $\sin^m$-transformation, when applied to both regular and singular integrals, produce results that compare very favorably with those obtained from IMT rules, despite the fact that the $\sin^m$-transformation does not satisfy (1.12). The approximations that were obtained for two such integrals are given in Section 5, and are seen to behave precisely as predicted by the results of Sections 3 and 4.

The results that we have obtained in this paper may suggest that variable transformations in $\mathcal{S}_m$ in general, and the $\sin^m$-transformation in particular, can be used very effectively in multiple integration in conjunction with lattice rules. We propose to study the theoretical and practical aspects of this usage in a future publication.

2 The $\sin^m$-Transformation

The $\sin^m$-transformation is defined by

$$\psi(t) = \frac{\Theta_m(t)}{\Theta_m(1)}, \tag{2.1}$$

where

$$\Theta_m(t) = \int_0^t (\sin \pi u)^m du, \quad m = 1, 2, \ldots. \tag{2.2}$$

The recursion relation

$$\Theta_m(t) = -\frac{1}{\pi m}(\sin \pi t)^{m-1} \cos \pi t + \frac{m-1}{m}\Theta_{m-2}(t), \quad m = 2, 3, \ldots, \tag{2.3}$$

with the initial conditions

$$\Theta_0(t) = t \text{ and } \Theta_1(t) = \frac{1}{\pi}(1 - \cos \pi t), \tag{2.4}$$

and the recursion relation

$$\Theta_m(1) = \frac{m-1}{m}\Theta_{m-2}(1), \quad m = 2, 3, ..., \tag{2.5}$$

with the initial conditions

$$\Theta_0(1) = 1 \text{ and } \Theta_1(1) = \frac{2}{\pi}, \tag{2.6}$$

can be used to compute $\psi(t)$ for any t in $(0,1)$ very efficiently and stably. It is easy to verify that $\Theta_m(1)$ can be expressed in terms of the beta function $B(\mu, \nu)$ as

$$\Theta_m(1) = \frac{2^m}{\pi}B\left(\frac{m+1}{2}, \frac{m+1}{2}\right) = \frac{2^m}{\pi}\frac{\left[\Gamma\left(\frac{m+1}{2}\right)\right]^2}{\Gamma(m+1)}. \tag{2.7}$$

Here is a short list of these transformations:

$$\begin{aligned}
\psi_1(t) &= \frac{1}{2}(1 - \cos \pi t) \\
\psi_2(t) &= \frac{1}{2\pi}(2\pi t - \sin 2\pi t) \\
\psi_3(t) &= \frac{1}{16}(8 - 9\cos \pi t + \cos 3\pi t) \\
\psi_4(t) &= \frac{1}{12\pi}(12\pi t - 8\sin 2\pi t + \sin 4\pi t)
\end{aligned}$$

2.1 A General Class of Variable Transformations

Now the $\sin^m$-transformation is a representative of a more general class of variable transformations that we shall denote $\mathcal{S}_m$. If $\psi(t)$ is in $\mathcal{S}_m$, then it has the following properties:

(a) $\psi \in C^\infty[0,1]$, increases on $[0,1]$, and satisfies $\psi(0) = 0$ and $\psi(1) = 1$.

(b) $\psi'(t)$ is symmetric with respect to $t = 1/2$, i.e., $\psi'(t) = \psi'(1-t)$. Consequently, $\psi(1-t) = 1 - \psi(t)$.

(c) $\psi'(t)$ has the asymptotic expansions

$$\begin{cases} \psi'(t) \sim \sum_{i=0}^\infty \epsilon_i t^{m+2i} & \text{as } t \to 0+ \\ \psi'(t) \sim \sum_{i=0}^\infty \epsilon_i (1-t)^{m+2i} & \text{as } t \to 1-. \end{cases} \tag{2.8}$$

where $\epsilon_0 > 0$. (Note that the second of the expansions in (2.8) is actually a consequence of the first and of property (b).)

By the nature of the function $\sin \pi t$ it is easy to show that the $\sin^m$-transformation possesses all three properties.

Note that all of the variable transformation $\psi(t)$ mentioned in Section 1 share properties (a) and (b). Property (c) implies that $\psi'(t) = O(t^m)$ as $t \to 0+$ and $\psi'(t) = O((1-t)^m)$ as $t \to 1-$ if $\psi \in \mathcal{S}_m$, and this is true also for the polynomial transformation of Korobov in (1.10). From this we conclude that, for $\psi \in \mathcal{S}_m$, the $\psi\left(\frac{j}{n}\right)$ and $\psi\left(\frac{j-1/2}{n}\right)$ will have the same kind of a distribution near the end points $t = 0$ and $t = 1$ as those of the Korobov transformation. It is in this sense that the $\sin^m$-transformation or any transformation in $\mathcal{S}_m$, and the Korobov transformation of (1.10) are analogous. It must be emphasized though that property (c) above is *not* shared by the Korobov transformation and is the most important and useful property of transformations in $\mathcal{S}_m$.

It would be interesting to know whether there are further variable transformations in $\mathcal{S}_m$ that can be expressed in terms of elementary functions as the $\sin^m$-transformation. So far we have not been successful in constructing such a transformation.

Before we close this section, we show that some kind of a "closure" property with respect to composition is satisfied by variable transformations in the classes $\mathcal{S}_m$, $m = 1, 2, \ldots$.

Lemma 2.1: *Let $\psi_i \in \mathcal{S}_{m_i}$, $i = 1, 2, \ldots, r$, and define $\Psi = \psi_1 \circ \psi_2 \circ \cdots \circ \psi_r$ by $\Psi(t) = \psi_1(\psi_2(\cdots(\psi_r(t))\cdots))$. Then $\Psi \in \mathcal{S}_M$, with $M = \prod_{i=1}^{r}(m_i + 1) - 1$. Also M is even if and only if all m_i are even.*

Proof: The assertion follows from the definition of $\mathcal{S}_m$ for $r = 1$ and is easy to prove for $r = 2$. The result for general r can be completed by induction. We leave the details to the reader. □

3 Euler-Maclaurin Expansions for Transformed Integrals of Regular Functions

Let the function $f(x)$ in (1.1) be differentiable on $[0, 1]$ as many times as needed. Let also $\psi(t)$ be a variable transformation in $\mathcal{S}_m$. In this section we will explore the properties of the Euler-Maclaurin expansions associated with the transformed integral (1.6). As the results for even m are very different from those for odd m, we shall treat the two cases separately.

From the results below it will become clear that much better results are obtained for even m.

3.1 The Case m =odd integer

Theorem 3.1 : *Let $m = 2k-1$ with k a positive integer. Then, for $f \in C^{2p+1}[0,1]$, $p \geq k$, we have*

$$I[f] - \hat{Q}_n[f] = \sum_{\mu=k}^{p} c_\mu [F^{(2\mu-1)}(1) - F^{(2\mu-1)}(0)] n^{-2\mu} + O(n^{-2p-1}) \text{ as } n \to \infty, \tag{3.1}$$

where $F(t) = f(\psi(t))\psi'(t)$ is the transformed integrand and c_μ are as in (1.4). Thus the error in $\hat{Q}_n[f]$ is at worst $O(n^{-m-1})$ as $n \to \infty$.

Proof: Let us analyze the behavior of $F(t)$ for $t \to 0+$ and $t \to 1-$. Since $\psi'(t) = O(t^m)$ as $t \to 0+$ and $\psi'(t) = O((1-t)^m)$ as $t \to 1-$, and since $f(\psi(t)) \sim f(0)$ as $t \to 0+$ and $f(\psi(t)) \sim f(1)$ as $t \to 1-$, we have that $F(t) = O(t^m)$ as $t \to 0+$ and $F(t) = O((1-t)^m)$ as $t \to 1-$. Consequently, $F^{(i)}(0) = F^{(i)}(1) = 0$, $i = 1, 2, ..., m-1$. In addition, in general, $F^{(m)}(0) \neq 0$ and $F^{(m)}(1) \neq 0$. Since m is odd and $f \in C^{2p+1}[0,1]$, (1.4) becomes (3.1). □

Let us now apply Theorem 3.1 to the constant function $f(x) = 1$. From the proof of this theorem it is easy to see that $I[f] - \hat{Q}_n[f] \sim -2(m!)c_k\epsilon_0 n^{-m-1}$ as $n \to \infty$ when $m = 2k-1$, $k \geq 1$. A much better result is obtained when m is even, as we show in Lemmas 3.2 and 3.3 below.

3.2 The Case m =even integer

Lemma 3.2: *Let m be an even integer and let $f(x) = 1$. Then*

$$I[f] - \hat{Q}_n[f] = O(n^{-\mu}) \text{ as } n \to \infty, \text{ any } \mu > 0. \tag{3.2}$$

Proof: The transformed integrand now is simply $F(t) = \psi'(t)$. Since m is even, the asymptotic expansions of $\psi'(t)$ for $t \to 0+$ and $t \to 1-$ contain only even powers of t and $(1-t)$, respectively, as can be seen from (2.8). This implies that $F^{(2i-1)}(0) = F^{(2i-1)}(1) = 0$, $i = 1, 2, 3, ...$. Thus all the terms in the Euler-Maclaurin expansion vanish. This completes the proof. □

A substantial improvement results in Lemma 3.2 if the variable transformation $\psi \in \mathcal{S}_m$ is taken to be the $\sin^m$-transformation. This is considered in Lemma 3.3 below.

Lemma 3.3: *Let the variable transformation $\psi(t)$ in (1.6) be the $\sin^m$-transformation with $m = 2k,\ k \geq 1$, and let $f(x) = 1$. Then, provided $n > k$, we have*

$$\hat{Q}_n[f] = I[f]. \tag{3.3}$$

Proof: We start by observing that for $m = 2k$

$$\begin{aligned} \psi'(t) &= \frac{(\sin^2 \pi t)^k}{\Theta_m(1)} = \frac{(1-\cos 2\pi t)^k}{2^k \Theta_m(1)} \\ &= \sum_{j=0}^{k} \delta_j \cos 2\pi j t \quad \text{for some constants } \delta_j. \end{aligned} \tag{3.4}$$

That is to say, $\psi'(t)$ is a trigonometric polynomial of degree k on $[0,1]$. Thus

$$\hat{Q}_n[f] = \sum_{j=0}^{k} \delta_j Q_n[\cos 2\pi j x], \tag{3.5}$$

where Q_n stands for T_n or M_n depending on whether $\hat{Q}_n$ stands for $\hat{T}_n$ or $\hat{M}_n$, respectively. But

$$Q_n[\cos 2\pi j x] = I[\cos 2\pi j x] = \begin{cases} 1 & \text{if } j = 0, \\ 0 & \text{if } j \neq 0, \end{cases} \tag{3.6}$$

provided $n > j$. This completes the proof. □

Corollary 3.4: *The results of Lemmas 3.2 and 3.3 remain unchanged when $f(x) = A + Bx$, where A and B are constants.*

Proof: We can rewrite $f(x)$ in the form $f(x) = (A + B/2) + B(x - 1/2)$. From symmetry considerations, we have both $I[x - 1/2] = 0$ and $\hat{Q}_n[x - 1/2] = 0$. Thus $I[f] = (A + B/2)I[1]$ and $\hat{Q}_n[f] = (A + B/2)\hat{Q}_n[1]$. The result now follows by invoking Lemmas 3.2 and 3.3. □

Note: The result of Lemma 3.2 is very similar to the known analogous result for the IMT rules, in that in both cases the constant function is not integrated exactly, but the corresponding Euler-Maclaurin expansions are empty, suggesting rates of convergence that are better than $n^{-\mu}$ as $n \to \infty$ for any $\mu > 0$. The result of Lemma 3.3, however, says that the constant function is integrated exactly if $\psi(t)$ is the $\sin^m$-transformation with even m.

We now turn to the analysis of the Euler-Maclaurin expansion for arbitrary integrands $f(x)$ that are differentiable on $[0,1]$ as many times as needed. This expansion has a very interesting structure as we show in Theorem 3.5 below.

Theorem 3.5: *Let $m = 2k$ with k a positive integer. Assume that $f \in C^{2p+1}[0,1]$ for $p \geq q(m+1)$, where q is the smallest integer ≥ 1 for which*

$$f^{(2q-1)}(1) - f^{(2q-1)}(0) \neq 0. \tag{3.7}$$

Then

$$I[f] - \hat{Q}_n[f] = \sum_{\mu=q(m+1)}^{p} c_\mu [F^{(2\mu-1)}(1) - F^{(2\mu-1)}(0)] n^{-2\mu} + O(n^{-2p-1}) \text{ as } n \to \infty, \tag{3.8}$$

where $F(t) = f(\psi(t))\psi'(t)$ is the transformed integrand and c_μ are as in (1.4). Thus the error in $\hat{Q}_n[f]$ is $O(n^{-q(2m+2)})$ as $n \to \infty$.

Proof: As in the proof of Theorem 3.1, we need to analyze the behavior of $F(t)$ for $t \to 0+$ and $t \to 1-$.

First, $f \in C^{2p+1}[0,1]$ implies that

$$f(x) = \sum_{j=0}^{2p} \frac{f^{(j)}(0)}{j!} x^j + O(x^{2p+1}) \text{ as } x \to 0+, \tag{3.9}$$

so that

$$F(t) = \psi'(t) \left\{ \sum_{j=0}^{2p} \frac{f^{(j)}(0)}{j!} [\psi(t)]^j + O([\psi(t)]^{2p+1}) \right\} \text{ as } t \to 0+\,. \tag{3.10}$$

Next, since $\psi \in \mathcal{S}_m$, we have from (2.8)

$$\psi(t) \sim \sum_{i=0}^{\infty} \frac{\epsilon_i}{m+2i+1} t^{m+2i+1} \text{ as } t \to 0+\,. \tag{3.11}$$

This, with $m = 2k$, implies that

$$\phi_j(t) \equiv \psi'(t)[\psi(t)]^j \sim t^{j(2k+1)+2k} \sum_{i=0}^{\infty} \epsilon_{ji} t^{2i} \text{ as } t \to 0+ \tag{3.12}$$

for some constants ϵ_{ji} with $\epsilon_{j0} > 0$. Thus the asymptotic expansion in (3.12) contains only even powers of t when j is even and only odd powers of t when j is odd. Since odd order derivatives of $t^{2\nu}$, $\nu = 0,1,2,\ldots$, are all zero at $t = 0$, we have $F^{(2j-1)}(0) = F_0^{(2j-1)}(0)$, where

$$F_0(t) = \psi'(t)\left\{\sum_{j=1}^{p} \frac{f^{(2j-1)}(0)}{(2j-1)!}[\psi(t)]^{2j-1} + O([\psi(t)]^{2p+1})\right\} \text{ as } t \to 0+. \tag{3.13}$$

Treating the other end point $t = 1$ similarly, we obtain $F^{(2j-1)}(1) = F_1^{(2j-1)}(1)$, where

$$F_1(t) = \psi'(1-t)\left\{\sum_{j=1}^{p} \frac{f^{(2j-1)}(1)}{(2j-1)!}[-\psi(1-t)]^{2j-1} + O([\psi(1-t)]^{2p+1})\right\} \text{ as } t \to 1-, \tag{3.14}$$

and we have invoked $\psi'(1-t) = \psi'(t)$ and $\psi(1-t) = 1 - \psi(t)$.

Since $\psi'(t)\psi(t) = \epsilon_{10}t^{4k+1} + O(t^{4k+3})$ as $t \to 0+$, we see that $F^{(i)}(0) = 0$, $i = 0,1,\ldots,4k$, but $F^{(4k+1)}(0) \neq 0$ in general. Similarly, $F^{(i)}(1) = 0$, $i = 0,1,\ldots,4k$, but $F^{(4k+1)}(1) \neq 0$ in general. This implies that the Euler-Maclaurin expansion is at worst $O(n^{-2m-2})$ as $n \to \infty$, and this is the result of (3.8) with $q = 1$ in (3.7).

To obtain the result for arbitrary $q \geq 1$ we need a more careful analysis of the odd order derivatives of $F(t)$ at $t = 0$ and $t = 1$.

From (3.13) and (3.14) we have

$$F^{(2\mu-1)}(1) - F^{(2\mu-1)}(0) = \sum_{j=1}^{p} \frac{f^{(2j-1)}(1) - f^{(2j-1)}(0)}{(2j-1)!}\phi_{2j-1}^{(2\mu-1)}(0), \quad \mu = 1,2,\ldots,p. \tag{3.15}$$

By (3.7), the summation on the right hand side of (3.15) actually begins with the term $j = q$. From (3.12),

$$\begin{aligned} \phi_{2j-1}^{(\nu)}(0) &= 0, \quad \nu = 0,1,\ldots,2j(2k+1)-2, \\ &\neq 0, \quad \nu = 2j(2k+1)-1. \end{aligned} \tag{3.16}$$

Consequently,

$$\begin{aligned} F^{(2\mu-1)}(1) - F^{(2\mu-1)}(0) &= 0, \quad \mu = 1,2,\ldots,q(2k+1)-1, \\ &\neq 0, \quad \mu = q(2k+1), \end{aligned} \tag{3.17}$$

and

$$F^{(2\mu-1)}(1) - F^{(2\mu-1)}(0) \propto f^{(2q-1)}(1) - f^{(2q-1)}(0) \text{ for } \mu = q(2k+1) = q(m+1). \tag{3.18}$$

The proof of the theorem can now be easily completed. □

Remarks:

1. It is very instructive to compare the results of Theorems 3.1 and 3.5. Under appropriate conditions on $f(x)$, we have for $\psi \in \mathcal{S}_m$

$$I[f] - \hat{Q}_n[f] = \begin{cases} O(n^{-m-1}) \text{ as } n \to \infty, & m \text{ odd}, \\ O(n^{-2m-2}) \text{ as } n \to \infty, & m \text{ even}, \end{cases} \tag{3.19}$$

in general, which shows the superiority of the transformations in $\mathcal{S}_m$ with even m.

2. It is also instructive to compare the errors in $\hat{Q}_n[f]$ obtained from a variable transformation $\psi \in \mathcal{S}_m$, namely, (3.19), with those obtained from the polynomial transformation ψ of Korobov given in (1.10). For the latter it can easily be shown that

$$I[f] - \hat{Q}_n[f] = \begin{cases} O(n^{-m-1}) \text{ as } n \to \infty, & m \text{ odd}, \\ O(n^{-m-2}) \text{ as } n \to \infty, & m \text{ even}. \end{cases} \tag{3.20}$$

It is clear that also for the Korobov transformation better results are obtained for even m. In addition, transformations in $\mathcal{S}_m$ and the Korobov transformation produce the same rate of convergence for odd m, but the former are much better than the latter for even m.

4 Treatment of Transformed Integrals of Singular Functions

In the previous section we assumed that $f(x)$ was differentiable a sufficient number of times on $[0,1]$. We now turn to the case in which $f(x)$ is differentiable on $(0,1]$ a sufficient number of times, but has an algebraic singularity at $x = 0$ of the form x^α, where $\alpha > -1$ is not an integer. We write the function $f(x)$ in the form

$$f(x) = x^\alpha g(x), \tag{4.1}$$

where we also assume that $g \in C^{2p+1}[0,1]$ and $g(0) \neq 0$.

By a result due to Navot (1961), the Euler-Maclaurin expansion of $I[f] - \hat{Q}_n[f]$ has different contributions coming from the end points $t = 0$ and $t = 1$ in the transformed integral (1.6). By taking $\psi \in \mathcal{S}_m$ in (1.6), we shall now give a brief analysis of these contributions.

First, we have

$$f(x) = \sum_{j=0}^{2p} \frac{g^{(j)}(0)}{j!} x^{j+\alpha} + O(x^{2p+\alpha+1}) \quad \text{as } x \to 0+. \tag{4.2}$$

As a result, for some $a \neq 0$, $a \propto g(0)$,

$$F(t) \equiv f(\psi(t))\psi'(t) = at^{(m+1)\alpha+m}(1 + O(t^2)) \quad \text{as } t \to 0+. \tag{4.3}$$

Therefore, the dominant term in the contribution to the Euler-Maclaurin expansion of $I[f] - \hat{Q}_n[f]$ from $t = 0$ is of the form $bn^{-(m+1)(\alpha+1)}$ at worst, for some $b \neq 0$, $b \propto g(0)$.

Next, we also have

$$f(x) = \sum_{j=0}^{2p} \frac{f^{(j)}(1)}{j!}(x-1)^j + O((x-1)^{2p+1}) \quad \text{as } x \to 1-. \tag{4.4}$$

As a result,

$$F(t) = \sum_{j=0}^{2p} \frac{f^{(j)}(1)}{j!}[-\psi(1-t)]^j \psi'(1-t) + O([\psi(1-t)]^{2p+1}\psi'(1-t)) \quad \text{as } t \to 1-. \tag{4.5}$$

Now the contribution from $t = 1$ to the Euler-Maclaurin expansion of $I[f] - \hat{Q}_n[f]$ is of the form $\sum_{\mu=1}^{p} c_\mu F^{(2\mu-1)}(1)n^{-2\mu}$ with c_μ as before. From (4.5) and (2.8), we see that, for p sufficiently large, the first odd order derivative of $F(t)$ that does not necessarily vanish at $t = 1$ is $F^{(m)}(t)$ when m is odd and $F^{(2m+1)}(t)$ when m is even. This implies that the lower limit in the summation above is $\mu = \mu_0$, where $\mu_0 \geq \bar{\mu}$ with $\bar{\mu} = (m+1)/2$ for odd m and $\bar{\mu} = m + 1$ for even m. Therefore, for p sufficiently large, the contribution from $t = 1$ to $I[f] - \hat{Q}_n[f]$ is at worst $O(n^{-2\bar{\mu}})$ as $n \to \infty$.

Finally, combining the different contributions from $t = 0$ and $t = 1$, we obtain

$$I[f] - \hat{Q}_n[f] = O(n^{-\omega}) \quad \text{as } n \to \infty, \tag{4.6}$$

where

$$\omega = \begin{cases} \min((m+1)(\alpha+1), m+1), & m \text{ odd} \\ \min((m+1)(\alpha+1), 2m+2), & m \text{ even}, \end{cases} \tag{4.7}$$

at worst.

The treatment above can easily be extended to functions $f(x)$ that are of the form $f(x) = x^\alpha(1-x)^\beta g(x)$, where $\alpha > -1$ and $\beta > -1$ are not integers, and $g(x)$ is differentiable on $[0, 1]$ a sufficient number of times. In this case (4.6) holds with

$$\omega = \min((m+1)(\alpha+1), (m+1)(\beta+1)), \tag{4.8}$$

at worst. This result follows again from Navot (1961) and also from Lyness and Ninham (1967).

Similarly, we can treat the case in which $f(x)$ has logarithmic singularities at $x = 0$ or $x = 1$ or both, by using the results of Navot (1962). See also Lyness and Ninham (1967). We shall not pursue this matter further, however.

5 Numerical Examples

We have applied the transformed rules, namely $\hat{T}_n$ and $\hat{M}_n$, in conjunction with the $\sin^m$-transformations to various integrals, both regular and singular. The numerical results of our experiments provide ample verification for the theoretical results of this work and the conclusions that we have derived from them.

The numerical experiments reported in this section were done in extended precision arithmetic on an IBM-370 computer at the Computer Center of the Technion.

Example 5.1: $f(x) = e^x/(e+1)$.

$I[f] = (e-1)/(e+1)$ for this case. Table 5.1 shows the errors $|I[f] - \hat{T}_n[f]|$ obtained by employing the $\sin^m$-transformation, $m = 1, 2, ..., 8$, and $n = 2^s, s = 1, 2, ..., 10$. It is easy to verify that (3.19) is satisfied exactly. That is, the $m = 1, 2, ..., 8$ columns tend to zero like $n^{-2}, n^{-6}, n^{-4}, n^{-10}, n^{-6}, n^{-14}, n^{-8}, n^{-18}$, respectively. The last column shows the errors obtained by using the IMT rules with $c = 1$ in (1.14), and was taken from Table 3 in Iri et al (1970).

Example 5.2: $f(x) = \sqrt{x}$.

$I[f] = 2$ for this case. Table 5.2 shows that errors $|I[f] - \hat{T}_n[f]|$ obtained by employing the $\sin^m$-transformation, $m = 1, 2, ..., 8$, and $n = 2^s, s = 1, 2, ..., 10$. It is easy to verify that (4.7) is satisfied this time. That is, the $m = 1, 2, ..., 8$ columns tend to zero like $n^{-2}, n^{-4.5}, n^{-4}, n^{-7.5}, n^{-6}, n^{-10.5}, n^{-8}, n^{-13.5}$, respectively, by the fact that $\alpha = 1/2$. Again the last column shows the errors obtained by using the IMT rules with $c = 1$ in (1.14), and was taken from Table 3 in Iri et al (1970).

Table 5.1

m \ n	1	2	3	4	5	6	7	8	IMT
2	1.1D-01	1.9D-02	6.0D-02	1.3D-01	1.9D-01	2.5D-01	3.0D-01	3.5D-01	1.2D-01
4	2.6D-02	1.1D-04	2.1D-03	2.6D-03	6.9D-03	8.1D-03	5.7D-03	2.3D-04	4.0D-03
8	6.5D-03	9.9D-07	1.5D-04	2.0D-07	1.5D-05	1.8D-07	8.4D-06	4.9D-05	3.5D-05
16	1.6D-03	1.5D-08	9.4D-06	4.9D-11	2.2D-07	9.4D-13	1.1D-08	4.0D-12	3.5D 06
32	4.0D-04	2.2D-10	5.8D-07	4.3D-14	3.4D-09	4.8D-17	4.0D-11	1.9D-19	4.5D-09
64	1.0D-04	3.5D-12	3.6D-08	4.1D-17	5.2D-11	2.8D-21	1.5D-13	6.1D-25	3.2D-13
128	2.5D-05	5.4D-14	2.3D-09	4.0D-20	8.1D-13	1.7D-25	6.0D-16	2.2D-30	4.2D-18
256	6.3D-06	8.5D-16	1.4D-10	3.9D-23	1.3D-14	1.0D-29	2.3D-18	8.5D-33	4.0D-26
512	1.6D-06	1.3D-17	8.9D-12	3.8D-26	2.0D-16	9.6D-33	9.2D-21	9.6D-33	
1024	3.9D-07	2.1D-19	5.5D-13	3.7D-29	3.1D-18	6.6D-32	3.6D-23	6.5D-32	

Table 5.2

m \ n	1	2	3	4	5	6	7	8	IMT
2	1.1D-01	4.0D-02	1.7D-01	2.8D-01	3.7D-01	4.6D-01	5.5D-01	6.3D-01	1.0D-01
4	2.6D-02	6.1D-04	2.2D-03	6.3D-04	3.1D-03	1.1D-02	2.4D-02	4.2D-02	3.9D-03
8	6.5D-03	2.1D-05	1.5D-04	1.2D-06	1.6D-05	9.9D-08	4.1D-06	2.1D-06	1.4D-05
16	1.6D-03	8.8D-07	9.3D-06	5.1D-09	2.2D-07	1.1D-10	1.1D-08	7.0D-12	2.6D-07
32	4.0D-04	3.8D-08	5.8D-07	2.7D-11	3.4D-09	7.0D-14	4.0D-11	4.6D-16	1.4D-11
64	1.0D-04	1.7D-09	3.6D-08	1.5D-13	5.2D-11	4.7D-17	1.5D-13	3.7D-20	3.7D-16
128	2.5D-05	7.4D-11	2.3D-09	8.1D-16	8.1D-13	3.2D-20	6.0D-16	3.1D-24	1.4D-22
256	6.3D-06	3.3D-12	1.4D-10	4.5D-18	1.3D-14	2.2D-23	2.3D-18	2.7D-28	8.0D-32
512	1.6D-06	1.4D-13	8.9D-12	2.5D-20	2.0D-16	1.5D-26	9.2D-21	6.0D-32	
1024	3.9D-07	6.4D-15	5.5D-13	1.4D-22	3.1D-18	1.1D-29	3.6D-23	8.5D-32	

References

1. M. Beckers and A. Haegemans (1992), Transformations of integrands for lattice rules, in *Numerical Integration: Recent Developments, Software and Applications*, (T.O.

Espelid and A. Genz, eds.), NATO ASI Series, Kluwer Academic Publishers, Boston, pp. 329-340.

2. P.J. Davis and P. Rabinowitz (1984), *Methods of Numerical Integration*, 2nd Edition, Academic Press, New York.

3. M. Iri, S. Moriguti, and Y. Takasawa (1970), On a certain quadrature formula, *Kokyuroku of Res. Inst. for Math. Sci. Kyoto Univ.*, vol. **91**, pp. 82-118 (in Japanese). English translation in *J. Comp. Appl. Math.*, **17** (1987), pp. 3-20.

4. N.M. Korobov (1963), *Number-Theoretic Methods of Approximate Analysis*, GIFL, Moscow. (Russian).

5. J.N. Lyness and B.W. Ninham (1967), Numerical quadrature and asymptotic expansions, *Math. Comp.*, **21**, pp. 162-178.

6. M. Mori (1978), An IMT-type double exponential formula for numerical integration, *Publ. Res. Inst. Math. Sci. Kyoto Univ.*, **14**, pp. 713-729.

7. I. Navot (1961), An extension of the Euler-Maclaurin summation formula, *J. Math. and Phys.*, **40**, pp. 271-276.

8. I. Navot (1962), A further extension of the Euler-Maclaurin summation formula, *J. Math. and Phys.*, **41**, pp. 155-163.

9. T.W. Sag and G. Szekeres (1964), Numerical evaluation of high-dimensional integrals, *Math. Comp.*, **18**, pp. 245-253.

10. I.H. Sloan (1985), Lattice methods for multiple integration, *J. Comp. Appl. Math.*, **12 & 13**, pp. 131-143.

11. I.H. Sloan and P.J. Kachoyan (1987), Lattice methods for multiple integration: theory, error analysis and examples, *SIAM J. Numer. Anal.*, **14**, pp. 117-128.

Author's address:

Prof. Avram Sidi

Computer Science Department, Technion - Israel Institute of Technology,

Haifa 32000, Israel.

Problems

One evening during the conference was devoted to a discussion, in which several open problems were presented. Nine of these problems have been formulated by the authors in written form and are given here.

Problem 1, proposed by B. Bojanov:

Optimal Formulas for Analytic Functions

Denote by H the class of all functions $f(z)$ which are analytic in the unit disc $D := \{z : |z| < 1\}$ and bounded by 1 on $\bar{D}$. Assume further that $f(z)$ is real for real z. It is known that for each N there exists a set of optimal nodes $\xi_1, \ldots, \xi_N$ for integration in H. In other words, there exist points $-1 < \xi_1 < \ldots < \xi_N < 1$ such that

$$R(\xi_1, \ldots, \xi_N) = \inf_{-1<x_1<\ldots<x_N<1} R(x_1, \ldots, x_N) =: R_N(H),$$

where $R(x_1, \ldots, x_N)$ is the error of the best quadrature formula in H with nodes $x_1, \ldots, x_N$,

$$R(x_1, \ldots, x_N) := \inf_{\alpha_1, \ldots, \alpha_N} \sup_{f \in H} \Big| \int_{-1}^{1} f(x)\,dx - \sum_{k=1}^{N} \alpha_k f(x_k) \Big|.$$

Now consider more general quadrature formulas which use N values of the function f or its derivatives:

$$\int_{-1}^{1} f(x)\,dx \approx \sum_{e_{ij}=1} \alpha_{ij} f^{(j)}(x_i)$$

defined by an incidence matrix $E = \{e_{ij}\}_{i=1}^{n}{}_{j=0}^{r}$ (E has only 0 and 1 entries) which contains N 1's. Denote by $R(x, E)$ the error of the best quadrature formula of this type in H.

Problem: Show that

$$R_N(H) \le R(x, E)$$

for each choice of the nodes $-1 < x_1 < \ldots < x_n < 1$ and the matrix E with N 1-entries.

Problem 2, proposed by B. Bojanov:

(0,2) Quadratures

P. Turan (1980) raised the question about existence of quadrature formula of (0,2)-type:

$$\int_{-1}^{1} f(x)\,dx \approx \sum_{k=1}^{n} A_k f(x_k) + \sum_{k=1}^{n} B_k f''(x_k),$$

with Algebraic Degree of Precision (ADP) greater than $2n-1$. Recently my student D. Dimitrov (1991) solved this problem constructing explicitly a (0,2) quadrature with ADP $= 2n+1$. Let MADP be the maximal ADP that can be achieved by a (0,2) quadrature formula. Clearly $2n+1 \leq \text{MADP} \leq 4n-1$.
Problem: Find MADP or improve the estimation

$$2n+1 \leq \text{MADP} \leq 4n-1.$$

References

P. Turán (1980), On some open problems of Approximation Theory. J. Approx. Th. 29, 23–85.
D.K. Dimitrov (1991), On a problem of Turan: (0,2) quadrature formula with a high algebraic degree of precision. Aequationes Math. 41, 168–171.

Problem 3, proposed by H. Brass and K. Petras:

Extremal Quadrature Rules

Let $w \geq 0$ be given and let $I[f] = \int_{-1}^{1} f(x)w(x)\,dx$. Then we define

$$\mathbf{Q}_m^+(I) := \{Q \mid Q[f] = \sum_{\nu=1}^{n} a_\nu f(x_\nu), \quad x_\nu \in [-1,1], \quad a_\nu \geq 0 \quad \text{and}$$
$$Q[p] = I[p] \text{ for all polynomials of degree less than or equal to } m\}$$

and consider a convex functional ϕ on $\mathbf{Q}_m^+(I)$, which means that

$$\phi(\alpha Q_1 + (1-\alpha)Q_2) \leq \alpha\phi(Q_1) + (1-\alpha)\phi(Q_2)$$

holds for all $Q_1, Q_2 \in \mathbf{Q}_m^+(I)$, if $\alpha \in [0,1]$. We call $Q_0 \in \mathbf{Q}_m^+(I)$ *extremal relative to* ϕ, if

$$\sup_{Q \in \mathbf{Q}_m^+(I)} \phi(Q) = \phi(Q_0).$$

Problem: For a given functional ϕ determine the extremal formula.

Solutions or at least partial or asymptotical solutions exist for the following examples, $\phi_1(Q) = a(Q,\eta)$ (coefficient of Q at the node η), $\phi_2(Q) = x_1$, $\phi_3(Q) = x_1 - x_n$, $\phi_4(Q) = \operatorname{Var} Q$ (variance of Q) and $\phi_5(Q) = K_r(Q)(\eta)$ (K_r is the rth Peano kernel), $\phi_6(Q) = \|K_r(Q)\|_1$ and $\phi_7(Q) = \|K_r(Q)\|_\infty$. Explicit solutions are in general not known for ϕ_4 to ϕ_7. Of particular interest are extremal formulae for functionals of the form

$$\phi(Q) = \sup_{f\in K} |I[f] - Q[f]|,$$

where K are subclasses of $C[-1,1]$.

We say that a quadrature formula is of Markov resp. of Gauss type, if its number of nodes, with nodes at -1 or 1 counted half, is less than or equal to $\frac{m+2}{2}$ resp. equal to $\frac{m+1}{2}$. The known solutions for the functionals mentioned above are of Markov or of Gauss type. We conjecture that this also holds for many further important functionals ϕ.

Problem 4, proposed by R. Cools and H.J. Schmid:

Fill the Gap

Consider the following integrals

$$\int_{\mathbb{R}^2} f(x,y)(x^2+y^2)^{\alpha-1} e^{-x^2-y^2}\, dx\, dy, \quad \alpha > 0, \tag{1}$$

and

$$\int_\Omega f(x,y)(x^2+y^2)^\alpha (1-x^2-y^2)^\beta\, dx\, dy, \quad \Omega = \{(x,y)\in\mathbb{R}^2 : x^2+y^2\le 1\}, \alpha,\beta > -1. \tag{2}$$

The integral (1) for $\alpha = 1$ is usually refered to as $E_2^{r^2}$ and the integral (2) for $\alpha = \beta = 0$ is denoted by S_2.

The number of points N in a cubature formula (c.f.) of degree $2k-1$ for the integrals (1) and (2) satisfies the lower bound (l.b.) by Möller (1976), i.e.,

$$N \ge \frac{k(k+1)}{2} + \lfloor \frac{k}{2} \rfloor.$$

Verlinden and Cools (1992) derived necessary and sufficient conditions such that this bound will be attained for the integrals under consideration for degree $4k+1$. We proved (in this volume) that these conditions do not hold for $k \ge 2$. Thus we obtain the bound

$$N \ge 2(k+1)^2 \quad \text{for degree } 4k+1,\ k\ge 2.$$

Remark that this bound is attained for $k=2$ for the regions S_2 and $E_2^{r^2}$.

In the following table we give an overview of the known results for the regions S_2 and $E_2^{r^2}$. References to the actual formulae can be found in Cools and Rabinowitz (1993).

degree	Möller's l.b.	new l.b.	known c.f.
1	1		1
3	4		4
5	7		7
7	12		12
9	17	18	18
11	24		25
13	31	32	34
15	40		44
17	49	50	61

It is proven that for the integrals (1) and (2), the minimal number of knots in a cubature formula of degree 11 is 25. Hence, for degree 13 and higher there is a gap between the highest known lower bound and the minimal number of knots in a known cubature formula.

Problem: Fill the gap between the lower bound and the best formula for (at least) degree 13. In other words: construct a cubature formula or prove that no cubature formula exists, with 32 or 33 knots for the regions we are interested in.

References

R. Cools and P. Rabinowitz (1993), Monomial cubature rules since 'Stroud', a compilation. To appear in J. Comput. Appl. Math.

H.M. Möller (1976), Kubaturformeln mit minimaler Knotenzahl. Numer. Math. 25, 185–200.

P. Verlinden and R. Cools (1992), On cubature formulae of degree $4k+1$ attaining Möller's lower bound for integrals with circular symmetry. Numer. Math. 61, 395–407.

Problem 5, proposed by K.-J. Förster:

A Problem in Chebyshev Quadrature for Ultraspherical Weight Functions

Let w be a nonnegative weight function on $[-1,1]$. We say that w admits *Chebyshev quadrature in the strict sense*, if for every $n \in \mathbb{N}$ there are n pairwise distinct numbers $x_{\nu,n} \in [-1,1]$ such that

$$\int_{-1}^{1} x^\mu w(x)\,dx = \frac{m_0}{n} \sum_{\nu=1}^{n} (x_{\nu,n})^\mu \qquad \text{for } \mu = 0,1,\ldots,n.$$

Chebyshev quadrature has been investigated for more than a hundred years (for a survey see Gautschi 1976 and the references cited therein). Here we are interested in *ultraspherical weight functions* w_λ,

$$w_\lambda(x) = (1-x^2)^{\lambda-1/2}, \qquad \lambda > -1/2.$$

In 1874 Chebyshev proved that w_0 admits Chebyshev quadrature in the strict sense. For every $\lambda > 0$ it is shown by Gatteschi (1963/64) that w_λ does not admit Chebyshev quadrature in the strict sense. For every $\lambda \leq \bar{\lambda} = -0.30056...$ such a negative result is proved in Förster (1986). For $\lambda \in (\bar{\lambda}, 0)$ the question is still open. By numerical examples given in Förster (1986), one may conjecture that there exists a $\lambda^* \in (\bar{\lambda}, 0)$ such that w_λ admits Chebyshev quadrature in the strict sense for every $\lambda \in [\lambda^*, 0)$.
Problem: Prove or disprove the above conjecture.

References

K.-J. Förster (1986), On Chebyshev quadrature for ultraspherical weight functions. Calcolo 23, 355-381 (MR 89g : 65021).
L. Gatteschi (1963/64), Su di un problema connesso alle formule di quadratura di Tschebyscheff. Univ. e Politec. Torino Rend. Sem. Mat. 23, 75–87 (MR 30 # 4386).
W. Gautschi (1976), Advances in Chebyshev quadrature. In: G.A. Watson (ed.), Numerical Analysis, Proc. Dundee Conf. Num. Anal. 1975, Springer, Lect. Notes Math. 506, Berlin, 100–121 (MR 57 # 7956).

Problem 6, proposed by W. Gautschi and S.E. Notaris:

For each $n \in \mathbb{N} = \{1, 2, 3, \ldots\}$, let

$$\int_{-1}^{1} f(t)w(t;\lambda)\,dt = \sum_{\nu=1}^{n} w_{\nu,n}^{G}(\lambda) f(t_{\nu,n}^{G}(\lambda)) + \sum_{\mu=1}^{n+1} w_{\mu,n}^{K}(\lambda) f(t_{\mu,n}^{K}(\lambda)), \qquad f \in \mathcal{P}_{3n+1},$$

be the $(2n+1)$-point Gauss-Kronrod quadrature formula for the Gegenbauer weight function $w(t;\lambda) = (1-t^2)^{\lambda-1/2}$, $-1 < t < 1$, $\lambda > -1/2$, i.e., $t_{\nu,n}^{G}(\lambda)$ are the zeros of the nth-degree Gegenbauer polynomial $\pi_n(\,\cdot\,;\lambda)$ and $t_{\mu,n}^{K}(\lambda)$ the zeros of the Stieltjes polynomial of degree $n+1$ orthogonal to all lower-degree polynomials relative to the oscillatory weight function $\pi_n(\,\cdot\,;\lambda)w(\,\cdot\,;\lambda)$. It is known from a result of Szegö (1935) that for each $n \in \mathbb{N}$, provided that $0 < \lambda \leq 2$, the Kronrod nodes $t_{\mu,n}^{K}(\lambda)$ are all simple, contained in $(-1, 1)$, and interlace with the Gauss nodes $t_{\nu,n}^{G}(\lambda)$, when ordered decreasingly,

$$i_n(\lambda): \quad -1 < t_{n+1,n}^{K}(\lambda) < t_{n,n}^{G}(\lambda) < t_{n,n}^{K}(\lambda) < \ldots < t_{2,n}^{K}(\lambda) < t_{1,n}^{G}(\lambda) < t_{1,n}^{K}(\lambda) < 1.$$

Algebraic considerations (Gautschi and Notaris 1988) lead to the conclusion that for each $n \in \mathbb{N}$ there exists a number $\Lambda_n > 2$ such that

$$\Lambda_n = \sup\{\lambda' : \quad i_n(\lambda) \text{ holds for } 0 < \lambda \leq \lambda'\}.$$

Problem: Prove (or disprove) that the sequences $\{\Lambda_{2m}\}$ and $\{\Lambda_{2m+1}\}$ both decrease monotonically. If this is true, their limits as $m \to \infty$ exist. Prove (or disprove) that

$$\lim_{m\to\infty} \Lambda_{2m} = \lim_{m\to\infty} \Lambda_{2m+1} \geq 4.$$

Numerical evidence from Gautschi and Notaris (1988, Tables 2.1 and A.1), plotted below in Fig. 1, lends considerable support to the validity of both conjectures, the latter perhaps even with equality sign. A lower bound strictly larger than 2 in the above limit relations would already be of interest as it would show that Szegö's result is not sharp.

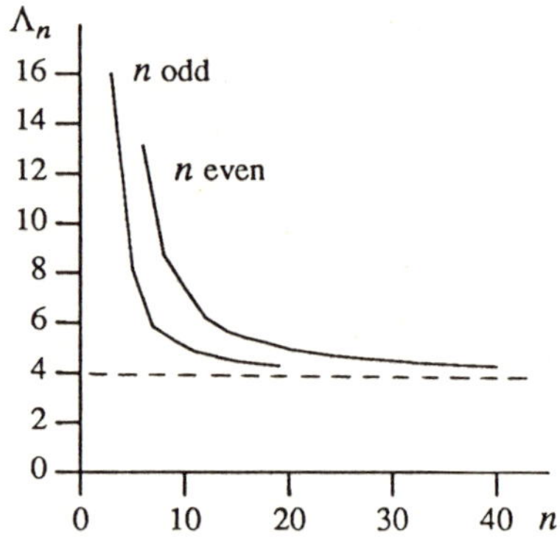

Figure 1. Λ_n for $3 \leq n \leq 40$, $n \neq 4$

References

W. Gautschi and S.E. Notaris (1988), An algebraic study of Gauss-Kronrod quadrature formulae for Jacobi weight functions. Math. Comp. 51, 231–248.
G. Szegö (1935), Über gewisse orthogonale Polynome, die zu einer oszillierenden Belegungsfunktion gehören. Math. Ann. 110, 501–513 (Collected Papers, R. Askey, ed., v. 2, 545-557).

Problem 7, proposed by G. Schmeisser:

Kernels of Quadrature Methods

A sequence of quadrature formulae

$$\int_a^b f(x)\,dx = \sum_{\nu=1}^{n} A_{n\nu} f(x_{n\nu}) + R_n[f] \qquad (n \in \mathbb{N}) \tag{1}$$

is called a *quadrature method.* For the familiar quadrature methods error estimates show that (within certain limits) high regularity of f implies rapid convergence to zero of the remainders $R_n[f]$ as $n \to \infty$. However, the opposite is not true. There may exist quite erratic functions f for which the remainders tend rapidly to zero. For the purpose of sharp results it becomes desirable to know about the set of all functions f for which all remainders of (1) vanish.

For a class S of functions for which (1) is applicable, we introduce

$$\mathcal{K}_S := \{f \in S : R_n[f] = 0 \text{ for all } n \in \mathbb{N}\}$$

and call $\mathcal{K}_S$ the *kernel* of the quadrature method (1) with respect to S. For the class of Riemann integrable, continuous and absolutely continuous functions, we replace S by R, C and AC, respectively.
Refering by superscript tr to the trapezoidal method on $[0,1]$, it is obvious that $R_n^{tr}[f]$ vanishes for all odd functions on $[0,1]$ as well as for all polynomials of degree at most 1 and hence for all functions from

$$K := \{f := c + u : c \in \mathbb{R} \text{ and } u(x) \equiv -u(1-x)\}.$$

It is implicitly shown in Brass (1978, Lemma 2) that $\mathcal{K}_{AC}^{tr} = K \cap AC[0,1]$. We may therefore say that $\mathcal{K}_{AC}^{tr}$ is *trivial*. However, $\mathcal{K}_C^{tr}$ and $\mathcal{K}_R^{tr}$ are non-trivial. For example, the function

$$f(x) := \sum_{n=1}^{\infty} \frac{\mu(n)}{n} e^{i2\pi nx},$$

where $\mu(\cdot)$ denotes the Möbius function introduced in Number Theory, belongs to $\mathcal{K}_C^{tr} \backslash K$. Further,

$$f(x) := \begin{cases} 1/q & \text{if } x = 2^{-1/2} + p/q \\ 0 & \text{otherwise,} \end{cases}$$

where p, q are relatively prime integers and $q > 0$, is an example of a function belonging to $\mathcal{K}_R^{tr} \backslash (\mathcal{K}_C^{tr} \cup K)$. Some partial results for the kernels of the midpoint method and the Simpson method can be found in Loxton and Sanders (1980) and Donat (1991).
Suppose that the quadrature formulae (1) are all precise for polynomials up to degree m but none of them is precise for $f(x) := x^{m+1}$. It then seems as if $\mathcal{K}_S$ is rather trivial if the functions of S are of much higher regularity than those of $C^m[a,b]$ and quite complicated if their regularity is much lower. A rigorous investigation is desirable, especially for compound Newton-Côtes formulae. Further, what can be said about the kernels of the sequence of Gaussian quadrature formulae?
Research on kernels should be encouraged and even non-trivial partial results may be of interest.

References

H. Brass (1978), Umkehrsätze beim Trapezverfahren. Aequationes Math. 18, 338–344.

R. Donat (1991), Fourier-Methoden in der Theorie der numerischen Integration. Diplomarbeit, Erlangen.

J.H. Loxton and J.W. Sanders (1980), On an inversion theorem of Möbius. J. Aust. Math. Soc., Ser. A 30, 15–32.

Problem 8, proposed by F. Stenger:

Let an $m \times m$ matrix $A = (a_{ij})$ be defined as follows. Let

$$\alpha_k = \frac{1}{2} + \int_0^k \frac{\sin(\pi x)}{\pi x}\, dx, \quad k = 0, \pm 1, \pm 2, \ldots,$$

and

$$a_{ij} = \alpha_{i-j}.$$

Problem: Prove that if λ is an eigenvalue of A, then $\operatorname{Re}\lambda > 0$.
(Mr. Stenger offers a reward of 50 DM for the solution of this problem.)

Problem 9, proposed by H. Woźniakowski:

Tractability Problem

Let F_d be a normed linear space of functions $f : [0,1]^d \to \mathbb{R}$. Let

$$e(n, F_d) = \sup_{\substack{f \in F_d \\ \|f\| \le 1}} \inf_{\substack{a_i, x_i \\ i=1,\ldots,n}} \left| \int_{[0,1]^d} f(t)\, dt - \sum_{i=1}^n a_i f(x_i) \right|.$$

Problem: For which $\{F_d\}$ do there exist C, p, q so that

$$e(n, F_d) \le C d^q \left(\frac{1}{n}\right)^p \qquad \text{for all } d \text{ and } n\ ? \tag{1}$$

If so, find a_i and x_i for which (1) holds.
Partial result: Let F_d be a reproducing kernel Hilbert space with kernel K_d. Then

$$e(n, F_d) \le \sqrt{\frac{1}{n} \int_{[0,1]^d} K_d(t,t)\, dt}.$$

Hence, if $\sqrt{\int_{[0,1]^d} K_d(t,t)\, dt} \le C$ for all d, then (1) holds with $q = 0$ and $p = 1/2$. Unfortunately the proof is not constructive.

ISNM

A series with a long-standing reputation

Since its foundation in 1963 more than 100 volumes have been published by Birkhäuser Verlag in the **International Series of Numerical Mathematics.**

John Todd's *Introduction to the Constructive Theory of Functions*, published as Volume 1, was a remarkable start. Proceedings volumes and further monographs such as Fenyö/Frey, *Moderne mathematische Methoden in der Technik*, Ghizzetti/Ossicini, *Quadrature Formulae*, Todd, *Basic Numerical Mathematics* (two volumes) and Heinrich, *Finite Difference Methods on Irregular Networks* followed, always presenting the state of the art in exposition and research.

Originally the Editorial Board consisted of Ch. Blanc, A. Ghizzetti, A. Ostrowski, J. Todd, H. Unger, A. van Wijngaarden. Despite a number of changes, it has shown long years of continuity; Prof. Ostrowski and Prof. Henrici, for instance, were members of the Board all their lives.

At present the series is being edited by
Karl-Heinz Hoffmann, München, **Hans D. Mittelmann**, Tempe, **John Todd**, Pasadena.

As in the past, we do not intend to restrict the series a piori to certain subjects. The series is open to all aspects of numerical mathematics with emphasis on mathematical content. At the same time, we wish to include practical applications in science and engineering, with emphasis on mathematical content.

Some of the topics of particular interest to the series are:
Free boundary value problems for differential equations, phase transitions, problems of optimal control, other nonlinear phenomena in analysis; nonlinear partial differential equations, efficient solution methods, bifurcation problems; approximation theory.

If possible, the topic of each volume should be discussed from three different angles, namely that of Mathematical Modelling, Mathematical Analysis, Numerical Case Studies.

The editors particularly welcome research monographs; furthermore, the series is to contain advanced graduate texts, dealing with areas of current research interest, as well as selected and carefully refereed proceedings of major conferences or workshops sponsored by various research centers. Historical material in these areas would also be considered.

We encourage preparation of manuscripts in LaTeX or AMSTeX for delivery in camera-ready copy which enables a rapid publication, or in electronic form for interfacing with laser printers of typesetters.

Titles previously published in the series

INTERNATIONAL SERIES OF NUMERICAL MATHEMATICS
BIRKHÄUSER VERLAG

ISNM 93 **R.E. Bank, R. Bulirsch, K. Merten (Eds.):** Mathematical Modelling and Simulation of Electric Circuits and Semiconductor Devices, 1990 (3-7643-2439-2)

ISNM 94 **W. Haussmann, K. Jetter (Eds.):** Multivariate Approximation and Interpolation, 1990 (3-7643-2450-3)

ISNM 95 **K.-H. Hoffmann, J. Sprekels (Eds.):** Free Boundary Value Problems, 1990 (3-7643-2474-0)

ISNM 96 **J. Albrecht, L. Collatz, P. Hagedorn, W. Velte (Eds.):** Numerical Treatment of Eigenvalue Problems, Vol. 5, 1991 (3-7643-2575-5)

ISNM 97 **R.U. Seydel, F.W. Schneider, T.G. Küpper, H. Troger (Eds.):** Bifurcation and Chaos: Analysis, Algorithms, Applications, 1991 (3-7643-2593-3)

ISNM 98 **W. Hackbusch, U. Trottenberg (Eds.):** Multigrid Methods III, 1991 (3-7643-2632-8)

ISNM 99 **P. Neittaanmäki (Ed.):** Numerical Methods for Free Boundary Problems, 1991 (3-7643-2641-7)

ISNM 100 **W. Desch, F. Kappel, K. Kunisch (Eds.):** Estimation and Control of Distributed Parameter Systems, 1991 (3-7643-2676-X)

ISNM 101 **G. Del Piero, F. Maceri (Eds.):** Unilateral Problems in Structural Analysis IV, 1991 (3-7643-2487-2)

ISNM 102 **U. Hornung, P. Kotelenez, G. Papanicolaou (Eds.):** Random Partial Differential Equations, 1991 (3-7643-2688-3)

ISNM 103 **W. Walter (Ed.):** General Inequalities 6, 1992 (3-7643-2737-5)

ISNM 104 **E. Allgower, K. Böhmer, M. Golubitsky (Eds.):** Bifurcation and Symmetry, 1992 (3-7643-2739-1)

ISNM 105 **D. Braess, L.L. Schumaker (Eds.):** Numerical Methods in Approximation Theory, Vol. 9, 1992 (3-7643-2746-4)

ISNM 106 **S.N. Antontsev, K.-H. Hoffmann, A.M. Khludnev (Eds.):** Free Boundary Problems in Continuum Mechanics, 1992 (3-7643-2784-7)

ISNM 107 **V. Barbu, F.J. Bonnans, D. Tiba (Eds.):** Optimization, Optimal Control and Partial Differential Equations, 1992 (3-7643-2788-X)

ISNM 108 **H. Antes, P.D. Panagiotopoulos:** The Boundary Integral Approach to Static and Dynamic Contact Problems. Equality and Inequality Methods, 1992 (3-7643-2592-5)

ISNM 109 **A.G. Kuz'min:** Non-Classical Equations of Mixed Type and their Applications in Gas Dynamics, 1992 (3-7643-2573-9)

ISNM 110 **H.R.E.M. Hörnlein, K. Schittkowski (Eds.):** Software Systems for Structural Optimization, 1992 (3-7643-2836-3)

ISNM 111 **R. Burlisch, A. Miele, J. Stoer, K.H. Well:** Optimal Control, 1993 (3-7643-2887-8)